Jürgen Haffer

**Ornithology, Evolution, and Philosophy**

The Life and Science of Ernst Mayr 1904–2005

Ernst Mayr in his office at Harvard University, Cambridge, Massachusetts, ca. 1960 (AMNH Library photographic collection, negative no. 334102)

Jürgen Haffer

# Ornithology, Evolution, and Philosophy

## The Life and Science of Ernst Mayr 1904–2005

With 71 Figures and 4 Tables

Dr. rer. nat. Jürgen Haffer
Tommesweg 60
45149 Essen
Germany

j.haffer@web.de

Library of Congress Control Number: 2007937237

ISBN 978-3-540-71778-2 Springer Berlin Heidelberg New York



**Springer is a part of Springer Science + Business Media**
springer.com



Editor: Dr. Dieter Czeschlik, Heidelberg, Germany
Desk editor: Ursula Gramm, Heidelberg, Germany
Cover design: WMXDesign GmbH, Heidelberg, Germany
Cover illustration: The photograph, taken in 1995, shows Ernst Mayr in his house in Cambridge, Massachusetts. Source: Oliver Rüther, © www.oliver-ruether.de
Typesetting and production: LE-TEX Jelonek, Schmidt & Vöckler GbR, Leipzig, Germany
31/3180/YL – 5 4 3 2 1 0 – Printed on acid-free paper

# Preface

Ernst Mayr (1904–2005), eminent naturalist-systematist and ornithologist, "architect" of the Synthetic Theory of Evolution, leading evolutionary biologist and influential historian and philosopher of biology, was asked repeatedly by friends, associates, and peers to write an autobiography, but he always declined such suggestions because he considered those of other people he knew as inadequate. Moreover, he could not very well talk about himself without bragging, which did not appeal to him. Who else has had a private dinner with the Emperor and Empress of Japan? Who else has had honorary degrees from the Universities of Oxford, Cambridge, Harvard, the Sorbonne, Uppsala, and Berlin? And who else has been able to celebrate the 75th anniversary of his PhD? He also published several autobiographical accounts, dictated copious autobiographical notes, and annotated his bibliography. I had the privilege of using these notes and, in addition, of asking him numerous questions, which he answered patiently in many personal letters. He supported the plan of a biography wholeheartedly from the beginning and reviewed a first draft in detail. In many ways he should be considered a coauthor of this book.

As a high school student and eager birdwatcher in Germany, I read Kramer's (1948) extensive review of Mayr's *Systematics and the Origin of Species* (1942e) and was impressed with the general evolutionary relevance of the discussions and conclusions in this book based, in large part, on his ornithological studies. In 1952, I took a course on the evolution of man and the evolutionary synthesis offered at the University of Göttingen by Gerhard Heberer (1901–1973), editor of *Die Evolution der Organismen* (1943), who had just translated into German G. G. Simpson's *Tempo and Mode in Evolution* (1944). Direct communication with Ernst Mayr began during the early 1960s, when I was studying the bird fauna of Colombia, South America, and with my first visit to him at Harvard University in 1968. In 1961, I returned to Europe for a break from my work as an exploration geologist in South America and had chosen the route across the Pacific Ocean to visit the Hawaiian, Samoan and Fiji Islands as well as New Guinea on whose bird faunas Ernst Mayr had published many detailed studies. Across the Bay of Manokwari, I could see the Arfak Mountains, which rise 2,500 m in the Vogelkop region of northwestern New Guinea, where, 33 years earlier, Ernst Mayr had set out in search of certain "rare birds of paradise." He had also explored the bird fauna of the precipitous Cyclops Mountains further east along the coast, where I had been a few days before, traveling comfortably in the open country around Lake Sentani. Greatly influenced by his publications, I applied Mayr's ideas on biological species and speciation in my studies of the bird faunas of tropical South America and northern Iran (Haffer 1969, 1974, 1975, 1977) and in research on the history of the biological species concept (Haffer 1992, 2006). We exchanged letters occasionally during the 1970s and 1980s and also met at several ornithological congresses. During the 1990s, when I edited the correspondence between Ernst Mayr and his mentor and friend, Erwin Stresemann, and published (with two coauthors) a biography of the latter, Ernst Mayr contributed much information to both of these projects and also reviewed parts of the final texts.

Note: All quotes from Mayr's publications, personal letters and autobiographical notes as well as from other letters are rendered in English (if the original is in German, his translations are unmarked, whereas my translations are marked "transl.").

# Contents

## Part IV Appendices

## Part IV Appendices

# Introduction

Like Charles Darwin during the 19th century, Ernst Mayr worked indefatigably for a better understanding of the central importance of organic evolution, and he fought for the recognition of the independence and autonomy of biology among the natural sciences. His research career comprised several branches of biology and spanned 80 years. His extraordinary scientific contributions total over 750 articles and 21 books (plus about 120 book reviews). His books, and many of his articles, manifest a rare ability to critically synthesize the knowledge gained in distant fields of research. In the sense of emergence, these synthetic works are more than the sum of their parts integrating successively (a) systematics and genetics, (b) evolutionary biology and (c) the history and philosophy of biology.

Above all, Ernst Mayr was a naturalist who endeavored to comprehend the living world in all its relations with respect to diversity, populations, and evolution. He watched birds since his youth in Germany and continued this activity later in North America. He was interested in the behavior of birds, their diverse ecological relations as well as their environments, and he included in his studies other groups of animals as well, like Charles Darwin and Alfred Russel Wallace had done a hundred years earlier.

**Ornithology, systematics and zoogeography—Mayr's first synthesis.** I distinguish three periods of Mayr's career which, of course, were broadly transitional. The first, as an ornithologist, systematist, and zoogeographer, encompassed his work as an assistant curator at the Museum of Natural History in Berlin (1926–1930) and as curator of ornithology at the American Museum of Natural History in New York (1931–1953). This transition, his familiarity with European and North American ornithological research, was decisively important. Dr. Erwin Stresemann (1889–1972) was his teacher in Berlin and conveyed the knowledge of modern ornithology ("New Avian Biology") and systematics to Mayr (see Stresemann 1927–1934) and had arranged for expeditions to New Guinea and the Solomon Islands (1928–1930). The results gained during these expeditions were the foundation of Mayr's later systematic studies at the museum in New York where he analyzed the variation, distribution patterns, and speciation of the birds of Oceania, especially of New Guinea, Melanesia, and Polynesia. He also established the basic tenets of island biogeography and forged a synthesis of systematics, natural history, genetics and evolution with the publication of his book, *Systematics and the Origin of Species from the Viewpoint of a Zoologist* (1942e), which became one of the founding documents of the new synthetic theory of evolution and explained a whole set of phenomena well known to systematists and naturalists but not to the geneticists such as species and speciation, the effects of natural selection on natural populations, the nature of geographical variation, and the role of species in macroevolution. Mayr influenced ornithology by encouraging amateurs in the New York region to undertake biological studies of local birds, by publishing M. M. Nice's pioneering population study of the Song Sparrow, by establishing bird family relations when cataloging and arranging the Rothschild Collection as well as by editing the continuation of J. L. Peters' *Check-list of Birds of the World*. Mayr's work as an ornithologist formed the foundation of his studies as an evolutionist.

**Evolutionary biology—Mayr's second synthesis.** The second period of Mayr's career as an Alexander Agassiz Professor of Zoology, Harvard University (Cambridge, Massachusetts), beginning in 1953, witnessed the development of modern evolutionary biology. Some topics of his studies were the emergence of evolutionary novelties, the nature of isolating mechanisms, the biological species concept, the significance of ecological-geographical separation of populations, and the dual nature of evolution: "vertical" phyletic evolution (adaptive change through time) and "horizontal" evolution in geographical space (speciation, origin of diversity). The climax of this work was the publication of his books, *Animal Species and Evolution* (1963b) and *Populations, Species, and Evolution* (1970e), masterly summations of species and speciation, and magnificent syntheses of population genetics, variation of populations, the origin of species, and adaptive specialization.

**History and philosophy of biology—Mayr's third synthesis.** The third period of Mayr's career was devoted to the history and philosophy of biology with new considerations on the basis of systematics and evolutionary biology. It began during the late 1950s but mainly followed his "retirement" as director of the Museum of Comparative Zoology at Harvard University in 1970 and as a university professor in 1975. He discovered in the history of science an evolution of certain themes, concepts, problems, and ideas similar to that of organisms. He wrote on Darwin and his time (the "First Darwinian Revolution") and on the development of the Synthetic Theory of Evolution during the 1940s (the "Second Darwinian Revolution"). The core of the history of science is the evolution of ideas and concepts. Mayr used the theories of natural selection and population thinking as theoretical models within the framework of historical biological studies. He suggested that various competing paradigms may exist side by side and more or less pronounced "revolutions" may occur in different fields from time to time. Changes of concepts have a much stronger effect on the development of biological sciences than the discovery of new facts. Mayr was the first to emphasize the role of biopopulations, thereby pointing out the basic difference between "population thinking" and typological essentialism. Population thinking takes into consideration the uniqueness of each individual and unlimited variation of populations that may lead to the development of new species. On the other hand, typologists assume that the unchanging essence of each species determines variation and fixed limits of variation preclude speciation from occurring except through saltation. The genetic program of organisms is the result of selection, which acted upon millions of generations and therefore is a causal factor that differs fundamentally from physico-chemical causes. He emphasized the significance of immediate (direct, proximate) causes and ultimate causes in biology. Functional biology deals with direct, proximate causes, which concern the phenotype and poses "How?" questions, whereas evolutionary biology investigates the history of the genotype of organisms and poses "Why?" questions. In his recent books, Ernst Mayr united systematics, evolutionary biology, and the history of biology with the theoretical and philosophical foundations of the biological sciences. The titles of these books are *The Growth of Biological Thought* (1982d), *Toward a New Philosophy of Biology. Observations of an Evolutionist* (1988e), *This is Biology. The Science of the Living World* (1997b), *What Evolution Is* (2001f), and *What Makes Biology Unique? Considerations on the Autonomy of a Scientific Discipline* (2004a). Mayr decisively influenced the fields of systematics, evolutionary biology, and history of biology and he worked toward the foundation of a modern philosophy of biology. In this process, he developed a new vision of modern biology as the *Leitwissenschaft*, the guiding science, of the 21st century, when functional and evolutionary biology jointly contribute to further progress.

"When it comes to philosophical questions that specifically relate to man, his well-being and his future, it is the science of biology that will be most suitable as starting point for all analysis, rather than the physical sciences" (1969g: 202). "Overpopulation, the destruction of the environment, and the malaise of the inner cities cannot be solved by technological advances, nor by literature or history, but ultimately only by measures that are based on an understanding of the biological roots of these problems" (1997b: XV).

**Historical chance events.** In both publications and oral presentations, Ernst Mayr repeatedly emphasized the historicity of the course of life of most people and that much of what we find in the world around us has come about in consequence of historical accidents, not natural inevitabilities. Chance and coincidence played an especially important role in his life "just as in Darwinian evolution" determining the course of his unfolding career during the 1920s and 1930s (Bock 1994a; 2004c). In this metaphor, "necessity" through natural selection, the second factor of Darwinian evolution, is the manner in which Mayr utilized these chance events each of which gave a new turn to the course of his life and career. Most important was that in each of these new turning points, it was up to him and his hard work and dedication in which way he made the best of the new opportunities that these chance events opened up for him. These events will be mentioned in the context of his developing career in the following chapters, but are summarized here in view of their biographical significance:

(1) The observation of a pair of rare ducks near his hometown of Dresden (Germany) in March 1923 led to his first contact with Dr. Erwin Stresemann in Berlin, the leading ornithologist in the country, who later persuaded Mayr to become a biologist instead of a medical doctor (pp. 22, 32).
(2) Plans for ornithological expeditions to Peru and Cameroon in 1926 and 1927 did not materialize; instead Mayr went on an expedition to New Guinea

(1928–1929) to replace Lord Walter Rothschild's collector in that region who had a stroke and had to return Europe (p. 49).

(3) When Mayr was about to return to Germany from New Guinea in the spring of 1929, the leader of the Whitney South Sea Expedition (WSSE) of the American Museum of Natural History in New York (AMNH), Rollo H. Beck, retired after 8 long years of service and the expedition badly needed the assistance of an ornithological expert and therefore Mayr joined the expedition to the Solomon Islands (July 1929 to February 1930); pp. 74–75.

(4) The ornithologist Ernst Hartert, director of W. Rothschild's private museum in Tring, near London, retired in early 1930 and W. Rothschild planned to have Mayr as his successor (p. 96, footnote). However, at that time, Rothschild was in financial difficulties and within 2 years had to sell almost all of his bird collections to the AMNH.

(5) At this same time, an expert especially on the avifaunas of the southwest Pacific and New Guinea was needed in New York, and Mayr had already attracted Dr. Sanford's attention, a trustee of the AMNH, who arranged (on Dr. Stresemann's recommendation) for him to work in New York on the material of the WSSE during temporary assignments in 1931 and 1932 (p. 96).

(6) The transfer of Rothschild's collection to New York at the "right time" led to Mayr's employment as curator of the Whitney-Rothschild collections at the AMNH (p. 98).

(7) At a symposium on speciation in December 1939, Mayr gave a splendid lecture, his first theoretical paper, by observing and learning from his preceding speaker what not to do. Mayr's clear presentation led to his invitation by Professor L.C. Dunn to give (with E. Anderson) the famous Jesup Lectures at Columbia University, New York, in March 1941 (p. 190).

(8) The shift of the research interests of his botanical colleague E. Anderson after their joint lectures induced Mayr to enlarge his contribution and to prepare his seminal book, *Systematics and the Origin of Species* (1942e), pp. 193–194.

(9) When the "Committee on Common Problems of Genetics and Paleontology" was founded in early 1943, the head of the Eastern Group's genetics section, Th. Dobzhansky, left for Brazil for 1 year and Mayr was placed in charge. From this position, developed his leadership in evolutionary studies in the United States during the late 1940s and 1950s (p. 234).

# Part I
# The Young Naturalist in Germany

Ernst Mayr has been an observer all his life. At a young age, he became a naturalist and learned to watch and intimately study the numerous kinds of birds inhabiting the parks and rural surroundings of the cities where he grew up. As a boy, he ventured into the fields, the forests, and along the lakes and streams nearly every free minute possible, watching wildlife and observing the habits of birds, locating their nests and, in some cases, following the hatching of their eggs. He thus developed very early a general idea of what animal species are—sharply separated breeding communities at particular localities, as he would say many years later, each with particular ecological requirements for food and living space. Throughout his career, Mayr continued to watch birds and other animals as well as to study the plants in the areas where he lived and traveled (Lein 2005). He still made observations even at the age of 100 on the birds around his home in Bedford, Massachusetts, and followed the change of the seasons, longer than anyone who has ever lived. His theorizing had a broad basis in his birdwatching activities during his youth and in his later taxonomic studies. One has to be an observer, he said; if a person does not observe and see what is going on in nature, then he or she has a good deal of difficulty really understanding it. The term naturalist refers to a person who studies plants or animals in a natural environment rather than in a laboratory. He or she is a person who is "fascinated by biological diversity [...] that excites our admiration and our desire for knowledge, understanding, and preservation" (Futuyma 1998: 2). The evolutionary synthesis during the 1940s was a period when the mathematical geneticists and the naturalists including paleontologists and systematists began to speak the same language. All of the "architects" of this synthesis in North America had been "young naturalists", except the paleontologist G. G. Simpson. As Mayr remarked, "this was Simpson's handicap who never had a feeling for what a species is." The same is true for many philosophers, other paleontologists, and even certain systematists. Mayr referred to them as "armchair naturalists" who interpreted species either as constant, nonvariable entities of nature or as evolutionary lineages because they had never watched and studied the animals or plants in the areas where they had gone to school.

# 1 Childhood and Youth

## The Family

Ernst Walter Mayr was born on 5 July 1904 in Kempten, at the northern foot of the Alps, located in southern Germany, and he was the middle of three brothers. His father, Dr. jur. Otto Mayr (23 July 1867–1 July 1917) was a successful jurist from a peasant family in Oberegg near Kempten (Fig. 1.1). The family name "Mayr" is fairly frequent in Bavaria. Ernst's great-grandfather, Johann Evangelist Mayr, was born in Westenried near Unterthingau on 16 December 1807 and died in Altusried (near Kempten) on 28 June 1864. He and a second great-grandfather (Bonifaz Müller) were the first physicians in the family and established a long medical tradition that is still alive in the family today. Johann Mayr had two sons, Dr. med. Otto Mayr (OM 1, born in Altusried on 8 August 1833, died in Lindau on 17 January 1912), Ernst's grandfather, and Eugen Adolf Mayr (born on 16 December 1839, died in Nürnberg after 1920). Dr. Otto Mayr (OM 1) studied medicine in Munich. He was at first a physician in Oberstdorf in southwest Germany (where Ernst's father was born) and later in Lindau am Bodensee (Lake Constance) where he became a doctor at the town hospital and served as surgeon for the poor. He was a naturalist, apiculturist, and hunter, and he participated on a regular basis in the annual meetings of the German Society of Naturalists and Surgeons. He married Wilhelmine ("Minna") Müller, a member of the distinguished Gruber family, respected landowners in this region. Minna Mayr lived from 3 October 1847 to 22 December 1928. They had two sons, Otto and Hermann, and one daughter Dora. When a little over 70 years old, the grandfather retired, and built a house in the outskirts of Lindau, with a large garden, where Ernst and his brothers spent several summer holidays. They received much attention especially from their grandmother because they were her only grandchildren. Uncle Hermann in Heidelberg and Aunt Dora had no children. Like her brother Otto, Aunt Dora was very interested in philosophy and the family considered her brilliant. Occasionally, the grandfather or Aunt Dora would take them for a walk and then for a piece of cake at a cafe. After the grandfather died in 1912, grandmother Minna and Aunt Dora (who later married August Selle) continued to live in the house in Lindau. Ernst was in New Guinea, when his grandmother died in December 1928. The house in Lindau-Reutin (Rennerle 7) still belongs to Dr. Jörg Mayr, one of Ernst's nephews. His brother Dr. Otto Mayr (OM 4) maintains a comprehensive family archive there.

Uncle Hermann Mayr (1873–1914) was a medical doctor and his older brother Otto Mayr (OM 2, 1867–1917; Ernst's father) became a jurist in the Bavarian court system. He had a brilliant career and, although he was only 49 years old when he died from cancer of the kidneys on 1 July 1917, was to be appointed to Germany's supreme court in Leipzig. His library comprised several thousand volumes, particularly the subjects of history, philosophy, and the classics. On Sundays he would read Homer in the Greek original, without a dictionary.

"He was a gymnast," Ernst Mayr recalls, "belonging to the local sports club, was one of the pioneers of skiing in Germany (he bought his skis in Norway) and he was an enthusiastic mountaineer. He served in the Bavarian Guard Regiment and later was a reserve officer in the army, but when war broke out in 1914, he was too old and ill to serve, much to his regret. He was not at all severe with us children, indeed I have the feeling that my mother was more or less running the show."

Ernst Mayr's mother, Helene Pusinelli, was born of German parents in Le Havre, France, on 22 July 1870 and died in Bad Boll, Germany, on 31 May 1952. She was a member of a well-known Dresden family. An Italian ancestor, Anton Pusinelli (1790–1828), had come to Germany from Nesso, a village on the eastern shore of Lake Como (northernmost Italy). In 1809, he settled in Dresden where his older brother Carlo owned a wine restaurant, which Anton took over after Carlo's death in 1812. Anton married Caroline Brügner (1790–1853) from Torgau and they had six children, four girls and two boys, of whom the older one, Carl Anton Pusinelli (1815–1878), became a rather wealthy pediatrician to the Saxon court in Dresden and a close friend of the composer Richard Wagner (1813–1883). Carl Anton treated quite successfully the children of the Saxon princes and had the title "Hofrat," but was not an employee of the court. His younger brother was Carl Louis (1820–1879), Ernst Mayr's grandfather. When Carl Louis was 23 years old, he left Dresden and established an import-export business in Le Havre, France, during the late 1840s. Upon a visit to his hometown in 1852, he married Camilla Leonhardt (1833–1891), the daughter of his former boss in Dresden, and they had 11 children. The next-to-youngest was Helene Pusinelli, Ernst Mayr's mother. She was only a few weeks old when the Franco-Prussian War (1870–1871) broke out. The family was expelled and returned (via Rotterdam and Hamburg) to Dresden where, through his family connections, Carl Louis Pusinelli became the director of a local bank. He had owned a profitable wholesale business in France and had suffered substantial losses because of the expulsion. However, eventually he became quite affluent through considerable compensations from reparation payments. He died in 1879 of stomach cancer. His daughter Helene, her four sisters (another one had died early) and five brothers grew up in Dresden and in part lived there later. Her uncle and a cousin were medical doctors. However, all the family fortune, invested in government securities at the beginning of World War I, was lost due to hyperinflation during the 1920s.

The early Mayrs were members of the Catholic Church. Because the grandfather Otto Mayr had married the Protestant Minna Müller, their three children became members of a Protestant church, too. The Pusinellis in Dresden belonged to the reformed Protestant church. They too had originally been Catholic.

How did the Mayr lineage in Bavaria and the Pusinelli lineage in the distant Saxon city of Dresden meet? The best friend of Dr. Otto Mayr (OM 2) in Bavaria

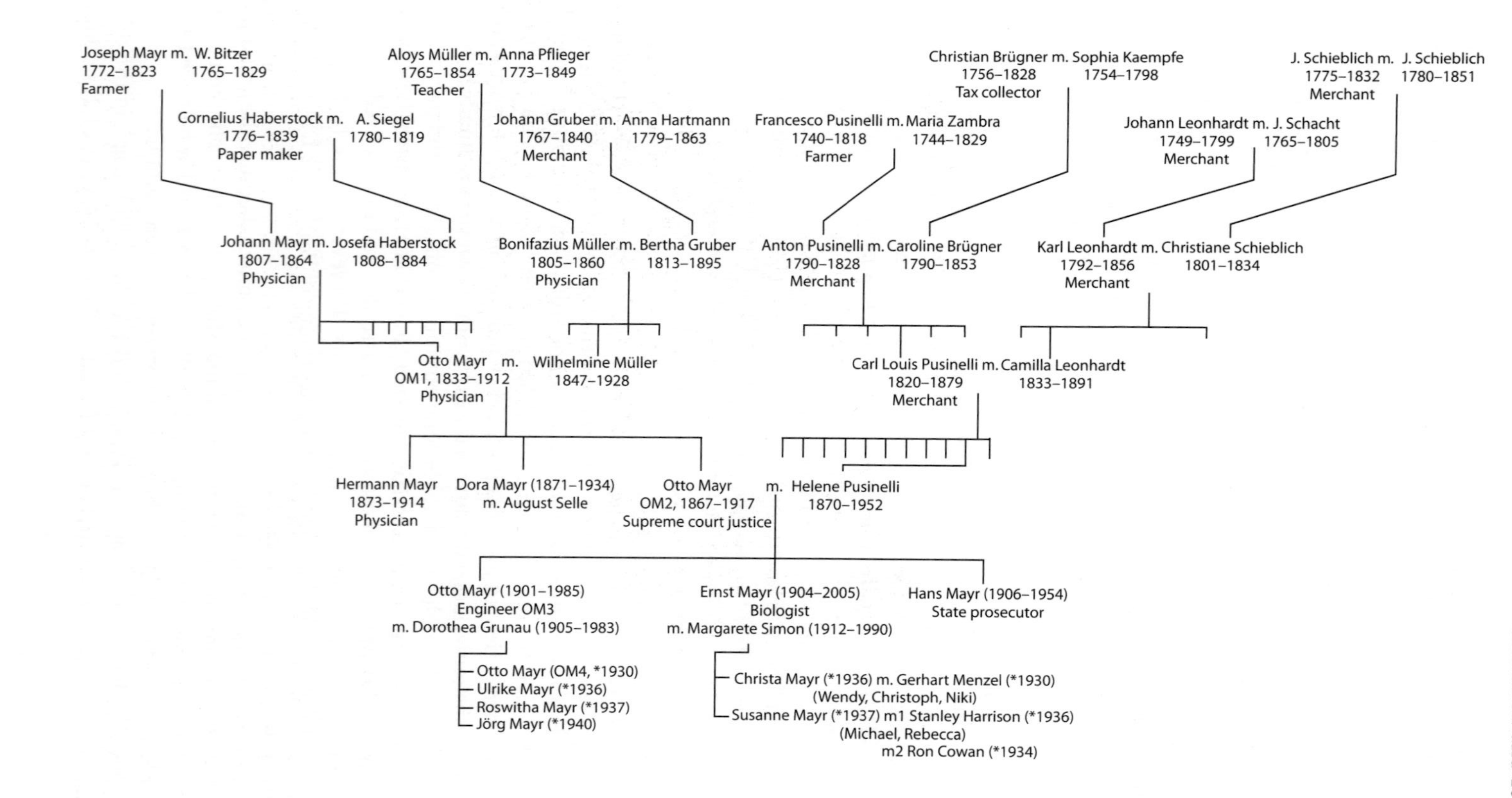

Joseph Mayr m. W. Bitzer
1772–1823 1765–1829
Farmer
Cornelius Haberstock m. A. Siegel
1776–1839 1780–1819
Paper maker
Aloys Müller m. Anna Pflieger
1765–1854 1773–1849
Teacher
Johann Gruber m. Anna Hartmann
1767–1840 1779–1863
Merchant
Francesco Pusinelli m. Maria Zambra
1740–1818 1744–1829
Farmer
Christian Brügner m. Sophia Kaempfe
1756–1828 1754–1798
Tax collector
J. Schieblich m. J. Schieblich
1775–1832 1780–1851
Merchant
Johann Leonhardt m. J. Schacht
1749–1799 1765–1805
Merchant
Johann Mayr m. Josefa Haberstock
1807–1864 1808–1884
Physician
Bonifazius Müller m. Bertha Gruber
1805–1860 1813–1895
Physician
Anton Pusinelli m. Caroline Brügner
1790–1828 1790–1853
Merchant
Karl Leonhardt m. Christiane Schieblich
1792–1856 1801–1834
Merchant
Otto Mayr m. Wilhelmine Müller
OM1, 1833–1912 1847–1928
Physician
Carl Louis Pusinelli m. Camilla Leonhardt
1820–1879 1833–1891
Merchant
Hermann Mayr
1873–1914
Physician
Dora Mayr (1871–1934)
m. August Selle
Otto Mayr m. Helene Pusinelli
OM2, 1867–1917 1870–1952
Supreme court justice
Otto Mayr (1901–1985)
Engineer OM3
m. Dorothea Grunau (1905–1983)
Ernst Mayr (1904–2005)
Biologist
m. Margarete Simon (1912–1990)
Hans Mayr (1906–1954)
State prosecutor
Otto Mayr (OM4, *1930)
Ulrike Mayr (*1936)
Roswitha Mayr (*1937)
Jörg Mayr (*1940)
Christa Mayr (*1936) m. Gerhart Menzel (*1930)
(Wendy, Christoph, Niki)
Susanne Mayr (*1937) m1 Stanley Harrison (*1936)
(Michael, Rebecca)
m2 Ron Cowan (*1934)

**Fig. 1.1.** Family genealogy of the Mayr-Pusinelli lines, the ancestry of Ernst Mayr. The chronological sequence of unidentified brothers and sisters is from left to right; m.–married

was Dr. Oscar Mey (1865– ca. 1940). His wife Gretchen, a close friend of Helene Pusinelli from her youth in Dresden, took things into her hands. She decided that Otto Mayr, at 32 still a bachelor, and Helene Pusinelli in distant Dresden, also still unmarried in 1899, were the perfect match. She invited her friend to visit her in Bavaria and then brought the two together. Mrs. Mey's instinct had been sound. The "eligibles" agreed that they were suitable for each other and were married on 24 June 1900 (Fig. 1.2)[1]

Otto and Helene Mayr first lived in Kempten (Allgäu) where he was a magistrate of the city and here their three sons Otto, Ernst, and Hans were born (Fig. 1.3).[2]

**Fig. 1.2.** Ernst Mayr's parents, Helene Pusinelli and Dr. Otto Mayr in 1900. (Photograph courtesy of O. Mayr.)

[1] Ernst Mayr and his brothers did quite a bit of genealogical research in old church records across southern Germany during the 1920s. The results indicated—as is also evident from the above account—that there was no Jewish ancestor in their father's or mother's family back at least to the 17th century. Ernst Mayr's emigration from Germany to the United States in 1931—two years prior to the Nazi takeover of the German government—was not motivated by political (but strictly professional) considerations. He had, however, no sympathy for the Nazi regime at all (see p. 300).

[2] Ernst Mayr became the longest-living member of the entire family. The life spans of 62 (of a total of 64) nearest ancestors are known. Among these eight (13%) have reached an age of 80 or more years including his mother who was almost 82 years old when she died.

**Fig. 1.3.** The three Mayr brothers in Kempten (Allgäu, Bavaria), August 1907. From *left* to *right*: Otto (*1901), Hans (*1906) and Ernst (*1904). (Photograph courtesy of O. Mayr.)

When Ernst was nearly 4 years old his father was transferred as District Prosecuting Attorney to Würzburg on 1 May 1908 and, in December 1913, to Munich as an Associate Justice at the Supreme Court of Bavaria.

Although a jurist by profession, the father was an enthusiastic naturalist and "took the family out on a hike just about every Sunday. We usually took the train somewhere away from Würzburg and then walked cross-country to some other train station or to the terminal of the electric tram near Würzburg. It was on these trips that we collected flowers, mushrooms, fossils in some quarry, or did other natural history studies. When my father heard of a heron colony north of the Aumeister near Munich we also visited it. Otherwise, perhaps more through the interests of my mother, we visited old towns, castles, and villages on weekends and during vacations. The family attitude was rather academic and very much that of upper middle class Germans that one should never stop trying to add to one's 'Bildung'."

Through these nature walks, Ernst became a naturalist at an early age, and his father's interests in history and philosophy broke through in later years when Ernst turned to the study of the history and the philosophy of biology. Holistic philosophies associated with the ideology of "Bildung" (learning and a general knowledge of culture) formed the basis of the educational system of upper middle class Germans who sent their children to the "Humanistisches Gymnasium" (high school)

None of the ancestors reached an age of 90 or higher, the average age at time of death was 63 years.

where they learned Latin and classical Greek followed by a modern language. This reflected a commitment to intellectual synthesis rather than an early specialization (Harwood 1994: 17). The gymnasiums to which the Mayrs sent their sons in Würzburg, Munich, and Dresden were elite schools, the best available in these cities. This education prepared Ernst to deal with complex issues of evolutionary theory and philosophy in later years.

The family was quite well-to-do before the early death of Ernst's father in 1917 and the German hyperinflation during the 1920s. They used to have a servant and at times even a second one. However, they always lived fairly modestly in a large apartment in the city, not in their own house. Like most other people in the country, they had no telephone or gaslight, no running hot water or refrigerator. Food was kept from spoiling during the summer in a big ice box cooled by thick slabs of ice; and, of course, they had no car. For any long distances in the city one used the electric tram.

Ernst was not interested in technical things, as many other boys were, and in school his worst subjects were drawing and singing. Otherwise, however, he was a first-class student with top grades. He did not collect stamps, rocks, or insects but he was an early observer. At the age of about two or three, he knew, like most children, to discriminate all the colors.

"One day my mother gave a tea to her sisters and girlfriends and wanted to brag about my achievement. It was spring and one of the aunts asked, What color are trees? To my mother's horror I answered, Pink. When pressed, What trees are pink? I immediately answered, Almond trees. They were pink flowering just at that time in several neighboring gardens, and mine and my mother's prestige was rescued. I think my enthusiasm for birdwatching was in part due to my capacity for making observations." One day grandmother Minna remarked to Otto, Ernst's elder brother: "Watch out well, little Ernsty will eventually pass you."

In an early notebook, Ernst recorded that the family went to the Munich Hoftheater to see Friedrich Schiller's "Wilhelm Tell" in 1914. During 1916–1917, Ernst took piano lessons and played several pieces for the family on Christmas Eve as he reported to Aunt Marguerite (one of his mother's sisters in Dresden) on 21 January 1917. He had recently attended a church concert and was reading G. Freytag's *Soll und Haben*. However, since he never practiced in spite of his mother's urging, the piano lessons were dropped. On 25 March 1917, the family attended Mozart's *Magic Flute*. Ernst noted the cast of this his first opera in detail and how the soprano voice of Maria Ivogün as the Queen of the Night had enchanted him.

After her husband's death from cancer in Munich on 1 July 1917, when Ernst was not yet 13 years old, Mrs. Mayr took her three sons to live in Dresden (Saxony), where she still had two brothers and four sisters. The Mayr boys finished the Staatsgymnasium ("Royal Gymnasium" until 1918) in Dresden-Neustadt and went to college on scholarships, some contributions from Aunt Gunny, and what little their mother could spare from her rather meager pension. Ernst always felt a tremendous admiration for his mother.

"She was an exemplary representative of the best Protestant ethics: generous, frugal, hardworking, full of ideals, and with a wonderful sense of humor. I think

**Fig. 1.4.** Ernst Mayr, 12 years old (*left*), and his brother Hans in Munich on 24 July 1916. (Photograph courtesy of O. Mayr.)

she always thought that it was important that one 'did one's job.' If science was one's job, one had to devote oneself to science. If the cataloguing and classifying of the Rothschild Collection was one's job, this is what one did without complaining."

The three brothers represented rather different personalities. Otto (OM 3, 1901–1985), 3 years older than Ernst, felt since the death of their father he had to give advice to his younger brothers, which was not received with gratitude. Ernst always wanted to excel over his older brother and worked very hard as a student, even finishing before him. After his early natural history interests, Otto turned to historical studies but became an engineer with several mining companies in northwestern Germany. He married and had four children (Otto [OM 4], Ulrike, Roswitha, Jörg) who are carrying on the Mayr tradition in Germany. However, Ernst and his younger brother Hans (1906–1954) were very close (Fig. 1.4):

"He was only a year and three-quarters younger than I and my inseparable companion. As far back as I can remember, Hans and I always did things together. We played together. We spent several vacations together in Lindau, and in the last years of my university studies in Berlin, we shared a room together. He studied law, following the footsteps of my father, and was particularly interested in criminal law. As a person, Hans was quite different from me. He was easy-going, but since he was very bright, studying was no problem for him. He was always in good humor, ready to make jokes, and to laugh at any adversity. At his bar exam (or the German equivalent), he ranked first among 133, and [later in 1936] was given the position of prosecuting attorney of eastern Saxony in the town of Bautzen. [At the outbreak of World War II in 1939] he was immediately drafted into the army. Somewhere at the Russian front, he developed the severe symptoms of multiple sclerosis, and was shipped home in a Red Cross train which, even though conspicuously marked, was heavily bombed. However, he made it home safely, but never left the hospital until he died in 1954. Even in his last years, he never complained, but somehow managed to find something good in every situation in which he found himself. He couldn't rave enough about his fortune to have been placed in such a wonderful hospital, etc."

After moving to Dresden in 1917, the family owned no longer the means to rent an apartment in a summer resort during school holidays but had to wait for invitations from relatives or friends. One or two summer vacations were spent at Uncle Karl Pusinelli's country place in "Switzerland of Saxony," the Elbsandstein Mountains, that marvelous area of sandstone cliffs not far southeast of Dresden, and their father's friend, Uncle Oscar Mey, invited them to Gargellen in Vorarlberg (Alps) where he owned the hotel "Madrisa" at an altitude of 1,500 m. The Mayr brothers were deeply impressed by the most beautiful landscape and alpine flowers. Aunt Gunny invited Ernst and Hans to visit her for a month in 1924 at her summer place in the Ammer Valley (Vorarlberg).

## Birdwatching

When he was 9 years old, Ernst learned (with the help of brother Otto) the more common birds in the large park of the Residenz in Würzburg. Otto also owned an aquarium and took Ernst along to catch sticklebacks and aquatic insects in the backwaters of the Main River. While Ernst's interests remained strongly centered on natural history from an early age on, his younger brother Hans never developed a serious interest in animals or plants. The parents subscribed to the well-known natural history magazine for amateurs, *Kosmos*, and also provided them with all the popular volumes by Wilhelm Boelsche and other authors in the *Kosmos Library* series. Certainly, this reading furthered Ernst's tendencies and he acquired a great speed in flying diagonally across a page for contents rather than looking for literary style (see also p. 33). This way, he consumed most of Karl May's adventure stories and often managed to read an entire 400 to 500-page volume in a single evening. He devoured all the books he could get on Arctic and Antarctic expeditions and on travels to unknown parts of South America, Africa, and Asia. In particular he was fascinated by the travels of the Swedish explorer Sven Hedin to Tibet. He read on different topics including geology, anthropology, and art history. In fact, there was a time at the gymnasium when he thought of studying art history as his life's subject. Ornithology remained strictly a hobby.

At about 14 years of age, he started a "most active ornithological period stimulated, I am embarrassed to admit it, by competition with another student in my class in the Gymnasium, who bragged about his knowledge of birds. That was more than I could endure. However, I was also the friend of a forester's son in Moritzburg, with whom, for instance, I tried to observe a badger. In Dresden, if weather was reasonable, I went birdwatching almost every single day. As soon as I had a bicycle, I also traveled to the Lausitz, to Moritzburg, and to mountainous areas east and south of Dresden. Once I got word that a nightingale was singing about twenty kilometers from Dresden, I immediately got on my bicycle and was more than elated actually to hear the bird singing."

During the winter of 1919–1920, the young birdwatcher checked almost daily on the birds in the Grosse Garten, a large city park. He observed that the Bohemian Waxwings were rather common there feeding on mistletoes in the large Linden trees and he found out which of the old hollow trees was the home of the Tawny Owl. In spring, he checked on the nests of Blackbird *Turdus merula* and Songthrush *T. philomelos* using a mirror with a long handle to see the contents of their nests. However, there was nobody to advise him about some constructive research or the literature. His guide was Alwin Voigt's *Excursion Book for the Study of Bird Voices*. It offered a system of signs representing bird calls and songs. This seemed to him more useful than Bernhard Hoffmann's book (1919) where bird songs and call notes were described in musical symbols. At school, he enjoyed most the biology course and did a lot of reading on biological subject matter. He also took a course in local flowers and learned how to identify them with the help of identification keys. He was a good high school student but judged by his classmate, Karl Baessler, when asked in 1990, there was nothing particular pointing to his later success in science.

In 1920 and 1921, he entered in his notebook 25 detailed bicycle routes that he took around Dresden for birdwatching, carefully noting the distances and times spent (ranging between 23 km in 2 h and 178 km in 10 h).

"I used to be an enthusiastic bicyclist myself while I was a bird student in Germany. The poor bike certainly suffered very much. I used to take it through swamps and over the beach, through woods and over fences. It is really much handier for birding than the automobile, and after a while one gets so good at it that one can ride almost anywhere. Besides it is good exercise" (letter to J. P. Chapin, 1 December 1936).

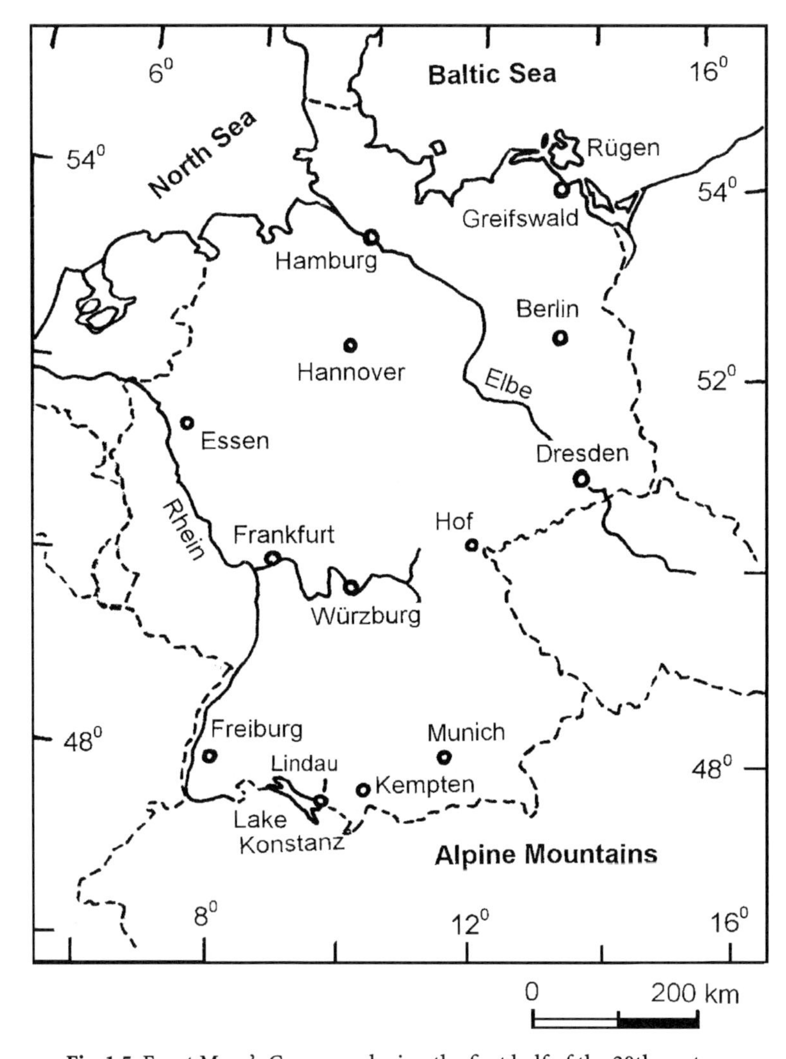


**Fig. 1.5.** Ernst Mayr's Germany during the first half of the 20th century

From 6 to 14 September 1925, he rode his bicycle across southern Germany over most of the distance between Hof (Bavaria) in the northeast and Lindau on Lake Constance in the southwest (Fig. 1.5), only occasionally using the train for short hops when the weather was really bad. He wrote a detailed diary in 4 days. On the first evening, he slept in a barn on straw and was on his way again at 4 o'clock in the morning after a cold night:

7 September (transl.): "It had stopped raining but a strong western storm was raging. It was pitch black but the old alley trees marked the way. Konradsreuth was still completely quiet, only some dogs were barking. The clouds rushed very low over the road. However, as a happy promise, the morning star then appeared

in a gap between the clouds. Bicycling was heavy going, strong contrary wind and bottomless [muddy and unpaved] roads, in addition they went continuously up and down. The landscape was very pretty, mountain meadows, and spruce forests. When I stood on top of a hill drawing breath, the moon broke through the rushing clouds, a wonderful setting. Dawn commenced a quarter off five o'clock and at six o'clock I was in Münchberg. [...] A freezing wind–Fichtel Mountains–in addition the fear that it might start raining again. Finally I reached Gefrees. From there on the riding became easier. To the left one might have had a nice view toward the 'Schneeberg' but everything was cloud covered. Finally, the descent began toward Berneck, a long-missed pleasure, I even had to use my backpedaling brake. I was very delighted with Berneck and its environs. To continue toward Bayreuth would not have been difficult, if I hadn't been completely exhausted. So I put up just before the last ascent, drank half a liter of milk and ate several slices of bread. Meanwhile it started to rain. When it stopped after an hour, I rode via Bindlach to Bayreuth. Because it rained again I decided to take the train to Pegnitz. [...] There it cleared up. I jumped on my bike and rode over deeply muddy roads through the romantic 'Switzerland' of upper Franconia to Pottenstein. When I got there it poured again for a change and I had to look for cover. The land lady asked me: 'Where are you headed?' 'For Lindau.' 'Oh, there I know the hospital doctor, Dr. Mayr. Yes, I was employed by his son in Heidelberg,' etc. She was Elisa Radl, the former housekeeper of Uncle Hermann. We were both quite surprised. Over even more deeply muddy roads I reached Muggendorf where I stayed overnight.

8 September. At 8 o'clock in the morning I continued and reached Forchheim riding along the less interesting lower valley of the Wisent River passing through Streitberg and Ebermannstadt. The road defied any description; on occasion the mud was so deep that bicycling was impossible." From Forchheim he took the train south to Neumarkt (via Nürnberg) "and then had a wonderful ride downhill with back wind along the Danube-Main canal. I was enthused with Berching where Fritz Gösswein[3] was born and reached the Altmühl Valley at Beilngries." Following the Altmühl upstream to the village of Arnsberg he turned southward up onto the plateau where Böhmfeld is located arriving after dark: "I had a strange feeling when I entered the birthplace of my great-grandfather.[4]" On the following day (9 September) Ernst checked the church books and found the data he was looking for. On 11 and 12 September he stayed in Unterthingau near Kempten (Allgäu) where his other great-grandfather J. E. Mayr was born and where he certainly also checked the church records. These activities were part of the genealogical research that the Mayr brothers conducted during the 1920s (see p. 11, footnote). Regarding the continuation of Ernst's trip from Ingolstadt to Lindau via Augsburg

[3] F. Gösswein was Ernst Mayr's classmate and friend at the gymnasium in Munich for three and a half years. They lost contact when the Mayrs moved to Dresden in 1917. His father was a blue collar worker (running a locomotive engine) and Fritz was living proof that the saying, children of blue collar workers were not accepted at the humanistic gymnasium, was wrong.

[4] Bonifazius Müller (1805–1860), father of Wilhelmine Müller (p. 8), was Royal Bavarian District Physician in Lindau at least from 1837 until his death.

**Fig. 1.6.** Ernst Mayr (on *right*) on a field trip with members of the Saxony Ornithologists' Association, April 1925. From *left* to *right*: R. Köhler, E. Dittmann, P. Bernhardt, H. Förster, E. Mayr. (Photograph courtesy of E. Mayr.)

and Kaufbeuren only a list of expenses is preserved but no detailed diary. In early October he was back in Berlin, where the German Ornithological Society celebrated its 75th anniversary (p. 22, footnote). The Mayr brothers also covered long distances on foot; 35–40 km a day was the average, but occasionally they walked up to 50 km (see also p. 50).

Shortly after it had been founded in April 1922, Ernst, while still a high school student, joined the Saxony Ornithologists' Association (Verein Sächsischer Ornithologen) and attended its annual meetings (Fig. 1.6). The majority of these local ornithologists pursued special aspects of the bird fauna and none of them was the kind of list chasers ("twitchers") like the birders of New York whom Mayr would meet 10 years later. Once he had bicycled a long distance to participate in an annual meeting because he could not afford the train fare. When the weather turned bad a kind dentist from Chemnitz (Johannes Keller) paid the train fare for him and his bicycle back to Dresden. Another member, Paul Bernhardt was specialized on the birds of the lakes at Moritzburg; W. Salzmann, the society's secretary, made faunistic observations around Leipzig; H. Foerster studied the Peregrine Fal-

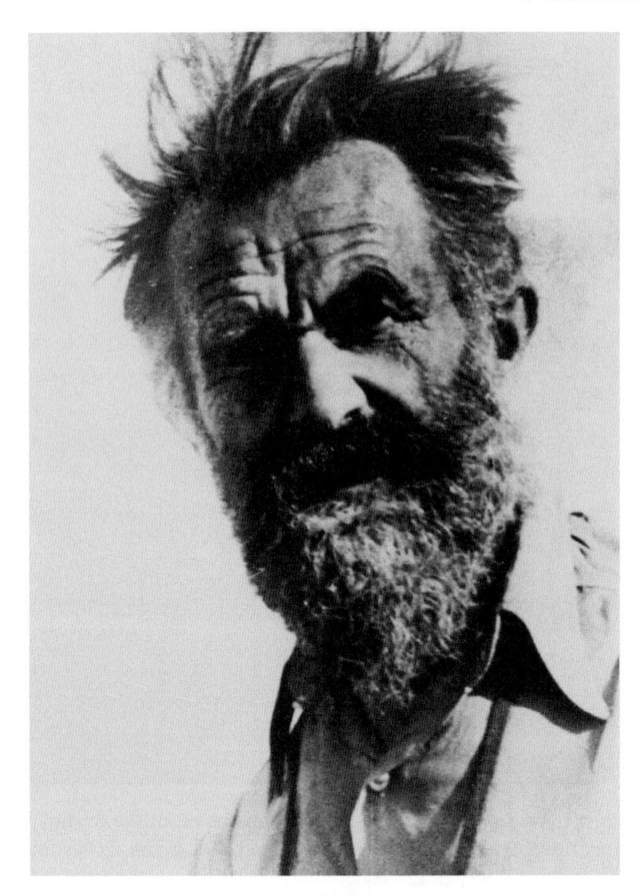

**Fig. 1.7.** "The lovable eccentric"—Rudolf Zimmermann; photograph taken in 1942 (Archive, Museum of Natural History Berlin, Orn. 158,1)

con, E. Dittmann and R. Köhler observed other birds near Dresden. At one of the meetings, Mayr was introduced to Rudolf Zimmermann (1878–1943), editor of the association's publication. With him, our young ornithologist learned more than from anyone else, particularly that one had to know the relevant literature and that one had to ask biological questions. Through Zimmermann, Mayr also met the dean of the Saxon ornithologists, Richard Heyder (1884–1984) in Oederan whom he visited for the last time in 1973. Their correspondence ended when Heyder died about 6 months before his 100th birthday. Ernst Mayr wrote very personal reminiscences of Rudolf Zimmermann (Fig. 1.7):

"He was my chief ornithological mentor during my high school days, and furthermore, one of the most extraordinary, lovable eccentric people one can imagine. He lived in Dresden in a one-room apartment largely filled with books and periodicals. He had no kitchen, only an alcohol burner on which he made his simple meals if he happened to have enough money to buy some food stuff. If not, he simply smoked cigars. He had a long, dark brown beard and was always dressed for the outdoors. I believe just about all of his income came from writing popular nature articles for various magazines and from his nature photography. He was a very good photographer, even in the technical matters of developing his films or plates.

Zimmermann never had any formal education. I presume that he finished some kind of a high school, but that was all. Everything he knew (and he had an enormous amount of knowledge) was acquired by reading. Why he engaged in all that is quite a puzzle to me. He was, of course, a bachelor; perhaps his speech defect—particularly when a little excited he stammered badly—contributed to his solitude. This was, of course, the reason why he couldn't earn his living as a teacher or lecturer.

I often visited him in his room because he had so many interesting books and could tell so many interesting stories about birds and their lives. He never ceased emphasizing how important it was to study the living bird, that one needed patience, that one should sit by a nest or watch a displaying bird by the hour and only thus could one get the exact details of what was going on. Most importantly, he took me along on some of his excursions to the Lausitz region [east of Dresden] with its innumerable ponds and marvelous bird life. Here he showed me how to find the nests of all sorts of birds, and determined which ones were apt to be parasitized by the cuckoo. He had a splendid knowledge of its habits, and he could tell the cuckoo egg from the own eggs of the birds, etc. One of his weaknesses, however, was that he did not carefully record everything he saw; so much of his immense knowledge of bird life died with him. He was just as much the opposite of the American birdwatcher as one could possibly think. He was not concerned with records and rarities but with the living bird and its behavior.

I remember on one of the trips we came to a fish pond that had just been emptied in order to catch all the carp that were in the pond. However, in one of the drainage ditches Zimmermann spotted a carp that had been forgotten. He picked it up and put it somewhere, presumably in his pocket. After much more birding, we finally caught the last train back to Dresden. It was too late for me to go home, I think it was about midnight, so I went to Zimmermann's room. There he cooked the carp on the alcohol burner and we had a delicious meal at about 1:00 am. Then I laid down on the floor and slept soundly until about 6:00 am. I said goodbye to Zimmermann and walked to my own home, where my mother was rather astonished when I appeared ringing the bell at 7:00 am.

The lessons Zimmermann taught me at my most impressionable age have stayed with me all my life. Alas, in New Guinea, the collecting of birds took so much of my time that there was no opportunity, really, for making good life history observations.

As I have mentioned already, Zimmermann had a splendid library. It included such rarities as the major book by Pernau [1720]. Alas, I am told, that all this, including a valuable series of periodicals, was destroyed during the infamous Dresden bombing [in February 1945].

Zimmermann died in his 60s from cancer of the throat, which I attribute to his life-long smoking cigars."[5]

## The Duck with a Red Bill

As a reward for just passing the high school examination (Abitur), in February 1923, Ernst's mother gave him a pair of binoculars. For several weeks, he went on daily excursions into the hilly neighborhoods of Dresden, to the lakes of Moritzburg (a hunting chateau of the Saxon kings), or to the gravel banks of the Elbe River, where he worked out the differences in call notes of the two common plovers (Mayr 1925a). It was on one of the lakes of Moritzburg, the Frauenteich, on 23 March 1923, that the first major historical accident in his life occurred: He discovered with his new binoculars a pair of ducks, the male with a red bill that was totally unknown to him and that he determined at home with the help of his bird books as Red-crested Pochard (*Netta rufina*). This bird had not been seen in Saxony since 1845 and, therefore, arguments over the reliability of the identification of these ducks arose among the members of the local bird society in Dresden. To settle the issue, Dr. Raimund Schelcher (1891–1979), at that time a pediatrician in Dresden, suggested that Mayr visit his former schoolmate Erwin Stresemann during a stop-over on his way to Greifswald where Mayr was to begin medical studies. Schelcher wrote a letter of introduction, thereby establishing the fateful link between these two scientists. Dr. Stresemann, already a leading ornithologist in the country, after a detailed "cross examination" and an analysis of Mayr's field notes accepted his observation as valid and published a brief note on it (Mayr 1923a).[6] Stresemann was so taken by the enthusiasm and knowledge of his visitor that he invited him to work between semesters as a volunteer in the ornithological section of the museum: "It was as if someone had given me the key to heaven," Mayr (1997d: 176) recalled this event many years later. The contact with Stresemann was to change the course of his life very soon.

"In the next university's vacation, I worked at the museum. On the very first day, Dr. Stresemann handed me two or three trays full of little brown treecreepers. There are, in Europe, two species which are so similar that until a generation ago all the leading ornithologists insisted that they were only one. It was not until their songs became known that their specific distinctness was recognized. The specific differences between the two are so minute that it requires a lengthy

[5] Rudolf Zimmermann and Ernst Mayr appear standing side by side in the group photograph taken on the occasion of the annual meeting of the German Ornithological Society in Berlin in October 1925 (see Haffer et al. 2000, p. 434). An obituary of R. Zimmermann was published in the *J. Ornithol.* 92, 1944; (see also H. C. Stamm, Mitt. Landesverein Sächsischer Heimatschutz 1999, pp. 57–61; Mayr 2003a). Ernst Mayr is mentioned repeatedly in the letters exchanged between R. Zimmermann and R. Heyder from the 1920s to the 1940s which were published recently (Stamm, H. C. & J. Hering, Mitteilungen des Vereins Sächsischer Ornithologen, vol. 10, Sonderheft 1, 2007).

[6] The year of the observation was 1923, not 1922, as incorrectly given in the published note (pers. comm.).

study of each specimen before it can be placed. Actually, to an experienced worker there are some intangible differences which greatly facilitate the identification. The beginner might require two or three days to carefully identify 150 specimens and even then there is a good chance that he has made about thirty or forty per cent misidentifications. To make a long story short, I came back to Dr. Stresemann after half an hour and had, with one or two exceptions, every specimen identified correctly. It turned out that I had realized and recognized the intangible differences in about one glance, and, therefore, had little difficulty in doing the job. 'You are a born systematist' he exclaimed, and [in early 1925] he prevailed upon me to give up my medical studies and enter in the museum career. This is what I eventually did and I do not think that I have ever been sorry for it. The one point that I believe is brought out in this story is that there is something like a born systematist. Of course, anybody with intelligence and discrimination can do fairly good work. But there is no question that certain authors have been able to straighten out one taxonomic 'mess' after the other that had stumped all the preceding workers" (E. Mayr to Professor E. Anderson on January 28, 1941; Harvard University Archives, E. Mayr Papers).

## University Student in Greifswald and Berlin

In the spring of 1923, Mayr entered the University of Greifswald as a medical student (Fig. 1.8). Recorded in an early notebook, the thought had been continuously in the back of his mind: Could his father's early death not have been prevented by

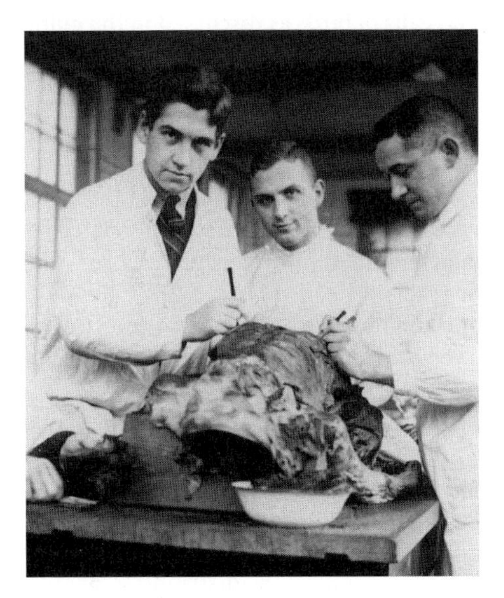

**Fig. 1.8.** Ernst Mayr (*left*) as a medical student in Greifswald, 1924. (Photograph courtesy of E. Mayr.)

a diagnosis of his illness in time? This thought and the family tradition on both his father's and mother's side led him almost automatically to choose a medical profession. He had selected Greifswald on the Baltic Sea not because of its academic reputation but because of the excellent birding areas nearby, and the Darss spit and the islands of Hiddensee and Rügen not far away (1980n). He went birdwatching in the forests and along the beach of the Baltic Sea almost every day, mostly alone or in company with his friends Herbert Kramer (1900–1945), a student of zoology, or Wilhelm Bredahl and Werner Klein, fellow medical students. They banded lapwings and dunlins in the marshy Rosenthal area and found the Red-breasted Flycatcher (*Ficedula parva*) commonly breeding in the beech forests of the Elisenhain in Eldena (Mayr 1923b), saw the fairly rare Middle Spotted Woodpecker (*Picoides medius*) and watched Little Gulls (*Larus minutus*) at the seashore (Mayr and Klein 1924a). Hans Scharnke (1931), a younger schoolmate from Dresden, later published many of Mayr's and Kramer's records together with his own field observations in the Greifswald region. Mayr had a glorious time, summer and winter and felt truly exhilarated on his birding excursions, especially those in the early morning. Some of his discoveries he still remembered with excitement 80 years later, e.g., the nest of a snipe in a bird-rich marsh, nests of both treecreepers (*Certhia familiaris* and *C. brachydactyla*) permitting a careful comparison of these sibling species, vast flocks of wintering geese on unused fields, and there were many more! Also, he enjoyed his studies at the university and his complete freedom tempered by a sense of responsibility.

Mayr's considerations about bird study are well reflected by a preserved list of 28 points. This list was probably written during the first half of 1924 (definitely after September 1923; see point 16). Evidently he was particularly occupied at that time with the songs and calls of birds as discussed in the publications of the Saxon ornithologists Rudolf Zimmermann, Bernhard Hoffmann, Bernhard Hantzsch and Alwin Voigt (1921). These were his questions:

(1) Is the call of the European Jay (*hiäh*) species-specific, or an imitation of that of the buzzard?
(2) Which species of birds can imitate?
(3) How different are the vocalizations of closely related species?
(4) How good is the memory of birds?
(5) Song and mating calls.
(6) One should not claim that species have expanded their range when they had not been observed previously. They might have been overlooked (nocturnal mammals; see Zimmermann).
(7) Conversely, a previously observed species has not necessarily become extinct if it is no longer observed (nutcracker).
(8) One must know the literature before one can make statements about the status of a species.
(9) Which are the latitudinal and altitudinal species borders?
(10) Which species of plants control the distribution of animals?
(11) How far does a nonmigratory bird roam? (see Bacmeister, *J. Ornithol.* 1917, II).

(12) Test by aviculture whether the two species of treecreepers, the willow tit and the marsh tit, the sprosser and nightingale, etc., can be crossed in captivity (like the horse and donkey).
(13) Investigate the degrees of relationship among species of a single genus, for instance *Charadrius dubius* and *hiaticula*.
(14a) Do only such species hybridize in which the female alone raises the young (Capercaillie, etc.)? Or also such others that are very similar in their whole reproductive biology?
(14b) How helpful is microscopic analysis of feathers in order to distinguish between juvenile, female, and male plumage?
(15) Investigate the distribution of *Muscicapa hypoleuca, semitorquata, albicollis.*
(16) How far north does the osprey range? Migrant ducks from further north or east passing through Moritzburg, were afraid of it on 12 September 1923.
(17) Is it legitimate to make anthropomorphic descriptions of birds, as done by B. Hoffmann?
(18) How independently has the beauty (a relative term!) of bird songs evolved?
(19) Closely related species of birds often differ strikingly in the length, structure, and elaboration of their songs (B. Hoffmann).
(20) Different species of birds differ strikingly in the tonal quality of their songs.
(21) What is the psychological significance of the various call notes of a species (fear, pleasure, etc.)?
(22) [A repeat of question 5] B. Hoffmann, p. 21
(23) How differently do different nations represent the vocalizations of the same species of birds?
(24) To what extent do females sing? (Hantzsch) Which calls are shared by male and female?
(25) If the song of the male serves to attract females (Hoffmann, p. 93), why do birds sing also long after the period of pairing and frequently even in the autumn? (*J. Ornithol.* 1917, II, F. Braun)
(26) Can birds really count up to three? (B. Hoffmann, p. 120).
(27) When searching for bird nests, always record how far away from the nest the male was singing.
(28) Find out how many clutches per season the willow tit raises, and search for their nest. Willow tits tend to spend the night in their nest cavity.

Despite the birds, Mayr attended all his medical courses very conscientiously except for his stay at a farm for a couple of weeks during the first semester to help with haying in order to get something to eat (in 1923 the inflation in Germany was rampant and food scarce in the cities). After some difficulties he had found a furnished room in Burgstrasse 17. He had to go down a staircase and cross a yard to get to a toilet. There was, of course, no heat, no warm or running water, only a wash basin (where the water sometimes froze in the winter) and a pitcher with water to fetch somewhere from a faucet. His lunch in the mensa, the student dining hall, consisted mostly of a rather thin soup.

Mayr disliked the student fraternities and their activities like dueling and drinking and turned down their attempts to enroll him. Eventually he joined the Deutsche

Hochschulgilde Sankt Georg. This local Gilde had developed from the Youth Movement: no drinking, but much hiking, singing folksongs, etc. Originally quite unpolitical, the Gilde at some universities became nationalistic in later years and some of them even Nazi.

"My intimate knowledge of the countryside around Greifswald led to a noteworthy incident in my life. A famous professor of paleontology at the university, O. Jaeckel, was an ardent nationalist and at the same time involved in a youth group somewhat like the boy scouts. One of their activities was to have war games. How I came to get involved in this, I no longer remember. However, I was chosen to be the leader of the blue party. The red party had a fort in the middle of the Elisenhain (a forest region, my favorite birding locality) and the assignment of the blue party was to enter the woods on the main road and to take the fort. I saw at once that the whole scheme was set in such a way that the red party (with Professor Jaeckel) had to win. How could I thwart this plan? I knew enough about battles and wars to know how often a battle is won because one army did something totally unexpected by the other. So I decided to surprise the red party. From my birding excursions I knew that there was a small road into the Elisenhain from the opposite direction. A road that could be approached by using service roads through the fields. So I entered the Elisenhain with my blue army from the back and of course had no trouble at all taking the fort by surprise. In the ensuing review the military experts decided that I had not won because I had not followed exactly the instructions. All of my arguments that such a policy would be the ideal education for losing wars were of course rejected. Also I was never again asked to lead an army in such war games. And that was the closest I ever was involved in war."

During short vacations, Mayr explored nearby areas on the Baltic seashore, the Darss spit and the island of Rügen, and on one occasion he visited his older brother Otto in Danzig (Poland today) where the latter was taking an engineering degree. Near Danzig he watched his first Common Rose Finch (*Carpodacus erythrinus*) and heard its distinctive clear whistle. Most importantly, however, he spent several weeks between semesters at the Museum of Natural History in Berlin as a volunteer with Dr. Stresemann. There he also assisted visiting ornithologists like R. C. Murphy from New York and the famous Russian ornithologist M. Menzbier from Moscow. On one occasion, Stresemann assigned to him the identification of a recently received collection of birds from Java. He could have done this himself in less than an hour, but he knew that this assignment would greatly widen Mayr's horizon.

*Early scientific views on species and evolution.* The roots of Mayr's later contributions to the evolutionary synthesis reach far back to the early 1920s and his seemingly abrupt appearance as an "architect" of the synthetic theory of evolution during the early 1940s had a long history which, however, must be reconstructed from unpublished letters, notebooks and articles in little known journals.

In a letter to Dr. Erwin Stresemann dated 12 May 1924, he reported from Greifswald not only details of his ornithological observations in the field but also reflected on theoretical matters concerning the ecological and historical origin of geographic variation in a bird species and on the evolution of species themselves!

Evidently he had followed up on certain topics discussed with Stresemann during earlier visits to the Berlin museum. Most of this letter is here included because of its historical interest (my translation; letters refer to the notes at the end; see also Haffer 1994b, 1997):

> *Dear Dr. Stresemann!*
> *[...] Now I would like to ask you to undertake several studies. You told me in Berlin that [A. B.] Meyer mentioned in his* Birds of Celebes *[A] that everything in ornithology will soon be accomplished with the help of mathematical formulas. How about you setting up principles of an ornithological mathematics? This would save us from a quaternary nomenclature (which appears necessary from a purely theoretical point of view). (1) Intensity index. Compare* Parus atricapillus rhenanus *and* subrhenanus, Motacilla flava rayi *and* thunbergi, Carduelis l. linaria *and* cabaret. *(2) a geographical, respectively climatic, factor (desert, steppe, polar climate, island, humid coast line, etc.), (3) the individual variation needs to be taken into account, and (4) the nomenclature of intermediate forms (Fig. 1.9)*
> *Five strongly differentiated forms are to be expected in this region where one has to distinguish between the intermediate forms between 1 and 2, 2 and 3, 3 and 4, 5 and 6 on one hand and between 4 and 5 on the other hand. For instance, the transition from 5 to 6 will be gradual so that a [b, c, etc.] will be rather uniform. However, hybridization will take place between 4 and 5 and therefore a strong individual variation will occur (see Long-tailed Tit [B]). Therefore intermediate forms between 5 and 6 have to be labeled 6–5, by contrast those between 4 and 5 should be designated 4 × 5, etc [C]. Isolation needs to be taken into consideration too. If a form does not continuously receive fresh blood [gene flow] from the parent form, it will enter into a totally aberrant evolution. This, however, probably cannot be expressed mathematically. So many subspecies have been described within recent years that the time is ripe to write a* "general Hartert" (or "The Theory of Geographical Variation and of the Species") *[emphasis added].*
> *The facts mentioned above will have to be discussed as well as modifying factors, i.e., active factors (climate, etc., partly based on Görnitz [D]) and passive factors (color, size, biology, etc.). In addition the following topics need to be treated: (1) How is it possible that members of the same Formenkreis overlap their ranges without hybridizing? and: (2) May similar and geographically representative species nevertheless be members of different Formenkreise? An example: If* Ficedula hypoleuca *and* F. albicollis *would exclude each other geographically, they would surely be included in the same species. Conversely, someone may say: I can no longer recognize the forms of the Yellow Wagtail only as subspecies, although they mostly exclude one another geographically; just as the delimitation of genera is more or less a matter of taste.*
> *Furthermore the phylogenetic connections of the different species must be clarified. With the help of mutations and several representative forms like Sprosser* [Luscinia luscinia] *and the Short-toed Tree-creeper (*C[erthia] brachydactyla*)*

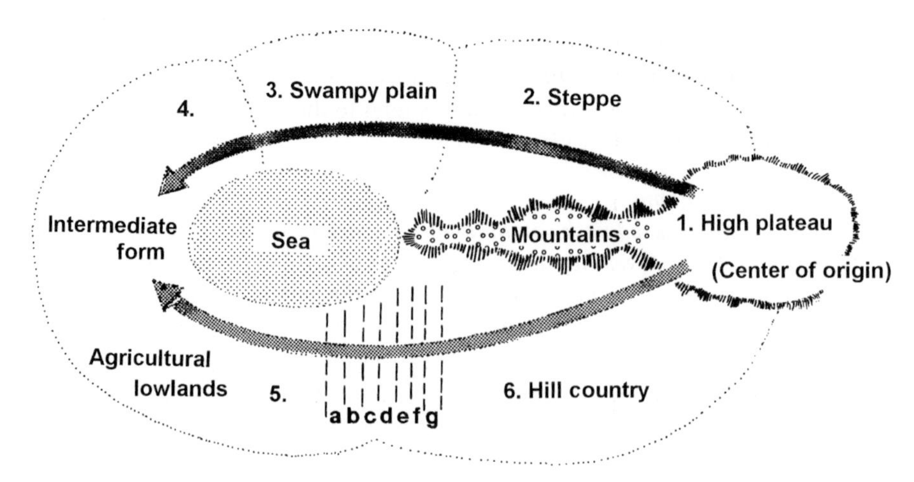


**Fig. 1.9.** Dynamic-historical interpretation of the origin of geographical variation in a bird species as sketched schematically by the medical student Ernst Mayr in a letter to Dr. E. Stresemann (Berlin) dated 12 May 1924 (redrawn and terms translated). Arrows indicate range expansion of populations from a high plateau ("center of origin"). An intermediate form (*Zwischenform*) results from secondary contact of populations circumventing barriers like mountains and the sea. Various ecologically different regions and their respective subspecies are indicated as follows: 1. High plateau, 2. Steppe, 3. Swampy plain, 4.–5. Agricultural lowlands, 6. Hilly region; a–g symbolize stages in a cline between the subspecies inhabiting areas 5 and 6

*a theory may be elaborated [E]. This is necessary to replace Kleinschmidt's dogma which does not advance science [F]. I am an adherent of Lamarckism (despite the theory of genetics, Baur's modifications, etc.). Each organism has a large number of equal possibilities of development and this explains the phenomena of convergence.*

*An attempt should also be made to find out for all bird species where they originated based on certain characteristics (Berajah [G]). The comparative morphology of immature plumages would probably play a decisive role in such an endeavor. That form whose immature plumage most resembles the adult plumage, perhaps represents the ancestral type. In searching for the center of origin one must avoid the mistake (as happened before, particularly in ethnography) and select that region from where the least amount of material is available. One thought has not yet been incorporated into the ornithological theory of colonization which (perhaps more than justified) is prevalent in ethnography, namely the thought that a center of origin sends out again and again entire waves of animals. According to what I have read so far ornithologists seem to think that the centers of expansion are always located along the periphery of the distributional range. However, they may equally well have been located in ecologically favorable areas [within the ranges].*

*Another interesting question is the rate of differentiation ("Polish Serin" [H]). Supporters of the mutation theory should answer this question: In which way does the established form disappear from its range? Possibly the mutant is more strongly expansive. The peculiar phenomena of convergence which caused so many errors in systematics need to be taken thoroughly into consideration. Very many ornithologists would appreciate a comprehensive treatment of these problems. Surely such a study would suit you. [...]*
*With ornithological greetings. Sincerely yours, Ernst Mayr.*

*Notes*:

[A] In cases of clinal or stepped clinal geographical variation, A. B. Meyer and L. W. Wiglesworth (1898) and L. Wiglesworth (1898) had proposed to assign subspecies names only to the endpoints and to selected intermediate stages in the character progression.
[B] This is a reference to the hybridization between the white-headed and stripe-headed forms of the Long-tailed Tit (*Aegithalos caudatus*) in central Europe.
[C] Mayr distinguished here between primary and secondary intergradation.
[D] Görnitz (1923: 498) had concluded "that the majority of the geographical subspecies are not due to the effect of natural selection but originated through the effect of climatic factors."
[E] This is a reference to two pairs of sibling species (*Luscinia luscinia/L. megarhynchos* and *Certhia brachydactyla/C. familiaris*) whose members probably originated in geographical isolation from their respective common ancestors and are today in secondary contact along a narrow (nightingales) or broad (treecreepers) zone of overlap in central Europe.
[F] The Protestant pastor and ornithologist Otto Kleinschmidt (1870–1954) had proposed a typological species concept ("Formenkreis") and the independent origin (creation) of all animal species.
[G] In his excellently illustrated monograph series "Berajah" (1905–1936) Kleinschmidt had discussed geographical variation and range expansion of numerous species.
[H] Several subspecies of the Serin (*Serinus serinus*) had been described from regions (Germany, Poland), which this bird had colonized only during the last century. In his dissertation on "The expansion of the Serin" Mayr (1926e) was able to prove that the Serin finch does not differ in these latter areas taxonomically from the populations in the Mediterranean region.

When I sent a copy of the above letter to Ernst Mayr in 1992, he was totally surprised and answered (transl.):

"I am terribly amused at my forwardness with Stresemann. No doubt this is part of the reason why in the Stresemann family I was always referred to as the fresh ('freche') young Mayr. I admire the patience with which Stresemann tolerated me and my letters. He must have appreciated my genuine deep interest in science. I had completely forgotten all about this letter and even reading and re-reading it now does not trigger any remembrance."

The three-volume *magnum opus* of the renowned ornithologist Ernst Hartert on the birds of the Palearctic fauna (1903–1922) had just been completed. However, Hartert was mainly a practical taxonomist. The above letter of the 19-year-old Ernst Mayr shows his early interest in theoretical analyses of taxonomic data and his ability to synthesize critically the results derived from studies in widely different fields, capacities which characterize many of his publications in later decades. It is evident that already in the spring of 1924 he was familiar not only with writings of the leading ornithologists Stresemann, Hartert, Kleinschmidt, etc., but also with the basic concepts of evolution, genetics and systematics through a close reading of textbooks and specialized articles, as shown by his reference to Erwin Baur's "modifications" and the theory of inheritance of acquired characters. His concern with genetics probably goes back to the volume on human heredity (*Menschliche Erblichkeitslehre*, 1923) by E. Baur, E. Fischer and F. Lenz which greatly impressed the student in Greifswald (notebook). Mayr's course in genetics was rather traditional consisting largely in exercises demonstrating the "Mendelian Laws"; the emphasis was on mutation and physiological genetics. The connection between genetics and evolution was not dealt with at all.

Heavily occupied with the manuscript for his large volume on *Aves* and many other projects, Stresemann was not able to follow up on Mayr's suggestions. However, he was so much impressed by his enthusiasm and knowledge that he wrote to his fatherly friend Ernst Hartert a few weeks later (12 July 1924):

"I have discovered [...] a rising star, a young Studiosus med[icinae] by the rare name of Mayr, of fabulous systematic instinct. Unfortunately, he will probably have to wither away as a medical doctor. If only one could always place the right man in the right position!" (transl.)

At this moment, neither Stresemann nor Hartert could have imagined that this young man, less than 8 years hence, would be proposed to be Hartert's successor at Lord Walter Rothschild's private museum in Tring (see p. 96, footnote) and actually became the curator of the Rothschild Collection at the American Museum of Natural History.

In 1925, Mayr entered the following remarks into his notebook:

"Theses regarding the dispute between Darwinism and Lamarckism.

1. The controversy in evolutionary theory today does not concern the question of selection. Selection is also acknowledged by the Lamarckian.
2. The controversy relates to the cause of variability (which Darwin accepts as given).
3. Varieties [heritable] originate according to De Vries by random mutations. The mode of life of an organism is determined by the structures thus originating. This seems to be the view of some evolutionists, particularly of experimental zoologists and geneticists.
4. The Lamarckian, on the other hand, claims that new variants originate under the influence of the mode of life [Lamarck] or of the environment [Geoffroy].
5. The Lamarckian has the right to interpret the laws and findings of genetics in his sense; this refers both to mutations and Dauermodifications; cumulative aftereffect–Alverdes.

6. While the experimental biologist states that mutations are undirected (random), the Lamarckian asserts mutations on the way to an adaptation are directed[7]. It would make no sense to believe that the destabilization of the germ plasm (caused by a mutation) would be without influence on subsequent mutations–Orthogenesis.
7. Changed conditions of the environment influence the reaction of the body plasma [soma] (modification of the phenotype). If this influence continues for a lengthy period, the germ plasm will also be influenced, the modifications become heritable, they become Dauermodifications, which after return to the normal environment will disappear only after many generations (cumulative aftereffect). We must assume such Dauermodifications in many geographical races.
8. There are no proper arguments against Lamarck's claim that organs deteriorate by lack of use. It is in line with the economy of the organism that of the available 'fund' (Hesse) [resources] particularly those organs will be endowed which are very much in use (Roux, the Struggle of Parts in the Organism). On the other hand, it is reasonable that in the organs which are used most actively corresponding to the degree of use, mutations will occur, the maintenance of which will be controlled by selection.
9. The Lamarckian theory [in its modern version] is not teleological."

Additional entries in his notebook read (transl.):
"Darwin distinguished between sudden and gradual changes too."
"The genetico-darwinists always claim that a very small mutation is sufficient gradually to prevail through natural selection. However, there are sufficient examples that at least a predisposition for mutations is triggered by biological processes."
"Certain doubts appeared whether everything can be explained by mutations as, e.g., Baur hints at. These mutations have been derived from experiments. It is questionable that this is a secure basis, because the conditions during experimental work often are not normal, or better, not natural. There is another path, because we encounter frequently natural experiments of speciation that originated under natural conditions. An example is Stresemann's 'mutations', [...] and borderline cases of the doctrine of Formenkreise."

Mayr (1980n: 413) later stated, "I have no recollection of when I first learned about evolution." However, it is true that he had an early interest in evolution and genetics if only coincidental to his enthusiasm for birds. Even though he was inscribed as a medical student in Greifswald, he was first and foremost an ornithologist.

Wide-ranging biological interests are documented by a notebook of 1925 with hints to what Mayr considered open problems and topics suitable for possible future research (transl.):

[7] "I am here in agreement with Professor Buchner in Greifswald with whom I discussed this point."

- Convergence versus phylogenetic relationship in birds: (1) significance of food, bill, etc. (2) significance of subsoil and climate (pigmentation), (3) significance of mode of life (sense organs, legs, claws, wings, etc.);
- Biology and relationship; nest building, feeding, voice, courtship, sociology, birds and plants, diurnal and nocturnal mode of life, migration instinct, number of broods and eggs, molt, wintering areas, direction of migration, proportion of males and females during migration, care of young, ecology, race formation, variability, differences of males and females in size and coloration;
- Influence of age, temperature, weakness, irritants, poison, etc. on the determination of sex (in plants);
- Fragrance of flowers: (1) Which flowers of our flora are fragrant? (2) During which months and at what time of the day are they visited by which insects? (3) Definition of fragrance (4) Repellents (mercaptane, ether), (5) Organs producing odor, (6) Elimination of all optical markers, e.g., cut off flowers.

Originally, it was necessary to study five semesters before one could take the cand. med. examinations. Mayr had planned to stay in Greifswald for three semesters and do the next two semesters at Tübingen University. While he was in his second semester, a new regulation permitted taking the cand. med. only after four semesters. Therefore he decided he should finish in Greifswald. In February 1925, Mayr passed his pre-clinical examinations with straight "A"s in all six subjects, a rare achievement. When he again visited Berlin and his "beloved Zoological Museum" (entry in an early notebook), Dr. Stresemann persuaded him to switch to zoology and to major in ornithology, partly by promising to place him on an expedition to the tropics later on. This was a temptation Mayr could not resist, particularly because by that time, certain doubts had been growing in his mind regarding medical practice as his lifelong occupation. Within the field of medicine, he could see himself only as a researcher in one of the basic medical sciences (early notebook). Stresemann gave him at once the topic for his dissertation: "The range expansion of the Serin finch *Serinus serinus* in Europe." He started work during his last semester in Greifswald, now registered as a student of zoology. In the fall, he transferred to the University of Berlin where, in October 1925, he participated in the annual meeting of the DOG (see the group photograph in Haffer et al. 2000: 431). Stresemann remarked in a letter to O. Kleinschmidt on 19 August 1926, "I am placing great expectations in his further scientific development." Without Mayr's chance observation of the pair of rare ducks at Moritzburg, which led to his encounter with Dr. Stresemann, he probably would have become and "withered away" as a medical doctor (p. 30) somewhere in Germany, perhaps known only to the local community of birdwatchers.

The biweekly meetings of the DOG comprised a lecture, usually with slides, or, alternately, a special session (Fachsitzung) when recent literature was reviewed. Most of the new books and monographs were introduced by Stresemann himself, others by his graduate students. When Mayr was asked to discuss a recently published avifauna of a region in Bavaria, authored by a most distinguished local ornithologist, the medical doctor J. Gengler, he simply presented a long list of all the mistakes he had found in the book, for his work on the Serin finch had made him quite familiar with the faunistic literature. When he finally sat down after his review, the president of the DOG, Herr F. von Lucanus, and other members protested against such a young student criticizing the master. Heinroth and Stresemann had to quiet the troubled waters, but Dr. Stresemann later took him aside and advised him always to say something favorable at the beginning (advice that Mayr passed on to his students in later years).

The graduate students had noticed that Stresemann occasionally reviewed a book that he had had no time to read. However, while introducing the author and the title of the book, he studied the table of contents and made a few remarks on it. Meanwhile he opened a few pages that appeared interesting. Reading a page "diagonally" with one glance, he then picked out several important sentences, especially those that he could criticize. Someone who had studied the book from cover to cover could not have done better. Mayr continued (pers. comm.):

"I was an equally quick reader as Stresemann, and bold as I was already in those days, I bragged to my fellow graduate students that I could do the same. When I was assigned again a book for review, (probably) Kattinger seized it immediately, before I could take a look at it. He sat next to me during the following session and gave me the book the moment that Stresemann called me up. My heart was beating to my neck but, strictly following his example, I managed to do the review and Stresemann didn't notice anything. We never confessed to him."

The student home in Borsig Street run by the widow of a Protestant minister was rather primitive to say the least. There was no real vacancy when Mayr arrived but he and 6 or 7 others were permitted to sleep in dormitory-style "housing" directly under the slanting roof in the attic. Each student had a camp bed separated from the next one by a hanging sheet. Several months later he got his own room, but it was alive with bedbugs so that he could hardly sleep. When he complained, he got another room and the first one was fumigated. He stayed in this place until his PhD examination and during one semester shared the room with his younger brother Hans who studied law. Since they always got along splendidly, this was a very enjoyable time.

## Friendships at School and University

At the gymnasium in Dresden, his classmates Robert Hensel and Karl Baessler were Ernst's best friends. They remained in contact throughout their lives. Mayr visited Baessler 60 years later in Bamberg where, after having left Leipzig at the end of World War II, he had established a publishing company, and they corresponded until shortly before Baessler's death in 1990. Hensel became a metallurgist and later went to America where he lived in Indianapolis. However, they saw each other only once or twice after his arrival in the States. In Greifswald, during his first semesters, Ernst met one friend for life, Martin Hennig, and commented: "It is somewhat ironic that the best friend of my student days should be a theologian. But except for a few Christian dogmas, our thinking about man's obligations, about ethics, about all practical philosophy and wisdom were very much the same." They were still corresponding and visiting each other at the age of 90. However, none of these friends shared Ernst's interest in natural history.

Among the PhD students he met at the Museum of Natural History Berlin was Emil Kattinger. They saw each other daily. He was a great help in Ernst's preparation for the PhD examinations: "Every morning in the lab I asked him about all the novel things he had learned the previous evening and night and with my own good memory I stored it in case it should be asked in my final exam." There were no friendships with girls, although he did enroll in dancing classes as usual in society at that time.

In a survey of his friendships during his youth, Mayr felt that none of them had a particularly decisive impact on his development.

# 2 The Budding Scientist

## PhD Thesis and Examination

When, in February 1925, Erwin Stresemann persuaded Mayr to switch from medicine to zoology, the former knew that an assistantship would open up at the museum on 1 July 1926. Therefore he chose a thesis that Mayr could finish in a little more than a year provided he worked very hard. Since about 1800, the Serin finch (*Serinus serinus*) had expanded from the Mediterranean region northward into and through central Europe, but some authors claimed it had only been overlooked previously. Mayr's main work consisted of searching through literally hundreds of local natural histories from France to Poland for statements on the occurrence of the Serin finch (see also p. 165). After his thesis had been published, nobody questioned the steady spread of this bird any more.

The Museum of Natural History and the Zoological Institute of the University of Berlin, although located next to each other, were administratively separate entities and only the director of the Museum, Dr. Carl Zimmer, was a full professor (*Ordentlicher Professor*) at the university giving lectures and courses. He was entitled to have graduate students and to accept their PhD theses, but not the curators of the museum. At that time, they could not submit a habilitation thesis to become regular professors at the university.[1] Through personal agreement with the director of the museum Dr. Stresemann who, in 1930, was awarded the title of professor, was able to accept graduate students and to supervise their work, but officially Professor Zimmer presented the theses of Stresemann's students to the university and was legally their supervisor. This situation caused some problems for his graduate students, when it came to finding a job in the German university system, because their actual supervisor was not part of that system.

By working day and night Mayr accomplished all the course requirements, finished his thesis on time and passed his PhD examination *summa cum laude* on 24 June 1926. He also obtained the position at the museum starting work a few days later on 1 July. His monthly salary was 330.54 Reichsmark.

Mayr remembered many details of his PhD examination: "The director of the museum Carl Zimmer asked me one specific question after the other for a full hour and he later said that I was the only PhD candidate he ever had who had been able to answer correctly every single one of his questions. I had a fabulous

[1] This was the reason why Bernhard Rensch (1900–1990) left the Museum of Natural History Berlin in 1937 and went to the Natural History Museum in Münster where he was able to habilitate at the university and eventually became a full professor of zoology.

memory in those days and Kattinger and I exercised every day all sorts of out-of-the-way questions and answers. I think I somewhat embarrassed Zimmer because sometimes I answered as follows: "According to the standard textbooks the answer is–such, but there was a recent article in the Biologisches Zentralblatt in which it was pointed out that this answer is not correct but that the real answer should be such and such." I think Zimmer never knew whether I was simply pulling his leg or was really presenting new information. He was the director of the museum and was far too busy with administrative duties to keep up with the modern literature. Where I had potential trouble was with the general zoology examination, which lasted half an hour. I was examined by the newly arrived Professor Richard Hesse (successor of Karl Heider) who had never before given a course in Berlin nor examined anybody. No one knew what kind of questions he would ask. Only later did I learn that he was very much interested in the structure of the eye and this is indeed what he asked me. He asked for the names, description, and function of the glia elements in the retina. I can swear that no zoology student in Berlin could have answered that question but, after all, I had been a medical student before and in my histology classes in medical school I had learned all about these glia elements. I, so to speak, in my mind opened up my histology textbook to the appropriate page and then presented to Hesse all the information.

In philosophy I was examined about positivism because fortunately students who were not philosophy concentrators could suggest the subject on which they wanted to be examined. I think I answered everything satisfactorily. The end was that I passed with an A in all parts of my exam and was awarded the degree of *summa cum laude*. This was on June 24, 1926, when I was still 21 years old. I later learned that during the past 10 years my record had been equaled once before and who was it? My good friend Curt Stern (p. 249).

With my *summa cum laude* [and Stresemann's support] I had little trouble in getting the coveted job. The person who actually had to make the decision was Walther Arndt, quite a wonderful person who became a good friend. During World War II he was executed by the Nazis because he had the courage to say in 1944 that continuing the war would do far more damage to Germany than stopping it[2]. Arndt first assigned me to the library and until I went to New Guinea I was in charge of everything connected with the library, although I had a clerical assistant. I developed a new catalogue of the periodicals of the library which used entirely new principles useful for people who know nothing about library science and which my colleagues found very practical. It was published by my friend Wilhelm Meise during my absence in New Guinea (Mayr and Meise 1929)."

The director of the Zoological Museum, Professor Carl Zimmer, justified the selection of Ernst Mayr for the assistantship with a letter to the ministry dated 15 July 1926 (transl.): "Dr. Mayr struck me already as a student because of his eagerness, his thorough knowledge and his sense for scientific problems. He submitted an

[2] Since the late 1990s, the Museum of Natural History and the Natural History Society of Berlin jointly invite every year a noted scientist to hold a "Walther Arndt lecture." Ernst Mayr was this lecturer in June 2001, when he spoke on "The autonomy of biology." In the introduction he mentioned his relations to W. Arndt during the late 1920s.

excellent dissertation and passed his examinations several weeks ago. The test went excellently and, by general agreement, he received the rarely given mark *summa cum laude*. I am most interested to keep this very promising young man who is leaning toward museum work at the Berlin museum" (see also Landsberg 1995: 124).

Several knowledgeable zoologists now advised Mayr to switch to Entwicklungsmechanik (experimental research in embryology), if he were to choose academic zoology as his career: "Spemann [1869–1941] fills all the vacant chairs," they told him. His field had established a virtual monopoly. However, Mayr's goal remained to work at the Berlin museum and to prepare himself for an expedition to the tropics.

## The Influence of Teachers

Ernst Mayr remembered fondly his grammar school teacher in Würzburg and several others at high school in Munich and Dresden. Mr. Löwe, their natural history teacher, was beloved by every pupil in the class. He told them they could be as lazy as they wanted to while in the gymnasium, as long as they passed on to the next higher grade but afterwards, at the university, they had to work like hell. Mayr followed this advice with great success. Their superb teachers of Greek and Latin had actually traveled in Greece and Italy and collected personal experience with nearly everything described in the classical literature. In natural history they used the text book by Kraepelin in which ecology, adaptation and behavior featured prominently. What these and other teachers achieved quite splendidly was to give a broad general education as a basis for university studies. Late in his life, Ernst Mayr looked back at his high school education evaluating it as follows:

"I received a classical education at a German gymnasium where I had 9 years of Latin, German, and mathematics, 7 years of Greek and history, 4 years of French, and no English whatsoever; also a great deal of geography, together with 1-year classes of various science subjects. Now, 75 years later, how do I evaluate such a strongly classics-based education? I still think it was very valuable, but I must admit that it crowded out some subjects that would have been even more important. Although I had lots of history, it was mostly dynastic history, and I had no courses in the social sciences, about democracy and citizenship or some other subjects valuable for daily life. But ignorance is met wherever we look, not only in Germany. What struck me most when I came to the United States in 1931 was the incredible ignorance of most Americans, including college graduates, about the rest of the world" (1997i: 287).

At the University of Greifswald, Ernst Mayr noted, "physics was rather old-fashioned. Chemistry was an excellent course taught by Pummerer, a specialist in polymer chemistry, soon called to one of the big universities. Physiology was taught competently but somewhat old-fashioned although I bought the latest book in biochemistry in order to broaden my horizon. The teaching in anatomy was first-rate. There were three professors who taught different aspects including histology and embryology and supervised the dissections. There was no shortage of bodies and I was able to dissect every part of the human body. I spent every day several hours in the dissecting room and only in retrospect did I become aware of how quickly one became adjusted to work on preserved human bodies. Very soon we were virtually oblivious to the fact. The chief professor of anatomy was Karl Peter [1870–1955], who had a strongly adaptationist concept of body structures. He published a book and a book-length monograph on the adaptationist aspects of ontogeny [*Die Zweckmäßigkeit in der Entwicklungsgeschichte*, Springer, Berlin, 1920]. As everyone in Germany at that time, he was of course a Lamarckian, and this perhaps was the reason why his splendid interpretative analysis has been so completely ignored. Peter was very much impressed by my knowledge of and interest in anatomy and tried to persuade me to become an anatomist. He would give me a thesis and there would be no problem about a splendid future. I was flattered and slightly tempted but not very much. However, I greatly revered Peter and was in touch with him again after the war when he had retired to a small village at the foot of the Alps and studied polymorphism in insect populations. This was quite typical for him. He was not one-sided but his interests ranged from natural history to the most philosophical aspects of form and function.

The botanist was Buder who soon afterwards moved to one of the large universities. He gave me a solid foundation in this field based on the standard text of Strasburger revised by his successors. The old professor of zoology had just retired when I got to Greifswald. His temporary successor, F. Alverdes, was very much interested in behavior and sociology. Also he was a prominent Holist, having even published a book on the subject [*Die Totalität des Lebendigen*, Leipzig, 1935]. We students were not too sure at the time that he was on the right track because Holism to us sounded too much like metaphysics. However, no doubt he stimulated our thinking. With the fall semester came Paul Buchner, from Richard Hertwig's school. In contrast to nearly all the other zoology professors appointed at that time, he was not an Entwicklungsmechaniker but a specialist of cytology, chromosomes and particularly of intracellular symbiosis. His wife was Italian, the daughter of a well-known professor of zoology, and their house was a place of real culture and the cultivation of art. We admired Buchner for his splendid anatomical drawings on the blackboard, since he was able, as some other masters of this art, to have chalks of different color in his two hands and to draw with both of them simultaneously. Buchner was a convinced Lamarckian and strongly defended his views in his otherwise excellent textbook of zoology published in 1938 [*Allgemeine Zoologie*. Leipzig, Quelle and Meyer 1938, 372 pp.].

For people like Peter and Buchner the Mendelian mutationism was contrary to Lamarckism. For them, it seemed impossible to explain gradual evolution by mutations. Since they were quite convinced, and rightly so, that most evolution was gradual, they were forced into opposing mutationism and to adopt Lamarckism and naturally so did I. In the botany department I actually took a special course in genetics but the teacher strictly followed one of the standard texts and most of the class consisted of Mendelian exercises. It certainly helped nothing to improve my understanding of modern genetics or the relations between genetics and evolution. Of course I also took some courses outside of my curriculum, for instance one on

European archeology, a description of all the various cultural levels represented by different kinds of ceramics, etc. I may have taken some course in history of literature but have no recollection. In my last semester in Greifswald, to get a real feeling of medicine, I attended classes in small surgery and pathology occasionally.

After I arrived in Berlin, I left medicine entirely behind me. As a graduate student, I was given a table in the room for PhD candidates in Erwin Stresemann's department. With him I primarily discussed my thesis. He contributed mainly references to relatively unknown local faunas. I don't think I discussed species and speciation with him at all in this period. Neither did I do with Rensch at this time, although after I came back from New Guinea and after he had published his 1929 book, I got more from him than I did from Stresemann.

In Berlin, I had the good fortune to take the famous Karl Heider's last lecture course on Vermes. It was a superb analysis of structure, comparative anatomy and phylogeny of most of the lower invertebrates. The Big Course was given by Ernst Marcus. He was a delightful Berlin Jew, patriotic as any German, who owing to his conspicuous courage had received the Iron Cross First Class in the First World War. He was married to the granddaughter of Du Bois Reymond. His lectures and presentations were lots of fun because he always spiced his explanations with some typical Berlin witticism. However, all the teaching was strictly classical in the style of the best of comparative anatomy. Adolf Remane, who had recently received his degree in Berlin, showed up in the institute occasionally and we chatted with him. The Big Lecture in Special Zoology was given by Carl Zimmer and it was rather a disaster. He simply paraded endless facts before us, practically a recital of the big Claus-Grobben textbook. I don't think I learned much from either Marcus or Zimmer that couldn't have been found in a traditional textbook. The botany course was given by Kniep but I had to travel all the way to Dahlem for his lectures. An outstanding researcher he had discovered the life cycles of many fungi and lower plants. Finally, I had to take philosophy to satisfy the Berlin requirements for a PhD I took the major systematic philosophy course of Dessoir, a mass course with about 800 students, very often only standing places being available. But I also took some additional courses and a seminar on Kant's *Critique of Pure Reason*. Somehow, I rather suspect that at that time I was not able really to connect what I learned in my philosophy courses with what I was doing and thinking" (see also Mayr 1980k and 1980n).

## Erwin Stresemann—Teacher and Friend

Erwin Stresemann (1889–1972) was curator of ornithology at the Museum of Natural History of Berlin, secretary general and later president and honorary president of the Society of German Ornithologists from 1921 to 1972, and one of the outstanding ornithologists of the 20th century (Fig. 2.1). During the 1920s, he initiated the transformation of former ornithology that had been primarily systematic and faunistic in scope, into a branch of modern biological science, a "New Avian Biology," and influenced a large circle of contemporaries (the "Stresemann revolution"). He

**Fig. 2.1.** Ernst Mayr and Erwin Stresemann at the XIVth International Ornithological Congress in Oxford, July 1966 (phot. E. Hosking; reproduced from the Proceedings of the XIVth IOC)

forged links, directly and indirectly, between ornithology and genetics, functional anatomy, physiology, ecology, and ethology, when he published his seminal volume *Aves* (1927–1934) in W. Kükenthal's *Handbook of Zoology*. These accomplishments were of greater general importance and had broader consequences than his ideas on species, speciation and biogeography. He also supervised a large number of PhD students including Ernst Mayr as one of the first. He organized biweekly meetings of the DOG in Berlin, edited the *Journal für Ornithologie* and the *Ornithologische Monatsberichte* which, during the 1930s, were the best among ornithological journals in the world, and he has been President of the VIIIth International Ornithological Congress in Oxford in 1934 (Haffer 2000, 2001a,b; Haffer et al. 2000; Bock 2004a).

The problems of the theoretical species concept and the difficulties of the delimitation of species taxa in birds as well as the questions of speciation were at the center of Stresemann's studies. As one of the first zoologists to do so, he discussed, around 1920, the biospecies concept based on genetic-reproductive isolation between groups of populations. Such isolation can only develop in relatively small and geographically separated populations, he said. He described numerous species and subspecies of birds from his own collections made in the islands of the Moluccas and from the material of other expeditions that he received for study: from Balkan Peninsula, northern Iran, New Guinea, China, Sikkim, and Burma (Myanmar). His monograph on the birds of Celebes or Sulawesi (1939) was largely responsible for the new ecological approach to the old science of zoogeography. During the 1920s and 1930s he cooperated closely with his friends Ernst Hartert (1859–1933) in Tring, England and Leonard Sanford (1868–1950) in New York and New Haven.

In an appreciation of Stresemann's influence Mayr (1997l: 853) wrote:

"I was only 12 years old when my father died and in the course of time Stresemann had become a replacement father figure for me. As the average son would turn to his father for advice, so I always turned to Stresemann when in need. He greatly honored me in 1930 (after my return from the expedition) by, so to speak, moving me up to the rank of the younger brother. [From then on they used the personal 'Du' in German and addressed each other in their numerous letters with the Malay words 'adek' (younger brother = Mayr) and 'kaka' (older brother = Stresemann).] Yes, there were only 15 years of age difference between us and indeed I became more and more like his younger brother. Sometimes even to the point where there were the little competitive jealousies between us as there are between two brothers. He was my closest friend, indeed I had no other close friend. Yes, in those years I regularly went into the field with Gottfried Schiermann but this was an entirely different thing. Stresemann clearly was my closest friend until my friendship with Dobzhansky developed. For Stresemann I was not only a friend but even more so a disciple. He was delighted when I was invited to come to New York and work on the collections because it would mean that his influence would now begin to spread in America. In a way, of course, he must have been slightly jealous of the wonderful opportunities which I had in New York, but he never mentioned it. Fortunately, at least in the early 1930s, he had the Heinrich collections to work on. It was at this time, particularly in connection with his work on the *Aves*, that his interest shifted from avian species systematics to the morphology and physiology of birds. With no more collections coming in, such a shift in interest had become a necessity. I certainly had become his disciple. And virtually all the ideas I had in the field of systematics I had acquired from Stresemann, including the biological species concept."

When Mayr first came to see him in April 1923, Stresemann had immediately recognized the young student's scientific abilities and "systematic instinct." Therefore, he treated him from the start almost like a colleague sending him long letters written in a rather leisurely style despite his busy schedule. Mayr became deeply influenced by Stresemann's thinking, more by reading his publications than through discussions. He acquired from Stresemann (and also from Rensch) the principles of "the new systematics," including the biological species concept and the model of allopatric speciation, as well as the ideas of a dynamic faunal zoogeography. Mayr could not have had a more charismatic, erudite, and scholarly teacher with wide-ranging interests (including the history of ornithology) than Erwin Stresemann and they became close personal friends, as Mayr confirmed on a return trip to New York from a visit to Germany:

"It is no exaggeration when I say that you are my best friend. Nobody else takes the trouble to point out to me my mistakes, and nobody knows better than you to stimulate my imagination and ambitions. I miss such a person in New York now and bitter experiences must replace my friend's advice" (6 October 1934; transl.).

Throughout his life Mayr remained grateful to his mentor thanking "his fate that granted me to become your disciple. This was the basis on which my entire career rests" (19 November 1946). The publication of Stresemann's book, *Ornithology from Aristotle to the Present* (1951, 1975) made him "again proud to be your disciple" (5 August 1951) and "This again gives me an occasion to say how much I owe you in this respect. I would hardly have been able to do so much if I wouldn't have had your standard always in front of my eyes. It is so much easier to be lazy!" (18 July 1964, transl.). Conversely, Erwin Stresemann followed with great delight Mayr's brilliant career in the United States. When the latter assumed the directorship of the MCZ at Harvard University, Stresemann wrote: "We are very glad with you that you already climbed up again another step on the steep ladder of your successes and that you now sit where once Alexander Agassiz, Tom Barbour and other men of their caliber had been enthroned" (22 April 1961; transl.).

The line of influence went not always from the older to the younger partner. When Stresemann had arrived at an interpretation of polymorphism in wheatears conflicting with current genetic explanations, Mayr invited him to become co-author of a paper on polymorphism in *Oenanthe* (Mayr and Stresemann 1950) compatible with genetic conceptions. This, so to speak, gave Stresemann an opportunity to revise his earlier incorrect explanation. He was an orthodox Darwinist and acknowledged the importance of natural selection. He felt, however, as did most other German evolutionists of the time, that this mechanism was insufficient to explain the origin of complex adaptive structures. Until the end of his life he was searching for an additional evolutionary "factor X" (Haffer et al. 2000).

Their extensive correspondence from 1923 until Stresemann's death in 1972 consisting of about 850 letters documents the strong emotional bond between them, it demonstrates the way in which this bond grew stronger over several decades, how it withstood difficulties and temptations and thus became a beautiful testimony of the friendship between two scientists. Beyond their scientific contents and historical significance these letters have in part literary qualities of a culture of communication which, compared to our age of telefax and e-mail messages, already seems to belong to another time (Haffer 1997b, 2007b).

Erwin Stresemann is relatively little known in English-speaking countries, because he published most of his books and journal articles in German. The top international ornithologists knew his *Aves* volume (1927–1934) and his decisive influence on the advanced modern contents of the *Journal für Ornithologie*. They elected him as President of the International Ornithological Congress in 1930, when he was only 40 years old. The English translation of his excellent book, *Ornithology from Aristotle to the Present* (1975) eventually made his name more widely known internationally, at least among historically interested ornithologists. Stresemann was one of the key figures of ornithology who, from 1921 onward, merged systematic ornithology and field natural history in the New Avian Biology (Haffer 2007a).

## Assistant Curator at the Museum of Natural History in Berlin

As a general museum assistant Mayr, like the three other assistants, was assigned special tasks. His assignment was the Museum's main library where he decided which books to buy, how to classify them, which departments should get incoming reprints, but most of all how to arrange a planned new catalogue of the Museum's 1750 journals. He developed a system of three key words: (1) A single name, e.g., proceedings, annals, bulletin, etc.; (2) the publishing institution (museum, society, university, etc also a single name); and (3) the place of publication (London, New York, Berlin, etc.). By contrast, the standard catalogues alphabetized every single word in the title and if the user did not realize that the word "Royal" was in the title he would have great trouble finding the title wanted. This catalogue was not yet finished, when Mayr left for New Guinea. His colleague Wilhelm Meise completed and published it under both names in 1929. It was Mayr's first book. Many scientists praised such a simplification. It remained in continuous use until the advent of the computerized catalogue at the Berlin Museum in 1992 made it obsolete.

Mayr was also a lecture assistant for the course in systematic zoology. "It was about as old-fashioned a course as one could imagine. Apparently it was Zimmer's ideal to make the student familiar with the total diversity of the animal kingdom. For each class the lecture assistants had to set up scores of preserved specimens, stuffed birds and mammals or boxes with insects as the course moved from protozoans to mammals. There were sometimes as many as 60 or more containers with spirit specimens. The assistant had to find them in the teaching collection and roll them on small carts to the lecture hall. After the lecture he had to take them back again. It was a tedious and senseless activity—my only connection with teaching while I was at the Berlin Museum. When I came to the American Museum in New York, no one in the Bird Department did any teaching. And so, for many years, I experienced no teaching at all."

The senior assistant Bernhard Rensch (1900–1990) was now in charge of the public exhibits, where he did a pioneering job in presenting biologically relevant topics. Mayr also appreciated his publications and remembered:

"Rensch came to the Berlin Museum while I worked there for my PhD. He was 4 years older, but considerably more mature than myself. He had very broad interests, and was at that time almost as much interested in psychology and philosophy as a student of Ziehen[3] as he was in biology. In biology, likewise, although he had a considerable interest in birds and bird study, he was interested in all sorts of organisms. He was later on given the position of curator of mollusks at the Berlin Museum, and contributed quite a few taxonomic additions, and had PhD students working in mollusk taxonomy. Rensch, in a way, was a rather shy person. He was quite pleasant but not outgoing, and a bit difficult to get into a conversation. We always lunched together nearby at the lunchroom of the Ministry of Labor, with Stresemann, Rensch, Professor Neumann, and had lively conversations in which Rensch usually was the least active one. Neumann was a great person for telling jokes, particularly dirty jokes. And he was always amused in what sequence people started laughing at one of his far fetched dirty jokes, and, much to my disgrace, I must admit that I was the one who always laughed first, and Rensch always was the one who laughed last. Obviously, he was very innocent and somewhat naïve and he had one weakness: He had no sense of humor at all. If you tried to tease him,

[3] Theodor Ziehen (1862–1950) was Professor of Philosophy at the University of Halle.

he always took it seriously. Even though he was not terribly impressive as a person, I admired him greatly for a number of reasons. One was his breadth of knowledge, secondly, he had ideas—he had plans; he thought of the future. He was placed in charge of the exhibition halls of the museum, and organized perhaps the most modern exhibits any museum had. It always illustrated basic biological principles and did not just show interesting exceptions or particular things. From his 1929 book, I undoubtedly learned more than I ever actually learned from Stresemann. And I often have said that in a way, at the Berlin Museum, Rensch was perhaps more my teacher than Stresemann. Later on he published a manual of taxonomy (1934), which I thought was really quite admirable, and it inspired me to write my own textbook one day, which actually later on I did, although it was finally combined with a parallel manuscript of Linsley and Usinger (1953a).

One of Rensch's interests was biogeography. He wrote a whole book on the biogeography of the Sunda Islands and eastern Indonesia (1936), which I think was inferior to his other work. He was an inveterate land bridge builder. When I did my papers on Wallace's Line and on Timor, I disagreed with almost all of his conclusions. Later on in life, he wrote a good deal on philosophy. Here again, I rather disagreed with him. Since he did not accept emergence, he had to assume that such phenomena as mind and consciousness occurred already at the lowest levels (molecular) and I suspect that his philosophical writings found few followers.

Rensch was quite an artist. He painted excellently, and many of his paintings are in German art galleries. He also could write poetry, and wrote a very nice autobiography. He had a rather tough time during the Nazi period, because he was anything but a Nazi. He had married the daughter of the president of the province of Brandenburg in Germany, who was a social democrat, and, of course, his daughter probably as well. So when the Nazis came to power, he was going to be dismissed a few years later, but Stresemann and the director of the museum interfered successfully. Nevertheless Rensch left the museum, when he had found a position at the museum in Münster owing to the great generosity of the local Nazi leader, who was a good man and did not take the Nazi laws too seriously.

Rensch continued to publish almost up to his death. He died a couple of weeks after his 90th birthday."

After Mayr's PhD examination, Stresemann introduced him to systematics and suggested that he study certain groups of Palearctic songbirds such as accentors (*Prunella*), snowfinches (*Montifringilla*) and Rosy-Finches (*Leucosticte*). Mayr described a couple of new subspecies of *Prunella* (1927e) and wrote a detailed revision (1927f) of the other two groups of songbirds mentioned above which Hartert (1910) in his great work on birds of the Palearctic fauna had combined in one genus. However, Mayr was able to show that these two groups differ conspicuously in their molt patterns. The species of *Montifringilla* change not only the body feathers but also wing and tail feathers during their first (juvenile) autumn molt ("complete molt"), whereas the young of *Leucosticte* species change only their body feathers at that time retaining wing and tail feathers for another year ("incomplete molt"). Two different and unrelated genera are involved; *Montifringilla* is related to the sparrows (*Passer*) and *Leucosticte* to the finches (*Carduelis* group). Peter Sushkin

(St. Petersburg, 1868–1928) having studied skull and skeleton in these groups fully confirmed in a personal letter to Mayr the latter's results based on molt patterns (see also Bock [2004b] who also supported Mayr's earlier conclusion). Mayr ended his article with an analysis of the history of dispersal of *Leucosticte arctoa* documenting his early interest in zoogeographic questions. The detailed description of geographic variation and speciation of the genus *Leucosticte* shows how familiar Mayr was already at that time with the principles of geographic speciation. This needs emphasis because it was later claimed that Mayr had learned about geographic speciation from Sewall Wright's 1932 paper (Ruse 1999: 118).

His efforts to elucidate the genetic basis of geographical variation and speciation and to establish ties with genetics now found public expression for the first time. In the general discussion of his paper on snow finches, Mayr (1927f: 611–612) laments that the geneticists attempt "to analyze the factors of speciation without taking into consideration the examples offered by nature." He further deplored "how little geneticists and systematists cooperate even today" and "that the geneticists still today apply the Linnean species concept which is by now 170 years old (and in many respects outdated)," the systematists had abandoned it long ago. His justified criticism referred primarily to the mutationists among the classical geneticists, for he (like most other naturalists) was unaware of the recent publications of population geneticists emphasizing small mutations and they still adhered to Lamarckian views explaining gradual evolution.

During the mid-1920s all of Mayr's colleagues at the Museum of Natural History in Berlin were deeply concerned with evolutionary questions and fighting against the saltationist views of the "mutationists" like De Vries, Bateson and Johannsen (as well as R. Goldschmidt and O. Schindewolf during the 1930s) who interpreted evolution through macromutations (saltations). All of these latter scientists were typologists who thought new species originated with major mutations and all of them rejected natural selection. The Berlin zoologists at the Museum of Natural History studied in detail the phenomenon of geographical variation in animals leading to an emphasis on environmental factors. As long as mutations were believed to cause large phenotypic changes, their only alternative was a Lamarckian interpretation of the gradual (clinal) geographic variation they observed in numerous continental species. They realized—like Darwin and Wallace previously—that a thorough study of variation as well as adaptation in natural populations was an indispensable precondition to understand the problems of evolution and speciation. They all agreed that speciation was a slow process and that macromutations à la de Vries and Morgan's "freaky" *Drosophila* flies with yellow body color or crumpled wings had nothing to do with the development of new species. These naturalists divided characters into Mendelian (particulate) ones, which they considered evolutionarily unimportant, and gradual or blending ones which, following Darwin, should be the material of evolution. So the important thing that had to happen, and which indeed happened during the 1920s (at first unnoticed by the naturalists), was that the geneticists completely rejected the saltationist views of the early Mendelians and showed that genetic changes could happen through very small mutations which in the long run could be of great evolutionary importance.

Erwin Stresemann emphasized that extant birds provide a number of borderline cases between species and subspecies and that the taxonomic rank of geographically separated (allopatric) taxa as subspecies or species is to be determined by inference on the basis of several auxiliary criteria (Stresemann 1921: 66):

(1) Similarity or dissimilarity in morphological and other biological characteristics (ecological requirements, voice, etc.),
(2) overlap or nonoverlap of individual variation in several characteristics,
(3) comparison with other congeneric forms that are in contact and either hybridize (subspecies) or overlap their ranges without hybridization (species).

Stresemann had established some ties between ornithology and genetics through a series of publications. Based on the work of Seebohm, Berlepsch, Hartert, Kleinschmidt, Hellmayr and their extensive discussions of the species problem, he had initiated a theoretical broadening of the Seebohm-Hartert tradition in the direction of the "new systematics." This conceptual modernization was continued by Bernhard Rensch and, in particular, Ernst Mayr (pp. 204–206).

During the late 1920s Mayr also reviewed several publications for Stresemann's *Ornithologische Monatsberichte*. Much of his spare time he devoted to bird excursions around Berlin discovering, e.g., that the Willow Tit (*Parus montanus*) was far more common in Brandenburg than recorded in bird books; probably it had been often confused with its sibling species, the widespread Marsh Tit (*P. palustris*). A report of his observations on this species appeared in the *Journal für Ornithologie* (1928). He also published notes on the nesting of the Chaffinch (1926a) and the House Martin (1926c), on the occurrence of the Waxwing (1926g), the calls of the Bittern (1927c), on snake skins as nest material (1927d) and he often accompanied Gottfried Schiermann (1881–1946), whose population studies at that time were pioneer efforts continued in Germany only many years later.

Mayr reported: "My best friend in Berlin, except of course for Stresemann, was Gottfried Schiermann, a superb field ornithologist. I met him through Stresemann and through the meetings of the DOG. Schiermann was so much older than I, he could have been my father. He had no academic background but had been an ardent egg collector in his younger years. The bird censuses and ecological studies he published in the 1920s and 30s are pioneering. I loved to go in the field with him because he was such an acute observer. One day, just going through the woods together, he discovered 54 occupied bird nests. This included such difficult to find nests as that of the Wood Warbler (*Phylloscopus sibilatrix*). He was a specialist in finding the nests of birds of prey, most of whom use abandoned crows nests as their first foundation. Hence, in suitable woods he mapped in winter all the old crow nests and old hawk nests and then checked them systematically during the season for occupancy by hawks. I once made a memorable trip with him to the lower Spreewald where he showed me nesting Black Storks, and woodland White Storks, as well as many other exciting things. (We were nearly eaten up by the mosquitoes). He took me out to the nesting place of the Rohrschwirl (Savi's Warbler, *Locustella luscinioides*), where I was lucky enough to find a nest of this species (Fig. 2.2).

**Fig. 2.2.** Gottfried Schiermann at a nest site of Savi's Warbler (*Locustella luscinioides*), Kremmener Luch near Berlin, 29 May 1927 (phot. H. Siewert; courtesy E. Mayr)

He was such a modest, warm, friendly person that it was always a pleasure to be with him. He never bragged about the enormous knowledge he had. I am told that during the bombing of Berlin he was quite heroic in extinguishing fires and making himself otherwise useful. His only son was killed on the Russian front and Schiermann was apparently quite unable to cope with this loss. To obtain food in the immediate days after the fall of Berlin must have been very difficult, and he was too modest to push himself. He died of malnutrition and illness on September 10, 1946. I rank him among the highest of all the human beings I have ever been fortunate enough to meet."

All the time Mayr and the PhD students had friendly personal relations with Dr. Stresemann and his family. Occasional invitations at his home were celebrated in high spirits. On 6th of December Mayr appeared as St. Nicolaus at Stresemann's door and surprised his three children. Usually he attended the biweekly meetings of the German Ornithological Society (DOG) and those of the Society of Naturalists (Gesellschaft Naturforschender Freunde zu Berlin) where representatives of different disciplines exchanged their views. He listened to presentations on faunistic observations in China (R. Mell) and on ethnological research in Peru (G. Tessmann) and had also contacts with the Mammalogical Society, the Zoological Institute of the University and their representatives.

Mayr now lived in a furnished room in Berlin-Hermsdorf with the Schneiders, a house right at the edge of the woods. Hermsdorf was on the same electrical transit line as Stresemann's home in Frohnau, one stop beyond Hermsdorf. Otherwise, Mayr did not enjoy much social life while he was an assistant curator at the museum. However, several students enlarged Stresemann's group of pre-docs. Mayr joined some young university staff members playing volleyball (Faustball) and persuaded Stresemann to do so on occasion, but he dropped out soon. There was still no girlfriend.

*History of bird migration.* In Mayr's paper on the evolutionary origins and development of bird migration (written in 1927 when he was 23 years old and completed by his colleague W. Meise after Mayr's departure for New Guinea; Mayr and Meise 1930c) he is convinced that most of these phenomena cannot be explained completely through evolution by natural selection "which can eliminate but not create anything new" (a view he would vigorously attack later). This is the last anti-selectionist statement found in Mayr's publications. Because of his Lamarckian views he was not able to discuss, much less answer, valid questions that he posed, e.g., "In which way originate genetically determined changes of migratory routes?" Mayr and Meise (l.c.) endorsed Thomson's four-fold division of complementary causes of migration (function, origin, physiology) but restricted themselves to a discussion of the evolutionary-zoogeographical development of bird migration (Beatty 1994).

In their opinion the route of migration is originally a backtracking of the route of immigration. However, subsequent phenomena, such as route abbreviation and route prolongation complicate any historical interpretation of the original home range of migratory birds. The major reason for route prolongation in species which expand their breeding area northward seems to be a strengthened physiological apparatus leading to a southward displacement of the wintering area. Mayr ends his theoretical discussion with the remark that without hypotheses, scientific progress is not achieved—as emphasized in the hypothetico-deductive method of testing previously conceived hypotheses. This theoretical paper was reviewed in detail by Mayr's colleague at the AMNH John T. Nichols (1931) who emphasized the point that the migratory drive, once started, tended to increase, "causing the bird to swing annually pendulum-wise over an ever increasing course." The original home of the species may then lie neither in the present breeding or wintering areas, but at some intermediate location.

## Expeditions to New Guinea and the Solomon Islands (1928–1930)

After Mayr had started his assistantship at the Zoological Museum Berlin in July 1926 Dr. Stresemann, mindful of his earlier promise, attempted to place him

**Fig. 2.3.** Ernst Mayr, 1927, a few months before leaving Germany for New Guinea (Archive, Museum of Natural History Berlin, Orn. 103, 1)

on an expedition. Plans of travels to Cameroon (with the ethnographer Günther Tessmann) or to Peru (with the geologist Harvey Bassler) failed. At the International Zoological Congress in Budapest (October 1927), Stresemann introduced Ernst Mayr to Lord Walter Rothschild and Ernst Hartert and convinced them that he would be the best man to continue A. F. Eichhorn's work in New Guinea (this collector for Lord Rothschild had retired owing to illness).

This was going to be Mayr's great chance. He was 23 years old now (Fig. 2.3). Back to Berlin he wrote to E. Hartert on 22 October 1927 (transl.):

*Dear Dr. Hartert!*
*You can hardly imagine how overjoyed I was by your and Lord Rothschild's assent. For years it has been my loftiest goal some time to be able to conduct a scientific expedition and to become acquainted with those birds in life whose*

*skins have impressed us in the museums as very special rarities. Hopefully I will not have any difficulties to get used to the high Papuan diversity, because I have a very sharp eye sight and an excellent ear for bird voices. So I hope that my journey will have the success which you expect. [...] The steamer of the North German Lloyd will depart from Genoa on 7 February. [...] I hope to be able to collect extensively in the moss forest and around the tree limit. I believe this is where the best things occur. I am a rather persevering climber. For example, I've made excursions in the Dolomites without special fatigue recently: 10 km approach [on foot], then the ascent from 1,000 m elevation to 2,400 m followed by the descent and a march of 15 km. Last Sunday, I walked 40 km and would have been able to continue easily for another 10–15 km. I mention this only to show you that I am not afraid of the mountains, as has happened in the case of certain collectors. [...] Cordially greeting you, your gratefully obedient, Ernst Mayr.*

Mayr's first goal was the Arfak Mountains in the Vogelkop region (NW New Guinea) to look for certain "rare birds of paradise." The expedition was financed jointly by the Rothschild Museum and the American Museum of Natural History in New York. Hartert informed Dr. Sanford, Trustee of the AMNH, on 25 October 1927 as follows:

*Dear Dr. Sanford,*
*[... Instead of Dr. H. Snethlage who has taken a position at a museum and is unavailable] I have found another man, probably even better, in the person of Dr. Mayr. He is attached to the Berlin Museum as an assistant, but the authorities are willing to give him a year's leave. The expedition will not cost more than L 1000 and probably less. Dr. Mayr is an ornithologist and a young, very active and decent man whom I know personally, in fact he was also at Budapest for the Congress. I have arranged for him to come to Tring for ten or fifteen days to get full instructions and to make the acquaintance of some of the especially rare birds from that region. I have also seen Dr. Dammermann, Director of the Buitenzorg Museum in Java, and he is willing to let him have at least one of their Malay skinners and will assist him with recommendations, permission to stay free of charge in government stations etc. There is one condition, however, that is that they get some duplicates of the birds collected for the Buitenzorg Museum, that is that after the division into two lots, one for Tring and one for your museum, the Buitenzorg Museum is to get a third pick. It will not affect the first two lots in species, only diminishing the numbers of specimens. The help of the Dutch officials is so important that we have decided to agree to this condition and I trust that you will also agree. [...]*
*With kind regards from Lord Rothschild, Believe me, Yours sincerely, [E.H.]*

In late November Mayr went to Tring, England:

"My instruction sessions with Hartert went fine with one exception. One time he took a gun, and gave me a gun, and said, let us go out and shoot some pheasants.

To be honest, I virtually never before had a gun in my hands, and of course had never shot at a flying bird. The result could have been predicted. I missed every pheasant that got up in front of me on Chiltern Hills while Hartert, in spite of his age, brought down most of those that I had missed. I rather suspect that Hartert began to doubt my success as a bird collector, but it was really too late to reverse history.

On that occasion and again when, after New Guinea I had gone to the Rothschild Museum to study birds, I frequently encountered Lord Walter Rothschild. I was surprised how shy he was. When Hartert and I were talking and looking at some specimens and Rothschild arrived he stayed at the door until Hartert asked him to come in and join us in looking at the specimens. He had an unbelievably good memory. In my New Guinea collection I had a specimen of a *Poecilodryas* species, which Hartert had never seen. Rothschild said, oh this was illustrated by J. Gould in his *Birds of New Guinea*, plate 84. Hartert got the volume out of the library, and lo and behold, Rothschild was right, it was on plate 84[4]. I am sure he could have told us for every other plate number what species it was.

He was a mountain of a man, well over 6 feet tall and extremely heavy. His mother had a special comfortable chair built for him, and every morning Rothschild went to the library where the chair stood, backed against it and then let himself drop into this comfortable seat. It had special steel reinforcement because whenever Rothschild sat on another chair it would collapse. This had happened in the Hartert household. Therefore also Mrs. Hartert offered a special chair constructed with steel reinforcement, on which Rothschild sat whenever he visited the Harterts."

In early February, 1928, Mayr left Germany all by himself to lead—as it turned out eventually—a three-partite expedition to New Guinea and Melanesia of over 2 years. During this entire period he was on leave from the Museum of Natural History without salary payments from that institution (receiving only about 200 Marks per month from the expedition funds). He returned to Berlin in the first days of May 1930 resuming his duties as an assistant curator at the Museum of Natural History. These three expeditions were administratively independent undertakings and explored the following regions (Fig. 2.4):

(1) Papua Province, Indonesia, or Irian Jaya, former Dutch New Guinea (Arfak, Wandammen and Cyclops Mountains), for the Rothschild Museum in Tring, England, and the American Museum of Natural History in New York; February 1928–October 1928; financed in equal measure by these institutions; for reports see Mayr (1930f, 1932e) and Hartert (1930);
(2) Papua New Guinea, the former German Mandated Territory (Saruwaget and Herzog Mountains) for the Museum of Natural History in Berlin; November 1928–June 1929; supported by the German Research Foundation (Notgemeinschaft der Deutschen Wissenschaft); for a report see Mayr (1931l)

[4] This was probably plate 6 of Part XVI illustrating *Poecilodryas bimaculata* (Black-and-White Flycatcher), which is currently placed in the genus *Peneothello*. Mayr (1931l: 680) had collected a female of this species at "Sattelberg."

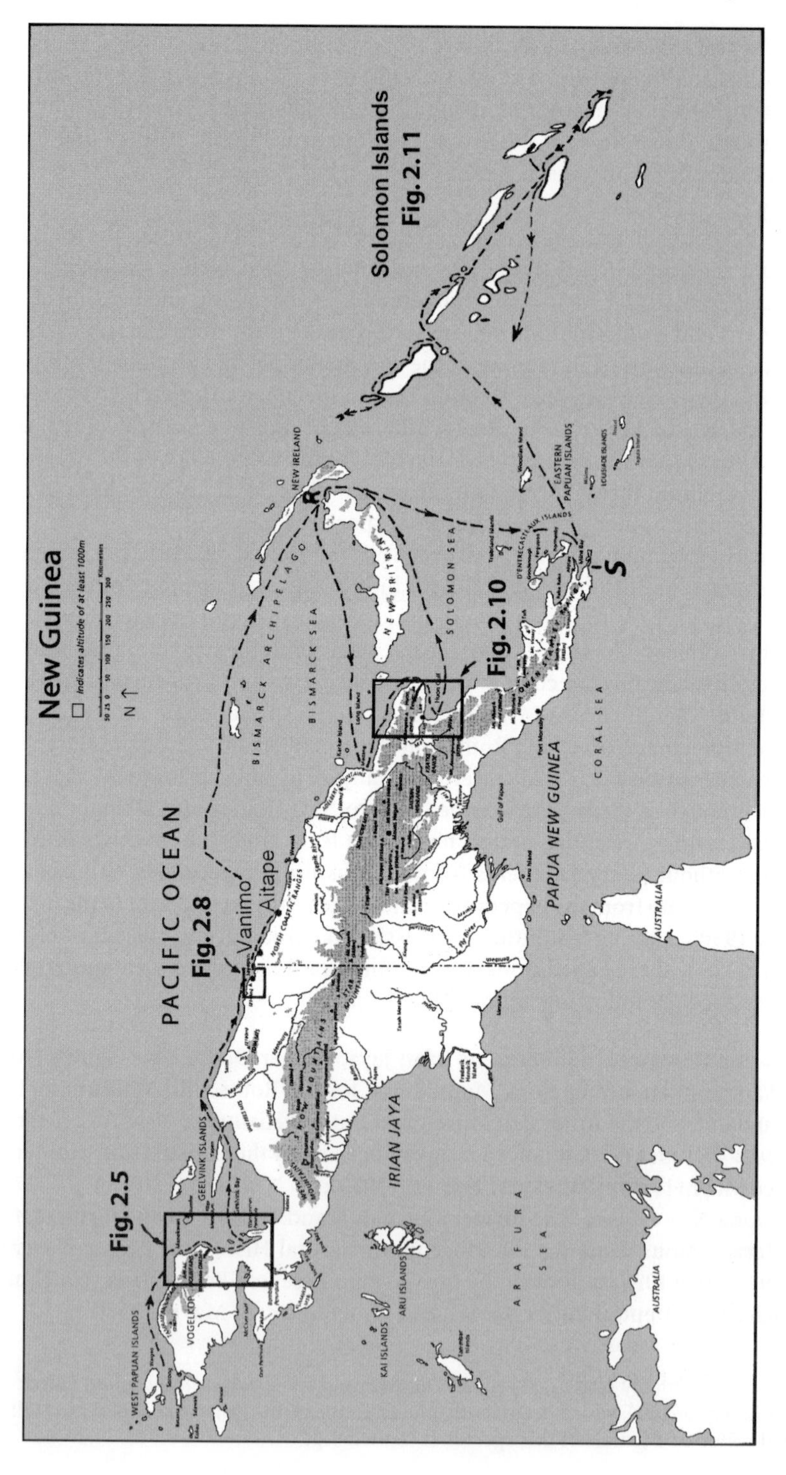
New Guinea
indicates altitude of at least 1000m
Kilometers
N
Solomon Islands
Fig. 2.11
PACIFIC OCEAN
Fig. 2.8
Vanimo
Aitape
Fig. 2.5
Fig. 2.10
BISMARCK ARCHIPELAGO
BISMARCK SEA
NEW BRITAIN
NEW IRELAND
SOLOMON SEA
D'ENTRECASTEAUX ISLANDS
EASTERN PAPUAN ISLANDS
CORAL SEA
PAPUA NEW GUINEA
IRIAN JAYA
ARAFURA SEA
ARU ISLANDS
KAI ISLANDS
GEELVINK ISLANDS
VOGELKOP
WEST PAPUAN ISLANDS
NORTH COASTAL RANGES
AUSTRALIA
AUSTRALIA
R
S

**Fig. 2.4.** Ernst Mayr's expeditions routes (*dashed lines*) in New Guinea and the Solomon Islands (1928–1930) and the location of Figs. 2.5, 2.8, 2.10, and 2.11. R Rabaul, S Samarai. Shaded areas in New Guinea–elevations at least 1,000 m. Base map of New Guinea from Beehler et al. (1986).

(3) Solomon Islands, Melanesia, for the American Museum of Natural History, New York (part of that Museum's Whitney South Sea Expedition); July 1929–March 1930 (for a report see Mayr 1943h).

Since no detailed description of Mayr's expeditions in New Guinea and in the Solomon Islands has ever been published, besides his own brief popular accounts, I present an overview on the following pages and illustrate his expedition routes on four geographical maps.

## Papua Province, Indonesia (Irian Jaya, Former Dutch New Guinea)

### (1) Departure from Germany and Voyage from Genoa to New Guinea

While Mayr prepared for his departure, he received from Ernst Hartert "15 commandments for a zoological collector" giving instructions based on his own long collecting experience several decades earlier, for example "(1) Always try to collect series (population samples) of individual forms or species, (2) Make nice but at least acceptable skins, (3) Always take enough cotton along to fill the mouth of the collected bird, (4) On longer excursions always be accompanied by a second man for help in case of an accident," etc., but also "(7) The collector should never let himself be infatuated by sirens or Venuses." Mayr answered on 5 January: "I just received the 'commandments' and shall put them under my pillow. I was greatly amused by your wording and shall take the contents to heart" (transl.); see Haffer (1997b: 416–418) for a complete list of these commandments.

On 4 February 1928, E. Stresemann and his doctoral students accompanied Mayr to the railway station in Berlin. New Guinea still had a very ominous reputation and when they said goodbye Mayr could see that Stresemann was wondering whether he would ever see him again. The thought of not returning had crossed his own mind, of course, and in a letter to his mother from Genoa he wrote:

"I told myself all the time that if on this trip something really should happen to me it would be in the midst of a beautiful experience and in the midst of the most important phase of my profession, that is, in free research and that under such circumstances death would be a glorious end to my life and all the incidental accompanying circumstances would be quite unimportant. I enjoy every day on which I can work and make use of the experiences of the past days. But I am not terrorized by the thought of death after a life which, up to now, has been *innerlich* so happy. Actually, I am quite convinced that I will return in full health, and so it will be. I have penned down the previous sentences only in order to tell you what my own principles are in this situation" (5 February 1928).

The "Fulda" of the North German Lloyd with 160 passengers on board left Genoa on the afternoon of 7 February in beautiful weather and a calm sea. On the 10th they passed Crete arriving in Port Said, Egypt, on 12 February. During the entire voyage Mayr learned Malay with the help of several young passengers living in Java (and also picked up smoking, as he told Stresemann in a letter dated 23 February, which habit he gave up again in August 1939). He wrote a detailed travelogue to his mother about the food on board, the passengers, the flying fish, porpoises and birds observed and also about the activities on board. He was elected spokesman of the passengers in the tourist class and when a concert was arranged, he was the one to invite the captain to attend. He won the first prize (a watch) in one of the sports activities on deck, an "obstacle run," and described to his mother the various other sporting events in which he participated. He was very competitive and emphasized when his team or he himself had won. The ship called at Colombo (Sri Lanka) where he visited several Hindu and Buddhist temples and at Singapore before arriving, on 4 March, at Jakarta (Batavia), Java. Dr. Dammermann and Dr. Siebers of the Museum in Bogor (Buitenzorg), south of Jakarta, treated him generously for two weeks and sent along with him three native museum assistants or mantris: Two preparators and one insect collector. They were good collectors and bird skinners and knew how to organize a camp in the jungle. It was they who educated the totally inexperienced Mayr to become a highly successful explorer.

The voyage to New Guinea commenced on 20 March. The steamer called at Semarang, Soerabaja, Buleleng (Bali), and Makassar (Sulawesi) from where Mayr continued on the SS "Van Noort" to Manokwari (Vogelkop). Traveling through the northern Moluccas was the most beautiful and impressive part of his trip particularly the narrow strait west of the island of Batjan (Bacan). The volcano of Ternate was smoking and covered with dense vegetation almost to the top (1700 m). The "Van Noort" called at Waigeo and Sorong at the western tip of New Guinea and arrived on 5 April at Manokwari, a village on the northeastern coast of the Vogelkop. Viewed across the bay to the south, the towering Arfak Mountains rise abruptly to an altitude of 2,500 m. In those forested mountains Mayr was to search for the "rare birds of paradise" and to assemble a representative collection of the avifauna during the next several months.

### (2) Arfak Mountains (Vogelkop)[5] (Fig. 2.5)

Itinerary[6]:

| | |
|---|---|
| 5 April 1928 | arrival in Manokwari |
| 12 April 1928 | en route from Manokwari to Momi |
| 13–15 April 1928 | at Momi |
| 16–17 April 1928 | ascent to Siwi |
| 18 April–24 May 1928 | at Siwi collecting |
| 25–27 May 1928 | en route from Siwi to Ditchi via Ninei |
| 28 May–8 June 1928 | at Ditchi collecting (climbing Mt. Mundi and Mt. Lehuma) |
| 9–10 June 1928 | en route from Ditchi (via Dohunsehik) to Kofo |
| 11–15 June 1928 | at Kofo (Anggi Gidji Lake) collecting |
| 16–23 June 1928 | en route from Kofo to Momi (via Dohunsehik, Ditchi and Siwi) |
| 24–25 June 1928 | en route from Momi to Manokwari |

After several days of preparations, Mayr and his crew of seven (3 mantris and four Christian Papuas) departed on the government boat, the "Griffioen," on 12 April to Momi, 90 km south of Manokwari and located at the eastern foot of the Arfak Mountains (Fig. 2.5). Obtaining the necessary porters and converting the baggage into many small portable loads took another four days until the expedition was ready to leave Momi and to enter the foothills of the Arfak Mountains in a northwesterly direction on 16 April. The general plan was to collect birds at different altitudes up to the shores of one of the two Anggi Lakes at 1,925 m. Considering the steep mountainous terrain and the poor condition of the trails, the porters refused to take loads weighing more than thirty pounds. About 60 porters were required to carry the equipment and supplies. They entered the rainforest soon after leaving Momi and new bird calls were heard everywhere. It was quite a sensation when Mayr saw the first brilliant bird of paradise (*Ptilorhis magnifica*) alive, a species which he had known so far only as a museum skin. On the second day the hunters brought two adult males of the Superb Bird of Paradise (*Lophorina superba*) and a female with large eggs of the Dwarf Whistler (*Pachycare flavogrisea*). Mayr and his mantris had collected around Manokwari several common species like *Meliphaga albonotata*, members of *Rhipidura*, *Cinnyris jugularis* and *C. sericeus*, *Aplonis cantoroides*, *Oriolus szalai*, *Halcyon sancta* and *H. albicilla*, and had listened to the incessant calling of *Philemon novaeguineae*.

---

[5] Mayr (1930f, 1932e) published two accounts of this expedition based on his detailed diaries which have been preserved and are deposited in the archives of the Staatsbibliothek Berlin (Mayr Papers). Moreover, this archive holds the letters he wrote to his family, to Ernst Hartert and to Erwin Stresemann (Haffer 1997b). I have used all of these documents in the preparation of this account of the expeditions to New Guinea and the Solomon Islands (1928–1930). I did not include the rich information contained in the diaries and letters on the natives and their customs, general considerations on colonialism and life in the tropics or on the many logistic problems regarding camp moves, porters, and transportation which fill many pages of these diaries.

[6] The dates are taken directly from Mayr's diary; a few of them differ slightly from those given by Vink (1965: 487).

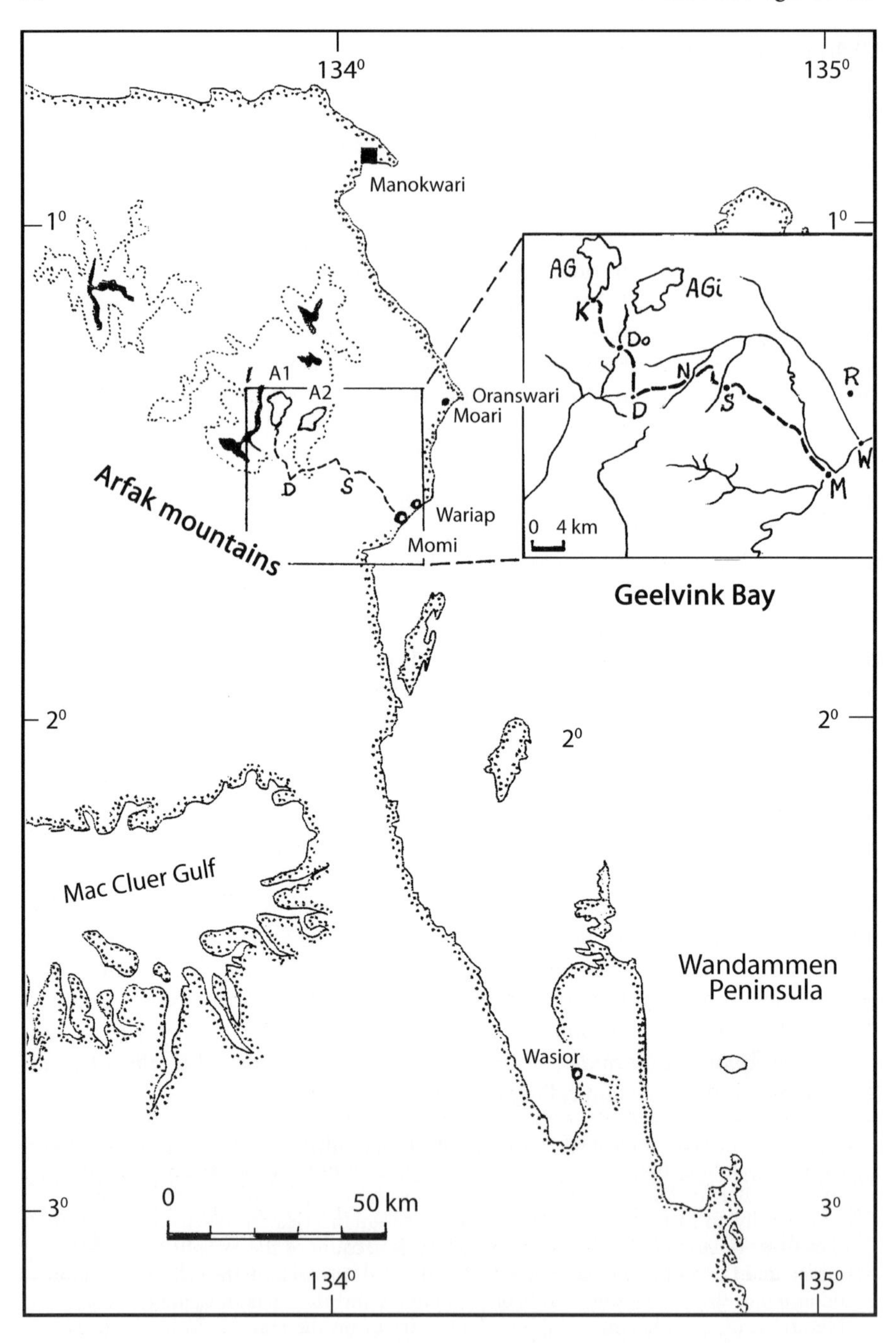

134⁰
135⁰
Manokwari
1⁰
1⁰
AG
AGi
K
Do
N
D
S
R
W
M
0 4 km
A1
A2
Oranswari
Moari
D
S
Arfak mountains
Wariap
Momi
Geelvink Bay
2⁰
2⁰
2⁰
Mac Cluer Gulf
Wandammen
Peninsula
Wasior
3⁰
0
50 km
3⁰
134⁰
135⁰

**Fig. 2.5.** Vogelkop region, NW New Guinea. Expedition routes (*dashed lines*) (1) from Momi on the coast to Anggi Gidji (Male Lake) in the Arfak Mountains and (2) from Wasior into the Wondiwoi Mountains, Wandammen Peninsula, April–July 1928. Stippled contour line–2000 m elevation; solid areas–mountains over 2500 m elevation. A1 Anggi Gidji (AG), A2 Anggi Gita (AGi, female lake), K Kofo, Do Dohunsehik, D Ditschi, N Ninei, S Siwi. M Momi, W Wariap, R Ransiki. For a location map see Fig. 2.4

Mayr felt absolutely fantastic when he entered the forests. He stepped from the shore into the tropical jungle and at that time New Guinea was virtually untouched. Walking into the interior, he came to villages where no white man had ever been. And to wake up in the morning and hear those tropical birds calling and singing around the camp was an overwhelming experience for him. While ascending the forested mountain slope on the first day, having heard many stories about the treachery of the mountain tribes, he became startled because:

"Our column had lengthened very much and I was alone at its head accompanied by a number of Papuans fully armed with knives, bows, and arrows. Suddenly a terrifying howling started at the end of the caravan, advancing toward me and getting louder and louder until it became a blood-curdling series of screams and yells. I had no idea what this could mean and had not read anything like it in the entire New Guinea literature. I was frightened, feeling certain that this was a signal to attack, and I expected every moment to feel the knives of the carriers in my back. As it turned out, it was really the war-cry of the Manikion tribe, but on this occasion it was uttered only to inspire the energy of the carriers."

The night was spent in Ingeni, a small campong (village) with three large houses. "I slept peacefully in my tent. Apparently everybody had gotten up early and there was an active camp life with much chatter and noise. Suddenly complete silence, when I had stepped out of the tent. If that doesn't give one megalomania, I don't know what will. I simply can't get used to playing the role of the tuan besar (big master). A little while later while I was supervising the breaking up of the camp the chief of the Papuas suddenly came to me and urged me to go to the head of the column because that is the place where the tuan should be."

On 17 April a camp was established at Siwi at about 800 m elevation after they had crossed a mountain range at over 1,300 m. Mayr climbed Mt. Taikbo (ca. 1,400 m) on the following day. A leg wounded from bathing on a coral reef at Manokwari had become worse during the climb but improved with medication and rest during the following days. The hunters collected special birds and everybody was busy skinning. Nothing would be wasted: The skins with the feathers were saved, but the bodies of somewhat larger birds went to the kitchen for dinner! Mayr ate more birds of paradise in those years than any other modern ornithologist. *Peltops* and *Halcyon sancta*, an Australian migrant, were shot near the camp, as were *Amblyornis* and *Rhipidura albolimbata*. There was much rain which, however, did not last for more than one or two days. On such occasions Mayr filled three notebooks with words of the native language with the help of several interpreters. The Papuas and the mantris proved to be faithful workers and sometimes they and Mayr had to stay up skinning at night until 11 p.m. Several local Papuas participated in hunting and some of them had a very good knowledge of animals.

"The people know the birds very well; every one has its own local name. When I describe one, they know immediately which species I have in mind. However, *Lamprothorax wilhelminae, Janthothorax mirabilis, Loborhamphus nobilis* and *Neoparadisea ruysi* [some of the "rare Birds of Paradise"] are unknown in this region" (to E. Hartert on 1 May 1928; transl.).

Some species with similar plumage had a group name, e.g., those of *Pachycephala, Graucalus, Manucodia* (including *Phonygammus*). The natives also knew quite a bit about the displays and life histories of *Parotia sefilata, Lophorina superba, Drepanornis, Diphyllodes, Paradisaea,* and *Phonygammus.*

On 28 April five police soldiers arrived from Momi sent by the governor for Mayr's "personal protection". However, they needed blankets and frightened the Papuas away from the camp. To dispatch the collections obtained so far Mayr marched down to the coast at Momi on 3 May and was back in camp at Siwi already on the 8th. Collecting had continued during his absence.

Erwin Stresemann stimulated Mayr's ambitions effectively by asking numerous questions on the biology of the birds of paradise and other species inhabiting the mountain forests and also gave specific advice, e.g., "Hartert joins me in the call: Do not overexert yourself and your people! Thou shalt sanction the holiday! Rest at least one day a week. [...] Don't measure your success with the scale. Don't be testy that I preach you this again and again, but I have done myself all those blunders against which I am cautioning you" (19 May 1928; transl.).

On 25 May, after Mayr had sent the five soldiers back to the coast, the expedition left Siwi for the villages of Ninei (800 m) and Ditchi (1,000 m). He was rather weak from diarrhea during the preceding days and had trouble climbing and descending the steep ridges, but was helped by the natives. The group stayed overnight at Ninei and then continued climbing Mt. Mundi where he observed *Pachycephala schlegeli, Myzomela rosenbergii,* and *Ptiloprora erythropleura* at 1,500 m. The descent toward Ditchi (1,000 m) was quite gradual following a long ridge. Camp was established in a newly built house on the opposite slope of the valley at 1,200 m elevation on 27 May. Montane birds and plants were collected in the luxuriant forests on the surrounding slopes during the following days, when they climbed Mt. Mundi (1,800 m) and Mt. Lehuma (1,900 m). Mayr made a list of 136 native names for the 137 species the birds in the area, and as soon as he received specimens he added their scientific designations. By this method he was certain not to miss any bird. Only two species were confused by the natives (Mayr 1963b: 17). To prevent his helpers to collect more specimens of very common birds he did not "pay" them. Otherwise they would get a little slip of paper that was worth one point, for better ones three or five, and for rarities ten slips of paper. When they had 25 points together, Mayr would give them an ax or a piece of loincloth or something like that. This system worked beautifully, and on the whole Mayr established excellent working relations:

"The Papuas always have great admiration for athletic feats. They are not at all impressed by the fact that I can write letters and that I have a precise idea of the value of each species of birds. This they take for granted. But when I lift a particularly heavy load with the little finger of my left hand and none of them can do that or when I toss a heavy rock a whole meter or more, further than even the strongest among them, then they really are full of admiration." An eclipse of the moon evoked no signs of interest or excitement and Mayr asked them if they had no myth about it. When he continued questioning them, one of the men slapped his shoulder and said soothingly: "Don't worry, master, it will become light again very soon." Mayr never again tried to acquire any information that was not given willingly.

Generally speaking, the habits of the natives varied regionally much more then Mayr had expected. Almost every valley was inhabited by tribes with a different culture. The natives, of course, were aware of Mayr's difficulties when crossing deep ravines on very steep trails or creeks on slippery tree trunks. There was always somebody waiting for the privilege of helping or carrying him across. When a river was not too wide and rather shallow he tried to cross it by jumping from rock to rock, greatly admired by the Papuas. Where they required five little steps he did it with one jump aided by a long stick (visible in Fig. 2.6).

In the interior of the Vogelkop he was alone for weeks and months with his three Malay mantris and New Guinea natives. Of course, he often felt depressed and lonesome. Stresemann's encouraging letters were a great help, praising him for his achievements, criticizing gently whatever he had done wrong asking numerous questions regarding the life history and courtship of many birds. Mayr was grateful for these letters "which I need badly. As you state quite correctly one gradually dulls against all tropical splendor, but such a letter shakes one awake again for a month" (late July 1928; transl.). To be addressed as "Dear researcher and friend" ... "pleased me very much. You will have noticed from my letters how much I consider you as my confidant and friend. After my return home your obligation as a friend will again be to educate me."

On 5 June, the 1,000th bird skin was made, quite a nice achievement. On the same day, a police force of two officers and about 20 soldiers and forty carriers arrived from Manokwari and that was the reason: The five soldiers dismissed at Siwi had told on their return all sorts of stories and rumors about the ambushed camp and murderous natives. Mayr had supposedly insisted on continuing into the interior and probably was no longer alive. The government therefore had sent a major police force into the mountains to rescue him or at least to punish the natives for his death. The five police soldiers were later court-martialed in Manokwari for all the lies they had told causing nothing but trouble.

On an excursion to Sohila (1,350 m) on the slope of Lehuma Mountain several new records entered the notebook and a display ground of *Amblyornis inornatus* was found: The carefully cleaned platform in front of the "house" was adorned with various red, grayish-green and white flowers (mostly orchids) and an empty snail shell. At 1,800 m the forest with taller trees and less moss covered was quite different from that on Mt. Mundi only 10 km away by air. Several days were devoted to collect plants and to label all bird skins.

The police force returned to the coast on 9th of June and, at the same time, the expedition departed for the mountain lake Anggi Gidji ("Male Lake") which was reached on the following day. Only one assistant accompanied Mayr and

**Fig. 2.6.** Ernst Mayr (*right*) and Sario, one of his Malay assistants, at Kofo, Anggi Gidji (Arfak Mountains), former Dutch New Guinea, June 1928 (Photograph courtesy of E. Mayr.)

a much reduced number of porters, while the other two mantris and crew members continued collecting around Ditchi. Actually, Mayr had intended to visit Lake Anggi Gita ("Female Lake") closer to Siwi (Fig. 2.5). But it was Basi, the local chief who provided the porters and chose the route. Why did he prefer this one? As Mayr later discovered, he wanted to get a bride for his son in which, however, he was unsuccessful. At the time Mayr suspected there might be hostile villages in the area or no trails. Sometimes at least he was in the hands of the natives and had to do what they told him.

The march from Ditchi via Dohunsehik consisting of two houses was very strenuous because several deeply incised valleys had to be crossed on slippery trails. At 1,500 m began the moss forest which was here as luxuriant as on Mt. Mundi. Lake Anggi Gidji is located at an elevation of 1,925 m. During the next 5 days, the expedition stayed in the village of Kofo with most friendly inhabitants. Mayr collected birds of a rather surprising Australian/Papuan faunal mixture: On the water *Anas superciliosa* and *Fulica atra*, in the marshes rails and *Acrocephalus* and in the grasslands *Megalurus* and a new species of *Lonchura* (*L. vana*). A few steps away in the forest he encountered such typically Papuan high mountain birds as *Machaerirhynchus nigripectus, Amalocichla incerta, Melidectes leucostephes*, etc.

Anggi Gidji had once before been visited by a botanical expedition in 1912, and the natives had a great respect for white people. On the fourth day of his stay, the chief made to Mayr a generous offer:

"He appeared before me leading a girl of about 12 years by his hand, his daughter. He proposed that I should marry this girl and become permanent resident of his village. They would build me a house, they would give me a piece of land for my 'garden,' they would plant my fields, and provide everything I need for living, and I wouldn't have to do anything, as I interpreted it. All I had to do was lend my prestige for his greater glory and that of the village of Kofo. As preposterous as the idea was, for a fleeting moment, I thought that if I accepted it, life for me would suddenly be easy, no more worries of any kind, but of course I knew it was impossible. Yet, if I declined the offer, I would insult him and I might get into immediate trouble. Most of all, I would not get the porters that I needed to get back to my base camp, so I gave the chief a very evasive answer stressing that half of my party was back at Ditchi and I had to take care of them and of my Malay mantris and I referred to the future rather vaguely. Fortunately, he accepted my answer and I rejected a carefree life as the guest of a proper mountain tribe."

**Fig. 2.7.** Ernst Mayr (*right*) with his assistant Darna in the government guest house at Momi, former Dutch New Guinea, labeling bird eggs, June 1928. (Photograph courtesy of E. Mayr.)

No trouble arose between Mayr and the Kofo people. In fact, on 16 June the whole village was lined up wailing about their visitor's departure. After the delay of a whole day when two hunters had shot a wild boar, they reached Ditchi at noon of the 19th and left for the coast at daybreak of the following morning. While they descended Mayr received the mail of several weeks. Overjoyed, he spent the evening in Ninei reading and rereading his 25 letters. On their arrival in Siwi the next forenoon they found the village empty since the people had left for the coast and for a distant valley to collect dammar[7]. By paying a whole load of salt to their chief, Mayr persuaded the Ditchi porters to stay with him. They arrived at Momi in the early afternoon of 23 June (Fig. 2.7) and, with a fairly old canoe, left again for Manokwari in the early afternoon on 24 June. They all paddled strongly for many hours until they reached their destination at 10 o'clock in the evening of the following day. Tired, unshaved, dirty, and sunburnt, Mayr immediately had to board the Dutch marine survey ship to tell about his adventures.

#### (3) Wondiwoi Mountains, Wandammen Peninsula (Fig. 2.5)

Itinerary:

| | |
|---|---|
| 28 June–5 July 1928 | en route to Wasior |
| 6 July 1928 | climbing Wondiwoi Mountain |
| 7–18 July 1928 | Wondiwoi Mountain collecting |
| 19–26 July 1928 | at Wasior |
| 27–29 July 1928 | en route to Hollandia |

When the mail ship reached Manokwari, Mayr received bad news: The two boxes from Java with the needed supplies including ammunition for their bird guns had not arrived. Transportation was always a problem. Going from one place to another, dividing the baggage into smaller shipments in order to make use of available boats, etc., all that caused many difficulties. After three days of preparations in Manokwari the expedition left again for Momi and Wasior (Wandammen Peninsula) on the evening of 28 June. The objective was to collect birds on the higher slopes and the top of the Wondiwoi Mountains which form the backbone of the peninsula. Paddling all night they had reached the base of the Arfak Mountains next morning and then moved slowly southward against a strong southeasterly wind towards Momi where they arrived in the evening of 30 June. By that time the mail canoe with the supplies and equipment was supposed to be there, but when they started again on the government ship "Griffioen" on 4 July Mayr was still waiting for his shipment. The next morning they unloaded their baggage in Wasior at the foot of the Wondiwoi Mountains. This was Mayr's 24th birthday; but instead of celebrating they had to prepare their departure on 6 July. Since hunting of birds of paradise had been forbidden and nobody had climbed up to the mountains, the trail was completely overgrown. The vegetation was wet and bamboo thickets

[7] The Malay name of dammar refers to resin collected from several different species of trees and is used to make nondripping torches.

conspicuous. On 7 July camp was established at 1,400 m. It started to rain the next day and never stopped while they stayed on the mountain. The Wondiwoi is a long crest with a number of ups and downs around 2,000 m elevation. There were several distinct subspecies of *Crateroscelis, Sericornis* and *Eupetes* which either tend toward their relatives in the Arfak Mountains or toward those of the Central Range of New Guinea. But otherwise bird collecting was poor. The firewood was soaking wet, only the kerosene lamp provided some heat. Moreover, they had run out of rice and no dry clothes were left. Everything was quite miserable. On 19 July the porters arrived and they returned to Wasior on the coast. Next day the weather changed to sunshine and wind. The natives commented that whenever the mountain hears gun shooting it covers itself in clouds. An interesting bird collected near Wasior was *Pitohui kirhocephalus* which in this area is intermediate between two strikingly different subspecies. Mayr was also pleased to get specimens of *Poecilodryas brachyura, Eupetes caerulescens* and on the last day one *Megapodius.*

The group left Wasior on 27 July on board a steamer where a big pile of mail was waiting for Mayr with good news: The German Research Council had granted 6,000 Marks for the exploration of the Huon Peninsula later this year.

#### (4) Cyclops Mountains (Fig. 2.8)

Itinerary:

| | |
|---|---|
| 31 July–15 August 1928 | at Hollandia |
| 16 August–14 September 1928 | Cyclops Mountains, collecting |
| 15–20 September 1928 | Ifar, Sentani Lake, collecting |
| 21 September–20 October 1928 | at Hollandia |

The ship was on its way to Hollandia (Jayapura today) on the northern coast of Dutch New Guinea, close to the eastern border. On the morning of 30 July they were traveling alongside the Cyclops Mountains, which drop steeply to the sea and soon entered Humboldt Bay. Until the Resident Officer had returned to issue the necessary permits for the mountains their excursions were limited to the vicinity of Hollandia. One collecting station was the village of Hol on a small coastal strip in Jautefa Bay where *Megacrex* and *Seleucides* were obtained.[8]

In Hollandia lived several Dutch and German planters, the latter displaced from former German New Guinea to the east after the end of World War I. Reflecting on "colonization" of foreign territories Mayr wrote in his diary: "Should one really interfere to such an extent with these natives, and deprive them of their old culture, replacing it by European civilization? I admit this is a very fundamental question which really amounts to the question: Should one do any colonizing at all? [...] A solution must be found to protect the rights of the natives, and, in spite of a certain strictness of regulations, to do the best for them."

[8] In his article on Mayr's bird collection Hartert (1930), thinking that Hol was an abbreviation for "Hollandia," added the rest of the word to Mayr's Hol labels. One can usually tell that it has been added but the Rothschild labels have "Hollandia" written out and one has to look at Mayr's labels. Mostly, this does not matter, but there are a few birds that Mayr obtained only at Hol (M. LeCroy, pers. comm.).

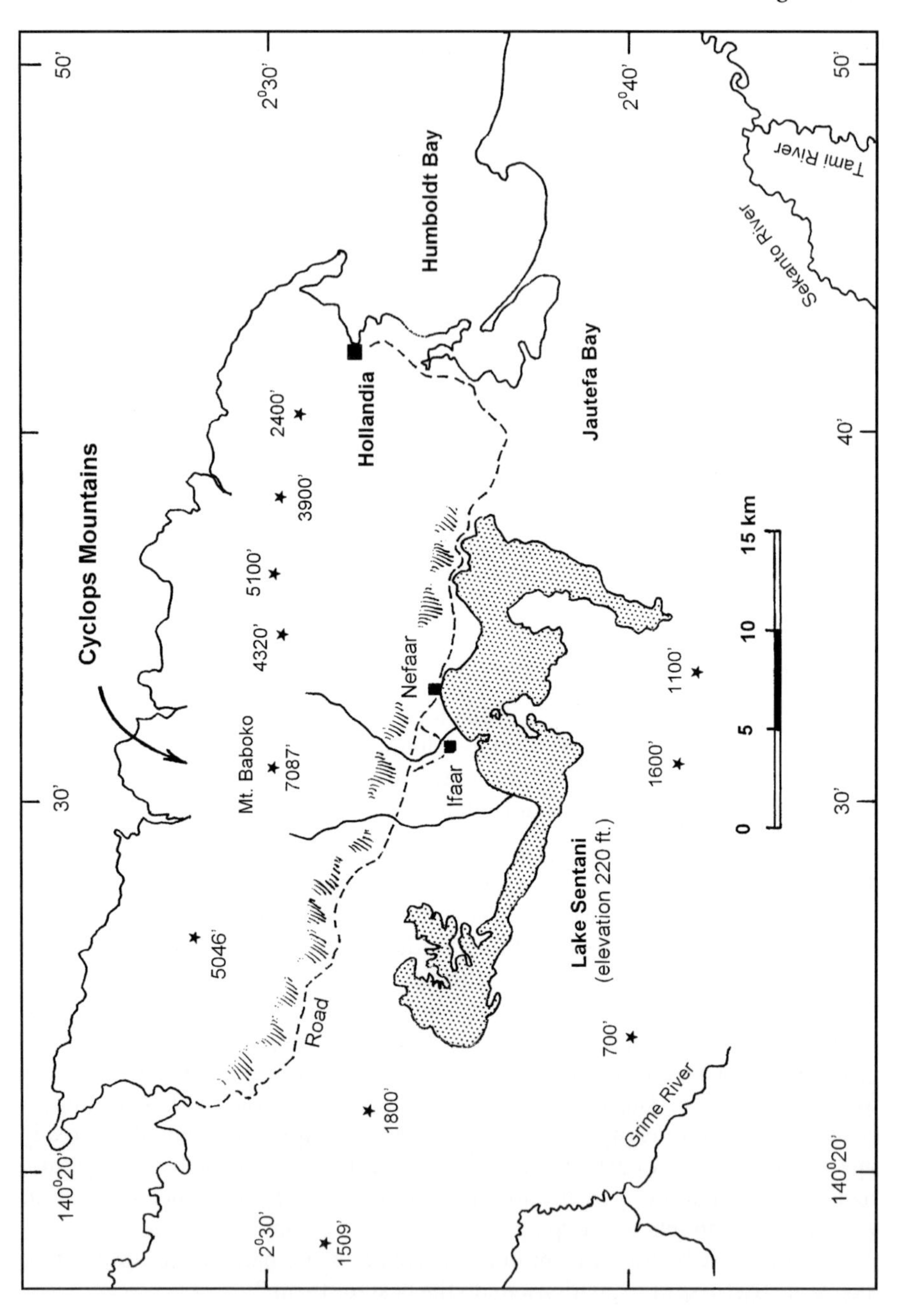

Cyclops Mountains
Humboldt Bay
Hollandia
Jautefa Bay
Nefaar
Ifaar
Mt. Baboko
7087'
4320'
5100'
3900'
2400'
5046'
1800'
1509'
700'
1600'
1100'
Road
Lake Sentani
(elevation 220 ft.)
Grime River
Sekanto River
Tami River
0
5
10
15 km
140°20'
30'
40'
50'
2°30'
2°40'

**Fig. 2.8.** The isolated Cyclops Mountains and Lake Sentani on the northern coast of Irian Jaya (former Dutch New Guinea). Elevations in feet. For a location map see Fig. 2.4

◄

After the Resident Officer had returned on 12 August they were ready to leave for the Cyclops Mountains on 16 August and on the same day reached Ifaar on Lake Sentani. A number of endemic subspecies of birds inhabiting this grassland region indicate that at least part of the grassland must be old and not man-made. Among the grassland birds were *Malurus, Megalurus, Saxicola,* and some *Lonchura* species. On 18 August a camp was established north of Ifaar at about 700 to 750 m elevation. Here they had found a spring, but no other water further up the slope. With several helpers Mayr concentrated for a few days on plant collecting, especially orchids were plentiful in this region. The forests were swarmed by chiggers and, not quite as bad, by leeches.[9] It rained a lot, but not as much as on the Wondiwoi Mountains a month earlier. Mayr found the lower limit of distribution of montane birds to be higher here than in the Arfak and Wondiwoi Mountains. He established that most of the montane avifauna was missing, particularly the birds of paradise. Therefore, the birds of the lower montane zones extend their vertical ranges all the way up to 2,000 m, e.g., *Phylloscopus, Pitohui dichrous,* and *Poecilodryas leucops.* Mayr suspected that the montane species never had colonized this isolated mountain range rather than being extinct there. This mountain range and others along the north coast of New Guinea probably have been islands cut off from the rest of New Guinea during the late Tertiary.

Hartert and Rothschild's idea that the Cyclops Mountains might harbor an unknown montane bird fauna had been based on a small bird collection supposedly from the "Cyclops Mountains" made by J. M. Dumas in the late 1890s. Rothschild (1899) had described *Melampitta gigantea* from "Mt. Maori west of Humboldt Bay." Mayr's (1930f: 24) statement that Mt. Moari (Mori) is actually located in the Arfak Mountains near Oransbari (Fig. 2.5) helped to solve many zoogeographical problems.

Around 2,000 m the forest was high and humid. While hunting birds Mayr looked at all the epiphytes in search of flowering orchids which were plentiful there. On 3 September he moved the camp to a site at 1,100 m which had been prepared during the preceding days and on 11 September he climbed the top of the Cyclops Mountains. A view from a high tree convinced him that this was indeed the highest peak. On the way back to camp he visited a cave inhabited by about 300 large fruit bats which, after his entry, cruised around screaming in the upper part of the cave.

On 14 September the expedition returned to the lowlands and to Ifaar at the shore of Lake Sentani. The grassland avifauna was studied for a week and on 21 September Mayr went alone to Hollandia to take care of his mail and to dispatch the bird collections to Bogor (Buitenzorg, Java). The steamer arrived on 24 September 1928 with rather unexpected news. Whereas Mayr expected his return to Germany

[9] As an example, after a march of 1½ hours, he removed 83 leeches from his shoes.

within a few months, Hartert suggested that after the completion of the Saruwaget (Huon) expedition, Mayr continue work in New Guinea, go to Biak Island for a rest and then on to the Weyland Mountains (although Mayr's contract signed for a collection of 3,000 bird skins had been fulfilled). Originally, Stresemann and Hartert had also planned that Dr. Rudolf Kuhk would travel to New Guinea in early or mid-1929 to join Mayr for the Weyland expedition. Both these projects, however, fell through when, in March 1929, the American Museum in New York approached Stresemann and the Museum of Natural History in Berlin proposing that Mayr join the Whitney South Sea Expedition. Stresemann later sent Georg Stein to northwestern New Guinea in 1931 to continue Mayr's work in that region.

During the first week of October 1928 Mayr and his crew finished searching around Lake Sentani and Ifaar and then returned to Hollandia on the coast. The collections from the Cyclops Mountains were packed and dispatched to Java. The time had come to part, to say 'Goodbye' and 'Thank you' to his faithful and hardworking mantris—Sario, Darna, and Soeab. They returned home to Bogor (Buitenzorg), on 20 October. It was they who deserved major credit for the success of Mayr's first expedition and he was fully aware of this. He also acknowledged gratefully Dr. K. W. Dammermann and Dr. H. C. Siebers of the Zoological Museum in Bogor, who so unselfishly and splendidly supported him. Mayr in the 1990s spoke very highly of this truly exemplary spirit of collaboration between the museums in Tring and Bogor. Drs. Dammermann (1885–1951) and Siebers (1890–1949) went out of their way to help. Not only did they make the three mantris available, but they also made a major contribution to the equipment of the expedition. In addition they received, opened and thoroughly dried the collections which Mayr sent in several shipments to Bogor. Subsequently, Dr. Dammermann had the consignments re-packed and forwarded to England. About 270 specimens of these collections were later returned to the museum in Bogor after Hartert had completed their scientific study.

## Papua New Guinea, Former Mandated Territory

### (1) Canoe Trip along the North Coast

In Hollandia, Mayr now prepared the continuation of his travels east to the Mandated Territory of New Guinea, the former German New Guinea. He was faced with a real dilemma: Although he was only about 200 kilometers away from Aitape, the nearest major port in the Mandated Territory to the east (Fig. 2.4), there was simply no connection between the Dutch and the Australian territories, except via Australia. The only possibility to reach Aitape was to rent a small fleet of canoes and paddle along the coast. Everybody told him that this was utter folly, but he felt he had no choice. Finally he found some people willing to take him to Vanimo, the nearest village to the east along the rocky coast and about 75 km away, from where he hoped to find some other way to continue toward Aitape.

On 21 October Mayr started with five coastal outrigger canoes each with a crew of three or four Papuas. Two former Malay bird of paradise hunters, Damar and

Sehe, accompanied him. A few hours out of Hollandia a strong wind sprang up, the waves became higher and higher and the sails had to be trimmed. In each canoe one person was steering and the others were busy bailing out water. The situation became increasingly serious, but after they had crossed Humboldt Bay and got close enough to the coast the sea smoothed down. Within a couple of hours they reached Vanimo harbor. After several days delay to find canoes for the trip to the village of Leitre (another 40 km east), they were ready to leave on the clear moonlit night of 27 October. But soon it turned stormy and again they bailed feverishly during the night and all of next morning. By one o'clock they reached Leitre and now faced a thundering surf. The two canoes with the lightest loads managed to get through rather quickly, whereas the cargo of the heavily loaded ones got wet. Finally all canoes had made it "safely" to the beach, i.e., with only minor damage. Mayr changed to a canoe of local people who had come out for help. When they approached the beach, they got over the first breaker, but on the second the canoe overturned. One of the Papuas pulled Mayr out of the boat sideways and dragged him to the shore "to dry out."

By now, they had covered just over half of the distance to Aitape which lay another 90 kilometers to the east. There was much trouble to obtain transportation to Serai, the next village. Three canoes left on 30 October and two were to follow later, while Mayr continued on foot walking along the shore with the most valuable part of the baggage. He had the greatest difficulties getting porters. Finally he threatened with his gun. They walked along the beach under the torrid sun, crossed small river mouths and suffered from sand flies. When they caught up with the three canoes which had left Leitre before their departure, they transferred the baggage to them and left several hours later. The wind picked up again but the boats landed safely in Serai in the afternoon of 31 October. The inhabitants were quite helpful. Several hours later the other two canoes also arrived and the group was reunited. They cooked a good meal and prepared their departure on the next day. The last distance from Serai to Aitape was covered mostly on foot (1–3 November), at first along the beach and later on a road through coconut plantations.

Mayr here reached the domain of a German Catholic Mission. Eager to meet a compatriot and to speak German again, he entered the missionary's house saying "Grüss Gott, Vater Franz!" The father almost fell off his chair asking "Where do you come from?" Apparently he knew nothing at all of his impending arrival. His astonishment was particularly great, since Mayr came from the west, that is, from the region of the Dutch territory, where no one ever came from. Naturally he was terribly pleased. During their conversation, it turned out that Father Franz was born and raised in Versbach, a little village near Würzburg, where Mayr had passed dozens of times as a boy hiking with his parents. The arrival at the mission station finished the most difficult trip Mayr had undertaken during his entire stay in New Guinea.

Near Aitape the song of the Varied Honeyeater (*Meliphaga virescens* = *Lichenostomus versicolor*) accompanied them. For part of the way they used canoes on inland waters. The last day porters were again employed.

## (2) From Aitape to Finschhafen

The steamer that had been scheduled to call at Aitape a week before their arrival had been delayed so much that Mayr and his crew were able to reach it just before its departure. They started for Rabaul, New Britain, on 6 November. The first stop was at Allison Island and then at Mal, one of the Ninigo Islands, to load copra. The next halt was at one of the hilly islands of the Hermit Archipelago where the largest island still had original forest. In Rabaul, Mayr had more time than expected because his steamer, and before continuing to Finschhafen, had to pull a sister ship off a reef. This permitted him to visit Father Otto Meyer (Fig. 2.9) on Vatom Island off the coast near Rabaul who studied the bird fauna of his island and had been in contact with the Berlin Museum of Natural History for over 20 years publishing the results of his valuable research in several scientific journals. His invitation ran: "You must come, because I just got 40 eggs of *Megapodius* which need to be examined by a scientist—afterwards as an omelet!" He received Mayr most cordially, and they celebrated this meeting in the South Sea with a glass of wine. They had the most lively conversation about ornithological topics, particularly of course about the nesting habits of the mound-building megapodes.

The steamer "Mantura" with Mayr and his crew on board left Rabaul on 3 December reaching Alexis Harbour and Madang on 7 December and calling at Finschhafen on 8 December. There they were received most friendly by the Protestant-Lutheran Neuendettelsauer Mission. After two days of preparations they went up to the

**Fig. 2.9.** Father Otto Meyer, Uatom Island, off Rabaul, New Britain (Archive, Museum of Natural History Berlin, Orn. 121, 5)

mission center at Sattelberg, where Mayr became totally immersed in a Christian atmosphere. He participated wholeheartedly because he was quite serious in testing once more his position toward Christianity. "However," he wrote in his diary, "I fear I simply cannot accept the Christian dogma. It demands too much sacrifice from one's reasoning."

### (3) Saruwaget Mountains (Fig. 2.10)

Itinerary:

| | |
|---|---|
| 3–8 December 1928 | en route from Rabaul to Finschhafen |
| 9 December 1928 | en route to Sattelberg |
| 10 December 1928–6 January 1929 | at Sattelberg collecting |
| 7–8 January 1929 | en route from Sattelberg to Junzaing |
| 9 January–1 February 1929 | at Junzaing collecting |
| 2–9 February 1929 | en route from Junzaing to Ogeramnang |
| 10 February–4 March 1929 | at Ogeramnang collecting |
| 5–7 March 1929 | climbing Rawlinson Range (Saruwaget Mts.) |
| 8–12 March 1929 | high mountains, collecting |
| 13–14 March 1929 | returning to Ogeramnang |
| 15 March–3 April 1929 | at Ogeramnang, collecting |
| 4–9 April 1929 | en route from Ogeramnang to Finschhafen |
| 10–27 April 1929 | at Finschhafen |

At first, about ten birds a day were collected which increased with growing experience of the two Malayan skinners. Mayr's plan was to search the surroundings of three stations at different elevations: (1) Sattelberg (700–1,000 m), (2) Junzaing (1,100–1,400 m), and (3) Ogeramnang (1,600–2,000 m). Christmas and New Year were spent at Sattelberg. On 7 January 1929 the expedition left for Nanduo, stayed there overnight, and arrived at Junzaing on the 8th. Interesting birds obtained at this locality included, e.g., *Pseudopitta (Amalocichla) incerta, Sericornis arfakianus*, and *Casuarius bennetti*. From 18–27 January Mayr visited Sattelberg and Finschhafen to send off part of his collections but was rather desperate because the ammunition which he badly needed had not arrived. Now he had to continue reloading his brass cartridges for the bird guns every evening.

Camp move to Ogeramnang was scheduled for 2 February. They crossed a ridge and descended into a deep valley reaching Joangeng in the afternoon. The village of Kulungtufu (1,529 m) was their next home for several days after they had crossed the deep valley of the Mongi River. Departing for Tobou on 8 February they reached Ogeramnang on the following day. A watershed between the Kuac and Burrum Rivers had to be crossed and just before reaching Ogeramnang (1,785 m) they had to descend once more into a deep valley. This village was the center of their collecting activity during the next several weeks. Much forest had been cut and now was interspersed with plantations. Most birds occurred along the forest edges, while hunting in the deep forest was poor. Mayr listed the names used by the

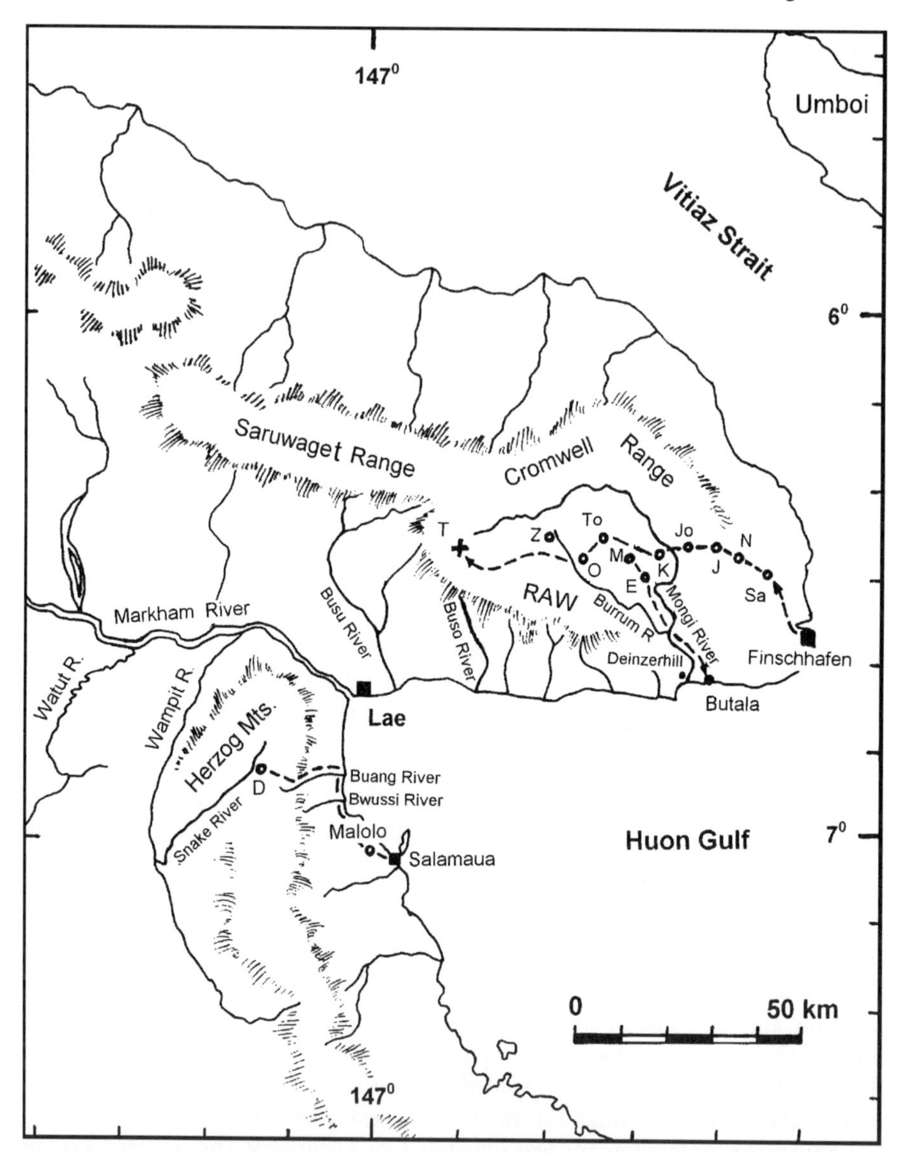


**Fig. 2.10.** Huon Peninsula, southeastern New Guinea. Expedition routes in the Saruwaget Mountains and the Herzog Mountains (*dashed lines*). Approximate locations of villages in the Saruwaget Mountains are based on a map by Wagner and Reiner (1986: 63). For a location map see Fig. 2.4. Sa Sattelberg, N Nanduo, J Junzaing, Jo Joangeng, K Kulungtufu, To Tobou, O Ogeramnang, Z Zangung, T Mt. Titaknan, M Mindik, E Ebabe, RAW Rawlinson Range, D Dawong

Burrum people and all they knew about bird life. On 22 February he had a severe malaria attack which he treated with quinine and the next day with plasmochine, an experimental drug offered by a pharmaceutical firm in Germany. It had never been licensed for it sometimes caused bad side-effects, particularly on the spleen. In his case it worked wonders. Mayr coped well with his situation, remembering that Erwin Stresemann also suffered with malaria throughout his expedition to the Moluccas in 1912. Although still weak, Mayr made a first short excursion on 25 February to watch one of the dance festivals of the natives some of whom wore large grass skirts. Each of the mountain tribes in this region had its own dialect or language and, besides it, the natives usually knew no more than one or two of the adjacent tribes. Therefore Mayr who spoke Pidgin English and Malay needed up to four local interpreters to communicate with certain groups of people.

On 5 March he left Ogeramnang for the high mountains, the Rawlinson Range, part of the Saruwaget Mountains (Fig. 2.10). Walking for him was still difficult and every step an effort. Ascending all day they reached a forest with tree ferns and unknown bird voices. The night was cold with pouring rain and only about 5°C. Marshy meadows, tree ferns, stretches of grassland along the water courses and woodland on the upper parts of the hills made it a very diversified landscape. The occurrence of grassland birds with endemic subspecies (*Megalurus* and *Anthus*) again indicated that much or at least most of the grasslands are old (see p. 65). On 8 March they reached a plateau with a lot of rhododendron and a high altitude bog. Here they found shelter in a hunting cottage of the natives. On the way up Mayr saw pipits (*Anthus*), the Alpine Thrush (*Turdus poliocephalus*), *Rhipidura brachyrhynchus*, high altitude species of *Crateroscelis, Sericornis*, etc. Above the tree limit on 9 March only *Megalurus, Turdus* and *Anthus* were observed. From the top of Mt. Titaknan (close to 4,000 m in elevation) they had a beautiful view over the Markham Valley to the southwest, the Huon Gulf in the south and toward the eastern part of the Rawlinson Range. The Saruwaget Mountains proper to the west were cloud covered. Collecting of birds and plants continued until 13 March, when they started the descent via Zangung, which was reached after dark; they were back in Ogeramnang the next day.

There was no mail, only disappointment, fatigue and exhaustion. Nevertheless the bird collection now comprised 1,000 specimens. Mayr here used again his system of paper money which, as in the Arfak Mountains (p. 58), was readily accepted by the mountain tribes. A very desirable bird was worth ten paper slips. A forest rail which Mayr later described as a new subspecies was at first such a most desirable bird, but after the natives had brought in more than ten specimens, he was forced to reduce its value (Mayr 2004g). Eventually, he had a total of 42 specimens of this montane rail which is shy, retiring and most difficult to observe. The explanation of why his natives had been able to collect so many specimens of this bird was their knowledge of how to catch them in their sleeping nests at night. This bird builds characteristic football-sized sleeping nests of moss and leaves in low trees in which several birds crowd together to spend the night (later confirmed by Mayr and Gillard 1954f). Thirty-five specimens of Mayr's series are still preserved in the Museum of Natural History of Berlin (Steinheimer 2004).

Finally the mail arrived in two installments on 26 and 30 March, the first in 2 months. Mayr's beloved grandmother in Lindau had passed away in December. Included in the mail was a box with Christmas cookies from his mother and a puzzling telegram from the Museum in Berlin: "Director favors Whitney suggestion." An explanatory letter was possibly on its way on the same April steamer with which Mayr intended to leave this region.

Collecting of birds, plants and insects around Ogeramnang continued until 4 April, when the expedition left this area and returned directly to the coast, this time along a different route than before, i.e., almost straight south via Mindik and Ebabe to Butala where they arrived on 6 April (Fig. 2.10).

"On the way we had to cross a torrent on a native bridge which essentially consisted of three rope-like lianas. One walked on the bottom liana, which was somewhat widened by additional vegetable material, and held on by the arms on the two other lianas. All the time the 'bridge' was swaying quite frighteningly. Even the natives were rather scared using it and the women in front of us at first refused to do so and wailed and cried. The men literally had to beat them to make them go. What a pity I didn't have a movie camera."

On the morning of April 7th, Mayr visited the mission station Deinzerhill across the bay from Butala and learned that the mission boat would pass by on 9 April. It brought him and his crew back to Finschhafen. A week later, on 16 April, he made an excursion to watch the displays of *Paradisaea raggiana*:

"We left at daybreak for the heights above Finschhafen. It went higher and higher and I was perspiring copiously. Finally the calls of *Paradisaea* electrified me and accelerated my ascent. We walked through a small piece of woodland and finally I see the orange-red feathers through the foliage. I sat down precisely under the tree which according to the natives is the favorite display tree. I waited here with my Malay. The birds are not yet very active since the real display time has not yet come. The plumage had been freshly molted and the feathers had not yet fully grown. There was a good deal of scattered calling on neighboring trees. Finally one male arrived on the display tree, my heart beating faster. The bird makes various intention movements and a moment later he begins to call, first slowly then accelerating and beginning with the display. The wings are spread and are beaten jerkingly, the plumes are raised. The bird during this display jumps excitedly on the branch on which he is standing and turns and rotates occasionally. The whole display lasts only a short time. Shortly afterwards two females appear which are vigorously pursued by the male but no copulation takes place even though the males at this time already have greatly enlarged testes. Copulation, according to the natives, takes place on the display tree and the activity is so vigorous that sometimes both of them together fall down from the display tree and can then be caught by the native standing under the tree."

A few days later, Mayr met Rollo H. Beck (1870–1950), leader of the Whitney South Sea Expedition from 1920 until late 1928 (p. 145). He had just returned from the Hompua-Wareo region, 5–10 kilometers north of Sattelberg, with a fine collection of mountain birds and was about to return to the United States.[10] He had mostly collected in the Finisterre, Cromwell and Adelbert Mountains of northern New Guinea. Beck demonstrated with pride a specimen of a new form of forest rail and Mayr did not have the heart to tell him that he had got a large number of specimens of this common, yet shy bird which he (1931l) later described as *Rallicula rubra dryas* (today placed in the species *Rallina forbesi*). Before Mayr could stop them, the natives had collected 43 specimens of this rail, because they caught them in rather conspicuous sleeping nests into each of which several birds crowded at night (2004g). Probably Beck also told Mayr details about the Whitney Expedition making him even more curious about the meaning of the cable received earlier.

*The Detzner incident.* While traveling in former German New Guinea, Mayr heard about an exciting book by Hermann Detzner entitled *"Four Years among Cannibals"* (1920, Scherl, Berlin). The author, an officer of the small colonial police troop of German New Guinea, had retreated into the interior in 1914, when the Australians occupied eastern New Guinea at the beginning of World War I. There he had hidden at the Neuendettelsauer Mission station Sattelberg, near Finschhafen, throughout the war. His book pretended to report his experiences and adventures during these years and his success in thwarting all Australian efforts to capture him. Instead he had been all the time under the protection of the German missionary Christian Keysser with whom Mayr discussed Detzner's "adventures," fights with the Australians, survival in the remote interior, discoveries of various mountain ranges, rivers, and untouched human tribes. Most of these stories were pure inventions! However, upon his return home after the war, the German Geographical Society of Berlin had honored him with the "Nachtigall Medal" for his alleged geographical discoveries; the University of Bonn had awarded him an honorary degree and the government the Iron Cross First Class. The Foreign Office had given him a good position, and a glorious report on his "researches" was published (Behrmann 1919). Detzner's book was reprinted three times in 1921, translated into Swedish in 1925 and into French in 1935.

Mayr was bothered that German science had become the laughing stock of the Australians who knew where Detzner had been during the war and mocked at the naiveté of the Geographical Society of Berlin. Mayr investigated the matter carefully and in discussions with missionary Keysser collected convincing material to prove that Detzner's book was nothing but a fairy tale. Upon his return to Germany in April 1930 Mayr informed the Geographical Society of Berlin but heard, after several months delay, that Detzner had been able to defend his stories. When Mayr then threatened to publish all of his evidence against the veracity of Detzner's "discoveries," the Society appointed a committee of three professors who investigated the case in more detail and interviewed Mayr later in New York. They came quickly to the conclusion that Detzner indeed had lied. He was obliged

[10] Mayr's expedition to the Huon Peninsula was financed by the German Research Council and all birds collected were going to be sent to the museum in Berlin. Probably for this reason the American Museum of Natural History had asked R.H. Beck to spend some time collecting in this region before returning home in 1929.

to publish a statement that his book of 1920 was meant to be literature rather than a scientific report (see *Zeitschrift der Gesellschaft für Erdkunde in Berlin*, 1932, pp. 307–308) and to resign as a member of the Society in order not to be expelled. The reputation of the Geographical Society of Berlin was restored.

Even after 1930 and into the 1960s Detzner's name and his "achievements" have been cited approvingly by several authors. Finally Biskup (1968) discussed in detail the contents of Detzner's book and challenged several aspects of his account growing suspicious about certain details, when he read Keysser's autobiography (1929). In an appendix to his article Biskup (1968, p. 21) cited a translation of "Detzner's statement" in *Zeitschrift der Gesellschaft für Erdkunde in Berlin* and the note indicating his resignation from this Society but, of course, he knew nothing about the background to Detzner's statement and how his resignation had come about.

#### (4) Herzog Mountains (Fig. 2.10)

Itinerary:

| | |
|---|---|
| 28 April 1929 | en route from Finschhafen to Salamaua |
| 29 April–6 May 1929 | at Malolo (mission station near Salamaua) preparing for inland trip to Herzog Mountains |
| 7–10 May 1929 | en route from Malolo to Dawong, Snake River |
| 11 May–2 June 1929 | at Dawong, collecting |
| 3–4 June 1929 | en route from Dawong to Malolo and Salamaua |

In Finschhafen, Mayr boarded a steamer for Salamaua on 28 April alone, without his Malay mantris for Europeans only were allowed to travel during the current mumps quarantine. Just before leaving, he had received his mail and learned more about the American suggestion to join the Whitney Expedition instead of returning to Manokwari. This sounded like an interesting project to him.

En route to Salamaua across the Huon Gulf he had a beautiful view of the Rawlinson Mountains to the north, the deep cut of the Markham Valley to the west and the heights of the Herzog and Kuper Mountains. Next day he met missionary Bayer in Salamaua and together they went to the mission station Malolo (Fig. 2.10). Several days were spent preparing for the trip north to the Buang River at the shore of the Huon Gulf. Here, through arrangements made by Mr. Bayer, Kademoi people from the Snake River Valley would meet him and carry his baggage west into the Herzog Mountains to the village of Dawong.

The expedition left Malolo on 7 May and met the Dawong carriers on the following day. They had been hindered by rains and a swollen river. Crossing the low mountain range on the 9th Mayr and his porters arrived in the Kademoi (Snake River) Valley and Dawong on 10 May. His Malay mantris were not able to catch up with him until 23 May. From then on, hunting and skinning improved a lot compared to the previous week, when Mayr had to employ local boys. On 29 May, a telegram arrived which read: "Are you willing to join Whitney Expedition?

Answer prepaid." It was sent from New York on 4 May! Even though he had more or less expected such a telegram for quite a while, he still was not sure what to do. Stresemann had advised him to accept this offer and even the museum director C. Zimmer in Berlin had telegraphed in late March that he favored the Whitney Expedition.

On 3 June the group returned to the coast in one day and Mayr even reached Malolo after dark, a distance that had taken him 3 days traveling to Dawong with all the luggage almost a month before.

### (5) From Salamaua to Samarai

The mission boat arrived at Salamaua in pouring rain on 10 June and Mayr received the eagerly awaited mail with final explanations of the various telegrams. Now he could make a decision. He telegraphed to New York two days later: "Accept your offer to join Whitney. Answer Samarai. Mayr." The contract was for one year. Consumed by homesickness after his long absence from Germany and suffering from loneliness during his long stays in the interior, he was more than ready to quit. On the other hand, Stresemann had added one tantalizing footnote to his telegraph, "joining the Whitney Expedition might be important for your future career." It was this footnote that finally led Mayr to accept the New York proposal. The Whitney Expedition would sail from Samarai toward the end of June.

Mayr and his mantris boarded the steamer at Salamaua on 12 June arriving at Rabaul on 14 June. Here he said goodbye to his two Malay boys who were going by steamer to Aitape to make their way back to Hollandia. In the morning of 16 June the SS Montoro with Mayr aboard was approaching the d'Entrecasteaux Archipelago with the high mountain of Goodenough Island clearly visible. Soon afterwards the ship entered the port of Samarai, a little island off the tip of southeastern New Guinea (Fig. 2.4). Mayr checked into a hotel to wait for the arrival of the Whitney Expedition's 77-ton motor schooner, the *France*. On the following day he received from New York an answer to his earlier cable: "Delighted. Report Hamlin, letter follows. Murphy." The *France* did not arrive from Port Moresby until 3 July. A new engine had been installed at Samarai in March–April of that year.

## Solomon Islands

Itinerary (see Fig. 2.11):

| | |
|---|---|
| 3–9 July 1929 | at Samarai, Papua New Guinea–Mayr joined Whitney South Sea Expedition (WSSE) |
| 10–14 July 1929 | en route from Samarai to Faisi, Shortland Island |
| 15 July 1929 | at Faisi, Shortland I. (polio epidemic) |
| 16–19 July 1929 | at Kieta, Bougainville Island |

(1) Buin region, southern Bougainville (20–27 July 1929), collecting

| | |
|---|---|
| 28–29 July 1929 | back in Kieta, Bougainville |

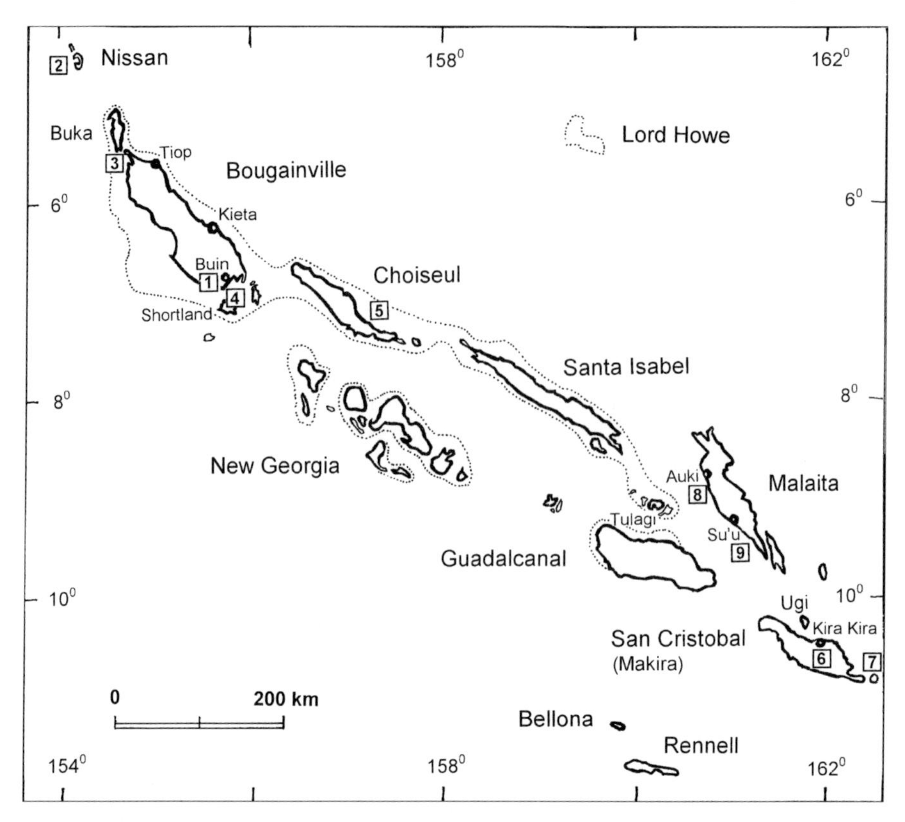


**Fig. 2.11.** Map of the Solomon Islands, southwest Pacific Ocean. Islands visited by the Whitney South Sea Expedition (from NW to SE) Nissan (2), Bougainville (1, 3), Shortland (4), Choiseul (5), Florida (Tulagi), Malaita (8–9) and San Cristobal (Makira, 6–7). *Dotted line* follows 200 m water depth. For a location map of the Solomon Islands to the east of New Guinea see Figs. 2.4 and 4.1

| | |
|---|---|
| 30 July–3 August 1929 | en route from Kieta via Buka Passage to Nissan Island |
| 4 August 1929 | visiting with inhabitants of Nissan Island |

(2) Nissan Island (5–10 August 1929), collecting

| | |
|---|---|
| 11–14 August 1929 | en route from Nissan Island to Soraken Harbor (Surikan) |

(3) Soraken, Bougainville Island (15–20 August 1929), some collecting inland

| | |
|---|---|
| 21–29 August 1929 | en route from Soraken to Kieta |
| 30 August–2 September 1929 | at Kieta; Messrs. W. Coultas and W. Eyerdam joined the expedition |
| 3–6 September 1929 | en route from Kieta to Faisi, Shortland Island, arriving 6:00 a.m. |

(4) Faisi, Shortland Island (6–13 September 1929) some collecting along the shore

| | |
|---|---|
| 14–15 September 1929 | en route from Faisi to Choiseul Island |

(5) Choiseul Island (15 September–4 November 1929), collecting inland

| | |
|---|---|
| 5–16 November 1929 | en route from Choiseul Island to Tulagi, Florida Island |
| 17–24 November 1929 | at Tulagi; "France" in dry dock for repairs (Hamlin stayed with her in Tulagi, when the expedition left for Kirakira) |
| 25–26 November 1929 | Mayr, Coultas, Eyerdam on steamer "Ranadi" en route from Tulagi to Kirakira, San Cristobal Island |
| 27–28 November 1929 | at Kirakira, San Cristobal Island |
| 29–30 November 1929 | en route from Kirakira to the interior of the island, where camp is established at Hunogaraha |

(6) Hunogaraha, 1900 ft. elevation, San Cristobal (1–21 December 1929), collecting

| | |
|---|---|
| 22–23 December 1929 | en route from Hunogaraha to Kirakira |
| 24–27 December 1929 | at Kirakira, some collecting |
| 28 December 1929 | en route from Kirakira to Santa Ana Island, off the east tip of San Cristobal Island |

(7) Santa Ana Island (29 December 1929–8 January 1930), some collecting

| | |
|---|---|
| 9–12 January 1930 | en route from Santa Ana Island to Tulagi, Florida Island on M/V "Hygeia" |
| 13–23 January 1930 | at Tulagi aboard "France," dispatching collections and making preparations for collecting on Malaita Island. Hamlin turned over leadership of the expedition to W. F. Coultas |
| 24–25 January 1930 | "France" en route from Tulagi to Auki, Malaita Island |

(8) Auki, Malaita Island (26–28 January 1930), some collecting

| | |
|---|---|
| 29–30 January 1930 | en route from Auki to Su'u, Malaita Island |

(9) Su'u, Malaita Island (31 January–16 February 1930), collecting inland

| | |
|---|---|
| 17–25 February 1930 | at Tulagi after overnight run from Su'u arriving early morning 17 February; Mayr left expedition. The "France" returned to Malaita on 25 February to resume collecting |
| 5 March 1930 | Departure from Tulagi for Marseilles, France on a French freight steamer arriving during the last days of April |

Another telegram from Dr. R. C. Murphy (1887–1973) in New York informed the members of the Whitney South Sea Expedition (WSSE, see p. 144) that Hannibal Hamlin (1904–1982) was to remain the official leader until replaced by William F. Coultas (of similar age) who was already on his way, together with Walter Eyerdam

**Fig. 2.12.** Schooner *France*, ship of the Whitney South Sea Expedition, anchored in the Solomon Islands (AMNH Library photographic collection, negative no. 117406)

**Fig. 2.13.** Schooner *France.* (AMNH Library photographic collection, negative no. 273099)

(1892–1974), and would meet the "France" (Figs. 2.12 and 2.13) in the Solomon Islands. Mayr was surprised since the preceding communication had implied that he was to be the leader of the expedition.[11] In retrospect (1992) he commented on his impressions as follows:

"Perhaps I should explain my annoyance after joining the 'France.' From the correspondence and cables I had, perhaps erroneously, concluded that I was to be the leader of the WSSE, or at least a co-leader. It turned out that I was merely considered an additional bird skinner and not involved in any decision making. Although later on, particularly after Coultas had joined, I became a good friend of Hannibal Hamlin[12], I never could get used to Hamlin's frequent indecision and even more frequent changes of decisions. One day Choiseul was our next objective, the next day he told me we would go to Kieta on Bougainville. The frequent changes of his plans were responsible for the fact that I received no mail for about 6 months (or more). Also Hamlin was often moody and not communicative. After all, he was very inexperienced, if not immature, just a recent Yale B.A. His knowledge of birds was almost nil. I think he had inferiority problems and his way of compensating for them was rather hard on me. And, owing to his inexperience, his judgment often was less than the best, as became evident again and again. For instance, when the natives of Choiseul, who knew the ground dove *Microgoura meeki* well,

[11] When Dr. R. C. Murphy at the AMNH in New York did not receive from Mayr a reply to his cable in May (see p. 74) and needed to fill the position of the expedition leader, he offered it to William F. Coultas, who accepted (LeCroy 2005). This, of course, was unknown to Mayr. Coultas joined the expedition on 30 August 1929 and took over the leadership from H. Hamlin on 23 January 1930.

[12] H. Hamlin (1904–1982) was the greatgrandson of A. Lincoln's first vice-president also named Hannibal Hamlin. Later he became a neurosurgeon in Providence, R. I., and an attending surgeon of neuro-surgery at Massachusetts General Hospital in Boston.

were unanimous that cats had exterminated the bird, Hamlin insisted that we stay weeks and weeks longer, a complete waste of time. At the same time he greatly admired me for my knowledge of birds, and when the time came for me to return to Germany he did everything he could to persuade me to stay; if I were to leave, he would also, he said (but this was about 8 months later).

The incompetent running of the expedition was not the only shock to me. The boat was the other. The 'France' was not a luxury expedition ship, but a converted copra carrier. Her greatest virtue was her sea worthiness. After I had left, she stood a taifun (cyclone) in Micronesia, which blew her a thousand miles out of her way. But she did not sink. Comfort she had none. There was not even a working flush toilet on board. Unless prevented by weather we had to sit on the railing to do our daily business. Whenever it was sunny, it became insufferably hot below deck and we had to sleep on top. But this being the tropics there was a downpour virtually every night and we got soaked. There was an awning, but it was old and leaked so that it was not of much help.

The ship was heaven for cockroaches, they bred in the bilge but at night crawled and flew all over the cabin and in large numbers. I don't know how often in my sleep I hit my face by reflex, when a cockroach crawled over it, while I slept, and squashed it on my face; not a pleasant experience.

Food was terrible. Usually rice and some canned goods, like canned Alaskan salmon (at that time the cheapest canned good) or corned beef. Hardly ever, except when we were on land, fresh fruit or vegetables; no dessert.

What I did enjoy was that I had to work like a sailor, hoisting the sail, heaving up the anchor, etc. There was, of course, no machinery for such tasks. I also had to take my regular turns at the wheel.

Ultimately, it was my participation on the WSSE that permitted me to place my foot, so to speak, in the door to America. But I was a rather disappointed young man in the first months after joining the 'France.' The situation was aggravated by a jaw operation and dengue fever.

The trip to the highest peak of Choiseul and later work on San Cristobal and Malaita eventually made me forget my disappointments. Lack of any news from home, of course, had made matters worse."

When collecting in the Buin region with Hamlin in mid-July for a week, Mayr suffered from dengue fever and in Kieta, on 28 July, a wisdom tooth was removed which had bothered him for several weeks. In order to get at it, the dentist extracted also the healthy neighboring tooth. In the process he broke off a piece of the upper jaw bone too, very painful. Mayr was unable to speak or eat for a while with his mouth half closed by the swelling and he felt most miserable.

During July and August 1929 H. Hamlin and E. Mayr were the only expedition members but they were joined in Kieta on 30 August by W. Coultas and W. Eyerdam who had arrived from the States (Fig. 2.14). William F. Coultas was about the age of Ernst Mayr and a professional collector. He had a bachelor degree from Iowa State University and retired later to a small farm in the Midwest which he had bought from his savings. Tragically, it later burned down together with his ethnographical collections and diaries. Walter Eyerdam (1892–1974) was also

**Fig. 2.14.** The members of the Whitney South Sea Expedition in the Solomon Island region (1929). From *left*: William Coultas, Ernst Mayr, Walter Eyerdam, and Hannibal Hamlin. (AMNH Library photographic collection, neg. no. 273081)

a professional collector primarily of mollusks and plants, but also of birds and mammals. He made five trips to the Soviet Union and the Siberian steppes between 1925 and 1931.

None of the members of the WSSE knew anything specifically about the birds of the Solomon Islands. However, Mayr had the advantage of his New Guinea experience; many Solomon birds have close Papuan relationships. Moreover while collecting on the islands he soon applied his well-established methods of questioning the natives about the birds and listing all their local names. Selected men were then employed as collectors on the basis of a premium payment to obtain the desired specimens.

The movements of the expedition between the islands of the Solomon Archipelago are listed in the "Itinerary" above which I prepared on the basis of H. Hamlin's diary (see also Fig. 2.11). I summarize below the collecting activities during the period when Mayr was a member of the expedition between July 1929 and February 1930 using the notes prepared by all members (on file at the Department of Ornithology, American Museum of Natural History, New York).

**(1) *Southern Bougainville Island.*** After stops at Faisi (Shortland Island) and Kieta (Bougainville) collecting was done in the southern part of this island (Buin region) during the period 20–27 July. Here Hamlin and Mayr visited Father Poncelet, a naturalist chiefly concerned with insects, at a place 8 miles inland. They went on bicycles provided by the patrol officer.

**(2) *Nissan Island.*** The "France" sailed from Buin via Kieta, Tiop and the Buka Passage to Nissan Island, northwest of Bougainville, during the first half of August.

Mayr and Hamlin collected 193 specimens of 24 species of birds on 6 days (5–10 August) and a list of 180 words of the Nissan language (Mayr 1931n). Seventy years later he still remembered many of them, for example tetjau (bird), vorvoron (breast), viliatun (war), kukuhalik (girl), kodan (liver). Salomonsen (1972) described a new subspecies of pigeon from their Nissan bird collection.

**(3) *Northern Bougainville.*** The motor of the "France" being out of order they sailed all the way back to the northwestern coast of Bougainville and arrived at Soraken (Surikan) on 14 August. During the period 15–20 August Mayr collected two species of paroquets in good numbers (a smaller green species and a larger red one) and observed a Peregrine Falcon which is a rare visitor to the Solomons and breeds in the New Hebrides and on New Caledonia. Arriving back in Kieta on 29 August they met W. Coultas and W. Eyerdam who had arrived there about 10 days earlier.

**(4) *Shortland Island.*** On several days, shorebirds were collected while the ship called at Faisi to purchase supplies in preparation for the inland trips on Choiseul Island (6–13 September).

**(5) *Choiseul Island.*** New York had instructed them to look for a good series of the endemic Choiseul Pigeon (*Microgoura meeki* Rothschild) that had not been found again since 1904, when A. S. Meek had collected several specimens for Lord Rothschild. The expedition arrived on 15 September anchoring at Choiseul Bay and found the usual lowland bird fauna but no *Microgoura*. Several natives well acquainted with this bird reported that previously it was easily caught provided they had found a low-branched tree where the pigeons roosted in groups of two to four birds by noting the droppings on the ground underneath. The natives were good naturalists as all inland people. They knew the bird represented on a colored plate of *Microgoura* and had a name for it but were unanimous in saying that it had become extinct, attributing its disappearance to introduced cats. An expedition member shot a large cat on the island which seemed to support the theory of cats being the cause of the extinction of Meek's Pigeon. However, Hamlin insisted that they continue the search.

Between 11 and 29 October the expedition split up in two groups. Hamlin and Eyerdam went to the Sasamunga region on the southwest coast to investigate the Kulambanara River Valley inland and later the southern tip of the island, whereas Mayr and Coultas were sent with the "France" to the mouth of the Wurulata River at the southeastern end of the island and close to a hilly region. Meek had discovered the species near this locality right at the coast. But again, the natives assured that it was extinct by cats. Mayr and Coultas collected for a couple of days and then went inland with the help of three porters establishing a camp at the base of the mountains, and the porters returned to the coast. On the way they had passed several abandoned villages. The inhabitants had either died from one of the introduced European diseases or after moving to this unhealthy coast. Mayr and Coultas climbed the mountain, with 550 m elevation the highest point in this part of the island, surely the first ascent by white people. All the birds were the same as in the coastal lowlands, no sign of *Microgoura*. On their return, and without porters, Mayr carried their canvas tent which was wet from frequent rains and gained further weight when they had to wade shoulder deep through the Wurulata River. Being very fit at that time, he managed to bring the tent all the way back to the coast.

For both W. Coultas and W. Eyerdam collecting on Choiseul was their first field experience in the tropics. They appreciated Mayr's advice and knowledge. As Coultas recorded: "Mayr has proven to be a great help to Eyerdam and me, both with bush lore and birds. To have someone with us who can identify material and direct the activities of the hunters is far more expedient than just trusting to luck" (23 September; p. 27) and again: "Mayr has been a great help in teaching me something of the bush and how to find one's way about without assistance. The average novice, when beginning his bush work, blazes Boy Scout trails a yard wide that will soon confuse him in an area if he makes enough of them. I have been told that one will develop a bush sense in time and will require few if any markers. Until that time, Mayr encouraged me to leave footprints in soft earth and always break twigs of uniform bushes on my left side as I went along. In that manner, I could later on determine the length of time since I had been over this or that trail and the direction in which I had been going. Crosstrails are always confusing but this method helped to overcome this difficulty. Mayr's early advice was well grounded. I never became definitely lost in all of my time with the expedition" (21 October, p. 40).

By now, the pigeon was so undoubtedly known to be extinct that the expedition shortened the stay on Choiseul and departed for Tulagi (Florida Island) on 5 November, however, leaving behind David, one of the three Polynesian hunters, to search several additional areas of the island. He was equally unsuccessful and joined the group again in Tulagi three weeks later.[13]

"The trip from Choiseul to Tulagi was nothing but disaster," Mayr wrote. "As we began to sail we sent two men from the crew to the mouth of a river with the fresh water tank to replenish our fresh water supply. They had strict instructions to taste the water to make sure they had pure fresh water, but once we were on our way and had our first tea, it turned out that they had disobeyed our orders and had indeed taken in brackish water. For the next twelve days we lived on brackish water. Our diet was equally unsatisfactory, canned, low-grade Alaska salmon and rice. The only redeeming feature was that the natives had discovered mangrove oysters and had filled a gunnysack with live oysters. We dragged this sack on a long rope behind the boat and pulled it up whenever we felt like a dish of oysters. They stayed fresh until we had consumed them all.

Since the "France" had no working motor we depended entirely on sailing. We had to sail essentially against the prevailing wind direction at this time of season, southeast. Whatever advance we made by tacking, we lost again in this clumsy tub by side-slipping and after 24 h we were often just where we had been the day before. Our nerves frayed, we fought each other. After we had arrived at Tulagi, we made a bee-line for the nearest pub and after about two Scotches we were again able to speak to each other."

[13] Further searches for *Microgoura* during the 1940s were also in vain. It had probably vanished during the 1910s.

***(6) San Cristobal (Makira) Island.*** The "France" went into dry dock for repairs at Tulagi harbor and H. Hamlin stayed with her, while Mayr, Coultas and Eyerdam left on the local steamer "Ranadi" for the island of San Cristobal (Makira) on 25 November. The Whitney Expedition had collected on this island before, but not extensively nor in the mountains. They landed again at Kirakira, Wanoni Bay, on the northern coast. From there they followed the Rayo River inland. On their two days' strenuous march they crossed this stream many times until they finally left it and climbed a steep trail to a mountain village, Hunogaraha, consisting of six houses at an elevation of about 1,900 feet, near the center of the island (15 miles inland from the coast at Kira Kira; 10°30'S, 161°55,' Times Atlas). They established their camp in "Charlie's house" for the next three weeks. Mayr asked the natives to build a small shed under which they could work and eat, with a table and benches exactly as he had become used to in New Guinea; all made from bush material. Later on a second shed was built so that they had an eating and a working room which was a much cleaner arrangement. From the beginning the sand flies were tormenting everybody. They gnawed at their legs, arms, hands, neck, and faces producing badly itching wounds. At times these flies became intolerable while they stayed in this camp.

At Mayr's suggestion, native hunters were hired. They knew their own district as well as the habits of the birds and were in some respects superior to the expedition members, particularly in the collecting of shy ground birds. The reward scale was as follows: three desirable common birds equals one stick of tobacco (three sticks cost one shilling); a rare bird equals one stick; a great rarity (rail, ground thrush, owl, black hawk) equals three sticks of tobacco (or one shilling). The average good hunter brought back birds to the value of about one shilling per day, but one of them was so successful that they had to pay him four shillings. The natives accepted this system quite readily and there never was one instance of ill feeling between them. It was basically the same system of paper money which Mayr had used in the Arfak and Saruwaget Mountains of New Guinea (pp. 58 and 71).

Mayr here again made a list of the native names of birds which he identified with their scientific names, as specimens were brought into the camp. Having a fair knowledge of the bird fauna of this region, he was able to describe other species similar to the material at hand. By acquiring a working vocabulary of native bird names he was able to guide the activities of the hunters. W. Coultas who later led the WSSE for 6 years (1930–1935) acknowledged: "I have to thank Mayr for this lesson which was indispensable to me in the future and added many more species to the Whitney collections" (3 December; p. 63).

On the contrary, Rollo H. Beck seldom went up into the mountains where, however, he sent his assistants for short stays. They probably missed rare mountain birds of the higher altitudes. Most of his work while conducting the WSSE for 8 years was done within a day's walk from the schooner "France." He nearly always returned to the ship to skin his birds and seldom camped in the forest, but also purchased skins from local people.

After the first week in Hunogaraha the most qualified natives did essentially all the hunting while the three expedition members stayed in camp for skinning and labeling. Despite their tribal taboos on hawks, owls, doves and pigeons, the natives turned in examples of such species and a wealth of other material. Each hunter received eight large and ten small cartridges per day with the understanding that each man return his unexpended cartridge cases at the end of the day.

When on the late afternoon of 13 December one of the hunters arrived with his catch, Mayr noticed at once that it contained a bird totally unknown to him and which he considered to be very rare, perhaps a new genus of ground birds. Terribly excited he said: "Look, today is Friday, the 13th. Naturally, a day on which we would have good luck." He nearly fainted with delight and excitement and "had to lay off from work the rest of the day; the thrill was too much for his constitution which otherwise is generally quite sound" (W. Eyerdam, expedition notes, p. 11).

Alfred Russel Wallace had a similar experience with a most gorgeously colored butterfly (*Ornithoptera croesus*) on Batjan Island in the Moluccas in 1858: "The beauty and brilliancy of this insect are indescribable, and none but a naturalist can understand the intense excitement I experienced when I at length captured it. On taking it out of my net and opening the glorious wings, my heart began to beat violently, the blood rushed to my head, and I felt much more like fainting than I have done when in apprehension of immediate death. I had a headache the rest of the day, so great was the excitement produced by what will appear to most people a very inadequate cause" (Wallace 1869: 257–258).

William Coultas in his expedition notes commented the incident of 13 December: "It is true with people, who have been some time in tropical countries and subjected to malarial fever, that they will come down with nervous disorders and malarial fever after a few moments of unusual excitement. I have noticed this with both Hamlin and Mayr and later with myself" (p. 69).

The bird that had caused Mayr's "nervous relapse" was the first specimen of the endemic ground-living San Cristobal Thrush (*Zoothera margaretae*). It did represent a new genus for Mayr because he had not collected any member of this group in New Guinea. *Zoothera [Oreocincla] dauma papuensis* was known from Sattelberg, but not found there despite much effort (Mayr 1931l: 692). No wonder the bird representing this group on San Cristobal gave him such a thrill. During the following days, the natives collected four additional individuals so that there are five specimens in New York today (LeCroy, Bull. AMNH no. 292 [2005], p. 35). Mayr (1935h, 1936b) later named this new species of ground thrush for his young wife Margarete. The San Cristobal Thrush (*Zoothera margaretae*) is not a subspecies of *Z. dauma* or of *Z. talaseae*; the ratio between wing length and tarsus length characterizes it as a distinct species.

Greenway (1973: 316) and Olson (1975: 1) erroneously attributed Mayr's excitement on 13 December 1929 to the collecting of the forest rail *Edithornis [Pareudiastes] silvestris*, another outstanding discovery on this island. However, the only individual of this species obtained during the expedition was collected on 4 December, when nobody got overly excited about it. Also, it was not "Stupe" who brought into camp the new thrush, as Eyerdam stated in his diary but "Charlie" (on which point both Mayr and Coultas agreed in their notes). "Stupe" appeared to be the village idiot who, as a joke, was given a gun one day. To everybody's surprise he brought back a rare black hawk. He received two sticks of tobacco and paraded through the village like a king. From that day on he went out hunting every day. On 19 December, he arrived with a collection that beat everything anybody had achieved previously. He earned four shillings and his father, the chief, was happy like a child.

When the expedition broke camp and returned to the coast at Kirakira on 22–23 December, they had assembled nearly 400 birds. A splendid record, but at the expense of much sweat and fatigue. Some additional specimens from the coast brought the total to over 400 ("which is more than the entire expedition collected during the three preceding months," Mayr). The San Cristobal collection included 13 species new to science and 67 new subspecies; the most outstanding discoveries described several years later were *Edithornis silvestris, Zoothera margaretae* and *Vitia (Cettia) parens*.

"The trip to the mountains had started from the residence of the chief magistrate of San Cristobal, where we had left behind everything not absolutely necessary, and this included a box with beer, whiskey, and special food for Christmas. After our hard work in the mountains we were very much looking forward to a happy Christmas celebration with the food and liquor we had left with the magistrate. Alas, during our absence he ran out of his own liquor and 'borrowed' ours. When we returned it was all gone, so we had a dry and very un-Christmasy Christmas. The only music we had was produced by an old hand-operated Victrola (phonograph) and two or three records of popular music hall ditties. It surely was the worst Christmas I ever had in my life."

**(7) *Santa Ana Island.*** From Kirakira the expedition went on December 28th to Santa Ana Island off the eastern end of San Cristobal. They did some collecting, while the planter Mr. Kuper from Hamburg took care of them until they sailed to Tulagi on board the "Hygeia" 9–12 January 1930. Meanwhile the "France" had been repaired and H. Hamlin was in good spirits, when he rejoined the expedition. Mayr received a letter from E. Stresemann who more or less ordered him to return to Germany. He was quite ready to go but in view of his contract for one year (or until June 1930) he cabled back: "I wish to work the Carolines. I cannot leave the ship without permission from New York." He also felt he had a real chance to get a position in America. H. Hamlin who enormously admired his knowledge of birds had, over the past months, fully informed Dr. Sanford about Mayr's important contributions to the expedition. This reputation was in part responsible for the later offer from New York. Ernst Mayr talked in detail about Hamlin and his enthusiasm about Mayr's ornithological knowledge in an interview of November 2003 (Bock & Lein 2005, CD-ROM in Ornithological Monograph 58).

On January 1, 1930 Mayr wrote in his diary: "The first day of the new year. I wonder what it will bring me. If I'm not very much mistaken it may well turn out to be the most important year of my life—when it will be decided whether I can obtain a good position outside of Germany or remain an untenured assistant in Berlin without pension rights."

**(8) *Malaita Island.*** Former hostilities between the inhabitants of this island had prevented other expeditions to collect birds there. A visit of Malaita was therefore one of the special aims of the WSSE. They encountered no problem and obtained a splendid collection which included two new species in the genera *Zosterops* and *Rhipidura* as well as 15 new subspecies, particularly a very distinct form of *Pachycephala pectoralis* without a black breast band (26–28 January). When Mayr near Auki, their first port of call, saw a white-eye (*Zosterops*) he decided right in the field that this was a new species. He later described it as *Z. stresemanni.*

On January 29th they sailed along the coast southward to Su'u where to collect appeared more promising (January 31st–16 February). This was indeed the case. Two weeks later Mayr and Hamlin left with ten porters for the interior mountains. They crossed the first range at an altitude of about 1,600 feet and descended on the other slope where they camped for the night. After crossing a river another camp was established at the next mountain range. Then, however, they returned to the coast because this path did not lead further inland to the highest mountains of the island. It was necessary to search for another entrance to the central highlands of Malaita.

On 15 February Mayr received permission from New York to return to Germany. When, on 16 February, he got a letter from Berlin urging him to come back right away, they pulled up anchor to return to Tulagi where the "France" arrived in the early morning of 17 February. Mayr's cable to the director of the museum in Berlin read: "Wire immediately your opinion concerning my departure. Permission New York. Will leave if you advise. Reply immediately. Mayr." The answer, next day, was: "Return advisable." This settled the situation.[14]

Calculating the fares it turned out to be much too expensive to return to Europe via New York, as Stresemann had suggested in one of his recent letters. Hence Mayr could only take the direct route to France and Germany. A French copra boat, the "Saint Eloi," passed through Tulagi on 5 March and took him to Marseilles, where he arrived during the last days of April. A cable he had sent to Stresemann from Soerabaja (Java) on 24 March 1930 read: "Start my work at the museum on 1 May. Mayr." The ship was somewhat late and he made it back to his home in Dresden by the afternoon of May 1st (and to Berlin a few days later).

The WSSE resumed the survey of Malaita's bird fauna on 25 February. In early June 1930, H. Hamlin and W. Eyerdam also left the "France" and returned to the States. W. Coultas remained in charge of the expedition until 1935.

## Expedition Results

It is truly astounding that Ernst Mayr not only survived the expeditions but actually accomplished so much. His ornithological achievements may be summarized as

[14] The Berlin museum had granted Mayr a leave of absence for two years, i.e., until March 1930. On the other hand, he had agreed to join the WSSE for one year, i.e., until at least June 1930. Stresemann therefore had contacted Dr. Sanford in New York and on 15 February 1930 Mayr received permission from the latter to leave the expedition.

follows: (1) For comparative purposes, he collected good series of birds in the Arfak Mountains (which represent the type region for many New Guinea birds) and made a vain attempt at finding some of the "rare Birds of Paradise." (2) He assembled further collections in the isolated Wondiwoi Mountains, Wandammen Peninsula, and in the Cyclops Mountains at the northern coast. (3) Bird collections were also made on the ornithologically poorly known Huon Peninsula (Saruwaget Mountains) and in the nearby Herzog Mountains, and (4) on certain islands in the Solomon Group, east of New Guinea. Besides containing a large number of bird species and subspecies new to science, Mayr's collections, together with other museum material, formed the basis for his later taxonomic and zoogeographical work.

Mayr, excellently prepared prior to his departure, knew most of the birds of New Guinea and their approximate altitudinal distribution from his study of the literature and the collections in the Museums of Natural History of Berlin and Tring. He was able to judge whether a particular bird was rare or desirable and whether it showed any peculiarities of interest to science. Therefore his samples were of unusual quality, containing relatively few of the common and well-known birds so often found in the collections of inexperienced travelers.

During the expedition long letters from Hartert, Stresemann and the botanist Dr. Diels contained numerous suggestions indicating the interest with which his sponsors followed Mayr's work. He himself sent detailed reports to Berlin and also to Tring relating many adventures. Stresemann published some of these letters while Mayr was still away (e.g., Mayr 1929b). His rich collections of bird skins were worked on by Hartert (1930) who published on the birds obtained in Dutch New Guinea and by Mayr (1931l) himself who studied the birds from the Huon Peninsula (Mandated Territory). The fresh and well-labeled material with exact locality data was important for comparisons with birds from other parts of New Guinea which in many cases had only imprecise or no locality data at all. But the "rare Birds of Paradise" proved to be elusive. Mayr related their story, the great enigma of New Guinea ornithology, in a lecture on "the joy of research" many years later:

"Prior to 1920, it was the height of fashion for a lady to have the plumes of a Bird of Paradise on her hat. Ten thousands of these birds, which occur only in New Guinea, were collected each year by the natives and shipped to the big plume dealers in Paris and London. In addition to the common species, these collections included every once in a while some rare or even new species of Bird of Paradise from some remote mountain range. And this is where Lord Rothschild comes into the picture. He had a standing offer of one hundred pounds (in those days an awful lot of money) for any new species of Bird of Paradise found among the skins collected by the natives. Over the years he was able to describe quite a few of such new species. But this method of collecting was scientifically most unsatisfactory, because one knew nothing about these birds, not even the part of New Guinea from which they might have come. One scientific expedition after another had gone to various parts of New Guinea between 1870 and 1920 to discover the home of these species and to discover new species of other bird families. And they indeed succeeded in discovering the home of most of the so-called 'rare Birds of Paradise.' But there were about five or six species left which simply could not be rediscovered. What Lord Rothschild wanted me to do was to go to three particular mountain ranges in western New Guinea which so far had been insufficiently explored or not at all.

I would like to tell you all about my trials and tribulations in exploring these untouched and largely uninhabited mountain ranges all alone, except for some native assistants from Java, but my time only permits me to tell you that I obtained a fine collection of birds, mammals, insects, and plants on all three mountain ranges, but saw no hide nor hair of any of Lord Rothschild's rare Birds of Paradise. As far as its principal purpose was concerned, you might say my expedition was a failure. But wait a minute!

My failure to find these birds in some of the last unexplored mountains of New Guinea gave Professor Stresemann in Berlin a bright idea. 'Maybe there is something fishy about these species; maybe they aren't even good species,' he thought. So he examined them very carefully, and was able to prove eventually that each of the so-called species, the home of which could not be found, was actually a hybrid between two other well-known species of Birds of Paradise [Stresemann 1930]. You can imagine how pleased I was to have contributed to the solution of this great puzzle of New Guinea ornithology, and to know that my expedition had not been a failure after all" (1981b: 149–150).

Mayr (1945h) estimated that possibly only one out of about 20,000 birds of paradise is a hybrid each of which is due to an "error" of the birds because of their breeding habits. Male and female birds of paradise never form a pair bond. The mature female is attracted to the display ground of the males when she is ready for mating and leaves the male soon after. In such a system, especially young inexperienced females may rather easily commit an error and approach the display ground of males of the "wrong" species, particularly along the geographical and elevational range border of its own species, where she may have difficulty locating a partner of the "right" species (Frith and Beehler 1998: 501–502; see also Nicolai 1977: 124).

The main lessons that Mayr learned in the Solomon Islands were (1) that each island (except those connected with each other during low sea-level stands of the Pleistocene) had a very distinct fauna and (2) that there was spectacular geographic variation and geographic speciation not only of the mountain birds (as in New Guinea) but also in lowland species.

From the mammal collection consisting of several hundred skins three new taxa were described: (a) a species of mice, Ernst Mayr's Leptomys (*Leptomys ernstmayri*), inhabiting the montane rainforest of the Saruwaget Mountains (Huon Peninsula), (b) a subspecies of Doria's Tree-kangaroo (*Dendrolagus dorianus mayri*) of the Wondiwoi Mountains (Wandammen Peninsula), and (c) a subspecies of the Long-fingered Triok (*Dactylopsila palpator ernstmayri*); Rümmler (1932), Rothschild and Dollman (1933), Stein (1932); see also Flannery (1995). In addition, Mayr collected about 4,000 butterflies, 4,000 beetles, 1,000 grasshoppers, a few spiders, scorpions, crustaceans, snails, worms, lizards, and snakes. A number of transmittal lists of these collections are preserved in the Museum of Natural History Berlin (Historische Bild- und Schriftgutsammlungen, Bestand Zool. Museum, Signatur SIII, E. Mayr "Schriftwechsel und Sammellisten seiner Expeditionen nach Neuguinea").

The director of the Botanical Garden in Berlin-Dahlem, Ludwig Diels, sent him detailed suggestions as to which groups of plants to collect and what practical points to observe. The total plant collection comprised several thousand sheets with dried plants. Those from former Dutch New Guinea were sent to Leiden via Java and were used in several subsequent Dutch publications (see below). The rich material from the Huon Peninsula went to the Botanical Museum in Berlin-Dahlem where it was destroyed during World War II without anybody ever having studied the majority of these plants. This greatly saddened Mayr because he had invested an enormous amount of time and effort in these plant collections. No plants had been collected during the short stay in the Herzog Mountains, where Mayr was alone most of the time.

Two botanists studied specimens from Papua Province (Irian Jaya, former Dutch New Guinea): Burret (1933) published the descriptions of eight new species of palms and named three species of different genera "*Mayrii.*" Among the 78 species of orchids listed by J. J. Smith (1934) 26 species and 2 varieties were new to science. Three of them were named "*Mayrii.*" Among 12 species of Ericaceae four were new and one was named "*Mayrii*" (J. J. Smith 1936).

In the two articles authored by J. J. Smith (1934, 1936), 13 species are listed "without locality" because one of Mayr's field catalogues had been lost in Leiden. These numbers "without locality" were all collected on the Cyclops Mountains (Mayr, pers. comm.). His plant collecting activities in Irian Jaya (Papua Province) have been summarized in "*Flora Malesiana,*" ed. C. van Steenis, Spermatophyta, vol. 1 (1950), pp. 352–353 and the species names dedicated to him have been listed by Backer (1936). Later Vink (1965) and Van Royen (1965) mentioned Mayr's botanical researches in the Arfak and Cyclops Mountains, respectively, and Vink (l.c.) presented a sketch map of his expedition route.

Similar to Alexander von Humboldt, Charles Darwin, and Erwin Stresemann who had also traveled in the tropics as young men and never visited these areas again, Mayr's tripartite expedition to New Guinea and the Solomon Islands remained his only expedition experience which, however, formed the basis of much theoretical work in later years. He worked on the material from Oceania at the AMNH until the late 1930s, when World War II interfered with new expedition plans. It was only during the winter of 1959/1960 that he was able to travel again overseas visiting remote regions in Australia.

## Future Plans

Mayr's expedition notebooks reveal that he intended to write a book, *The Birdlife of Germany*. They contain detailed notes on topics to be included (migration, ecology, food and feeding, psychology, breeding biology, etc.) and procedures to be followed (e.g., cooperation with G. Schiermann; literature search and evaluation). Further plans referred to

(1) a "critical test of the Territory theory (Howard, Nicholson)" in several landscapes near Berlin,
(2) an "ornithological field manual",
(3) an article on "Ecological approaches to birdlife",
(4) "The distribution of birds",
(5) "Contributions to the ornithology of New Guinea", and
(6) "Thoughts on the establishment of a zoological (ornithological) research station in New Guinea on a high mountain range at ca. 1,500 m elevation; one 'pure' ornithologist and one entomologically trained ornithologist, duration 2 years; problems to be studied are molt, feeding, breeding biology, ecology: It is important that the scientists observe during all periods of the year."

Evidently Mayr's interests were by no means restricted to systematic research but comprised a broad spectrum of topics which he planned to pursue through extensive ecological fieldwork. However, this was not to happen, because a few months after his return to Berlin he received an invitation to come to the American Museum of Natural History in New York as a research associate and taxonomist for one year. This assignment developed into his career as a museum systematist and evolutionary biologist, but his science included the lessons he had learned in the field.

# Part II
# Ornithologist and Evolutionist in New York

# 3 The New York Years (1931–1953)

## Emigration to the United States and Life in New York City

Mayr's main research in Berlin right after his return from the Solomon Islands concerned the study of his bird collections from the Saruwaget and Herzog Mountains, New Guinea (Fig. 3.1) which included a lengthy visit to the Rothschild Museum in Tring, England to compare certain specimens. In June 1930 Mayr attended the VIIth International Ornithological Congress at Amsterdam where he met many of the luminaries in ornithology, including Frank M. Chapman, chief curator of the Department of Ornithology, American Museum of Natural History (AMNH) in New York City. A few months later Mayr was invited to come to New York for

**Fig. 3.1.** Ernst Mayr back from the South Seas has joined the "Stresemann circle" again; Museum of Natural History Berlin, 1930. From *left*: Hans Scharnke, Erwin Stresemann, Ernst Mayr, Max Stolpe, Georg Steinbacher (*hidden*), Hans Schildmacher, and Hermann Desselberger (Photograph courtesy of Henriette Desselberger, Giessen, Germany)

one year to work on the bird collections of the Whitney South Sea Expedition. He did not leave Germany until the beginning of next year, having finished the report on his New Guinea bird collection. Arriving in New York on 19 January 1931 he entered the museum on the following day.

Mayr came to the United States as an employee of the Museum of Natural History of Berlin on temporary assignment of one year. When the AMNH was able to purchase the Rothschild Collection in 1932, he accepted the position of Associate Curator at the AMNH responsible, over the next several years, for organizing, cataloguing and preserving the 280,000 specimens which filled a huge number of cases. His salary had to be raised annually from the Whitney family. There was no tenure. This meant that if the Whitneys should ever stop giving the money (which, of course, never happened), he would be without a job and probably forced to go back to Germany. In view of his fascination with his work and with the institution he gladly accepted this risk.[1]

I emphasize that Mayr's move from Berlin to New York in 1931 had the simple reason that his position there was better, scientifically, than any he could expect in Germany (pers. comm.). Moreover, Mayr was the youngest of four assistants at the Museum in Berlin and knew that he would have to wait many years before a curatorship might open up for him. The Nazi regime which came to power in Germany in 1933 (i.e., 2 years *after* Mayr had arrived in New York) had nothing to do with his emigration to the United States, although Mayr was outspoken in his denouncement of this regime (see also p. 300, footnote). In addition, there was no room for a second major ornithologist in Berlin and probably not in all of Germany, next to Professor Stresemann. Their careers might have "collided," as Mayr thought in retrospect (pers. comm.; Bock 1994a). Moreover, Mayr would have had little chance of surviving World War II had he stayed in Germany.

## Employment in New York

Based on a suggestion by Dr. L. C. Sanford, trustee of the AMNH who had consulted in this matter with Hartert and Stresemann[2], Dr. Frank M. Chapman (1864–1945), head of the Department of Ornithology of the AMNH, offered Ernst Mayr, in October 1930, a position in New York as Visiting Research Associate for one year

[1] The Whitneys paid Mayr's salary until he left for Harvard University in 1953. He received an annual salary of $2,500 in 1931, $4,000 in 1932–1933, $4,500 in 1934, $5,000 from 1935 to 1944 and $6,540 starting in 1945 (M. LeCroy, pers. comm.).

[2] In January 1930, while the Whitney Expedition was in the Solomon Islands, Dr. Sanford, forever making enthusiastic plans and fully informed, through H. Hamlin, about E. Mayr, had already written to E. Hartert (Tring) regarding a further expedition to New Guinea: "If we wait until later we might have the chance of bringing Mayr home, resting him and sending him back. Personally he appeals to me more than any one else. Talk the matter over with Lord Rothschild." At the same time, Rothschild had decided to offer the position of curator of his museum to Mayr as successor to Hartert who retired in April of that year (Mayr 1984f). This plan, however, had to be dropped when Rothschild was in financial difficulties (p. 116).

to commence upon completion of his expedition report. Mayr gladly accepted this offer, because he knew that large unworked collections from the Pacific islands had been assembled there over the past decade. He left Germany in early January 1931. After spending two days in Tring (England) to study Solomon Island birds and to finish some of his New Guinea work, he left Southampton by the "Bremen" on January 15th arriving in New York on the 19th. Prior to his departure, he had written to Dr. Murphy from Tring on January 8th: "I think it is better nobody comes to take me off the steamer, because he might have to wait too long. To pass through those immigration officers takes an awful long time, I heard." Mayr "had arranged for a room at the International House located on Riverside Drive close to Columbia University and forty blocks north of the AMNH. Before he left Germany, Mayr talked to German colleagues who had recently visited New York, and learned that he could walk four blocks from the pier in Brooklyn to the closest subway station. He did so carrying his two suitcases, found his way through the New York City subway system, which involved several changes of trains, and reached the International House safely. The next day he made his way to the AMNH, again by subway, and reported for duty at the Department of Ornithology, totally surprising all members of the department. Their surprise was greatly increased upon learning that he had not made use of taxis in getting from the pier to International House and the next day to the museum. Clearly anyone who is able to plan his arrival in New York in such detail and with such determination will be successful in his career" (Bock 1994a: 280). Mayr reported to Stresemann (transl.):

*Tabeh kakaku,*[3] *21 January 1931*
*Sitting next to [J. T.] Zimmer I announce my arrival at the American Museum. The "Bremen" will leave for Germany tonight, therefore I must hurry. Murphy is not around at the moment, will come in the afternoon, Chapman is on Barro Colorado [Island, Panamá] until April, Sanford also that long in Florida; this means bright prospects. Had already good conversations with Zimmer and Mrs. Naumburg.*
*The voyage was pleasant despite stormy weather (the wind fell below 6 only once, mostly 7–8). I was one of the few passengers who were not seasick.*
*I found good accommodation in the International House, which costs me about $32 per week. New York did not impress me much; not very different from other big cities, only dirtier. You cannot beat Berlin.*
*In a hurry cordial greetings adek.*

Mayr was completely free on which collections from the southwest Pacific he wanted to work and in which sequence. Such independence from supervision greatly enhanced his admiration both for the American way of doing things, and for his boss Dr. Chapman. Mayr's first article (on *Halcyon chloris*, 1931b) was published on March 31 and before Chapman and Sanford returned to New York. To extend Mayr's assignment Chapman wrote to Berlin a few months later: "Dr. Ernst Mayr is rendering us invaluable cooperation in the study of our large

[3] "Dear Kaka." Regarding the use of "kaka" (older brother, Stresemann) and "adek" (younger brother, Mayr) see p. 41.

collections of Polynesian birds, research for which he is especially qualified by his field experience while a member of our South Sea Expedition" (8 October 1931) and requested an extension of his leave of absence for another six months which was granted on 12 November. When the Rothschild collection began to arrive in New York in the spring of 1932, Mayr got the position of Associate Curator of the Whitney-Rothschild collections without limit of time and terminated his employment in Berlin, effective on 31 July 1932.[4]

As Mayr recalled (pers. comm.), he encountered a certain amount of jealousy among young American ornithologists who in the depression years were without a job and quite naturally resented a German "intruder." However, everybody more or less realized that he was indeed the person best qualified for this position. He was elected a Fellow of the AOU remarkably early (1937) and later never had any problems when organizing meetings, societies and journals in evolutionary biology. Mayr was a staff member of the AMNH until 1953, when he accepted an offer as an Alexander Agassiz professor at the Museum of Comparative Zoology of Harvard University (Cambridge, Massachusetts).

In an interview on the occasion of his forthcoming 100th birthday, Mayr related in detail how he came to be attached to the Whitney South Sea Expedition in the Solomon Islands, how he was offered a position in New York and which projects he worked on there; he also described his colleagues in the bird department and his influence on the development of American ornithology (Bock and Lein 2005; CD-ROM in Ornithological Monograph 58). Mayr also recounted his life story for Peoples Archive, a London-based company that filmed the reminiscences of famous scientists and artists (www.peoplesarchive.com).

## A Manager of Large-scale Ornithological Projects— Dr. L.C. Sanford

The transfer of the Rothschild bird collections from Tring to New York as well as Mayr's employment by the AMNH in the early 1930s were due to the efforts of Dr. Leonard C. Sanford (1868–1950), a wealthy physician in New Haven, an influential member of the New York upper class society, and a Trustee of the AMNH (Murphy 1951; LeCroy 2005). He was not only the family doctor of many prominent families, but also a welcome associate as a splendid tennis player, an excellent bridge player, a superb raconteur, and a good friend (Fig. 3.2). Nobody ever wanted to disappoint him.

Through his activities during the 1910s and 1920s the bird collections of the AMNH had become the richest in the world and its Department of Ornithology a global center of research. It was Dr. Sanford who, so to speak, offered to Ernst Mayr the collections which enabled him to carry out a comprehensive research program on geographical variation, zoogeography, and speciation. This fatherly friend for

[4] This explains why the year of Mayr's emigration to the United States is sometimes given as 1931 (when he started his temporary assignment in New York) and sometimes as 1932 (when he terminated his employment in Berlin).

**Fig. 3.2.** Dr. Leonard C. Sanford. (AMNH Department of Ornithology archives.)

20 years (1931–1950) was an important figure in Ernst Mayr's life, "the knight in shining armor of this tale" (Bock 2004a) and a "manager of major ornithological ventures." He initiated (1) the Brewster-Sanford Expedition to the coastal areas of South America 1912–1917, (2) the Whitney South Sea Expedition to the islands in the Pacific Ocean 1920–1940, and (3) several expeditions to New Guinea, Celebes (Sulawesi) and the northern Moluccas in 1928–1932. He raised the necessary funds through his connections with financial circles in New York, laid the basic plans for these long-term expeditions and finally arranged for qualified scientists to be employed to study the collections obtained. As detailed below, his motivation was not entirely science but friendly rivalry with his comrade Thomas Barbour.

Through his acquaintance with the Whitney family Sanford also raised, in 1929, the financial means for the construction of the Whitney Wing of the AMNH to house the large incoming collections as well as new public galleries (Murphy 1951; LeCroy 1989, 2005; Bock 1994a). Dr. Chapman, head of the Department of Ornithology and mainly interested in birds of the Americas, did not object, of course, to any of Dr. Sanford's unsolicited plans, particularly since his department profited immensely by these global activities. However, Chapman may not have been really happy with the purchase of the Rothschild Collection in 1932 (Mayr, pers. comm.). To make room for this huge collection, the fourth floor of the recently completed Whitney Wing that Chapman had intended as another exhibition floor, was converted into a floor for bird collections. He never complained about this but Mayr felt that all of this happened without him really wanting it. In a sense, Chapman was somewhat afraid of Sanford. There was never any joint planning between the two. This is why Sanford had turned to Ernst Hartert and Erwin Stresemann regarding the ornithological exploration of New Guinea and the Malay Archipelago on which areas they were the experts. Frequently, when Mayr talked with Chapman and mentioned some needs on the fourth floor, he would make comments as if this floor belonged to a different museum. Occasionally, Chapman did object to Sanford's plans (February 18, 1937): "As regards Chapin, I am determined to have him go to the Congo and Chapman is determined he shall not" (Stresemann Papers, Staatsbibliothek Berlin). As usual, Sanford had the last word.

Before the First World War, Dr. Sanford was on the surgical staff of the New Haven Hospital and the physician of the Yale Football team. Because of his hunting and conservation interests he was, at that time, also one of the leading personalities of the Connecticut Fish and Game Commission and came in close contact with Theodore Roosevelt while he was President of the United States. Sanford's activities regarding the AMNH had been triggered by his friend Dr. Thomas Barbour (1884–1946), curator and later director of the Museum of Comparative Zoology (MCZ), Harvard University, who had pointed out to him that the MCZ's collections were more complete than those of the AMNH. Thus, as a trustee of the AMNH since 1921, Sanford competed with Barbour and the MCZ (Mayr, pers. comm., and Bock 1994a). He was a born collector and particularly enjoyed possessing things which others did not have and partly *because* they did not have them. However, he wanted nothing for himself and took delight in building up the scientific treasures of the AMNH, particularly its Department of Birds. He was not personally a man of very large financial resources, but his acquaintance was wide, and his approach well-nigh irresistible. If he could acquire some rarity for the Bird Department, that was splendid; but if he also knew that his friend, Tom Barbour, could not get a specimen of that same species, then Sanford's joy was doubled. In 1928 he traveled by airplane (!) to Europe and, after visiting Hartert in Tring and Stresemann in Berlin, continued to St. Petersburg in an effort to arrange an exchange of specimens of several extinct bird species from certain islands in the Pacific Ocean (*Phalacrocorax perspicillatus, Aphanolimnas monasa, "Kittlitzia" (Aplonis) corvina*). In this case, he was unsuccessful.

After winning the "competitive race" during the 1920s, Dr. Sanford continued his support of the Bird Department of the AMNH probably because of his collecting compulsion (he himself owned a large collection of North American birds) as well as his satisfaction in carrying through major ornithological projects on an international scale-like a modern manager of large industrial companies. When he was unable to exchange with other museums specimens of some rare birds from a number of islands in the Pacific Ocean, he planned and obtained the financial means for an expedition to these and many other islands conducted by the AMNH, the Whitney South Sea Expedition (1920–1940). He was happy to see Ernst Mayr's stream of publications during the 1930s, when the latter was working on the rich material brought back to New York. Dr. Sanford more or less concealed under the mantle of a mere fondness for sport a sense of high purpose and a deep love of nature. In 1948 Mayr supervised the organization and installation of a large Museum exhibit, the Leonard C. Sanford Hall of the Biology of Birds, in whose dedication Dr. Sanford took part. His bronze bust held a central position in the Hall (which in 2000 has been incorporated into other exhibits). Sanford died on December 7, 1950.

Ernst Mayr reported: "Soon after I had arrived in New York in January 1931, Sanford visited me in my office and took great interest in the series of papers I published. He wanted to talk with me all the time about birds the museum still lacked and places that ought to be visited. During the football season he invited me several times annually to New Haven to watch a game and stay overnight at the Tennis Club. I am sure it was Sanford who insisted that my contract be renewed for a second year, and when, during that second year, the Rothschild Collection was purchased by Sanford (with Whitney money) it was he who insisted that I be made the curator.

In 1934 I developed a medical problem and on 13th of April my left kidney had to be removed. Sanford could not have been more solicitous if he had been my own father. He continued to inquire about me, and as soon as I was mobile he invited me to his home in New Haven to spend some days there to recover even more, also under the care of Mrs. Sanford, a lovely lady. After that he sent me to his trout fishing camp in the Catskills (Beaver Kill Brook) taking care of all my expenses, etc. And in 1944 it was he who went to the director finding out how strange it was that I had never been made a full curator when several younger people with less distinction had been promoted to that rank. Needless to say, I was likewise promoted within half an hour.

At the time Sanford died (1950) I had the full intention to stay at the American Museum for the rest of my life. However, when in 1953 I got the offer to go to Harvard and the MCZ it would have been most awkward and surely would have broken Sanford's heart if I had gone to the rival institution. Fortunately, this conflict did not arise. I will always have him in grateful memory.

It was quite well known in New York circles that Sanford was soliciting anybody to get money for the bird department of the American Museum. My colleague Murphy once had a cocktail with Sanford in the University Club when one of Sanford's buddies walked by and whispered to Murphy, 'Don't give him a cent more than $10,000.' Actually, Sanford never wanted anything for himself. His great ambition was to build up the collection of the American Museum and he was unbelievably successful in this ambition. Together with its existing treasures and the uniquely valuable Rothschild Collection, the American Museum had finally a bird collection that was not rivaled anywhere else in the world. In fact, other museums today may have more specimens, but as far as balance and richness in types, etc., is concerned, I think the American Museum collection is still unique."

## Life in New York City

In New York City, Mayr occupied a small room in the International House at 500 Riverside Drive in upper Manhattan, at about 123rd Street, where he soon met other young Germans. To the museum he took the Broadway subway south or, when weather was nice, he walked the 44 blocks along the Hudson River. In the evenings, he occasionally went with friends from the International House to Times Square, to Radio City Hall, to a movie theater or a concert. They also on occasion sponsored social events, plays, musical performances, etc. In one of the plays Mayr appeared as a Catholic priest. Soon after visiting Europe in the summer of 1932 he moved, with two companions from the International House, to an apartment located at 55 Tiemann Place (Apartment 69 on the 3rd floor) just one block away. Each of them had a bedroom and there was a large living room. One of the other two roommates loved to go shopping and also otherwise running the household including cooking.[5] In spite of some turnover of roommates Mayr lived very pleasantly in this setup until he got married to Margarete (Gretel) Simon in the spring of 1935. By then he was 30 years old (Fig. 3.3).

They had met in 1932 at a Christmas party in the International House. She was the niece of a well-known Long Island family, the Dreiers, and had come to the USA as an exchange student at Wheaton College, Massachusetts for one year. After she had returned to Germany in 1933, he proposed to her during his next visit to Europe in the summer of 1934 (when he also attended the VIIIth International Ornithological Congress in Oxford with Erwin Stresemann as President). He left on SS "Veendam" in June and returned to New York on SS "Statendam" on 14 October. Ernst was not certain that Gretel would say "yes" in view of his recent kidney operation (p. 301) and the fact that she would have to leave Germany. However, his doubts were unfounded. He traveled again to Europe on SS "Europa" on 26 April 1935 and they were married in Freiburg in Breisgau on 4 May, a beautiful spring day when everything was in flower. The wedding ceremony was officiated by Gretel Simon's brother Ludwig, who was a minister, and with her younger brother Frieder playing the organ in the church in which their father had been the minister. Following a few days of "honeymoon" in the Alps they had a pleasant

[5] The six German and one American young men who had shared this apartment at one time or other during the years 1932–1935 remained in contact with one another into the 1990s, and four of them (with their wives) met in North Bennington, Vermont in May 1986 to exchange memories and renew their friendship. Except for Gustav Stresow who became a publisher in Germany, all the others had remained in the United States. Stresow and Mayr were especially close because both had intellectual occupations and usually worked at home in the evenings. They corresponded into their nineties and occasionally met, when Mayr was in Europe.

**Fig. 3.3.** Ernst and Gretel Mayr's wedding in Freiburg, Germany, on 4 May 1935. (Photograph courtesy of Mrs. S. Harrison)

steamer crossing on SS "Bremen" back to New York where they met Gretel's uncle and aunt, George and Helen Simon, on 27 May. He was her father's elder brother and vice president of Heyden Chemical.

Ernst and Gretel found an apartment alongside Inwood Park, just above 200th Street at the north end of Manhattan (at 55 Payson Street) with easy access both to the Broadway subway and a ferry to New Jersey. Erwin Stresemann and Phyllis Thomas visited the Mayrs a few months later and spent Christmas Eve 1935 with them in their apartment. Ms. Thomas, Ernst Hartert's former secretary in Tring, had come to New York to assist at the museum with cataloging and integrating the Rothschild Bird Collection.

Almost 2 years later, in April 1937, after the birth of a daughter, the Mayr family moved from Manhattan across the Hudson River to Tenafly, New Jersey, where they had purchased a home at 138 Sunset Lane (Fig. 3.4 a, b). Friends who lived in Tenafly (Professor Franz Schrader at Columbia University) had recommended a real estate agent to them. When consulted, he asked how large a down payment Mayr could afford and how large a monthly mortgage payment he could make. When he had these figures he said "Okay, then let us go to the NW section of Tenafly, where the $8,000 houses are." And indeed they liked one of them and bought it. At that time Tenafly was still relatively rural. Mayr's house was the only one on that side of the street, and there was an old abandoned apple orchard where screech owls and flickers nested and bob-white quail walked around. They tried to have a vegetable garden, but there were lots of rabbits. In later years, the area was completely built up. They stayed in close contact with their families and Erwin Stresemann, and saw all of them on visits to Germany with their two daughters Christa (born in 1936) and Susanne (born in 1937) after World War II.

In the Tenafly area, Mayr did some serious birding. He noticed the nesting trees of barred owls, located display grounds of woodcocks, and studied redwinged blackbirds in a nearby marsh. He always took visiting ornithologists, e.g., Delacour, Lack, Tinbergen, Stresemann, and Lorenz, to watch the courtship flight of the woodcock. Occasionally a field trip was organized around New York City, one of them to the Ramapo Mountains in the early 1930s guided by Ernest G. Holt. He had collected birds for several museums in South America during the 1920s and, from 1933 until 1942, worked for the Department of Agriculture, Washington.

**Fig. 3.4a.** Mayr's home in Tenafly, New Jersey, in 1948 (Photograph courtesy of Mrs. S. Harrison)

**Fig. 3.4b.** The Mayr family, spring 1947, with Christa (*left*) and Susanne (*right*) (Photograph courtesy of Mrs. S. Harrison)

Mayr remembered: "Holt said he knew all of the trails there, and we would have a wonderful time. My friend Koch-Weser came along and we parked the car at the foot of the mountains. The height of the fall coloring was already past because it was after the middle of October, and it was quite sunny early in the morning. But toward noontime it clouded up and in the later afternoon it was obviously time to get back to the car. But Holt got confused with the trails and pretty soon was utterly at a loss which way to turn. We met another hiker, and as an old New Guinea hand (none of us were smoking), I borrowed some matches from this man. Well, eventually it turned dark and we still didn't know where we were. We stumbled along for another hour or two in the dark, but having lost the trail on a bare rocky outcrop and having fallen over trees, etc., we finally decided we had to stop and make a camp. In the meantime it had started to rain and we had quite a bit of trouble getting a fire started, with the matches I had fortunately begged. We spent a miserable night being hot from the fire on one side and freezing in the back. We had left all our warm clothing in the car. Poor Holt was terribly upset because he was married and he knew his wife would worry. After a long and miserable night, dawn finally came, and in due time we found a way out to a road and eventually to our car. The story, of course, got around that the two famous explorers, Holt of Brazil fame and Mayr of New Guinea fame had gotten lost 'in the outskirts of New York City' and this finally led to a write up of our adventure even in the *New Yorker*."

In 1938, Mayr spent again three months in Europe and visited various museums to study the types of New Guinea birds in conjunction with the preparation of his monograph, the *List of New Guinea Birds*, which was published in 1941. He sailed from New York on SS "Europa" on 3 May and returned on SS "Deutschland" on 5 August 1938. Mayr's first visit to Europe after World War II took place during the period April–August 1951, when he gave a series of lectures in Italy, Switzerland, France, Denmark, Britain and Germany having become famous through his 1942 book. He met many of his old friends, especially Erwin Stresemann in Berlin. In later years he has, of course, returned to Europe many times to participate in symposia and international congresses, and to visit his relatives.

Both daughters went to primary and high school in Tenafly and were already teenagers, when Ernst Mayr accepted an Alexander Agassiz professorship at Harvard University in 1953. At that time he sold his home in Tenafly to his old friend Reimer Koch-Weser, one of his former roommates at 55 Tiemann Place.

## Birding around New York City

On many weekends Mayr went out birding in the surroundings of New York City, usually with other members of the Linnaean Society, the local bird club which a group of naturalists had founded in 1878. There he met Charles Urner (1882–1938) and his New Jersey group and the members of the Bronx County Bird Club (BCBC) some of whom had cars and took him along on their trips, eventually accepting him as a member under the name of "Ernie" (Farrand 1991, Barrow 1998: 193). The major criterion for membership in the BCBC was one had to be born in the Bronx. Therefore Mayr was nominated "honorary member" (as well as Roger T. Peterson, author of the field guides, and William Vogt, later editor of *Audubon Magazine*). In May 1933 Mayr bought his own car, "a second-hand

**Fig. 3.5.** Mayr's first car, 1933 (Photograph courtesy of Mrs. S. Harrison)

Ford which I need for my excursions" (Fig. 3.5).[6] By that time he had acquired a substantial knowledge of the local bird fauna.

"I am very busy. One evening I count starlings at their roost, the next evening I sit at a Barn Owl nest through half of the night, the third night I don't sleep at all because I leave on a field trip before midnight. It is great fun, if one finds interested young ornithologists who take up all suggestions. I am now secretary of the Linnaean Society which doesn't mean much. It is 'The local Bird Club,' extraordinarily layman-like but those people know their birds. [...] Strangely enough so far there was no interest in breeding biology or in ornithological problems generally, but this may be changed through slow education" (letter to Stresemann, 12 May 1933; transl.).

At that time, he assisted Irv Kassoy in his barn owl study and suggested that electric light, mirror glass on one side of the nest box and a dimmer be installed. This permitted them to follow and photograph the growth of the nestlings. In particular Mayr liked the big census trip (at the height of the spring migration in the middle of May) in which he participated once and which covered most of northern New Jersey.

"I arrived around 8 pm at Charlie Urner's house in Elizabeth, and at 9:00 our whole party went to bed (I don't remember how and where we all slept). Shortly after midnight we got up again and left around 1:00. At once we started to record a number of species of owls that had previously been staked out at certain localities. Then on to Troy Meadows in New Jersey, the largest freshwater marsh in New Jersey,

[6] Mayr got quickly used to New York traffic but in July 1935 he was summoned to traffic court for making a wrong turn and fined $2.00.

where we got several species of rails, two bitterns, and other swamp birds. Around 9:30 we had a short coffee break, and checked a list of New Jersey birds to see what we still needed. Horrors, we had no English Sparrows yet, so on to the next town to fill that gap. By early afternoon we reached Brigantane near Atlantic City where we got terns, gulls, skimmers, and a great variety of shore birds. On the way home we added a nighthawk.

When we finally got back to Elizabeth it was nearing midnight. Urner took me to the train station, where I took the train to Manhattan and there I dragged myself to the Broadway subway. I was so tired I fell asleep almost at once. When I woke up we had just left the 125th Station where I should have gotten out, so I got out at 135th St. I now had to make the decision, should I risk taking the next train back to 125th St, where most likely I would fall asleep again, or simply walk back the 10 blocks. I finally decided it was better to walk."

On one of the excursions with Charlie Urner to Barnegat Bay they had parked the car at a parking place and, returning half an hour later, found that the car had been broken into. Among other things, Mayr's suitcase had been stolen with a Leica camera and his hardbound master ornithological notebook in which he had carefully recorded all of his observations since his gymnasium days. He would have spent a lot of money to get it back.

In the spring of 1936 he went with the Kuerzi brothers to the northwest corner of Connecticut where the Berkshires begin (Mayr et al. 1937j):

"I had bragged that I could find the nest of the brown creeper, which the Kuerzi's had heard singing in a little swamp. It was now up to me to establish the first breeding record of the species in Connecticut. They stopped the car on a dirt road near a swamp and pointed in the direction where they had heard the bird singing. I walked in that direction, and about 40 yards in I spotted a dead tree with loose bark. I called back to the Kuerzi's this would be the ideal location for a brown creeper nest. When I looked, sure enough there was the nest. I had simply applied the Schiermann method [p. 46]. Of course, it was enormous luck. I have since repeated the search for brown creeper nests where I have heard them singing, but I have never again found one in the United States."

While he stayed at the "Trout Valley Farm" in the western foothills of the Catskill Mountains in May 1934, Mayr took notes on the breeding birds. He had found the nests of 22 species (Mayr 1935f).

In retrospect Mayr fondly remembered the field trips with other young ornithologists: "In those early years in New York when I was a stranger in a big city, it was the companionship and later friendship which I was offered in the Linnaean Society that was the most important thing in my life" (1999j: 3).

After his kidney operation in April 1934 (see p. 301), he "received about 100 letters and postcards and my local friends literally had to line up at the hospital door because only a limited number of visitors were permitted in. This gave me the comforting feeling that I have made true friends here" (letter to Stresemann, 4 May 1934; transl.).

The intellectual level of the Linnaean Society meetings was poor, just records, rarities, life lists, arrival and departure dates. In comparison, the German Ornithological Society under E. Stresemann was "far more scientific, far more interested in life histories and breeding bird species, as well as in reports on important recent literature. Most members of the German society were amateurs like those of the Linnaean Society, but somehow a very different tradition had become established" (Mayr 1999j: 3).

The activities of birdwatchers in the New York area and beyond had been influenced by Ludlow Griscom (1890–1959), an assistant curator at the AMNH. He dominated the meetings of the Linnaean Society (until he moved to Cambridge, Massachusetts in 1927) and was the acknowledged leader in field identification of birds. He inspired R. T. Peterson to prepare his first field guide which appeared in 1934. Griscom introduced the game of correctly identifying a maximum number of bird species in a minimum of time based on a thorough knowledge of the diagnostic differences of all similar species of birds. Millions of people enjoy it. However, as Mayr (1995k) pointed out, "Griscomites hardly ever make any contributions to the serious study of birds. [...] List-chasing à la Griscom does not produce amateur naturalists of the tradition of Selous and Howard." The one thing "birders" in the Griscom-tradition never do is to watch a bird carefully and describe its behavior.[7] This was precisely what Ernst Mayr now asked the birdwatchers to do. He wanted to lift their hobby of birding above the level of list-chasing.

He arranged an Ornithological Seminar alternating with the formal Linnaean Society meetings and similar to Stresemann's "Fachsitzungen" where Mayr had participated in Berlin. At first, he himself reviewed "Some problems of bird migration" and other topics from the "biological" bird literature like papers by G. Schiermann and K. Lorenz, an article by himself on B. Altum and the territory theory and others by Eliot Howard and Edmund Selous: "The ornithological seminar appears to be quite successful. There were 17 people present the first time, and 22 the second time. The meeting is on the first Tuesday every month" (letter to F. Chapman, 13 March 1933). Attendance later decreased to a core group of 8–10 young birdwatchers. "Everybody should have a problem to work on," Mayr used to say, which meant that each one should work on a research topic that is on a gap in scientific knowledge that they as bird students should attempt to fill.

"Once a pattern had been established, I asked members of our Ornithological Seminar to review short papers from the American literature and very much encouraged them, eventually, to adopt themselves a 'problem' and try to solve it by field work. Dick Kuerzi undertook a life history of Tree Swallow on a colony he himself had founded [Proc. Linn. Soc. New York 53:1–52, 1941]. Irv Kassoy did a superb study on nesting Barn Owls [never published]. Bill Vogt did a life history of the Willet nesting in New Jersey [Proc. Linn. Soc. New York 49:8–42, 1938], Dick Herbert did a census of occupied Peregrine Falcon eyries in New Jersey and southern New York state [Auk 82:62–94, 1965], and Jack Kuerzi a faunistic survey of the rarer breeding birds of Connecticut [on which he published several short

[7] After Mayr had moved from New York to Cambridge (Massachusetts) he wrote to Stresemann (Berlin): "Ornithologically, the Boston region was quite deteriorated, particularly under Griscom's unfortunate influence. 'List-chasing' had become the main activity" (4 January 1958).

reports]. Joe Hickey did a careful census, à la Schiermann [p. 46] of a plot in Westchester County. Bill Vogt, when he became editor of *Bird-Lore* (later *Audubon Magazine*) was so enthusiastic about Hickey's breeding bird census that he introduced breeding bird censuses in his magazine as a counterpart to the winterly Christmas censuses.

The reviews of literature as well as the problems they undertook were eye openers to these young birdwatchers who had had no academic training. Hickey indeed was so impressed by the possibilities that he quit his job at Consolidated Edison and went back to College, eventually becoming a professor of wildlife studies at the University of Wisconsin. He also wrote an excellent *Guide to Bird Watching*, 1943 (to a considerable extent based on the literature we had reviewed in our seminar) and eventually was one of the most important voices in the fight against DDT and other pesticides." Like Stresemann in Berlin, Mayr was convinced that with proper guidance, serious birdwatchers could make important contributions to science (see also Barrow 1998: 194).

In his report for 1932–1933, the Secretary of the Linnaean Society, William Vogt, stated: "Perhaps the most noteworthy event in the year's history of the Society was the establishment, under the leadership of Dr. Ernst Mayr, of a monthly seminar for the abstracting and discussion of current papers concerned with field ornithology. The formation of the seminar evoked a gratifying reception from the Society's members and it offers an obviously welcome opportunity for more technical discussions than are desirable or feasible in the regular meetings." These seminars continued at least until 1938 (LeCroy 2005).

As editor from 1934 to 1941, Mayr upgraded the publications of the Linnaean Society, opening the *Proceedings* and the *Transactions* for large contributions. Charles Urner was the publisher of a journal in dairy and egg trade knowing nothing about science. Nevertheless, he was willing to supervise the printing of the Society's publications and taught Mayr all the tricks of editing and publishing. Mayr first persuaded Margaret Morse Nice (1883–1974) to write a two-volume monograph (1937, 1942) on the life history of the Song Sparrow (*Melospiza melodia*). Volume 1 is dedicated "To my friend Ernst Mayr." He had done a lot of editing on this manuscript of population studies. Numerous letters were exchanged between him and Mrs. Nice. He also mediated her meeting with Erwin Stresemann during her visit to Europe in 1932. When Niko Tinbergen (1907–1988) visited New York in 1938, Mayr solicited a manuscript from him on his observations of Greenland birds, especially the Snow Bunting in spring (published in the *Transactions* in 1939). Mayr provided valuable criticism on the manuscript and on some of his views of animal behavior also suggesting that Tinbergen expand his interest into the genetics of behavior.[8] At first the council of the Linnaean Society felt that this type of publications would not be read by the old-time birdwatchers. However,

[8] Tinbergen stayed again with the Mayrs for a while when he lectured at various universities in the United States in late 1946–early 1947, a trip Mayr had organized. The latter suggested that Tinbergen put into book form the manuscript for the six lectures he gave at Columbia University (New York). It was published later under the title *The Study of Instinct* (1951).

Charles Urner as well as Mayr's young friends of the Bronx County Bird Club (Joe Hickey, Jack Kuerzi, William Vogt, Irving Kassoy, and others) backed his proposal leading to the approval for publication by the council.

"The council thought that for this more general publication [by M.M. Nice] maybe we should print 200 copies, or perhaps even 250. Everyone thought I was totally mad when I insisted that we print 1,000 copies. Hindsight tells me that I was probably not too diplomatic in getting my way, but Charlie Urner printed 1,000 copies, even though he thought we would probably get stuck with an unsold supply of about 750 copies. Much to everyone's surprise, even my own, it took relatively few years before the whole edition was sold out, the Society not having lost its shirt, but actually making a profit" (Mayr 1999j).

As Mayr had predicted, these monographs became vast successes and eventually sold out. The song sparrow monograph even had to be reprinted many years later (Dover Publ., Inc., New York, 1964). This editorial training proved invaluable for Mayr when he founded the journal *Evolution* in 1946 (p. 238). Mayr's encouragement of amateurs in New York to undertake serious biological studies of local birds and of Mrs. Margaret M. Nice to write her monographic population study of the Song Sparrow was a major influence on American ornithology during the 1930s.

## Seeing America

Even though he worked hard at the museum, often until late in the evening, Mayr took time off for birdwatching on weekends and to attend the annual meetings of the American Ornithologists' Union (AOU), e.g., in Detroit (October 1931) and Toronto (August 1935). By 1933 he had already lectured at Yale and Princeton Universities. The first opportunity to see a different part of the States came in November 1931, when Richard Archbold, a research associate at the Department of Mammals, invited him to come along for a visit to Georgia, where his father had a quail hunting preserve at Thomasville. He paid all the expenses, including the airplane tickets. The small plane from Washington to Jacksonville, Florida, stopped at least once in each state. Beyond South Carolina, they persuaded the pilot to fly low over the coastal salt marshes to see the thousands of wintering swans, geese, and ducks. A limousine brought them from Jacksonville to Thomasville where they met Herb Stoddard, the director of the preserve, who took them on daily excursions. "With its sandy soil and extensive pine forests, the region reminded me of the surroundings of Berlin. Ornithological tidbits were *Aramus, Anhinga* and, at the Gulf of Mexico, pelicans and many interesting songbirds" he wrote to Stresemann (12 November 1931). When they traveled south from Thomasville to Shell Point on the Gulf Coast, they passed through Tallahassee, which in those days was a charming, sleepy old Southern town with huge oak trees and lots of Spanish moss. This wonderful experience also helped to cement Mayr's friendship with Dick Archbold.

In the early summer of 1933, Sterling Rockefeller invited Mayr to accompany him and make a census of all the birds of small Kent Island, in New Brunswick near Grand Manan Island, which he had bought and intended to give to Bowdoin College as a nature reserve and field station. Particularly surprising were the large numbers of nesting Eider Ducks along the coast and the colony of Leach's Petrels at higher elevations. It was the beginning of the breeding season and males had selected their nesting holes and were "singing" to attract a female. In the spruce forest they observed Tree-Creepers, Golden-crowned Kinglets, nesting juncos, and Myrtle Warblers. Arctic Terns also nested on the island and dive bombed the visitors. From Kent Island they went by boat to another island to visit one of the colonies of the American Puffin. They could have stayed longer on Kent, but Mayr was anxious to get back to New York because Gretel Simon was due to leave for Germany. Two years later she became his wife.

When the 6th Pacific Science Congress took place in San Francisco in August 1939, Ernst and Gretel Mayr crossed the continent by train from New York to California. In Santa Fe (New Mexico) they lunched with R. Meyer de Schauensee of the Philadelphia Academy of Sciences who, with his family, spent a vacation there. At that time, Mayr prepared, jointly with him, an account of the birds of the Denison-Crockett South Pacific Expedition. From here the train went to Arizona, where they changed to another one that took them to the Grand Canyon. They arrived in the dark and were immediately housed in a cabin. When they stepped out next morning they were only about fifty yards away from the rim of the canyon. Mayr always considered this one of the two most impressive sights in his whole life. The other was the cave of Lascaux in France.

In Pasadena they were to meet Th. Dobzhansky who, at that time, was an assistant professor of biology at the California Institute of Technology (Caltech). They asked a taxi driver to take them to a "moderately priced" hotel. Instead, he took them to the famous luxury hotel, the Huntington where, greatly amused, the reception clerk let them have a room for $5.00 (which, at the normal rate, was in the order of $30–50). When Dobzhansky came to pick them up the next morning, he would not dare, with his old car, to drive up to the Huntington, and instead left it around the corner. Much to the entertainment of the clerks, they carried their suitcases down the street to Dobzhansky's car. They stayed another night in a less expensive hotel and then continued to San Francisco and Berkeley. The Mayrs and Dobzhanskys had a splendid time together in Pasadena and renewed their friendship (p. 133).

Alden H. Miller expected them in Berkeley and, over the duration of the congress, took them repeatedly with his car into the surroundings for sight-seeing and birdwatching. The Pacific Science Congress was held in San Francisco and at Stanford University in Palo Alto. Over a weekend they visited Monterey Peninsula and Yosemite Park. On their way back to New York the Mayrs stopped at Denver, Colorado where Adolf and Gwendoline Ley accompanied them on several trips into the Rocky Mountains. He was the "founder" of the "community living" at 55 Tiemann Place in upper Manhattan in 1932 and had moved to Colorado after he got married. Ernst went also birdwatching near Denver with Alfred M. Bailey of the Natural History Museum. His next stop was in Iowa City to visit Professor Emil Witschi (1890–1971), an endocrinologist from Switzerland, who proved that the sexual dimorphism of *Passer* species and some other birds was not affected by male or female gonadal hormones or else that it is affected by a pituitary hormone, as in fowl, certain weaver birds and others. He was a good friend of Erwin Stresemann. The last stop was in Chicago where Mayr spent a few hours at the Field Museum. Gretel had gone home several days earlier because she was worried about their children who had been with friends for the duration of their parents' trip.

## Curator of Ornithology at the American Museum of Natural History

For over 10 years Mayr worked mainly on bird collections of the Whitney South Sea Expedition from Polynesia and Melanesia, on collections from many areas in New Guinea, the Malay Archipelago and from southeastern Asia. A steady stream of articles and books documented his research activities.

Mayr also assembled general data for a comprehensive analysis of geographical variation and speciation in birds for which purpose the large collections of the Whitney South Sea Expedition from the islands of Oceania and of various expeditions to New Guinea were unusually well suited. There was no better qualified ornithologist to work on this rich material than Ernst Mayr whose scientific interests had been directed toward taxonomic and evolutionary topics by Stresemann and Rensch at the Museum of Natural History in Berlin. Without realizing it, he carried out in New York the suggestions he had written to Stresemann in May 1924 (pp. 27–28).

The natural history museum of New York City is located along Central Park West, between 77th and 81st Streets (Fig. 3.6). Since the 1930s its bird collections had become the most comprehensive and most representative in the world and the Department of Ornithology the foremost center of research. Numerous scientists from all over the world came to study the collections in conjunction with their own projects. The department occupies an entire eight-story building called the Whitney Wing in honor of its patron Harry Payne Whitney (however, the Biology of Birds Hall has been replaced by a Geology Hall in 2000). In early 1929, Whitney had contributed stock worth half the cost of the building, or $750,000, with the understanding that New York City provide for the other half. As Mary LeCroy (2005: 39) related the story: "For once, the city acted quickly and okayed the matching funds by the summer. The museum's comptroller rushed out and sold the stock against the advice of many, and he had the cash in hand when the stock market crashed in the late summer of 1929!" Construction of the Whitney wing began in April 1931 (Fig. 3.7), the collections and offices were transferred to the new Whitney Wing during 1933–1935. However, the official opening and dedication of the building took place only after the completion of the adjacent "Rotunda," the Theodore Roosevelt Memorial, in 1939. After the purchase, in 1932, of the Rothschild Collection, the department held about 700,000 birds. Today the collections comprise ca 1 million specimens representing more than 99% of all known species of birds (Lanyon 1995; Vuilleumier 2001).

**Fig. 3.6.** The American Museum of Natural History in New York City, a red brick structure in the Victorian Gothic style built from 1874 to 1877. The Theodore Roosevelt memorial (the Rotunda) and the Whitney Wing (to the right) were added during the 1930s. Ernst Mayr's office from 1935 onward was located in the right hand (northeastern) corner of the Whitney Wing on the 4th floor (large windows). Aerial view over the Museum westward with the Hudson River in the background. Photograph taken in 1957 (AMNH Library photographic collection, negative no. 125037)

Mayr would have had several opportunities to go out again on expeditions, but F. M. Chapman, Chairman of the department, firmly pointed out to him, that he was paid from the Whitney Fund to work on the Whitney Collections. To compensate for these restrictions, Mayr in his free time did quite a bit of bird-watching and fieldwork around New York, especially during the late 1930s and in 1940. In 1932, he had contacted the group of geneticists at Columbia University (New York) in conjunction with his studies of avian plumages. These contacts and those with Th. Dobzhansky in California guided him to general systematic and evolutionary problems on which he began to lecture in late 1939 ("Speciation phenomena in birds") followed by the Jesup Lectures on evolution at Columbia University in 1941, the foundation of his major work: *Systematics and the Origin of Species from the Viewpoint of a Zoologist* (1942e). He became very active as a member and secretary of the Society for the Study of Evolution (1946), first editor of its journal *Evolution* (1947–1949) and president (1950). Now an authority on evolutionary biology he was invited by various universities as a lecturer or visiting professor (e.g., Philadelphia Academy of Sciences, 1947; University of Minnesota, 1949; Columbia University, New York 1950–1953; University of Wash-

**Fig. 3.7.** Groundbreaking for the Whitney Wing of the American Museum of Natural History, 17 April 1931. H. F. Osborn, President of the AMNH, uses a bronze shovel (now in the Department's archives) in the ceremonial onset of construction of the building. Members of the Department of Ornithology present are (to the *left* of the President) R. C. Murphy, Alice K. Fraser (secretary), E. Mayr, Katherine Johns (secretary), C. O'Brien, and J. T. Zimmer (AMNH Library photographic collection, a portion of negative no. 313534)

ington, 1952). Mayr turned down several employment offers by various universities, partly with regard to Dr. Sanford who had done so much for him. When Sanford died in 1950 and Harvard University offered him a research professorship in 1953, Mayr felt that he now was free to accept and joined the staff of the Museum of Comparative Zoology in Cambridge, Massachusetts. His Alexander Agassiz Professorship was not restricted to the Bird Department, but was a position in the Museum of Comparative Zoology as a whole.

## Museum Tasks

Before the Whitney Wing was completed the Bird Department in the eastern half of the north wing of the museum was very crowded. Chapman, Chapin, and Murphy had offices, whereas J. T. Zimmer and Ernst Mayr sat at small tables in a little empty space among the collection cases. Whenever they needed to compare series of specimens they had to take them to a larger table in Chapman's office. After the department had moved into the new quarters during 1935, Mayr had a spacious office at the north end of the 4th floor with large windows overlooking Central Park (Fig. 3.6).

One of his main tasks at the museum during the mid- and late-1930s (besides his study of the Whitney birds) was to supervise the unpacking, storing, and cataloguing of the Rothschild Collection comprising 280,000 bird skins. The sale of this collection in 1932 had come about by the blackmailing activity of a "charming, witty, aristocratic, ruthless lady" beginning in the early years of the century. She continued her activity so successfully that (it is rumored) Lord Rothschild was in debt around 1930 (Murphy 1932; Snow 1973; M. Rothschild 1983: 92, 139, 302; LeCroy 2005). The sale was kept a secret until the deal was concluded. "If Sanford was the shining knight in this tale, she was the dark lady, but just as important to the outcome of Mayr's career" (Bock 2004c), because her activity led to Mayr's employment in New York. In a sense we may say he owed his career in New York to this still anonymous aristocratic lady. Mayr attempted to find out her name for 50 years but was unsuccessful. Historically more important was, however, to what good use through hard work Mayr put the new opportunities at the AMNH.

The 185 wooden packing cases (76×76×152 cm) in which Rothschild's birds had been packed in Tring under R. C. Murphy's supervision arrived successively in New York during the course of 1932 and were stored in an unused hall at AMNH. They could be unpacked only during the summer of 1935 (Fig. 3.8). The approximately 30,000 New World birds of the AMNH had been transferred to the 5th and 6th floors, the other 250,000 to the 3rd and 4th floors of the Whitney Wing during the course of 1933 and 1934, on completion of this building. For about three months Mayr and several colleagues calculated for each family of birds how many cases or half-cases were required. The calculations turned out to be correct with one exception—the unexpected size of the giant Siberian eagle owls required some last minute changes. The catalogue of the 185 large packing cases prepared by Phyllis Thomas, Ernst Hartert's former secretary, and R. C. Murphy was useful; furthermore Ms. Thomas herself came to New York for several weeks to assist with the organization. The ornithology collection manager Charles O'Brien was in charge of the move, assisted by three helpers: H. Birckhead, T. Gilliard, and C. O'Brien's younger brother. They transferred the birds from the shipping boxes in the storage room to the trays of three different sizes, wheeled them on carriages over to the Whitney Wing, and then placed the trays into the proper cases where they would remain permanently.

"The unpacking of the Rothschild collection is progressing favorably. We are now working on the small birds, of which we can unpack only a case and a half

**Fig. 3.8.** Examining the newly arrived Rothschild Collection, 1935. *Left* to *right*: R. C. Murphy, E. Mayr, J. T. Zimmer, and President F. T. Davison (AMNH Library photographic collection, negative no. 314574)

per day. Some of these cases contain as many as 10,000 specimens. We have, so far, unpacked 61 cases containing 62,000 specimens. This was done in fifteen days, which means that we have done an average of four cases and 4,000 specimens per day. This is just about twice as much as I had thought we could do. The boys who are doing the unpacking really deserve a great deal of credit for their fine work" and "The moving into the new wing is proceeding at a fast rate. By the time you come back from fishing the greatest part of the collection will already be installed in the new wing" (E. Mayr to Dr. Sanford on 5 March and 21 June 1935).

Not surprisingly, these activities consumed most of Mayr's time causing a noticeable drop in the number of taxonomic papers published during 1934 to 1936. After several months of unpacking it took another 6–8 years to catalogue the entire collection. This required decisions which families to recognize and which sequence of genera within each family and which sequence of species within each genus to accept. The consequence of this new classification was a whole series of family revisions by J. Delacour, E. Mayr, D. Amadon, D. S. Ripley and also C. Vaurie's authoritative survey *Birds of the Palaearctic Fauna* (1959, 1965). Mayr's taxonomic work on the Whitney Collections from the southwest Pacific, of course, was immensely facilitated by the Rothschild Collection which provided valuable comparative material including numerous type specimens.

The new classification of the combined collections permitted a species count of the birds of the world, based partly on actual numbers, partly on estimates. The results were 8,500 species (Mayr 1935d) and 8616 species (Mayr 1946a). The stability of these figures would depend entirely on the future development of the taxonomic viewpoint (1935d: 23). This turned out to be an amazingly perceptive prediction, since the application of the concept of allospecies (and superspecies) during the 1940s and 1950s led subsequently to the upgrading of many peripheral and well-differentiated "subspecies" to the status of (allo)species so that without an appreciable increase in the number of newly discovered species, the present estimate is 9,500 to 10,000 species of the world (a "quiet revolution," Mayr 1980d).

In the reference 1946(a), Mayr doubted "that in the entire world even as many as 100 new species remain to be discovered" (p. 67). This turned out to be underestimated. 162 "good" species have been discovered between 1938 and 1990. Periodically, Mayr reviewed critically the taxa described as new species in the literature (Zimmer and Mayr 1943b; Mayr 1957a, 1971d; Mayr and Vuilleumier 1983f, 1987c; Vuilleumier et al. 1992m). In 1957, he felt that the bird fauna of the world was then so well known that probably no more than 20 species would be discovered during the next 10 years. This estimate was again surpassed. Thirty-five good species were described until 1965, a rate of 3.5 species per year. This rate has only slightly decreased to 2.4 species per year until 1990 (Vuilleumier et al. 1992m). Most of these new species have been overlooked for so long because they are sibling species or have exceedingly small ranges in regions of difficult access like, e.g., parts of the tropical Andes Mountains. Recently, Mayr and Gerloff (1994t) estimated the total of subspecies of birds as 26,206.

## Colleagues at the American Museum

"There were (in the 1930s and 1940s) an intellectual excitement and a level of professional competence and ornithological universality at the American Museum that had nowhere existed previously and that perhaps can never again be duplicated" (Mayr 1975c: 372). Chairman of the Bird Department was Frank Michler Chapman (1864–1945), an outstanding ornithologist with imagination and a broadly interested researcher (Fig. 3.9), who had published pioneering monographs on the birds of Colombia, Ecuador, the Peruvian Urubamba Valley and the Roraima Mountains (Mayr 1975c, 1980n; Lanyon 1995; LeCroy 2005; Vuilleumier 2005a). He had popularized ornithology in North America, promoted the cause of conservation and pioneered life history studies of tropical birds but now being in his seventies was no longer an innovator. Mayr felt that it was not to the best of the department to wait until Chapman, in 1942, would retire at the age of 78 years.

Robert Cushman Murphy (1887–1973) was employed in 1921 in order to write a report on the birds collected by the Brewster-Sanford expedition to the coastal

**Fig. 3.9.** Staff members of the Department of Ornithology, AMNH, photograph taken on F.M. Chapman's 70th birthday, 12 June, 1934. *Seated* (*left* to *right*) Ernst Mayr (Associate Curator of the Whitney-Rothschild Collections), R.C. Murphy (Curator of Oceanic Birds), F.M. Chapman (Curator-in-Chief), J.P. Chapin (Associate Curator of Birds of the Eastern Hemisphere), J.T. Zimmer (Associate Curator of Birds of the Western Hemisphere). *Standing*: C. O'Brien (Research Assistant), Katherine Johns (Secretary), Alice K. Fraser (Secretary), A.L. Rand (Research Assistant), Elsie M.B. Naumburg (Research Associate), Albert Brand (Research Associate) (AMNH Library photographic collection, negative no. 314442)

waters of South America. He became the world expert on seabirds, especially the tubenoses (Procellariformes). After much delay his excellent expedition report eventually appeared in 1936 in two volumes under the title *Oceanic Birds of South America*, but the planned monograph of the Procellariformes was never completed. During Mayr's first 2 years in New York, Murphy corrected his English and taught him quite a few language subtleties.

John Todd Zimmer (1889–1957) had collected birds in New Guinea and Peru and was contracted to produce a companion volume on Peru to Chapman's books on the ecology and biogeography of Colombia and Ecuador. Zimmer felt that at first the taxonomy of all these bird species had to be clarified and described literally hundreds of subspecies of Neotropical birds. Chapman is said to have written annually, around Christmas, a letter to Zimmer reminding him of the book on the Peruvian avifauna. However, Zimmer, who totally lacked Chapman's zoogeographical imagination, never even started it.

James P. Chapin (1889–1964) was a superb naturalist who had done graduate work at Columbia University under the most modern biologists of the period. He was, of course, fully familiar with the modern genetics of T.H. Morgan, who was in the same department of Columbia University. Chapin and Mayr had numerous conversations on evolution and he convinced Mayr of the importance of the findings about the effect of small mutations and of the invalidity of any belief in the inheritance of acquired characteristics. Most importantly, he (and Dobzhansky, p. 133) helped Mayr to abandon his early Lamarckian ideas and to see that the gradual evolution that the naturalists had insisted on could be explained by the new genetics of R.A. Fisher and the other modern geneticists (based on small mutations and recombination) and did not require any saltational interpretation of the early Mendelians. Other colleagues were Charles O'Brien, the collection manager, and Austin L. Rand who worked on birds from the New Guinea region. Mrs. E. Naumburg studied Brazilian birds and A. Brand experimented with bird song recording.

Among Mayr's colleagues from other departments whom he met in the staff lunch room and with whom he discussed technical and other matters were Frank Lutz, G.K. Noble, Charles Bogert, Ned (Edwin H.) Colbert, Mont Cazier, Frank Beach, Harry Raven (sometimes accompanied by his tame chimpanzee "Meshie" who dined with a fork and spoon at the staff table), Jack (John T.) Nichols (1883–1958), curator of fishes and founder of the journal *Copeia*, but also excellently informed about reptiles and birds, and G.G. Simpson. Unfortunately, Mayr never had any scientific conversations with Simpson in 23 years of lunches at the AMNH. Mayr had intensive evolutionary discussions only with Herman T. Spieth of City College whom, in 1946, he persuaded to give up mayflies in favor of *Drosophila* behavior. During two weeks in May, the time of warbler migration, Mayr, J.T. Nichols and J.T. Zimmer would go out to Central Park right after lunch for birdwatching. In the 1940s or early 1950s, Mayr founded the "Systematics Club" at the AMNH which met once a month for a presentation and discussion and continued to do so for many years after Mayr had left New York for Cambridge, Massachusetts.

Several younger colleagues joined the Bird Department and Mayr very much enjoyed mentoring them (Fig. 3.10). Dean Amadon (1912–2003) was engaged in 1937 for the egg collection and later became assistant curator. He did his PhD thesis on the Hawaiian honeycreepers. Eventually Amadon became Chairman of the Department (1957–1973) and was appointed the first Lamont Curator of Birds. E. Thomas Gilliard (1912–1965) had been hired as Chapman's assistant in 1932, when he dealt mostly with South American birds. Later he became Associate Curator of Birds and Mayr inspired him to focus his research on New Guinea birds, particularly the birds of paradise and bower birds. Hugh Birckhead (who was killed in France during the last stages of World War II) and Dillon Ripley (later Secretary of the Smithsonian Institution) were both influenced by Mayr in their research as well as was Charles Vaurie (1906–1975) who, as a teenager, had come to

**Fig. 3.10.** Staff members of the Department of Ornithology (AMNH), photograph taken on Dr. Chapman's 75th birthday, 12 June 1939. *Seated* (*left* to *right*) Elsie M. B. Naumburg (Research Associate), John T. Zimmer (Executive Curator), Frank M. Chapman (Curator), R. C. Murphy (Curator of Oceanic Birds), Ernst Mayr (Associate Curator of the Whitney-Rothschild Collections). *Standing* (*left* to *right*) Dean Amadon (Research Assistant), Charles Schell (Assistant), Ruth Bowdon (Secretary), Charles O'Brien (Assistant Curator), Mildred Feger (Secretary), Hugh Birckhead (Assistant), E. Thomas Gilliard (Research Assistant) (AMNH Library photographic collection, negative no. 291112)

the United States from France and became a dentist. He came often to the museum and Mayr introduced him to taxonomy from the late 1940s, at first working jointly with him on a revision of the drongo family (Dicruridae) and later supervising his long series of revisions of Palearctic birds. Eventually Vaurie was appointed to the staff and gave up dentistry becoming a full curator at the museum.

Known to Dr. Sanford were two wealthy young men, Richard Archbold (1907–1976) and John Sterling Rockefeller (1904–1988). The former invited Mayr to his father's quail shooting place in Thomasville, Georgia in November 1931 (see p. 111). During the 1930s, Dick Archbold financed and led three museum expeditions to New Guinea (Morse 2000). Three additional Archbold expeditions under the leadership of L. J. Brass explored the Cape York area of northern Australia (1947–1948), the outlying islands of southeastern New Guinea (1956–1957), and the central highlands of New Guinea (1958–1959). Sterling Rockefeller provided the funds with which E. Stresemann (Berlin) sent Georg Stein to Timor and Sumba in 1932. Rockefeller was supposed to work out this collection under Mayr's guidance, but by that time had lost interest in birds. The result was that Mayr himself studied this splendid collection analyzing at the same time the colonization of Australia by birds from the Lesser Sunda Islands (p. 181).

Other volunteers were Cardine Bogert (1937), who clarified the migrations of the New Zealand cuckoo *Urodynamis*, Eleanor Stickney (1943; later the collection manager of the Peabody Museum of Yale University), who published on the northern shorebirds collected by the WSSE in the Pacific and Martin Moynihan who, in 1946(c), published a paper with Mayr on the evolution of the *Rhipidura rufifrons* group (Moynihan later became the director of Barro Colorado Island, Panama, and made it into a world-class tropical research station), Daniel Marien, Karl Koopman (later a curator in Mammalogy at the AMNH and an expert on bats) and Kate Jennings (who made a major contribution to their joint paper on the variation in Australian bower birds, Mayr and Jennings 1952i). Staff members using increasingly the Whitney-Rothschild collections included Austin Rand, Dean Amadon and Thomas Gilliard. Rand was a Research Associate in the Department of Ornithology—an unpaid position. He was paid by Archbold Expeditions associated with the Mammal Department. In 1941 he went to his native Canada and then joined the Field Museum in Chicago in 1947.

Jean Delacour (1890–1985) came to New York after the defeat of France during World War II and was appointed technical adviser at the Bronx Zoo. This position left him time enough to work at the AMNH almost every day for several hours. Mayr and Delacour collaborated from the time of the latter's arrival in late 1940. They published several joint articles and the book, *Birds of the Philippines* (1946k) until Delacour moved to California in 1952 (Mayr 1986f). Their reclassification (1945e) of the duck family (Anatidae) had a wide distribution and was reprinted several times.

## AOU Politics

From 1931 onward Ernst Mayr attended all or nearly all annual meetings of the national ornithological organization, the American Ornithologists' Union (AOU). He observed the performance and oral contributions critically and compared their standards with those of the well-known German Ornithological Society. Together with several other young colleagues, especially H. Friedmann, R. Boulton, J. Van Tyne, also J. Grinnell and P.A. Taverner he soon set out to modernize the organization and ornithological research in North America generally. In 1933 already, Mayr was on the Program Committee for the 50th Anniversary meeting of the AOU held at the AMNH and tried very hard to put together a well-balanced program. He recommended, for example, that certain zoologists who had been using birds as their experimental material be invited to present papers at this meeting. Thus, Professor L.C. Dunn of the Department of Zoology, Columbia University (New York) spoke on "Heredity of Morphological Variation in Birds" and Mayr gave a review paper on "The Physiology of Sexual Dimorphism in Birds." After the meeting he worked on an unpublished analysis of "The trend of interest in American ornithology as demonstrated by the percentage of various subjects in the total of papers presented at the annual AOU meeting." He felt that in the future, such matters as nomenclature and faunistic lists should be eliminated from the program. The number of papers on taxonomic topics should be reduced in favor of life history, behavior and ecology. As long as T. S. Palmer was Secretary of the AOU, there seemed to be no chance for improvement because "he is neither a scientist nor an ornithologist; narrow and dictatorial." Mayr opposed the custom of paying one third of the total income of the Society to the Secretary, Treasurer and Editor for their "services" and insisted that the election to fellowship should be decided entirely on ornithological merit documented by relevant publications. Moreover, there should be definite time limits not only for the President, but also for the Secretary and Editor of the *The Auk*, the Society's scientific journal.

When, at the 1935 meeting, the "Washingtonians" were again able to elect one of their "buddies" into the only opening for Fellow, the reformers "became very active in AOU politics with the major aim of breaking the total domination of the AOU by the 'Washington crowd.' They were mostly staff members of the Biological Survey, later called Fish and Wildlife Survey. By using all sorts of parliamentary tricks the Washingtonians succeeded in electing most of the officers and other Biological Survey people as Fellows. One year when they pushed through a person with the name of Preble who had never been distinguished, and in fact had not published anything in ornithology in 15 or 20 years, my tolerance had reached an end. Mrs. Nice had also been on the slate, and she was by an order of magnitude more deserving than Preble. Fortunately, there were a few others who felt like me, particularly Herbert Friedmann who, although also a Washingtonian, shared my sentiments. The trick of the Washingtonians was that the first ballot for any office or honor was called a nomination ballot. The results of this first vote were put on the blackboard and all the Washingtonians had reached a consensus before the meeting whom to nominate. As a result, invariably their candidate had three or four times as many votes as any other nominee. We young Turks adopted their method. We established a carefully chosen slate of the people we considered most deserving of election and then went to everybody else (except the Washingtonians) urging them to vote for this slate. Since these were all good candidates we had little trouble persuading them. When the first election took place after we had started our campaign the Washingtonians were utterly astonished that all of a sudden for each opening there was one nominee who had considerably more votes than their candidate. It took only a couple of years before we had gotten the Washingtonians out of all the offices and since we maintained the system they never again had a chance to push an unworthy candidate through. It was at this time that through the influence of Grinnell and Alden Miller in Berkeley and of Van Tyne in Michigan American ornithology slowly shifted into an entirely different direction. The AOU was still backward as shown particularly by the contents of the *Auk*. For a while the journal of the Wilson Club, the *Wilson Bulletin*, was indeed a better journal than the *Auk*, and this is where I published my paper on the 'History of the North American Bird Fauna' (1946h) and where Delacour and I published our subsequently so famous classification of the duck family (1945e)."

A leading voice among those demanding change and modernization in North American ornithology was Ernst Mayr (Barrow 1998: 190–195). As shown above, he played a key role in furthering both ornithological practice and the rigor with which it was pursued. The editor of the *Auk* was replaced in October 1936 and Mayr immediately sent to the incoming new editor Glover M. Allen a list of suggestions for improving the journal. Several other changes were also approved at that time. In preparation of the next annual meeting Mayr circulated a list of suggestions to his colleagues (Friedmann, Grinnell, Boulton, Taverner, Griscom) entitled *Proposed Amendments to the Constitution and the By-Laws of the American Ornithologists' Union* (7 pages including his detailed reasoning): (1) Election of Fellows should be based exclusively on outstanding ornithological accomplishments, (2) Inactive Fellows (who have not published technical papers in five consecutive years) should be transferred to the status of Emeritus Fellows, (3) Appointment of an editorial committee to support the editor of the *Auk*, (4) Secretary and Treasurer should hold honorary positions without receiving remuneration (except reimbursement for traveling and other expenses); this would reduce the overhead and enable the editor to publish colored plates, distribution maps and possibly even monographs as supplements to the *Auk*, (5) Publication in the *Auk* of the balance sheets presented by the treasurer at each annual meeting, (6) Nomination of a Program Committee at each annual meeting to prepare next year's meeting and to balance technical and more popular topics, etc. Mayr here also suggested that at annual meetings, symposia should be organized on particular topics, e.g., "The Species Problem, Genetics and Taxonomy, the History of the American bird fauna, Climate and Distribution, the Life History of shore birds (or other bird families), the Ecology of North American habitats, Bird Migration and Climatic Factors, Bird Migration and recent physiological investigations, Bird Behavior, Bird Sociology, etc. [...] In short, the program for the annual meeting should be a work of art and not an accident!" (quoted from an attachment to a letter from E. Mayr to J. Grinnell dated 14 October 1937; archives of the Museum of Vertebrate Zoology, University of California, Berkeley).

Some of these proposals were accepted at the 1937 meeting and most of the rest in one form or another during the next 5 years. On 5 December 1937 Mayr wrote to E. Stresemann (transl.):

"I was afraid to be outlawed but instead half of my proposals have been accepted, Palmer was brought down and Mrs. Nice and myself elected as Fellows. Friedmann is President, Chapin and Peters Vice Presidents [...]. All in all very pleasing. I now work toward next year; I wish that Elliot Howard will be elected as Honorary Fellow

**Fig. 3.11.** Ernst Mayr as President of the American Ornithologists' Union (AOU), 75th anniversary meeting in New York (14–19 October 1958). Persons in the *front row* from *right* to *left* are: (NN), Hoyes Lloyd, Dean Amadon, Ernst Mayr, George H. Lowery, Eugene Eisenmann, Charles Sibley, Finn Salomonsen, Austin Rand, Vesta and Erwin Stresemann (AMNH Department of Ornithology archives)

THE AMERICAN ORNITHOLOGISTS' UNION
SEVENTY-FIFTH ANNIVERSARY MEETING
(SEVENTY-SIXTH STATED MEETING)
AMERICAN MUSEUM OF NATURAL HISTORY

and Lorenz, Meise and Tinbergen as Corresponding Fellows. Those are the ones who, in my opinion, deserve it most."

It is understandable that one or two of his colleagues misinterpreted Mayr's drive for improving the AOU and believed that this foreigner, with ample self-confidence and energy, was primarily interested in himself assuming one of the leading positions in the organization.

In 1937, President Friedmann established a Research Committee with various subcommittees of which E. Mayr headed the one for "Migration, homing and related phenomena." He as well as H. Friedmann, L.J. Cole and P.A. Taverner presented their reports at the 1938 annual meeting (see *Auk* 56:113, 1939). The reformer had been successful. Mayr remained involved in the affairs of the Society and introduced many changes in the time to come. For example, he submitted a proposal to regulate the election of vice-presidents and, in 1957–1959, he served as the President of the AOU. In this capacity he managed the planning and invitation of international guest speakers for the Society's 75th anniversary in October 1958 (Fig. 3.11).

Through his activities in the AOU and the Linnaean Society of New York Mayr contributed effectively to "biologize" North American ornithology. In later years, several of his PhD students at Harvard studied the behavior of herons, flycatchers, tits, and wood warblers (p. 262).

## Ecology and Behavior of Birds

As a curator at the AMNH Mayr had only little time for studies on the ecology and breeding biology of birds, although he was able to complete a number of projects (e.g., Mayr 1935e) and to discuss the territory theory in birds (1935c). Also, during the early 1930s, he directed the research of several members of the Linnaean Society of New York who, at his suggestion, investigated the life history of selected bird species in the surroundings of this city (pp. 109–110). In the spring of 1940 Mayr studied several small colonies of Red-winged Blackbird (*Agelaius phoeniceus*) in northern New Jersey to test an idea that colonies of social birds would start nesting the earlier the larger the colony, also to see to what extent there was polygamy, and how the territory borders change with the seasons. He went out every morning at about 5:00 am and was back home shortly before 8:00 to start the commuting trip to New York. His findings indicated that the small colonies comprising only two or three territories in the small potholes with bushes and trees were established earlier than the larger colonies in the big cattail marsh where the birds had to wait until this year's cattail stalks were tall enough to support the nests. Hence the rule established for sea birds did not apply to these inland colonies (1941l). Mayr also watched birds while crossing the North Atlantic on board a passenger ship (1938h) and visiting the Bahama Islands (1953g), he described anting by a song sparrow (1948f) and gulls feeding on ants (1948g). During the early 1940s he intended to study the nature of behavioral isolating mechanisms in birds experimentally but failed because of logistic difficulties (p. 228).

In the course of his efforts to introduce experimental biology, ecology and ethology into American ornithology, Mayr encouraged Margaret M. Nice (1883–1974) to write her two-volume monograph of the Song Sparrow, which made her famous. They had met at the annual AOU meeting in Detroit (October 1931): "Thus started a warm and enduring friendship that became exceedingly important to me," she wrote in her autobiography (Nice 1979: 109). Mayr was delighted to find an American "interested in more than faunistic records and pretty pictures" and started her reading the German *Journal für Ornithologie*. Mrs. Nice lived in Columbus, Ohio where she had no chance to discuss her studies with other naturalists for she was excluded from the strictly masculine Wheaton Club. Mayr was present when, on his suggestion, she visited Berlin and the Museum of Natural History for ten days during the summer of 1932. Stresemann, similarly impressed with her work, published a detailed progress report in German in *Journal für Ornithologie* (1933/34). Mayr considered her Song Sparrow monograph "the finest piece of life-history work ever done." As the editor, he offered to publish the final manuscript in the Transactions of the Linnaean Society of New York (1937, 1943). He also persuaded N. Tinbergen to send him his observations of the Snow Bunting in spring for publication. At the AMNH Mayr established a seminar for birdwatchers in the New York area (p. 109) and cooperated closely with G. K. Noble (1894–1940) who pioneered in behavioral experiments with free-living birds and other vertebrates (Mayr 1990h). Noble's sudden death was particularly shocking because he was such a vital person, but died within three days by a throat infection. Similarly, Ernst Mayr encouraged and influenced the work of David Lack, Konrad Lorenz and Niko Tinbergen in Europe, as pointed out by Burkhardt (1992: 300, 1994: 360). The Austrian ethologist Lorenz worked in Germany during most of his career. He supported Mayr's lifelong interest in animal behavior. In fact, about half of Mayr's PhD students did their theses in behavior rather than either evolution or systematics (p. 262). Mayr wrote in his autobiographical notes:

"In 1951, Gretel and I visited the Lorenz's at Buldern, Westphalia. There we had very long discussions, even controversies. At this period, Lorenz always talked about *the* Greylag Goose in a strictly typological sense. By contrast, I insisted that every Greylag Goose was different from any other one. 'If a Greylag Goose becomes widowed,' he said, 'he or she will never marry again.' I asked him on how many cases his statement was based, but he had only a vague answer. At any rate my insistence that every goose should be treated as an individual eventually resulted in Lorenz hiring a special assistant to keep track of the activities of every single individual in the flock. Each goose had its own card in the cardfile and all about each goose was recorded daily. Needless to say, within the first year he already had one or two cases of widowed geese remarrying. Many other sweeping statements about *the* Greylag Goose were likewise refuted by individual records. [...]

When Lorenz went to Seewiesen [in 1956] I visited him several times and gave seminars to his investigators. Our friendship, of course, continued after his retirement [1973] when Gretel and I visited the Lorenz's at Altenberg, Austria. With part of the money for his Nobel Prize Konrad had built a very large seawater aquarium with the most wonderful coral reef fishes. There he went after breakfast with me and we sat on a comfortable bench facing the aquarium. He told me exactly what every fish would be doing the next minute, and what this meant, and quite obviously he had an amazing understanding of the psychology and social behavior of these fishes. What bothered me was that he never kept a note. When he looked at the fishes it was mainly for aesthetic satisfaction.

Lorenz was a born naturalist and observer who was able to see things that no one else notices. He could look at a group of displaying ducks and find some details in their courtship that others, who also saw the ducks, never saw. There is a wonderful story of Lorenz missing a class he was teaching at the University of Münster during the early 1950s." When he was staying at castle Buldern with a large park, he went by motorcycle to reach the university. One day he left at the usual time. After quite a while, his assistant called up Mrs. Lorenz and said that Dr. Lorenz had not appeared and asked what was the matter. She said he left at the usual time. He must have had a motorcycle accident. So the assistant said he'd jump on his bicycle and she should jump on hers and they would meet. She hadn't left the park when she saw the motorcycle lying on the side of the road and Lorenz lying flat on his belly with his field glasses studying some ducks. She said, "You've forgotten your class." He said, "Oh yes I did! But this was so interesting, I just couldn't miss it." Mayr also remarked that Lorenz did not fully understand natural selection, because he repeatedly said that this or that was good for the species. However, the target of natural selection is not the species but only the individual.

Like most of us, Mayr was intrigued by the homing ability of birds, that is, their ability to return to a known goal over an at least partially unknown flight-route. He reviewed the results of experiments conducted by German ornithologists who transported swallows and starlings from their breeding sites up to several hundred miles away in different directions and released them. A high percentage of these birds returned home in a relatively short time (Mayr 1937d, 1944l; see also Mayr 1952j, 1953j).

The origin of the migration routes of several species which fly across vast ocean expanses to winter on small isolated islands of the central Pacific (e.g., the Bristle-thighed Curlew *Numenius tahitiensis* from western Alaska and the Long-tailed Cuckoo *Eudynamis taitensis* from New Zealand) is consistent with the geological assumption that large archipelagos were extant in the Pacific during the Tertiary as they are at the present time (Mayr 1954l). This conclusion has been substantiated later by studies in conjunction with the theory of plate tectonics. Without this assumption it would be "difficult to believe that such a migration as that of the Bristle-thighed Curlew could have evolved in the face of the tremendous dangers that the establishment of such a route of migration must have faced" (p. 392).

The surprising preponderance of males (70–80%) versus females in several species of the honeyeater genus *Myzomela* which Mayr had noticed in New Guinea and later in the museum collections directed his attention to the problem of varying sex ratios in birds (1939a). This paper brought together a large amount of widely scattered information most of which, however, has never been followed up. In two brief notes (1938e, 1938m) he tried to stimulate ornithologists to collect additional field data. A differential vulnerability of the sexes to certain dangers cause a deviation from the ideal 50:50 ratio in adults. A high mortality of females during the breeding season occurs probably in all those species in which the female carries the whole burden of incubation, particularly in ground nesting birds. Females outnumber males in polygynous species like the weaver birds (Ploceidae). Unequal sex ratios favoring the male or the female sex have been found to be correlated with peculiarities in the life histories of these birds; however, caution must be observed in the use of sex ratio data gathered by field observers.

In one of the few papers coauthored by Ernst and Gretel Mayr (1954g) they demonstrated that the smaller species of owls (wing less than 210 mm) usually molt their tail feathers simultaneous. The small species may require the tail less in flight than the larger ones. In these species (wing more than 230 mm) the tail molt proceeds usually from the outer rectrices inward.

The Sanford Hall of "The Biology of Birds," opened on May 25, 1948, was the first one at the AMNH dedicated to the biology of any group of animals. Mayr was chairman of the planning committee. The hall demonstrated the diversity of species including examples of extinct birds, the constructions of nests; it explained the principles underlying flight, peculiarities of feathers and body (airsacs) as well as migration and evolution (Mayr 1948b). The basic rule was that not a single exhibit should be placed in the hall that did not illustrate some biological principle or generalization. These concepts for the layout were derived from those developed by Bernhard Rensch in Berlin (p. 44). After the opening ceremonies of the Sanford Hall Mayr had serious physical problems like irregular heart beat, from which he suffered for more than 5 years. The cause may or may not have been the pressure to finish the job in time and to stay within tight budget limits see (p. 302).

Once again Mayr planned to write a comprehensive book, *Natural History of Birds* for which he signed a contract in July 1940.[9] The manuscript was to be submitted in 1942. However, this project was laid out too exhaustive and never completed because of other more pressing tasks. He confessed to his colleague Wilhelm Meise in Berlin on 4 January 1950, almost 10 years after signing the contract:

"Concerning my own book on the biology of birds, the trouble is that I always seem to have so many other jobs to perform. This spring I shall give courses at Columbia University and will have little time for my own research and writing. However, for your information, I will send you an outline of the book as far as I have planned it, also part of the more detailed outline concerning the egg, the care of the eggs, and the care of the young. None of this is, of course, final. It always

[9] At that time, L. C. Dunn (Columbia University, New York) finalized the program of the Jesup Lectures on "Systematics and the Origin of Species" (see p. 190–193) and Mayr wrote to him on June 21, 1940 in a vain attempt to postpone these lectures: "I have had an offer from Oxford University Press to prepare a textbook on ornithology. This means so much to me and to my future that I can ill afford to turn it down in favor of the systematics' project. Maybe I can accept your kind invitation to lecture at Columbia by changing the subject. I could possibly use the notes assembled for the preparation of my textbook on ornithology, to give a really serious course of ornithology. I might be ready for this in the winter of 1941 to 1942." However, preparations for the systematics lectures to be held in March 1941 were too far advanced for any changes to be considered.

happens that when one works on a book like this, one has to rearrange the material. I want to accomplish primarily two things: (1) a clear presentation of the more interesting facts concerning birds, and (2) a reference to the most recent literature throughout the world. Too many of the books on bird biology seem to be merely copies, slightly paraphrased, of some earlier book on the subject. The difficulty that I have found up to now is that it seems to be impossible to stay within the stated limits. There is such an infinite variety of information available on birds that one really doesn't know where to stop. Parts of my manuscript are already typed and I may send you some of the typescript for your information. I would prefer if we do not go beyond this at the present time because I really don't quite know yet what I can do during the next year or two. The contract which I have with the publisher, Oxford University Press, is very elastic and it would cause no difficulties to include a co-author. I am very much interested in the possibility of doing this together with you. I think the publisher would like to have the volume well illustrated but perhaps a little more in the American rather than in the German way. This means that they are anxious to have illustrations of high artistic value and popular appeal rather than of great scientific value."

Subsequently, Mayr turned all of his material for such a book over to W. Meise (then in Hamburg) who used some of it for his three-volume *Natural History of Birds* (1958–1966) published in coauthorship with Rudolf Berndt.

## Conservation Biology

Ernst Mayr lived in a conservation-minded environment all his life. His parents subscribed for him and his brothers to the educational natural history magazine, *Kosmos*, which emphasized conservation. In Saxony his mother subscribed to the journal of the Society for Heimatschutz, which of course also promoted the preservation of historical buildings, etc. His fatherly birding companion Rudolf Zimmermann was interested in conservation. It may well have been he who got Mayr excited about the rapid decline of the Great Bustard in northern Saxony. In 1923 at least 25 birds had been seen in the region of Grossenhain in the area suitable for that bird. In 1924 Mayr decided to make a careful census. By bicycle he crisscrossed the entire suitable area but was able to count only 11 birds. This was in early spring when the birds were very conspicuous on the fields. He wrote about his experiences (1924b), his fourth article on birds, and called attention to the plight of this species in Saxony due to the intensification of agriculture, the increase of human population and traffic as well as hunting pressure. But more than that he outlined a series of measures that would have to be taken to save the species in this region, e.g., hunting to be discontinued, destruction of nests and eggs to be heavily fined, and information for the local people through leaflets and lectures increased. The fact that he was able to advance a series of carefully argued measures indicates that his interest in conservation was more than casual. During his years as a student in Greifswald he was very much concerned about the preservation of the Rosenthal, a marvelous area of marshy meadows where dunlins, snipes, and lapwings were breeding.

In a conservation pamphlet Mayr (1937i) later spelled out the reasons why birds of prey should be protected ending his brief discussion as follows:

"It often seems to me that in the effort to persuade the public to give protection to the hawks too much emphasis is laid on the benefit of hawks to the farmer. To me, the true value of the hawks lies in their beauty and strength. There is little in nature that can compare with the forceful ease of the eagle's flight, the graceful courtship display of the harrier, or the incredible swiftness of the duck hawk. To lead others to appreciate the beauty of hawks is to establish the base from which to start the fight for hawk conservation."

At the American Museum and in New York in general there was strong conservation activity. Frank M. Chapman, the chairman of the Bird Department, was one of the founders of the National Audubon Society and founder of its journal, *Bird Lore*. R.C. Murphy, one of Mayr's colleagues (who was a close associate of Rachel Carson), was very active on Long Island in fighting the indiscriminate use of pesticides, particularly spraying from airplanes. He is often credited with being the person responsible for the termination of these practices on Long Island. In the Linnaean Society were several members who were active conservationists: Richard Pough, who later founded Nature Conservancy, J. Hickey, who led an active campaign against DDT by showing how it had led to a dramatic decline in the numbers of robins in the gardens and also such birds as Peregrine falcons and ospreys. Another member, Warren Eaton, was also very active in raptor protection. Bill Vogt, a very active conservationist, was one of Mayr's close friends in New York.

In 1945(l), Mayr discussed "Bird conservation problems in the Southwest Pacific" pointing out the dangers for island birds often endemic to one island or one island group only. The populations are not only small but mostly genetically very uniform and therefore not capable of adjusting to environmental changes or new diseases and parasites. Equally destructive is the introduction of dogs, cats, and rats. Particularly vulnerable are the faunas of the smaller islands, e.g., those of Micronesia, where many species of birds have been already or are on the verge of becoming extinct. Mayr ended this article with a plea to avoid needless destruction of trees, starting of fires, shooting or killing, introduction of animals, especially rats, and random scattering of DDT or other poisons. He underlined our duty to protect these precarious biota stating, e.g., "it is necessary to take steps for the protection of the threatened Polynesian island fauna in order to hand it down intact to posterity" (1933j: 323) or "We have an obligation to future generations to hand over these unique faunas and floras with a minimum of loss from generation to generation. What is once lost is lost forever because so much of the island biota is unique" (1967f: 374).

Frequently, Mayr was active behind the scenes. This includes financial contributions toward conservation programs and activities in the Hawk and Owl Society or, after moving to Cambridge, in the Massachusetts Audubon Society (MAS) where he was on the Board for many years. He supported William Drury (1921–1992) who was in charge of research at M.A.S. in Lincoln and continued to collaborate with him after Drury joined the faculty of the College of the Atlantic in Maine. Mayr was also the driving force in the publication of Drury's posthumous manuscript

*Chance and Change. Ecology for Conservationists* (1998) for which he wrote a Foreword (1998a) emphasizing that predictions in ecology and field natural history are probabilistic; naturalists are impressed by the uniqueness of everything. In a brief note on endangered coral reefs Mayr (1970d) pointed out the dangers of ecotourism, scuba diving and shell collecting for marine organisms. It is possible that the elimination of triton snails (*Triton*), the principal enemy of the crown-of-thorns starfish (*Acanthaster*), through collecting of triton snails contributed to the starfish's sudden population explosion and the destruction of many coral reefs, as this starfish feeds on coral polyps.

In general discussions, Mayr (e.g., 1984a) summarized the contributions of ornithologists to conservation biology. In many cases the prohibition of shooting birds was insufficient to halt the decline of certain rare species. In other cases supplementing the wild stock by captive-bred releases was the solution, as in the Hawaiian goose *Nene*. In many other cases it became evident that the study of the native habitat was a prerequisite for the adoption of sound conservation measures. Conservation research is an area in which ornithologists have been pioneers and Mayr was one of them.

As to the protection of subspecies he wrote: "The possibility that a subspecies carries ecologically relevant adaptations coupled with the potential to become a unique new species are compelling reasons for affording them protection against extinction. [...] The Hybrid Policy of the Endangered Species Act should discourage hybridization between species, but should not be applied to subspecies because the latter retain the potential to freely interbreed as part of ongoing natural processes" (O'Brien and Mayr 1991b).

Later in life Mayr donated his Japan Prize (1994) to the Nature Conservancy for the protection of a desert river in New Mexico and continued his annual contributions to a number of conservation organizations totaling, for instance, $6,550 in the year of 2001.

## Contact with Geneticists

Early in his work on South Sea island birds, Mayr encountered several cases of conspicuous geographic variation in sexual dimorphism and corresponded about hormonal and/or genetic control of bird plumages with Walter Landauer at the University of Connecticut at Storrs who worked on such problems with chickens. He introduced Mayr to L. C. Dunn, geneticist at the Department of Zoology of Columbia University (New York), in late 1931 or early 1932, and from then on Mayr participated at first occasionally and later regularly, in the genetics seminars. He attempted to include genetic research into his own work such as the article on the physiological-genetic determination of bird plumages (1933l), a presentation at the IOC in Rouen 1938 on sex ratio (1938n, 1939a), his lectures at Columbia University in early 1938 and a talk on speciation in birds at the AAA meeting in December 1939 discussing processes of isolation and divergence (Mayr 1940c). Hereafter L. C. Dunn invited him to give the Jesup lectures in March 1941, together with the botanist E. Anderson. Since 1943, Mayr attended regularly the summer meetings of geneticists at Cold Spring Harbor (p. 243).

Mayr and Th. Dobzhansky met for the first time in October–November 1936, when Dobzhansky visited New York to deliver a series of lectures at Columbia University[10]. Mayr invited him to the museum and showed him some beautiful examples of geographic differentiation and speciation in island birds he had studied (e.g., species of *Pachycephala, Monarcha, Rhipidura,* and others). When Dobzhansky came to New York again in late 1937, L.C. Dunn invited him, the Professor Schraders and the Mayrs for dinner. This indicates the fairly close ties Mayr had established with the zoology department of Columbia University. He described his personal and scientific relations with Th. Dobzhansky (1900–1975) as follows:

"For many years I had been unhappy about the neglect by the geneticists of the problems that confronted the taxonomists (see p. 45). Nowhere in the writings of the geneticists did I find any appropriate discussion of geographic variation, incipient species, and the completion of speciation. All they did, as I saw it, was to discuss what happened within a single gene pool. For this reason, I was quite excited when I read a paper by a person with the name of Dobzhansky who discussed geographic variation in ladybug beetles and the genetic basis of this variation. I was so enthusiastic that I did something I had never done before in my life, I sat down and wrote him a fan letter (p. 185). This was in 1935, and between that time and Dobzhansky's death in 1975 there has been continuous interaction among the two of us.

I became much better acquainted with him when, in 1936, Dobzhansky, who at that time was at Cal Tech in Pasadena, came to the East and worked at Cold Spring Harbor, visiting New York at regular intervals. At one or several of these visits he came to the American Museum of Natural History, where I demonstrated to him the marvelous geographic variation of South Sea island birds and the many cases of incipient species. This quite fascinated him, and I had the impression that it revived in him an interest in these kinds of questions, an interest that had been dormant while he was working on more or less physiological problems during the preceding 8 or 9 years at the Morgan Laboratory. Of course, I also attended his lectures at Columbia in 1936, and had then also occasion to talk with him. What Dobzhansky did for me primarily was to teach me the most modern evolutionary genetics. Even though by that time I had abandoned regular Lamarckism, that is, a belief in the inheritance of acquired characters, I still was, in a manner of speaking, fighting mutationism and held the widespread opinion, which of course Darwin also had had, that there are two sources of genetic variation, mutational and gradual ones. I think it was Dobzhansky who convinced me that by accepting very small mutations etc. one could bring both types of variation on a common genetic denominator.

[10] These lectures were not Jesup lectures and his text was originally meant to be a stand-alone text in evolutionary genetics. Several months later, in May 1937, L.C. Dunn back dated, naming Dobzhansky a Jesup lecturer and including his book manuscript as the first volume in the revived Columbia Biological Series (Cain 2002a).

My next contact with Dobzhansky was in 1939 when Gretel and I traveled by train from New York to Pasadena (p. 112). Very shortly afterwards, Dobzhansky, of course, assumed his professorship at Columbia University, and from then on for the next years until I moved to Harvard in 1953, we had very regular contact, sometimes almost daily. He very often invited Gretel and me for dinner, and these dinner parties were always memorable because there were always some foreign visitors also present, and the conversations were at the highest level, much superior to any conversations I had at that time with people in the ornithological circles. What we didn't like so much was that Dobzhansky absolutely played the 'pasha,' as Gretel and I used to call him. Like the Turkish Pasha he was the absolute king pin and his poor wife was only his number-one slave. After dinner she usually retired to the kitchen, not so much because this was necessary, but because she liked to have a quiet smoke and Dobzhansky absolutely objected to her smoking. She persuaded Gretel to come along, and when Dobzhansky smelled the smoke she always said, 'Oh that was Gretel who smoked, not me!'

I admired Dobzhansky enormously, even though I realized that some of his character traits were less than admirable. Dobzhansky clearly had charisma, something that is difficult to describe, but everybody knew it. I was not the only one to admire him, but many others, and even though he was my very best friend from about 1941 until long after I had left for Cambridge, he probably was considered their best friend by many others, such as Michael Lerner and Howard Levene, both of whom openly wept at the memorial service for Dobzhansky in 1975.

What I most admired in Dobzhansky was his 'Bildung,' his interest and reading in philosophy, literature, psychology, and many other fields usually totally ignored by the average biologist. I sometimes borrowed books from him which he had recommended to me, usually books outside of biology. One of his great interests was anthropology, and this later documented itself in his book *Mankind Evolving* (1962), for a long time perhaps the most useful book for a person interested in man but not wanting to read something that was too technical. Of course, what made him particularly attractive to me was that he had such a similar background, continental natural history. He had been in taxonomy before becoming a geneticist, and when I talked about my scientific problems he knew exactly what I was talking about. There was no other geneticist of whom this could have been said. Also, since he had been working with ladybug beetles which are often highly polymorphic, he realized that the usual morphological-typological species concept was not valid and that one should have a biological species concept. However, like Darwin many years earlier, he was not always consistent. I remember that I was quite shocked when he published a paper around 1937–1938 on the classification of some of his beetles in western North America and, as far as I was concerned, confused morphs (intraspecific variants) and different species.

Dobzhansky loved verbal arguments. He and I argued by the hour about all aspects of evolution. What he did not like, was any printed criticism by his friends. Carl Epling once made the mistake of doing that and this was the end of their friendship. I have the feeling that verbal argument for Dobzhansky was like playing a game, but printed criticism was like an insult. For Dobzhansky everything was either black or white. He was very positive about any and all opinions. His 'That is what I say' was a proverb at Columbia University. To be quite frank, I was sometimes quite upset by Dobzhansky's megalomania. When he didn't get his way in some controversy or administrative arrangement, he could become extremely difficult.

At that time there were distinctly two schools in population genetics; a reductionist one going back to R. A. Fisher, and a holistic one. Dobzhansky definitely belonged to the holistic one but was not nearly as concrete about it as either Michael Lerner or Bruce Wallace or, for that matter, as myself. In the 1940's and 1950's when I had my closest contact with genetics I benefited more from my conversations with Bruce Wallace than with Dobzhansky.

Dobzhansky loved to travel and he wrote the most wonderful letters to his friends which, eventually, Bentley Glass collected and published under the title *The Roving Naturalist* (1980; American Philosophical Society). In addition to traveling he was passionate about horseback riding and used every opportunity to do so.

In due time as he became more and more famous, Dobzhansky apparently was not too happy being just one professor in the Columbia Zoology Department. And this is why L. C. Dunn wanted to make a special genetics department for him. This caused great dissension at Columbia.

Dobzhansky was notorious for being an egotist. He avoided all social responsibilities and, for instance, never attended any committee meetings, not even the faculty meetings of his department. He was furious when they made decisions he didn't like, but nevertheless continued not attending. He never served as the secretary, treasurer, or editor of any society. The only office he was willing to accept was that of the president, and he loved to give a presidential address in this capacity.

Dobzhansky had only one child, his daughter Sophie. He rather definitely stated that more children would be a nuisance and would interfere with his work. Also, when the time came for Sophie to go to school and she and his wife Natasha wanted to move to one of the suburbs, Dobzhansky prevented it, so Sophie grew up so to speak on the pavement of Manhattan. She was rather resentful about it, as well as about being the only child and had herself, I believe, five children.

What was Dobzhansky's influence on me? Most importantly, perhaps, his sensible type of genetics reconciled me with genetics and geneticists whom I had been rather opposed to previously. Presumably, it was he who cured me of any last remnants of my Lamarckian past. On the other hand, I never followed him in his extreme adherence to Sewall Wright's ideas on neutrality. At a time when he still thought his chromosome arrangements were without selective significance I already was convinced they had and expressed this to him. Later on, when he and Epling published the work on the 'desert snow' (*Linanthus*), it soon became obvious to me that the distribution of white and blue color was not a strictly random matter, but to some extent at least, controlled by selection. Even Sewall Wright came around to this, but Dobzhansky only very slowly. He and I had quite a few arguments about human blood groups and I said all along they must have selective significance while Dobzhansky insisted that they were strictly neutral.

When in 1951 he published the third edition of his *Genetics and the Origin of Species*, he thought this was so to speak the capstone on population genetics. All sorts of open issues were rather concealed in this volume as if he were trying to sweep all difficulties under the rug. At that time he temporarily switched over to the study of human genetics and together with Dunn, even founded a human genetics institute at Columbia University, because he felt that population genetics was finished. He later returned to it in connection with his scientific feud with H. J. Muller as to the nature and amount of genetic variation in populations. I had the feeling that this was not a very promising issue, and, as a matter of fact, nothing really came of it and it prevented Dobzhansky from making any major contributions during the last 25 years of his life.

As a person, he was very important in my life, and as I said before, for many years I considered him my closest friend until the time when I felt I had more in common in my ideas and ideals with John A. Moore, then also at Columbia, later at Riverside [see p. 297–298]. I have considered the latter my best friend since the 1960's and that he still is in 1993. What united Dobzhansky and myself was not only our similar European scientific background, but also our being non-English scientific immigrants. In the time before and during the war there was a great Anglophilia in this country and anyone being a German, or worse, a Russian, was looked down on. One had to be at the receiving end of these evaluations to be able to feel them. Dobzhansky's reaction to all this was so strong in the last months of his life that he switched almost entirely to talking only with people who could speak Russian, particularly Michael Lerner."

## Contacts with German Ornithologists

As a high school student in 1922, Mayr became a member of the Saxony Ornithologists' Association in Dresden and, as a medical student in 1923, he joined the German Ornithological Society. Ever since that time he has been an ardent reader of the *Journal für Ornithologie*. His dissertation on the "Range expansion of the Serin finch" appeared in this journal in 1926 as well as several other studies during later decades. He was named Honorary Fellow of the German Ornithological Society after publishing his *List of New Guinea Birds* (1941f) at a rather inopportune time during World War II. Whenever feasible, he attended the meetings of German ornithological societies. In mid-September 1934 he lectured on "Ornithology in the USA" at the meeting of the Saxony Ornithologists' Association (where he also demonstrated examples of Albert Brand's birdsong records[11]) and, in the summer of 1953, he attended the meeting of the German society in Cologne reporting on the Zoological Congress in Copenhagen.

---

[11] Albert R. Brand (1889–1940), Research Associate at the AMNH 1933–1936, was a pioneer in recording the calls and songs of North American birds (see Auk 58: 444–448, 1941 and Living Bird 1:37–48, 1962); Fig. 3.9.

Throughout his career Mayr has been interested, and to some extent remained involved, in the affairs of German ornithology. In a letter to Stresemann dated May 25, 1934 he referred to minor competitive animosities in Berlin and Munich suggesting several remedies to settle these differences. When he visited southern Germany in 1938, he proposed a field station at Radolfzell (Mettnau) to study not only bird migration but also avian ecology and general biology (letter of July 9, 1938). He corresponded regularly with his fatherly friends R. Zimmermann, G. Schiermann and R. Heyder as well as Oscar Neumann, his co-student W. Meise and others.

World War II ended in Europe in May 1945. The postwar relief programs for relatives and fellow ornithologists in Germany occupied the Mayrs for several years. One program was organized in the Tenafly area where they lived. Families of German descent pooled their energies and resources, packing and mailing innumerable packages with clothing, shoes and food. The wives were the heroes at that time. Many families depleted their savings accounts. The second task was the great relief program of the American Ornithologists' Union. By late 1946 Mrs. F. Hamerstrom, Mrs. M.M. Nice, Joseph Hickey, Ernst Mayr and others had been very active arousing the interest of American ornithologists and, due to their energy, nearly fifty C.A.R.E. packages were sent to German ornithologists (CARE was the "Cooperative for American Remittances to Europe"). Gretel Mayr was active as a mediator and in translating handwritten letters. Until May 1947, more than 100 food packages and an equal number of clothing packages had been mailed. Mayr reported to Stresemann:

"The relief program of the American Ornithologists' Union is making good progress. Packages with clothes as well as food and C.A.R.E. packages are being sent every day" (31 March 1947) and "The enthusiasm of Mrs. Hamerstrom and the other members of the committee is the finest experience I have had in recent years. It really gives you hope for a better world" (15 April 1947).

When Rudolf Drost at Göttingen needed a dark suit as a lecturer, Mayr sent him his only dark suit. The American Relief Committee for German ornithologists continued to operate until about mid-1949. Over 3,000 packages had been sent by over 1,000 American donors to European ornithologists in 15 countries. The readiness of American people to help Europeans after the war was extraordinary. Some details of this relief work and a formal "Thank you" are published in the *Journal für Ornithologie* 133: 455–456, 1992.

Because of the complex political situation in Berlin after the end of World War II (where the German Ornithological Society, DOG, was legally registered), a group of its representatives founded another society, the Society of German Ornithologists (Deutsche Ornithologen-Gesellschaft, DO-G), in West Germany in late 1949. This DO-G carried on the tradition of the old (dormant) DOG until the latter ceased to exist, when it was fused legally with the DO-G in 2006.

When after a pause of 6 years the *Journal für Ornithologie* had started to appear again in 1951, Mayr suggested to E. Stresemann "that the German Ornithological Society should make more of an effort to get foreign subscribers" (20 March 1952) and included a rough draft of a letter to be sent to 30–50 North American ornithologists. During his regular visits to Germany he met, besides Stresemann, with many colleagues, especially Gustav Kramer (1910–1959) and Konrad Lorenz (1903–1989) who also visited him in the United States, with Fritz Frank (1914–1988), Günther Niethammer (1908–1974) and Klaus Immelmann (1935–1987). Immelmann had participated in an expedition with Mayr to the interior of Australia which Dom Serventy had organized in 1959 and Immelmann's interests in ethology were close to Mayr's. When the DO-G held its 100th annual meeting in Bonn in 1988, Ernst Mayr attended as the keynote speaker.

Mayr also established close ties with the German Zoological Society, the German Society for the Theory and History of Biology, and the German Society for Biological Systematics all of which elected him as an Honorary Member.

The July issue of the *Journal of Ornithology* for 2004 was dedicated to Ernst Mayr on the occasion of his 100th birthday which he enjoyed greatly: "What is particularly impressive is the range of interests of the contributors and the large number of new findings. German ornithology is obviously very much alive [...] A good omen for the future!"

## International Ornithological Congresses

As mentioned in previous chapters, Ernst Mayr attended, when possible, the meetings of local and national ornithological societies in Germany and in the United States and frequently gave lectures at these meetings. He was also a regular attendant of the International Ornithological Congresses (IOC) which take place every 4 years: Amsterdam (1930), Oxford (1934), and Rouen (1938; lecture on the sex ratio in birds). Between these congresses he carried on an active correspondence, e.g., with E. Stresemann and other leading ornithologists, about presidents to be nominated, about the place and organizational details of future congresses. Because of legal problems with his passport (p. 251), he did not attend the 10th IOC in Uppsala, Sweden (1950), but R. C. Murphy read his progress report on "Speciation in birds" (Mayr 1951l). At the 11th IOC in Basle (Basel, 1954) Mayr lectured on the bird fauna of the table mountains of southern Venezuela (with W. H. Phelps, Jr., 1955f) and in Helsinki (1958) he organized a symposium on adaptive evolution (1960e). In Ithaca (1962) he was the President of the XIIIth Congress and spoke on "The role of ornithological research in biology" (1963r). From that year on he was a permanent member of the International Ornithological Committee. He attended the IOCs in Oxford (1966), where six presidents posed for a photograph (Fig. 3.12) and Mayr received an honorary PhD from Oxford University, The Hague (1970; chairman of the symposium on "Causal zoogeography") and Canberra, Australia (1974; chairman of the symposium on "The value of various taxonomic characters in avian classifications"). At the 17th IOC in Berlin (1978) he gave the Stresemann memorial lecture on "Problems of the classification of birds" and at the 19th IOC in Ottawa (1986) he reported on "The contributions of birds to evolutionary theory." In view of his advanced age he did not attend the congresses in Moscow (1982) and in New Zealand (1990) but did travel to Vienna in 1994 (21st IOC), when

**Fig. 3.12.** Presidents of International Ornithological Congresses at the XIVth I.O.C. in Oxford, United Kingdom (1966). From *left* to *right*: David Lack (1966), Sir Landsborough Thomson (1954), Alexander Wetmore (1950), Jean Berlioz (1958), Ernst Mayr (1962), and Erwin Stresemann (1934) (reproduced from the Proceedings of the XIVth I.O.C.)

the British Ornithologists' Union awarded him the Salvin-Godman medal and the University of Vienna a honorary PhD degree. In later years he always made an effort to attend the annual meetings of the American Philosophical Society which take place during three days in spring.

# 4 Ornithologist and Zoogeographer

## Birds of Oceania

At the AMNH, Mayr studied primarily the bird fauna of Oceania (Fig. 4.1), the island area bordered by and including New Guinea, Palau and Marianas Islands on the west and the Tuamotus and Marquesas Islands in the east. Based on anthropological research, Oceania is usually subdivided geographically into three regions: Melanesia (New Guinea eastward to Fiji and New Caledonia), Micronesia (Marianas, Caroline, Marshall and Gilbert Islands), and Polynesia (a triangle of many islands, including Hawaii, New Zealand and Easter Island). Mayr's main interest was in the problems of geographic speciation, but he had never before encountered material documenting this process quite so graphically as these island birds. There was no widespread species that did not contain clear-cut cases of geographic speciation. Mayr followed a three-pronged research program in systematic and regional ornithology mainly founded on the collections of the Whitney South Sea Expedition:

(1) Revision and monographic treatment of the birds of Polynesia and Micronesia;
(2) Study and revision of the birds of Melanesia in preparation of a book on the birds of the Solomon Islands;
(3) Study and revision of all species and genera of New Guinea birds and preparation of a book on the avifauna of this island.

He was eminently successful in carrying out these programs (Figs. 4.2, 4.3). A continuous stream of research articles appeared from 1931 onward punctuated by the publication of book-length contributions like the "Birds of the 1933–1934 Archbold Papuan Expedition" (1937c, with A. L. Rand; 248 p.), his *List of New Guinea Birds* (1941f, 260 p.), the field guides, *Birds of the Southwest Pacific* (1945n, 316 p.), *Birds of the Philippines* (1946k, with J. Delacour, 309 p.) and the handbook, *Birds of Northern Melanesia. Speciation, Ecology, and Biogeography* (2001g, with J. Diamond as coauthor, 492 pp). The total of his taxonomic and regional ornithological work comprises about 3,500 printed pages (which includes publications on Australian and southeast Asian birds to be mentioned later). Of this total he wrote about half as a single author and the other half with one or more coauthors. Among these were A. Rand, S. Camras, D. Serventy, and R. Meyer de Schauensee during the 1930s, D. Ripley, D. Amadon, J. K. Stanford, M. Moynihan, J. Bond, C. Vaurie and J. Delacour during the 1940s, and T. Gilliard as well as a few others during the 1950s.

From the start, this taxonomic and regional work, however, was not an end in itself for Mayr but a means to go beyond it. Regarding his scientific perspectives he wrote to E. Stresemann on 24 March 1934 (transl.):

"Systematics: I also strive beyond it and often pose the question: Cui bono? or in American: 'What of it?' On the other hand, I consider systematic studies an excellent training and I am paid to do such research. I do not know how I could justify to occupy myself with other subject matter as long as 40–60 undescribed new forms collected by the Whitney South Sea Expedition lie in the drawers. I do not want to give careless descriptions; every careless work I did so far, I regretted bitterly afterwards. Either-or!

The list of references I sent you recently will have demonstrated that I remain in very close touch with the progress of biology. There are many good and capable biologists and physiologists, but only very few really great systematists. Germany has been leading in this field, at least lately (Hartert–(Kleinschmidt)–Hellmayr–Stresemann), should this tradition be discontinued completely? [...] Therefore my proposal to train a truly significant systematist! I shall also see to it that I can train here a 'successor.' For as soon as I shall have completed three additional tasks, I'll also withdraw from the field of 'taxonomy.'[1] These three tasks are: Monographic treatment of all interesting Polynesian genera, the New Guinea list, and a book on the birds of the Solomon Islands. *Afterwards I shall restrict myself to more general problems of 'taxonomy' like 'geographical variation of morphological characteristics'* etc. [emphasis added]. Not much news is to be expected in the field of species description anyway (please keep this as a secret!!). I just unpacked the Archbold collection [from New Guinea] that contains nice things such as *Daphoenositta*, but practically nothing new. I hope to find more in Coultas' collection from the Admiralty Islands."

Since the mid-1930s Mayr increasingly took into consideration general systematic, evolutionary, and genetic aspects, particularly through his contact with Th. Dobzhansky (see p. 133ff.). Certain museum tasks such as the organization and integration of the Rothschild Collection, his increasing involvement in evolutionary studies and activities as a lecturer at various universities interfered only to some extent with his taxonomic work. However, the publication of the book on the birds of the Solomon Islands was long delayed until it eventually appeared in coauthorship with Jared Diamond (2001g). The work on the above research programs will be presented in some detail in the following sections. The results formed the intellectual database for Mayr's later work in the field of evolutionary biology.

---

[1] Stresemann had written to him on 25 February 1934: "I begin to think that my occupation with collection-based taxonomy has been for me only a transitonal stage which, inside, I have overcome completely, but outside not quite yet. It is physiology with its wide perspectives which now attracts me mightily (although perhaps too late)." Similarly already five years earlier: "Actually my real interests have shifted to very different fields [than systematics, zoogeography, and speciation], in particular functional anatomy and physiology" (letter to Mayr dated 14 October 1929).

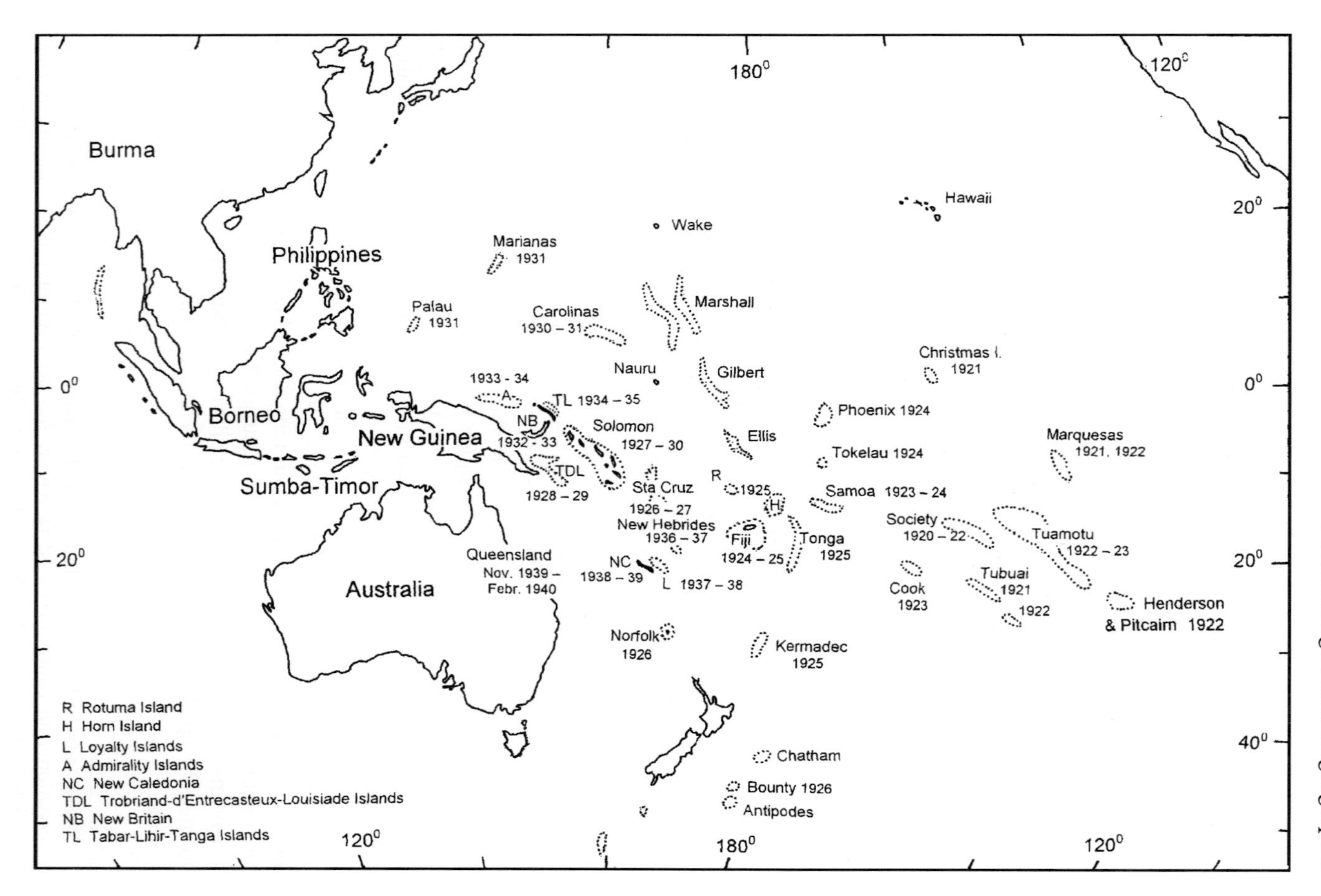

180°
120°
20°
0°
20°
40°
Burma
Hawaii
Wake
Philippines
Marianas 1931
Marshall
Palau 1931
Carolinas 1930 – 31
Christmas I. 1921
Nauru
Gilbert
1933 - 34
A
TL 1934 – 35
Phoenix 1924
Borneo
NB
Solomon 1927 – 30
Ellis
Marquesas 1921, 1922
New Guinea
1932 - 33
Tokelau 1924
TDL
R
Sumba-Timor
Sta Cruz 1926 – 27
1925
Samoa 1923 – 24
1928 – 29
H
New Hebrides 1936 – 37
Society 1920 – 22
Tuamotu 1922 – 23
Fiji
Tonga 1925
Queensland Nov. 1939 – Febr. 1940
NC 1938 – 39
1924 – 25
L 1937 – 38
Tubuai 1921
Australia
Cook 1923
1922
Henderson & Pitcairn 1922
Norfolk 1926
Kermadec 1925
R Rotuma Island
H Horn Island
L Loyalty Islands
A Admirality Islands
NC New Caledonia
TDL Trobriand-d'Entrecasteux-Louisiade Islands
NB New Britain
TL Tabar-Lihir-Tanga Islands
Chatham
Bounty 1926
Antipodes
120°
180°
120°

**Fig. 4.1.** Main geographical regions of Ernst Mayr's taxonomic and zoogeographical studies on the birds of Oceania, Australia, the Malay Archipelago, and southeastern Asia. The Pacific islands are schematically outlined and their names and years of visits by the Whitney South Sea Expedition indicated. The expedition ended with collecting activities in Queensland, Australia during World War II

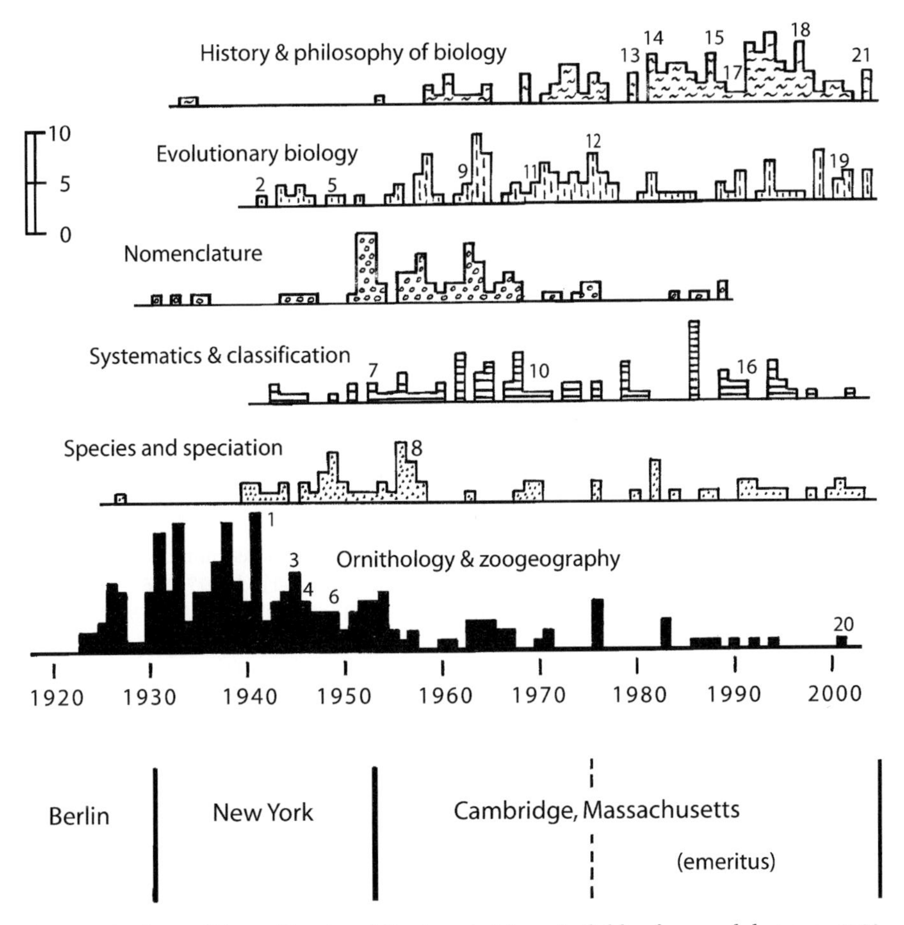


**Fig. 4.2.** Numbers of Ernst Mayr's publications in his main fields of research between 1923 and 2004. Figures refer to his books as listed on page 454. Not included are publications in various miscellaneous fields, his annual reports as Director of the Museum of Comparative Zoology, Harvard University (1961–1970) and over 120 book reviews. The drop in the number of ornithological publications during 1928–1929 and 1934–1936 was caused by Mayr's expedition to New Guinea and the incorporation of the Rothschild collection into that of the AMNH, respectively

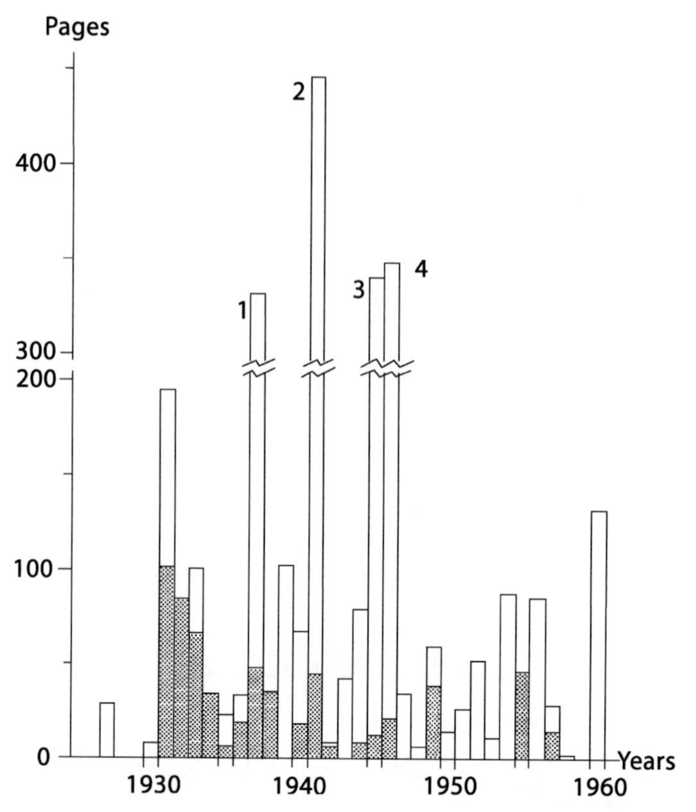


**Fig. 4.3.** Total of Ernst Mayr's taxonomic and regional publications (number of pages per year), mostly on the birds of Papua-Australasia, during the period 1927–1960. The tallest peaks include books as follows: (1) E. Mayr and A. Rand (1937) *Birds of the Archbold Expedition* (248 p.), (2) E. Mayr (1941) *List of New Guinea Birds* (260 p.), (3) E. Mayr (1945) *Birds of the Southwest Pacific* (316 p.), (4) J. Delacour and E. Mayr (1946) *Birds of the Philippines* (309 p.). Shaded–papers in the series "Birds collected during the Whitney South Sea Expedition."

## Pacific Islands

### The Whitney South Sea Expedition

Ernst Mayr's primary task, when he arrived at the AMNH in January 1931, was to study the collections of birds sent in by the Whitney South Sea Expedition (WSSE) which, at that time, continued field work in the southern Pacific Ocean and eventually developed into the longest, geographically most extensive and scientifically one of the most important ornithological expedition ever undertaken. Travelers were continuously in the field from 1920 until 1940 and R. C. Murphy at the AMNH functioned as general manager, but he never joined them anywhere during this entire period. He once got as far as California on his way to join the WSSE, but was summoned back by Dr. Sanford. Although several status reports have been published while the expedition was underway (Murphy 1922, 1924, 1938; Chapman 1935), no general overview as to where and when birds were collected over the 20-year period ever appeared after finishing in Australia in the middle of World War II (but see the recent brief account by LeCroy 2005). The "Introductions" to the two bound volumes of individual papers on the "Birds collected during the WSSE" (No. 1–25 and No. 26–50) prepared and published by Mayr in 1933(n) and 1942(f), respectively, include brief comments on the expedition and provide very useful lists of the individual papers, of bird species and problems discussed. Mayr (1942e: 12) summarized the accomplishments of the expedition in a few sentences and stated that a considerable part of his book was based on the magnificent material at hand. Rich collections were also made of mammals, reptiles, insects, mollusks and other groups. In 1969, Edwin H. Bryan, Jr., entomologist and curator of the Bishop Museum, Hawaii, from 1919 to 1968 and a member of the expedition in 1924, prepared an unpublished chronological summary of and guide to the manuscript journals and records of the WSSE until 1935 (on file in the Department of Ornithology, AMNH). However, the last years (1936–1940), when Lindsay Macmillan collected partly with the aid of Whitney funds are not included in Bryan's report. Macmillan's collecting stopped when he went into military service during World War II.

Plans for the WSSE probably originated through the friendly rivalry between Thomas Barbour and Leonard Sanford, both of whom wanted to help their respective museums, the Museum of Comparative Zoology and the AMNH, to be the richest in the world in rare species of birds (p. 100). The islands in Polynesia and Micronesia were famous for such species and Sanford apparently got the idea to launch an expedition funded by the Harry Payne Whitney family that would visit all the islands in the Pacific apt to have endemic species.

The first leader was Rollo H. Beck (1870–1950) who started field work in the Society Islands in 1920 and, over a year later, purchased a seaworthy two-masted schooner, the "France," to be able to visit small islands independently. This vessel was operated until July 1932, when it was sold. Beck was the most successful collector of seabirds in the world. He had participated in expeditions to the Galapagos Islands for Lord Rothschild in 1897 and 1901 and had led the Brewster-Sanford Expedition to the coastal waters of South America (1912–1917). Soon afterwards he began to prepare for the WSSE. He resigned from this task in 1928. From 1930 until his death he grew apricots, figs, and almonds in California (Pitelka 1986; Mearns and Mearns 1998). Besides several other persons, his longtime assistants were E. H. Quayle and José G. Correia. After Beck had retired, Hannibal Hamlin was in charge (1928–January 1930) followed by William F. Coultas (January 1930–1935) and L. Macmillan (1936–1940).

The WSSE explored most island groups in the Pacific Ocean (Fig. 4.1), all together more than 3,000 islands. Owing to the rapid changes in the South Sea region through habitat destruction, this was the last chance to make representative collections on the South Sea Islands. The typed and bound journals and letters of expedition members total 31 volumes. Two additional ones by Charles Richmond (Washington) contain geographical notes on most island groups in the Pacific prepared at the request of Dr. Sanford since mid-1918 and used during the detailed planning of the various stages of the expedition. This material is kept in the Department of Ornithology (AMNH). During World War II some of these volumes describing work of the expedition in certain islands of the southwest Pacific Ocean became important source material for US governmental agencies.

Starting in the east, birds of the Society, Christmas, Tuamotu and Marquesas Islands were collected in 1920–23, followed by those of the Cook, Samoa, Phoenix, Fiji, and Tonga Islands in 1923–25. The expedition sampled the avifauna of the New Hebrides (Vanuatu) and Santa Cruz Islands in 1926–1927 prior to collecting on some of the Solomon Islands and the islands off the southeastern tip of New Guinea during the period 1927–30. It then continued to the Carolinas, Marianas and Palau Islands in late 1930 and 1931, and, after the "France" had been sold, worked on New Britain in 1932–33 as well as on the Admiralty Islands and on the small islands off New Ireland in 1934–35. Additional work was done by L. Macmillan in the New Hebrides (Vanuatu, 1936–37), on the Loyalty Islands and on New Caledonia (1938–39). The expedition ended in mid-1940 in Australia where Macmillan collected in several inland regions of Queensland (east of Windorah, southeast of Boulia, eastern edge of Simpson desert, Birdsville area, and Dalby region) to provide comparative material for the Mathews type collection in New York.[2]

The only island groups not visited by the WSSE were the Hawaii Islands in the north and New Zealand in the south (both already well sampled in previous years) and the Ellice (Tuvalu), Gilbert (Kiribati) and Marshall Islands in the central tropical Pacific. No endemics were known from these latter island groups and none would be expected from there. These are low lying coral atolls and certainly have been flooded completely during high Pleistocene sea-level stands. No new subspecies have been described from there in the 70 years since the WSSE bypassed these islands. Ignoring them was deliberate.

The Whitney Hall of Oceanic Birds at the AMNH was designed with 18 habitat groups representing a range of Pacific island types from low lying atolls to high volcanic islands. To obtain the necessary photographs and other information for these habitat groups, three additional expeditions were undertaken by museum staff as guests on private yachts to particular Pacific islands in 1934, 1936, and 1940.

The results of taxonomic studies were published in the series "Birds collected by the Whitney South Sea Expedition" (American Museum Novitates) between

---

[2] The large private collection of G. Mathews was bought by W. Rothschild who had apparently promised Mathews that it would go to the British Museum of Natural History (London). Mathews was upset when his collection was sold to the AMNH along with the rest of the Rothschild Collection in 1932. One of the things Mayr had planned to do was to finish a list of Mathews' types, a task that Hartert had begun in Tring. However, Mayr was unable to complete this type list, but he left at the AMNH his card file of Mathews' taxonomic names which is still very useful today (M. LeCroy, pers. comm.).

1924 and 1957 by 14 authors and coauthors totaling 64 reports, 977 printed pages. More than half of this total (61.5%) was written by Ernst Mayr, curator of the Whitney and Rothschild Collections since 1932 and the first ornithologist able to devote full time to the study of these collections. 28 new species and 247 new subspecies have been described until 1953 (Murphy, unpubl. manuscript, 1953) plus nine subspecies described by Mayr (1955a, 1957b). Comparative material for Mayr's taxonomic studies of the Whitney birds was contained in the Mathews Collection of Australian birds at the AMNH and in the Rothschild Collection, particularly the rich material from New Guinea, the Bismarcks, and the Solomon Islands.

## Significance of the Whitney South Sea Expedition

Islands and island faunas represent thousands of experiments by nature of assembling animal communities (Mayr 1967f). There are many groups of fairly similar islands in the tropical oceans which differ in area, isolation and elevation. Islands have contributed decisively to the understanding of evolution and population biology. Following research on the Canary Islands in 1815, Leopold von Buch was the first to propose the theory of geographic speciation. The Galapagos Archipelago opened Charles Darwin's eyes to evolution and natural selection. The Malay Archipelago gave Alfred R. Wallace his insights into biogeography. The islands of the southwest Pacific and New Guinea taught Ernst Mayr geographic variation and speciation of animals. In recent decades islands stimulated biologists of the next generation to study ecology, behavior and conservation biology (J. Diamond, R. MacArthur, S. Olson, E. Wilson).

The most complete exploration of the Pacific islands by the WSSE provided the database for all subsequent studies of birds inhabiting the world's most extensive set of islands. This is the unique importance of this expedition. In particular it provided the basis for Ernst Mayr's analysis of variation and speciation in birds (Mayr 1942e, Mayr and Diamond 2001g, Schodde 2005). Among the phenomena discussed are many cases of simple and more complex geographical variation, primary and secondary intergradation of populations. Numerous well differentiated peripheral island forms represent borderline cases between subspecies and species and between species and monotypic genera demonstrating the continuity of the process of geographic speciation over time. Mayr realized that geographic variation provides the two components of speciation-divergence and discontinuity. Sympatry of distinct taxa indicates their status as species. Moreover, out of the biology and systematics of Polynesian birds Mayr (1940i) developed the basic tenets of "island biogeography" (p. 163). His detailed revisions of all subspecies and species of birds in Oceania also provided the basis for later conservation work in this region, particularly island endemics (Schodde 2005).

Not a single mountain bird was known from the Solomon Islands when the Whitney Expedition arrived there in 1927. No less than 17 mountain species, 11 of them new to science, were discovered. Of the 71 new taxa from northern Melanesia since 1940, most have been described from specimens collected earlier, only 23 from newly discovered populations (Mayr and Diamond 2001g: 34).

### Studies of the Whitney Collections

In January 1931, Ernst Mayr plunged into his work on the Whitney collections which had arrived in New York since 1921 but not studied in detail (including his own collections from the Solomon Islands). He had about ten papers published or in press by the end of that year.

"I work here really like a madman. The revisions of *Pachycephala* are in press; currently I study the Polynesian flycatchers. As soon as the revisions of the genera are finished, I shall prepare books on (1) Polynesia, (2) Solomon Islands and Bismarck Group and (3) New Guinea" (to W. Meise on 11 April 1932; transl.) and "I sit on two undescribed genera and many new species, among them an eagle. I would never have dreamed of something like that; I no longer describe 'dirty' subspecies unless for zoogeographical reasons" (to W. Meise, October 1931; transl.).

Mayr took notes on plumage colors and measured thousands of bird specimens (wing, tail, bill, tarsus) determining the individual variation of local populations and the geographical variation of subspecies and species. "He worked as a field naturalist but to laboratory standards of inference and proof–the essence of border practice" (Kohler 2002: 265). In some of his articles Mayr treated the avifauna of particularly interesting islands, in others he revised differentiated genera or species groups monographically. In the first paper (1931b) which appeared only two months after he had started to work at the AMNH, he introduced into the international literature the important concept of "superspecies" as an equivalent of the term "Artenkreis" coined by Rensch (1928, 1929). Mayr had read Rensch's book, *The Principle of Polytypic Species and the Problem of Speciation* (1929) upon his return from the Solomon Islands in 1930 and admired it greatly. He defined superspecies as "a systematic unit containing geographically representative species that have developed characters too distinct to permit the birds to be regarded as subspecies of one species" (1931b) explaining:

"I regard superspecies as a convenient compromise between diverse schools of ornithologists, the extremely modern ones, on one side, who want to put together as a species all geographical representatives, disregarding the most striking morphological differences, and the conservatives, on the other side, who demand perfect intergradation as a criterion" (p. 2).

This paper (1931b) is devoted to a study of the widespread Collared Kingfisher (*Halcyon chloris*). Mayr described several new subspecies from the New Hebrides, Santa Cruz Islands, and Rennell Island noticing the conspicuous differentiation of peripheral isolates: "The birds of the outposts, especially at the periphery of distribution, show advanced characters that lead to pronounced differences in appearance" (p. 1). "The sympatric occurrence of two taxa on the same island proves that they cannot belong to one species" (p. 3).

His second article (1931d) dealt with the isolated Rennell Island, a raised atoll 90 miles southwest of the Solomon Islands. Here "the bird fauna [...] has turned out to be one of the most interesting in the whole South Sea," because of its extraordinary endemism: 20 of about 33 native species were endemic species or subspecies. He described no less than three new species (one of them later reduced to subspecies rank) and 14 new subspecies. As an oceanic island, Rennell has received all of its fauna by dispersal across water barriers, from the Santa Cruz Islands and the New Hebrides in the east (30% of the native birds; 56% of the prevailing winds come from the SE or E), from the Solomon Islands in the north (48% of the native birds; 16% of the winds come from NW, N or NE), and from New Guinea and Australia in the west and southwest (5% of the winds come from the W or SW). The very frequent winds from the east compensate somewhat the greater distance of the eastern source region. Studies of particularly variable species like, e.g., the Golden Whistler (*Pachycephala pectoralis*; 1932c,d) and the Scarlet Robin (*Petroica multicolor*; 1934b), led to discussions of the genetic or hormonal control of bird plumages (1933l).

In monographic studies of particular species and species groups Mayr analyzed geographical variation on a regional scale, as illustrated by, e.g., the thrushes of the *Turdus javanicus–poliocephalus* group (Fig. 4.4). From Solomon Islands and New Guinea to the west they are high-mountain birds, usually not found below 2,000 m elevation, but some of the populations in southern Melanesia and in Polynesia occur in the lowlands or even on small coral islands. Subspecies from neighboring islands are sometimes more distinct (e.g., in the Fiji Islands) than those from opposite ends of the range. Other examples Mayr discussed include species of fruit doves, trillers, and fantails.

When the southwestern Pacific Ocean became a Theater of Operations during World War II, the demand for a field handbook or field guide to the birds of the islands was great. Ernst Mayr, the expert of this region, produced such a book in a relatively short time. It appeared in 1945(n) under the title *Birds of the Southwest Pacific*. The book covered the birds of Samoa, Fiji, New Caledonia, New Hebrides (Vanuatu), Banks and Santa Cruz Islands, Solomon, Marshall, Caroline, Mariana and Palau Islands, 803 forms in total (388 species and 415 subspecies). The layout is a happy compromise between a strictly systematic and a geographical treatment. Wide-ranging species like sea birds and shore birds, are presented in the first chapters. The land and fresh-water birds are grouped systematically in one section and geographically by seven island groups in Part II of the book. This organization permitted a separate treatment of all the endemic species of seven different island groups and yet avoided space-consuming repetition of the more wide-spread species. Mayr summarized the then available knowledge of these island birds and emphasized the many gaps particularly regarding life history, again and again stating "habits unknown" to encourage the observer in the field to gather more information. Lists of questions relating to the behavior and ecology of birds are included in the introduction. The book comprises this whole avian fauna in a manner which satisfied both the amateur observer in the field and the specialist in the museum. It also contains much unpublished data from personal research.

Three color plates by Francis Lee Jaques and numerous excellent line drawings by Alexander Seidel illustrate 55 species of birds[3]. A second edition appeared in 1949. When the copyright had expired, two different reprints were published immediately; one in Hawaii and the other in England (Wheldon and Wesley Ltd., 1968); still another one came out in the States in 1978. Ernst Mayr was delighted that his book got such wide distribution. Since taxonomically nothing of importance has happened in the area, *Birds of the Southwest Pacific* is still reasonably up-to-date. Of course, there are today several more recent field guides available, like *South Pacific Birds* (duPont 1976), *Birds of Hawaii and the Tropical Pacific* (Pratt et al. 1987), *Birds of the Solomons, Vanuatu and New Caledonia* (Doughty et al. 1999) and *A Guide to the Birds of Fiji and western Polynesia* (Watling 2001) with excellent color illustrations of all species and many subspecies of the birds. Rollin Baker (1951) published a comprehensive account of the avifauna of Micronesia based on collections assembled during World War II and comparisons with Whitney material at the AMNH.

Mayr was the bird specialist for the whole wide region from the Solomon Islands to Polynesia. Being *the* authority for a vast area gave Mayr some satisfaction, but he always regretted that he did not have competitors (pers. comm.). In many areas of the world various specialists usually have a great time disagreeing with each other. Here in Oceania, what Mayr said was correct. Only in the Papuan Region did he encounter occasional dissent, e.g., when he overlooked two sibling species, *Pachycephala melanura* and *Meliphaga orientalis.*

Several young ornithologists soon started to work with the Whitney material independently but under Mayr's supervision. These were Dean Amadon who published three papers in the Whitney series entitled "Notes on some non-passerine genera" (1942a, 1942b, 1943), Dillon S. Ripley and Hugh Birckhead who studied "The fruit pigeons of the *Ptilinopus purpuratus* group" (1942). Amadon stayed at the AMNH, later becoming the Chairman of the Bird Department; Ripley became

**Fig. 4.4.** Distribution and character variation of the Island Thrush (*Turdus poliocephalus* group). Map from Mayr (1942e: 58), illustrations of birds from MacKinnon and Phillipps (1993, Indonesia), Doughty et al. (1999, Solomons and Vanuatu) and Pratt et al. (1987; Fiji and Samoa)

[3] Whereas F. L. Jaques (1887–1969) was a well-known bird artist employed by the AMNH, A. Seidel remains unknown in ornithological circles. He was born in Germany where he studied art first in Munich and then, for five years, in Rome. His early work included murals for an industrial concern in Rüdersdorf and settings and costumes for a theater in Berlin. He came to the United States as a tourist in 1939 and decided to stay, when the war broke out. For some time he was a houseguest with Ernst and Gretel Mayr and from 1943 to 1961 a staff artist for the AMNH where he illustrated many ornithological books and scientific papers and painted murals of extinct birds, saurians, and primates. He has provided illustrations for Collier's Encyclopedia and the Encyclopedia Americana and has published two books for young people, about wild birds and water mammals (biography prepared by Steuben Glass, AMNH, March 1963).

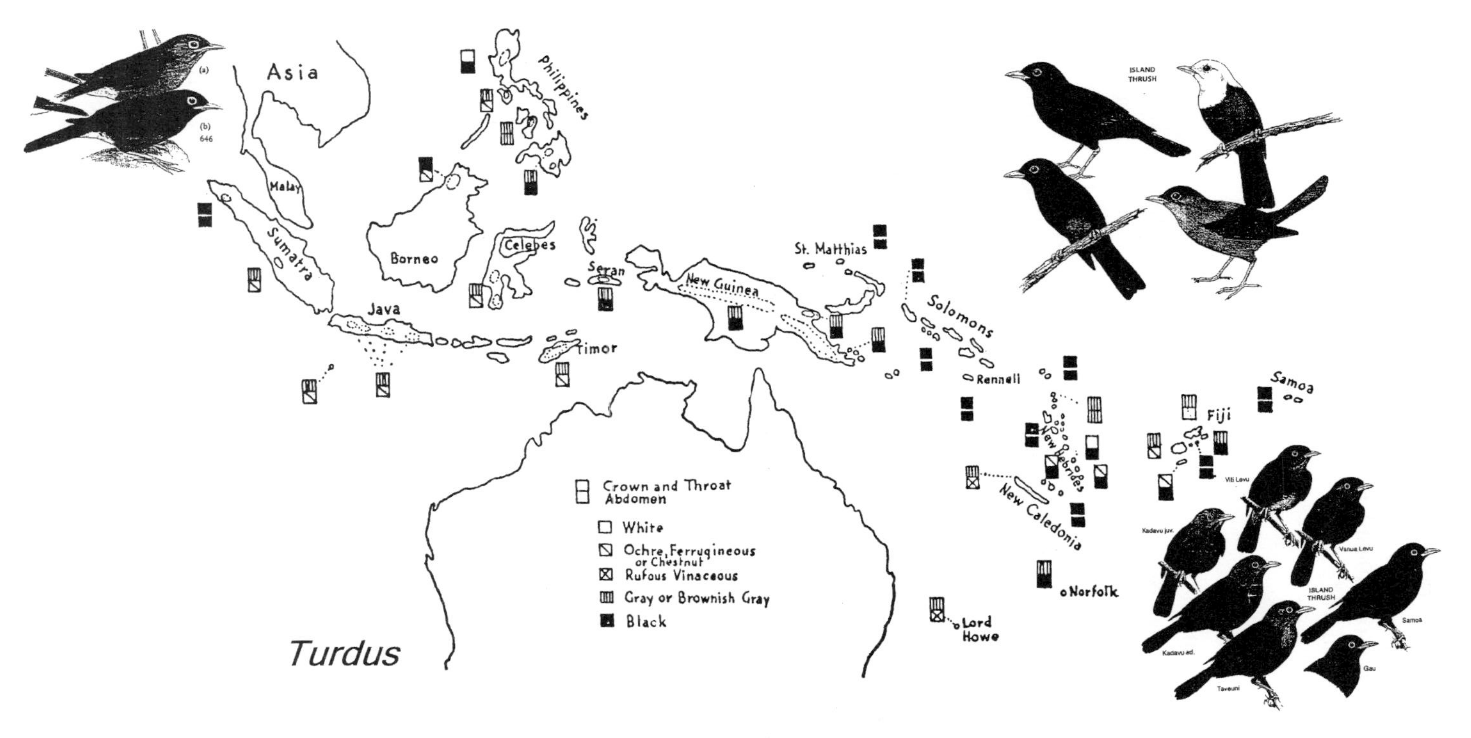
(a)
(b)
646
Asia
Malay
Sumatra
Java
Borneo
Philippines
Celebes
Seran
Timor
New Guinea
St. Matthias
Solomons
Rennell
New Hebrides
New Caledonia
Norfolk
Lord Howe
Fiji
Samoa
ISLAND THRUSH
ISLAND THRUSH
Viti Levu
Vanua Levu
Kadavu juv.
Kadavu ad.
Taveuni
Samoa
Gau
Crown and Throat
Abdomen
White
Ochre, Ferruginous or Chestnut
Rufous Vinaceous
Gray or Brownish Gray
Black
Turdus

an ornithologist at the Peabody Museum at Yale University and later the secretary (director) of the Smithsonian Institution; H. Birckhead was killed in action during World War II (Mayr 1945b); training of several volunteers in the Bird Department is mentioned above (p. 120).

**New Guinea**

During the 1930s and 1940s Ernst Mayr was also very much engaged to his first "love affair," the birds of New Guinea, with the aim of publishing a taxonomic overview of the entire avifauna of this large island. The last comprehensive account on New Guinea birds, T. Salvadori's *Ornitología della Papuasia* (1880–1882), as well as some reports on British and Dutch expeditions, were hopelessly out of date, the compilation of G. Mathews (1927, 1930) superficial and useless. The only modern paper on New Guinea birds using trinominals was Stresemann's (1923) on the birds of the German Sepik expedition.

The acquisition by the AMNH of the Rothschild Collection (in 1932), with its extensive series of New Guinea birds, much facilitated Mayr's work from 1935 onward (p. 116). In addition, he visited the museums in London, Paris, Leiden, Hamburg, Berlin, Munich, Dresden, Frankfurt, Stuttgart and Basel in 1930, 1932, 1934, and 1938 to examine doubtful type specimens. The types of the museum in Genoa were sent to Dresden. All this enabled him to complete the catalogue giving full synonymy lists and accurate descriptions of the range of each species and subspecies. About thirty specialized papers document the results of his taxonomic studies. He reviewed the subspecies of the Victoria Crowned Pigeon *Goura victoria* (with Berlioz, 1933m). From the collections of the 1933–34 expedition to southeastern New Guinea (Mafulu, Mt. Edward in the central range) financed, organized and guided by Mr. Richard Archbold of New York, Mayr and Rand (1935g, 1936c, 1936f) described 26 new subspecies and one new species (*Eurostopodus archboldi*). A massive report on the entire collection comprises 248 pages with detailed discussions of each species (Mayr and Rand 1937c). The birds of the 1936–37 and 1938–39 Archbold Expeditions were studied by A. L. Rand. Mayr continued to review the genera of New Guinea birds one by one, described several new subspecies (1936d) and determined the taxonomic position of certain isolated species such as the monotypic genera *Paramythia* (*P. montium*) and *Oreocharis* (*O. arfaki*) which he recognized as closely related to each other and as members of the flowerpeckers Dicaeidae (Mayr 1933g). Following Stresemann's example, he revised carefully some of the most difficult genera, such as *Sericornis* (1937a) and *Collocalia* (1937f). These were the most complicated groups of birds he ever tackled taxonomically with many exceedingly similar sibling species. Whenever he was interrupted in his work, he had to scrutinize them anew for several days to see the minute differences. With respect to *Collocalia* he wrote to Stresemann:

"On some days, and this is literally true, I sat there for 3–4 hours comparing a single specimen of *hirundinacea* with a single specimen of *vanikorensis*, or a single one of '*mearnsi*' with one of *germani*. Eventually I had developed such a clear 'mental picture' of them that the classification of the other forms became much easier [...] These difficult beasts must be looked at from the front, the back, from above and below for hours" (9 March 1937; transl.).

The article on the swiftlets gave him a chance to explain the changes in taxonomic concepts within several decades: "The literature on this genus [*Collocalia*] illustrates exceedingly well the trends of ornithological classification. We see in the earlier part of this century conscientious efforts to analyze the characters of the various geographical races without much of an effort to combine the many disconnected units into natural groups of related forms. Oberholser's papers were written in this analytical stage. In opposition to this trend the Formenkreislehre gained increasing influence during the twenty's, emphasizing the principle of geographical representation frequently with disregard of a thorough morphological examination of the treated forms. During this period (1925–1926) Stresemann proposed a classification of this genus, which grouped all the then known forms in six species [...]. A reaction to this ultra-synthetic trend was inevitable, and Stresemann himself was the first to suggest the breaking up of these large Formenkreise into smaller, but more natural species" (1937f: 2).

Most of the non-passerine birds of New Guinea extend their ranges considerably beyond this region. Revisions of these species had to take into consideration the material of the Rothschild Collection and the collections of the Whitney South Sea Expedition made in the Bismarck Archipelago, the Louisiade and d'Entrecasteaux Archipelagoes as well as the Solomon Islands and Polynesia. Therefore Mayr included the papers on non-passerine families entitled "Notes on New Guinea Birds I–VIII" in the series "Birds collected during the WSSE" (No. 33, 35, 36, 39–41, 43, 45; 1937–1941). In 1939(c,f,g), Mayr and Meyer de Schauensee reported in several installments on the collections made by S. D. Ripley during the Denison-Crockett Expedition to western New Guinea (Biak Island, Vogelkop and western Papuan Islands) in 1936–38. Here Mayr interpreted the speciation process for the first time. The bird fauna of Biak Island demonstrates all stages of increasing differentiation. Among 69 species 20 have not changed when compared to the New Guinea mainland population, either because they are exceptionally stable, or because they have reached Biak only recently, or because they belong to common mainland species which continue to swamp the Biak population. Some of the remaining 49 species are slightly different; others have developed into moderately or even conspicuously different subspecies. Partly, they might with equal justification be considered species or subspecies, and still others are so different (5 species or 8, if Numfor Island is included) that no taxonomist will hesitate to call them species (Mayr and Meyer de Schauensee 1939: 9).

By the end of 1940, Mayr had completed a taxonomic revision of every genus in preparation of his *List of New Guinea Birds. A systematic and faunal list of the birds of New Guinea and adjacent islands* to be published by the AMNH in 1941(f). Only the specialist is able to appreciate the quality of this work of which its author was justly proud. The *List* provided a sound basis on which later authors could build. Based on the Biological Species Concept, it gives an overview of the number, the systematic relations and the distribution of all the bird taxa of New

Guinea and surrounding islands (568 breeding bird species in 1,400 forms of the land and freshwater). Subspecies differentiation is conspicuous. For each form the name is followed by reference to the original description, a complete list of synonyms, and its range, and for each species there is a statement of habitat and altitudinal distribution. Species limits employed are broad to express relationships rather than differences among species and subspecies. Footnotes indicate which species together form a superspecies. A gazetteer of reference maps and collecting

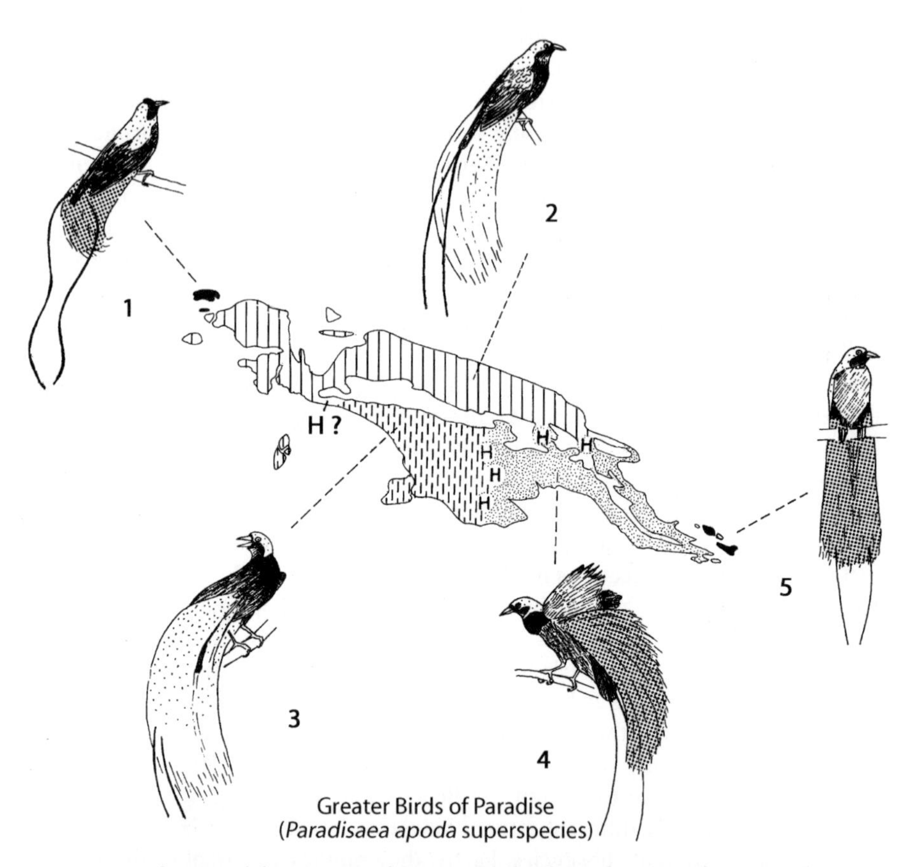


**Fig. 4.5.** Distribution of the crow-sized Greater Birds of Paradise, *Paradisaea apoda* superspecies of New Guinea (data from Gilliard 1969 and Cooper and Forshaw 1977). 1 *P. rubra,* 2 *P. minor,* 3 *P. apoda,* 4 *P. raggiana,* 5 *P. decora.* Adult males (illustrated) are maroon brown, breast mostly blackish, crown and nape yellow with two elongated central tail wires (or ribbons) and enormous flank plumes which are yellow (stippled) and white, orange or red (shaded); females are very different, smaller and lack flank tufts and tail wires. Ernst Mayr (1940c, 1942e) discussed geographical gradients in the color of back and flank plumes in *P. raggiana* of eastern New Guinea which are probably due to southeastward introgression of *P. minor* genes from the northcoastal lowlands. Hybridization (H) occurs in areas where the mainland species meet. Note strongly differentiated species on small islands off the coast of New Guinea near its northwestern and southeastern tips

localities in New Guinea at the end of the book is most useful to the user. The classification of the families used in this book follows Wetmore's proposal as far as the non-passerine groups are concerned and Stresemann's proposal for the Passeriformes.

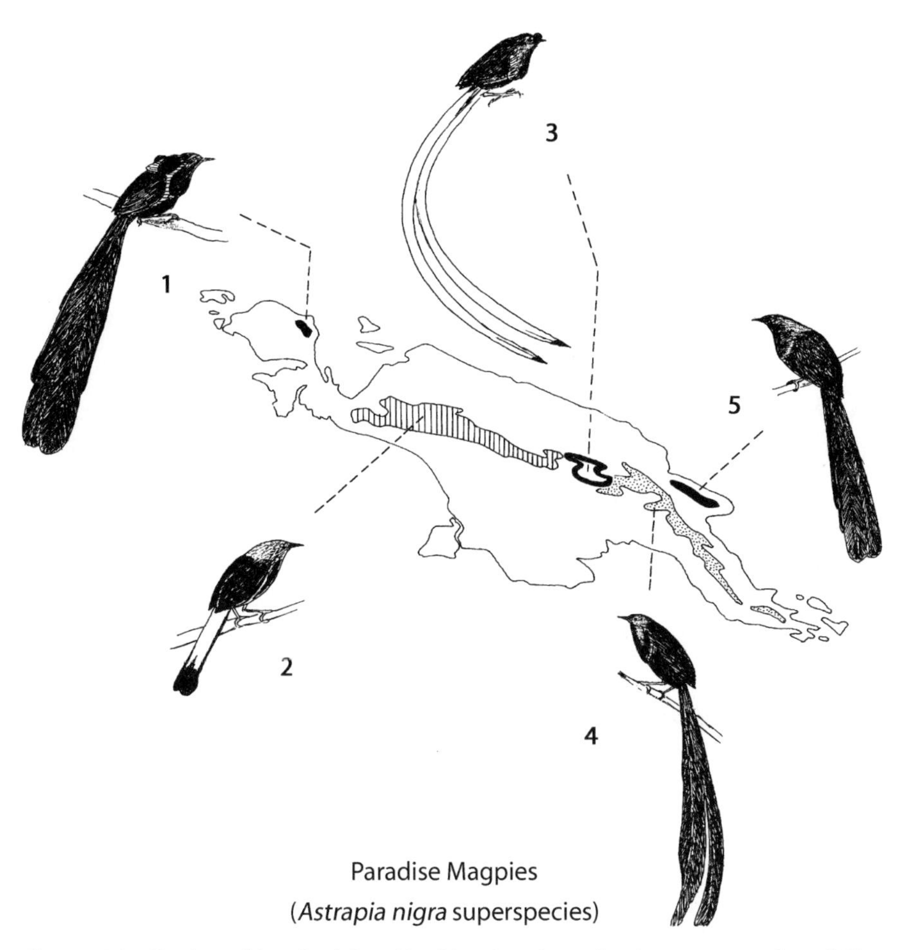


**Fig. 4.6.** Distribution of jay-sized Paradise Magpies, *Astrapia nigra* superspecies, of New Guinea (data from Gilliard 1969, and Cooper and Forshaw 1977). 1 *A. nigra,* 2 *A. splendidissima,* 3 *A. mayeri,* 4 *A. stephaniae,* 5 *A. rothschildi.* Adult males (illustrated) are mainly black with much green and purple iridescence, tail greatly elongated and black or black and white to white; females are more brownish with barred underparts. Hybridization has been recorded in the zone of contact between *A. mayeri* and *A. stephaniae.* Ernst Mayr (1942e, 1945h) used this group to discuss allopatric speciation: These five species, "descending from a common stock, have differentiated under conditions of geographical isolation. Each is restricted to a single mountain range, and none can exist in the lowlands. The differences acquired by these five species are a graphic illustration of evolution."

Examples of two superspecies of birds of paradise, Mayr's favorite birds[4], are illustrated in Figs. 4.5 and 4.6. Character gradients across hybrid zones occur in the Greater Birds of Paradise, geographical isolation and conspicuous differentiation is demonstrated in the Paradise Magpies. In a study of the Ribbon-tailed Bird of Paradise (*Astrapia mayeri*) Mayr and Gilliard (1952a) discussed the *Astrapia* superspecies. *A. mayeri* differs conspicuously from its representatives to the west (*A. splendidissima*) and to the east (*A. stephaniae*) but some gene flow exists between *stephaniae* and *mayeri* in the area where they have come into contact during recent geological time. These latter taxa are very close biologically. Partial overlap of closely related species occurs in the paradise kingfishers of the genus *Tanysiptera* (Fig. 4.7). During the Pleistocene *T. hydrocharis* was isolated on an island stretching from the Aru Islands to the mouth of the Fly River and thus separated from the mainland form *galatea* by a branch of the ocean. When this

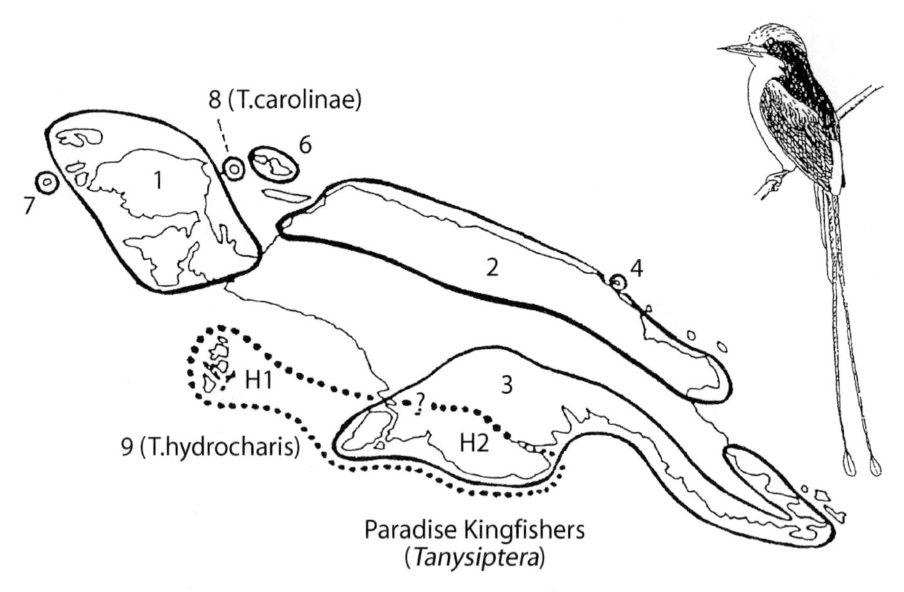


**Fig. 4.7.** Species and subspecies of the Paradise Kingfishers, *Tanysiptera hydrocharis-galatea* group. The subspecies 1, 2, and 3 of *galatea* on the mainland of New Guinea are exceedingly similar to each other. The subspecies *vulcani* (4) and *rosseliana* (5, on Rossel Island east of New Guinea, not shown) are much more distinct. The populations on Biak (6), Numfor (7), and Koffiao (8) have reached species level. The form on Aru Island, *hydrocharis* (H1), has also reached species rank and now coexists in southern New Guinea (H2) with a subspecies of *galatea* (3); slightly modified from Mayr (1942e, Fig. 15)

[4] In the foreword to Gilliard's book on the birds of paradise and bower birds, Mayr wrote: "Every ornithologist and birdwatcher has his favorite group of birds, whether they be nightingales or storks, hummingbirds or penguins. Frankly, my own are the birds of paradise and bower birds, [...] for in their ornamentation and courtship behavior birds of paradise are not surpassed in the whole class of Aves" (1969f).

strait fell dry the island joined with the mainland of New Guinea, and *galatea* invaded the range of *hydrocharis*, where the two species now live side by side without interbreeding and without obvious ecological competition. *T. galatea* displays no significant geographical variation in the vast area of New Guinea despite strong ecological contrasts. Yet each of the adjacent islands inhabited by this kingfisher has a markedly differentiated race or species even though they are in the same climatic zone as the neighboring mainland. Due to this observation and numerous similar examples in the literature Mayr (1954c) later proposed the theory of a relatively rapid genetic reorganization in small speciating populations (p. 219). Several sidelines to Mayr's taxonomic research on the birds of Oceania developed into additional publications on the avifaunas of Australia, the Malay Archipelago and Burma (Myanmar).

## Birds of Australia, the Malay Archipelago and Southeastern Asia

### Australia

A visit by Dominic L. Serventy (1904–1988) from Australia to New York in early 1938 led to a fruitful collaboration. So far Mayr had used the material of the Mathews Collection of Australian birds at the AMNH for the preparation of many papers on the birds collected during the Whitney Expedition. But the taxonomy of Australian birds owing to the publications of Gregory Mathews (1927, 1930) was chaotic. Serventy knew these birds in the field and Mayr, as curator of the Rothschild Collection, had access to all of Mathews' types. The first result of this new collaboration was their joint review (1938i) of the thornbills (*Acanthiza*), relatives of *Sericornis* and *Gerygone*. Each author contributed about equally to this revision. Eleven *Acanthiza* species are known from Australia and Tasmania and one (*A. murina*) from the mountains of New Guinea. In another article on "The number of Australian bird species" based on the concept of superspecies Mayr and Serventy (1944h) show that Australia, although ten times larger in area than New Guinea, harbors only a similar number of species. These joint papers aroused Mayr's interest in Australian birds.

After G. M. Mathews's reckless naming of subspecies in *The Birds of Australia* (1910–1927) "substantial progress in the modern treatment of genera, species and subspecies has been made by Dr. Ernst Mayr" (Serventy 1950: 267). His publications in *The Emu* (totaling 112 pages; Marchant 1972: 66) made him well known among Australian ornithologists and field naturalists some of whom he met when he visited Australia in the northern winter 1959–1960. He used this opportunity to observe in the field several "aberrant" Australian songbirds whose systematic position was unsettled and commented on their behavior (Mayr 1963c).

The recent "*Directory of Australian Birds. Passerines*" by R. Schodde and I. Mason (1999) is dedicated to the four founders of modern Australian variation studies: Ernst Mayr who initiated this new era, Allen Keast, Julian Ford, and Shane Parker.

## Malay Archipelago

Mayr also initiated taxonomic studies of birds from further west of New Guinea, based on collections made in Borneo, western China and northern Burma. This greatly added to his knowledge of species and literature. He studied the birds of prey of the Lesser Sunda Islands (1941m) and the birds of Timor and Sumba Islands (1944e) describing 19 new subspecies and analyzing the relationships within numerous species groups. Because of the dry climate and vegetation of these islands the bird fauna is comparatively poor and endemism is not strongly marked. Important zoogeographic considerations appended to these taxonomic discussions will be mentioned below (see p. 179ff.).

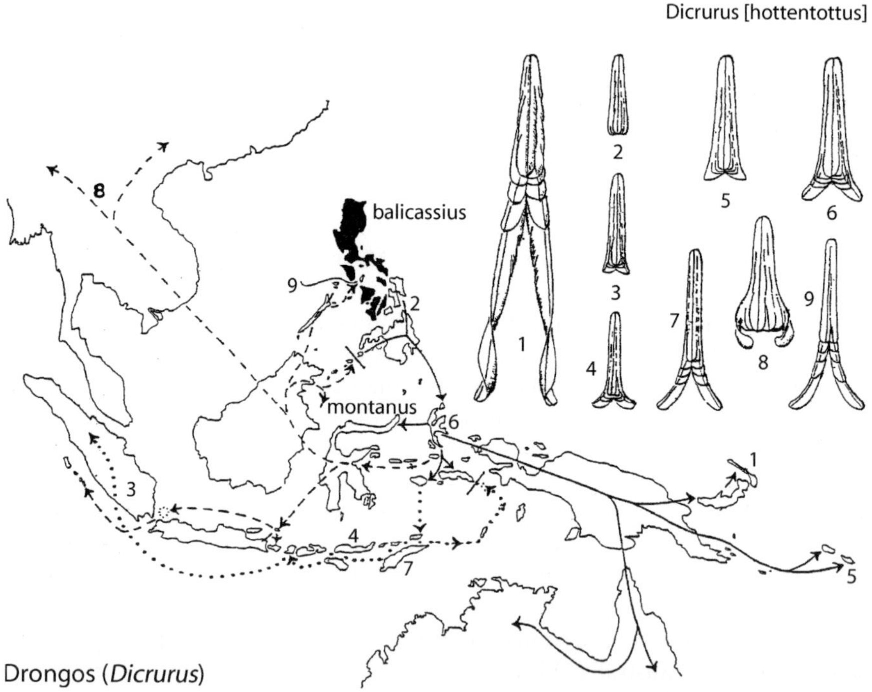

**Fig. 4.8.** The drongo superspecies *Dicrurus [hottentottus]*, Dicruridae. Tail forms and inferred dispersal routes of three branches of this highly variable superspecies are shown. Bizarre shapes of tail feathers developed in the geographically most peripheral populations. *Solid line:* the earliest branch to disperse; *dotted line:* the next branch; *dashed line:* the most recently dispersing branch. *D. balicassius*, a closely related species, occupies a range in the northern Philippines allopatric to *D. hottentottus* in the southern Philippines. The closely related *D. montanus*, the product of a double invasion of Celebes (Sulawesi) by *D. hottentottus*, now lives in the mountains of Sulawesi above *D. hottentottus* in the lowlands. The nine numbers on the map indicate the geographic ranges of nine taxa whose tails are depicted at upper right. For identification of these taxa and additional details see Mayr and Diamond (2001, map 52)

Study of the birds of the Philippines led to several taxonomic reviews, comments, descriptions of new subspecies, e.g., of the Honey Buzzard (*Pernis apivorus*, 1939d) and Tailorbird (*Orthotomus*, 1947b) and revisions of the classification and nomenclature of various other taxa (Delacour and Mayr 1945m). These articles were byproducts of a book, *Birds of the Philippines* by J. Delacour and E. Mayr (1946k) published to facilitate the identification of birds in the field. The work associated with this project was about evenly shared by both authors. Numerous excellent line drawings by E. C. Poole and A. Seidel illustrate many species. Keys permit the determination of species; subspecies are also briefly described. Remarks on ecology and habits of the birds point to insufficient knowledge and gaps to be filled. Zoogeographically, most Philippine birds show Malaysian affinities, but some have closer ties with Oriental and Palearctic elements.

Mayr's discussion of the "Evolution in the family Dicruridae" (Mayr and Vaurie 1948d) was based on a taxonomic revision of the drongos by Vaurie (1949) and therefore published in coauthorship. This family is widespread in the Old World tropics. Several of the 20 species are well differentiated in the Malay Archipelago (Fig. 4.8). Mayr emphasized here again that the more distinct subspecies and semispecies are geographically isolated and occur in peripheral parts of the species range, where they vary rather unpredictably. Those which readily cross water gaps formed widespread polytypic species and superspecies.

## Southeastern Asia

Mr. Arthur S. Vernay, a Trustee of the AMNH, sponsored several zoological expeditions to northern Burma (Chindwin region) during the period 1932–1936, with Major J. K. Stanford in charge of bird collecting. E. Mayr, who was asked to study the collections given to the AMNH published the results in *The Ibis* of 1938 and 1940–41 (with J. K. Stanford supplying field notes). Several British ornithologists had been rather unhappy that an inexperienced American colleague who had never done any work in this region should study this important collection. However, eventually they admitted that they themselves could not have done a better job.

An interesting example of discontinuous geographic variation of alternative plumage colors (white, gray, black) is presented by the Black Bulbul (*Hypsipetes madagascariensis*). The populations are fairly uniform, where they are isolated on high mountains (Fig. 4.9). However, hybrid populations develop where several of these forms come in contact with one another, which occurs from northeastern Burma into China south of the Yangtze River (Mayr 1941n, 1942e: 83–84).

Mayr's real triumph of working up the Burma collection provided the colorful minivets (*Pericrocotus*) (Fig. 4.10). Two exceedingly similar species were always confounded under the name *P. brevirostris* until Mayr (1940e) discovered a sibling species mixed in with the series of *brevirostris*. A name (*ethologus*) was already available for the newcomer. The two species co-exist over a wide range from Sikkim eastward. At first Mayr's finding was doubted by Ticehurst, Whistler, Kinnaer,

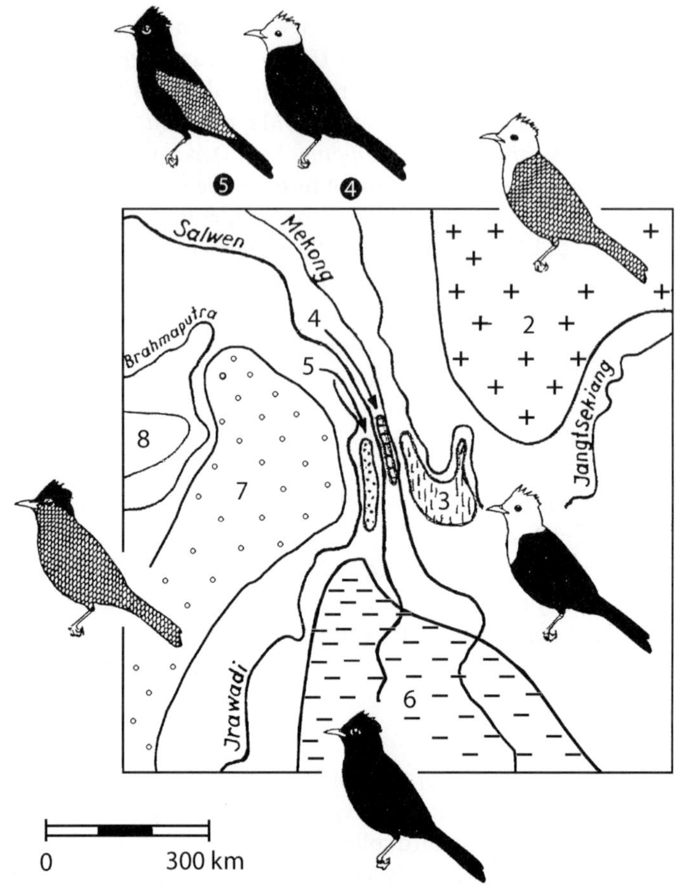


**Fig. 4.9.** Black Bulbul (*Hypsipetes madagascariensis*). Discontinuous geographical variation of alternative plumage characters in populations of northeastern Burma (Myanmar) and southwestern China. Rather uniform (stabilized) hybrid populations (3–5) inhabit isolated mountains between the ranges of four well-differentiated geographical subspecies (1 *leucocephalus* in S and SE China, 2 *leucothorax*, 6 *concolor*, 7 *nigrescens*, 8 *psaroides*); hybrid populations: 3 *stresemanni*, 4 *sinensis*, 5 *ambiens*). Plumage characters: Blank-white, shaded-gray, solid-black. Sketches of birds by R. T. Peterson. From Mayr (1941n, 1942e: 83)

and Baker, the experts on the birds of this region, but eventually they accepted it. E. Stresemann in Berlin was working on a collection of birds from Sikkim obtained by Ernst Schäfer in 1938–1939, when he received the letter from Mayr (dated 23 May 1940) describing his two sibling species of minivets. Stresemann had already determined Schäfer's specimens of *Pericrocotus* as belonging to one species (*P. brevirostris*) only and answered:

**Fig. 4.10.** Two sibling species of cuckooshrikes (Campephagidae), the Long-tailed Minivet (*Pericrocotus ethologus, left*) and the Short-billed Minivet (*P. brevirostris, right*) which are sympatric over a wide range from the Himalayas to western Yunnan. Ernst Mayr (1940e) discovered their distinctness as species. Illustrations of adult males from MacKinnon and Phillipps (2000)

"Your data convinced me that I hadn't noticed anything, which should not have happened! You are a tremendous fellow and have picked unfading laurel with this discovery. I could kick myself to have slept right through. The matter is crystal clear once it is understood. Schäfer collected in Sikkim 27 *brevirostris* and eight *ethologus* [...], *ethologus* breeds at 2,750 m, *brevirostris* in the subtropical zone, between 560 and 1,900 m" (19 June 1940; transl.; see Haffer 1997b: 517 and Stresemann, *Ornithologische Monatsberichte* 49 (1941): 9–10, footnote).

## Descriptions of New Species and Subspecies of Birds

The collections which Mayr himself assembled in New Guinea and in the Solomon Islands as well as those from the Pacific Ocean, Malay Archipelago, southeastern Asia and other areas which he studied at the AMNH came from many remote and ornithologically poorly known regions of the world. Therefore they contained an unusually large number of unnamed taxa of birds, new subspecies as well as new species, as I repeatedly mentioned above (Fig. 4.11). Among the total of 471 new taxa which Mayr named between 1927 and 1960 (and two nomina nova in 1986) are 26 new species (the description of one additional species was barely anticipated by Japanese authors) and 445 subspecies. Every new form was carefully studied and compared before it was named. If he had any doubts he walked over to John Zimmer, an expert on subspecies differences, who sat only 20 steps away

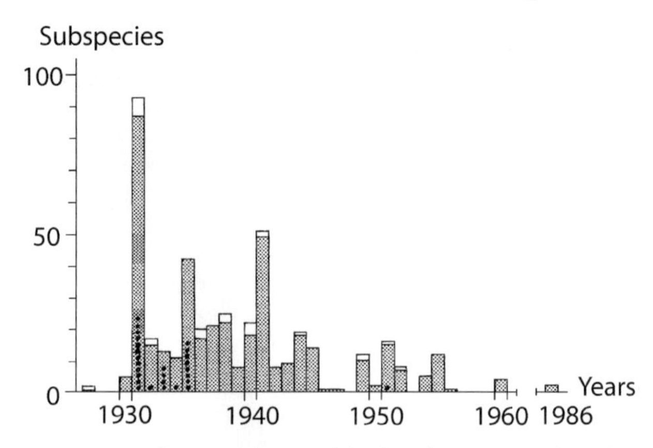


**Fig. 4.11.** Ernst Mayr named 26 new species (*dots*) and 445 new subspecies of birds (histograms) between 1927 and 1986. Subspecies accepted in Peters' *Check-list* are shaded, those synonymized are blank. For further details see text

from Mayr and he named only forms which Zimmer approved of. Admittedly, like the best taxonomists of the day, he was also a splitter. Several of Mayr's newly described subspecies, particularly among the insular forms, probably will be raised to species status when more details on their life history and behavior are known, e.g., *Phalacrocorax melanoleucos brevicauda, Eurostopodus mystacalis exul*, and others. Mayr described 376 taxa (including 26 species) alone and 95 in coauthorship with his colleagues A. Rand (33 taxa) during the 1930s, D. S. Ripley (15 taxa) in 1941, and T. Gilliard (29 taxa) during the 1950s. Other coauthors with whom he described one or several new subspecies were Birckhead, Greenway, Serventy, Meyer de Schauensee, K. Jennings, Gyldenstolpe, Van Deusen, and McEvey. From the birds which Mayr collected in former Dutch New Guinea, Hartert (1930) described three valid new species and 28 valid new subspecies.

Like many contemporary ornithologists (e.g., Hartert, Stresemann) Mayr also delimited polytypic species taxa rather broadly, in a few cases including in one species an allopatric representative as a new subspecies which, on the basis of additional evidence, is considered a full biological species by today's somewhat narrower standards. Examples are *Phalacrocorax melanoleucos brevicauda* Mayr 1931 (Rennell Island), possibly *Eurostopodus mystacalis exul* Mayr 1941 (New Caledonia), and also *Rallina rubra [leucospila] mayri* Hartert 1930 (northern New Guinea). The number of full biological species described by Mayr will probably increase when, in the future, genera of birds will be revised which comprise well-differentiated subspecies that he described. Of course, such taxa represent phylogenetic "species" under that "species" concept (p. 204).

The authors of J. L. Peters' *Check-list of Birds of the World* (1931–1987) accepted most of Mayr's new taxa. They and several authors of more recent monographs combined (synonymized) only 26 of Mayr's 445 subspecies with previously described forms. Among the authors of the *Check-list* is also Ernst Mayr himself who synonymized or accepted the synonymy of 12 forms which he had described in previous years (among which are five in *Pachycephala* and two *Amblyornis* bowerbirds). B. P. Hall synonymized a Mayr subspecies in *Pericrocotus ethologus*, G. Mees synonymized four other Mayr subspecies in the following species: *Rallina tricolor, Charmosyna placentis, Cinnyris sericea,* and *Zosterops novaeguineae.* C. Vaurie did the same with one form in *Cecropis striolata.* In one case (*Melanocharis striativentris albicauda*) an earlier available name was found. Seven forms, one in each of the following species, need additional study and are, at best, "weak" subspecies: *Butorides striatus, Rallina tricolor, Ducula aenea, Micropsitta bruijnii, Prunella rubeculoides, Staphida castaneiceps,* and *Melidectes leucostephes* (M. LeCroy, pers. comm.).

One reason for the high percentage of valid taxa in Mayr's work, besides his clear principles for recognizing subspecies, is the fact that he dealt mostly with island faunas composed of geographical isolates. Populations on islands often show more clear-cut geographical variation compared to continental faunas and to a lesser extent smooth and gradual variation; if so, it is in the form of a stepped cline. Smooth clinal variation of continuous continental populations often presents more problems for a subdivision into discrete subspecies than island faunas and, therefore, leads to more disagreement among workers about the validity and delimination of certain subspecies.

# Zoogeography

## Basic Concepts

Mayr's systematic studies of Pacific island birds soon led him to consider general zoogeographical topics, particularly in conjunction with the problems of dispersal, range disjunction, endemism, insular speciation, faunal turnover on islands and of continental speciation in ecological refuges during the Pleistocene. In later years, he made important contributions to historical-dynamic analyses of world faunas. He combined a series of miscellaneous papers on biogeography in his essay volume, *Evolution and the Diversity of Life* (1976m) which is a convenient summary of his major ideas on zoogeography, but he never published a book on this portion of evolutionary biology.

From his extensive work on Pacific island birds, Mayr (1933j, 1940i) established the basic principles of an equilibrium theory of biogeography, however in non-mathematical terms[5], as he stated himself in retrospect:

"My thesis that the size of an island fauna is the result of a balance between colonization and extinction was long ignored, but is now accepted, after it was transformed into graphs and mathematical formulae. It is the basic thesis of MacArthur

[5] Mayr (pers. comm.) had copious data on island sizes, distances from mainlands or other islands, number of species, etc. When he tried to determine relations among all these figures he got into mathematical problems and turned this material over to a graduate student with mathematical abilities. However, this student got sidetracked into other problems and this material was never exploited.

and Wilson's *Island Biogeography* (1967)" (Mayr 1976m: 616; see also Bock [1994a: 292] and Vuilleumier [2005b: 59, 65–66]).

Several dynamic zoogeographers had long ago realized that wherever there is biota, there is input (immigration, colonization) and output (extinction, emigration): The numerical values of MacArthur and Wilson's theory (1967) are "mere icing on the cake," as Mayr (1983e: 15) later put it. The discovery of numerous bird fossils on many Pacific islands (Steadman 2006)—unknown to Mayr and other early workers—indicates that during prehistoric times, much richer species communities inhabited many of these islands than currently found. Quantitative theories of island biogeography will have to be revised on the basis of inventories that describe the island avifaunas before human impact.

As to continental faunas Mayr (1965q: 484) emphasized that instead of a descriptive static definition ("A fauna is the totality of species living in an area") a dynamic definition is to be preferred: "A fauna consists of the kinds of animals found in an area as a result of the history of the area and its present ecological conditions" and "Instead of thinking of fixed regions, it is necessary to think of fluid faunas" (1946h: 5). Aspects of faunal studies are the analysis of endemism and of the faunal elements which differ in origin, age, and adaptation. The interest of modern zoogeographers centers on faunas rather than regions. Faunas originated (a) through continuous autochthonous differentiation in (at least temporary) geographical isolation from other faunas, during which time they acquired their special characteristics, (b) through continued single origin colonization (e.g., the Australian bird fauna), (c) through continued multiple-origin colonization (e.g., the bird faunas of Hawaii, West Indies, Pantepui, Tristan da Cunha), (d) through the fusion of two faunas (e.g., the faunas of the Malay Archipelago and of Central America).

Mayr also pointed out that adaptive radiation is not a phenomenon restricted to island faunas. There are examples on continents as well, e.g., the troupials (Icteridae) of the Americas. Another problem is the varying dispersal characteristics of individual taxa, even within a group of ecologically comparable families. Among grassland finches, the Estrildidae are widespread and richly differentiated in Africa, the Oriental Region and Australia, whereas the Ploceidae were successful only in Africa. The sunbirds (Nectariniidae) have successfully colonized the wide area from Africa to New Guinea, while the honey-eaters (Meliphagidae), so rich in species in the Australo-Papuan region, did not enter the Sunda Islands. The analysis of such differences has not yet progressed very far.

Continental drift is known to have occurred which, however, does not mean that we can or must explain all distribution patterns in terms of drift. Mayr (1972h) discussed faunal relations consistent with classical land bridge interpretations (Bering Straits bridge, Panama bridge, the relations between the Oriental and Australo-Papuan faunas). Other faunal relations are best explained by the theory of continental drift (e.g., North America–Europe).

Mayr was a prominent representative of the "Center of origin concept" or "Dispersal paradigm" whose basic tenets are: (1) Many animal groups differentiated in well-circumscribed regions of the world (e.g., marsupials in Australia); (2) some members of such groups colonized other continents or islands where they radiated conspicuously (e.g., meliphagids in New Zealand, drepanidids in Hawaii); (3) faunas are built up by waves of colonization from outside sources superimposed on original stocks; (4) geographical speciation occurs in populations separated by physical barriers (due to continental movements and climatic-vegetational fluctuations); (5) secondary range expansion from the areas of origin led to sympatry of representatives of different faunas.

Mayr (1982h) severely criticized an exclusive explanation of zoogeographic patterns by the vicariance paradigm but agreed (1983e) that it is no longer sufficient simply to develop an explanatory scenario. Hypotheses and predictions should be tested. On the other hand, he pointed out that not all problems can be settled in this manner, e.g., the placement of borders between biogeographical regions (Wallace's or Weber's lines versus "Wallacea" in the Malay Archipelago). In quantitative studies one must avoid treating all species of certain animal groups as exhibiting uniform biogeographical characteristics. Island birds are not a random sample of mainland birds with equal dispersal facilities. Among all kinds of animals there are good and poor dispersers. Mayr (1983e) emphasized again that the elements of faunas have had different histories and mean values are often meaningless. Mayr and Diamond (2001g) discussed these aspects in great detail with respect to the avifauna of northern Melanesia.

To explain discontinuous ranges, many authors favored an "either-or" choice, either secondary (vicariance) or primary (dispersal across a barrier). However, it is becoming clearer that in poor dispersers many if not most discontinuities are secondary, while in good dispersers they are primary. Also, the extant pattern of distribution was, at least in part, established in that geological period in which a group had its major evolutionary development, hence earthworms or mayflies have other patterns than mammals and birds. Single-track explanations are unable to cope with this pluralism and are, therefore, bound to be misleading.

Summarizing, spatial events and climatic-vegetational fluctuations shape faunas and species from their early history. Separation of populations through dispersal and vicariance, secondary contact and overlap led to speciation and faunal differentiation. The early history of faunal lineages was influenced by continental drift, i.e., the break-up and contact of continental blocks–processes which had more pronounced effects on some groups, like freshwater fishes, than others like birds.

## First Steps

The ecology and distribution of birds fascinated Mayr already as a student (see pp. 28, 35). In his dissertation prepared under Erwin Stresemann as thesis advisor, Mayr analyzed a zoogeographical problem, the range expansion of the Serin (*Serinus serinus*), a small greenish yellow finch which, spreading northward from the Mediterranean region, had colonized large portions of Europe since the early 19th century (Mayr 1926e; see Vuilleumier 2005b). He described the process of range expansion and discussed the ecology of this and other expanding species emphasizing that ecologically optimal areas were colonized first and less favorable areas later. The settlement of an expanding species is often followed by a population explosion, so the Serin finch became, within a few years, one of the most common songbirds of its new areas. Chance factors played an important role in the observed irregularities of the colonization process. Mayr speculated that possibly "a hereditary alteration" or "ecological mutant" was favored by natural selection thus triggering the range expansion (1926e: 653). His interpretation is still considered in this and other cases of rapid range expansion (e.g., Collared Dove, *Streptopelia decaocto*). His work on the Serin finch led Mayr to think about ecological aspects of biogeography and whether this species' genotype provided various potentials, an aspect emphasized by modern ecology. As an offshoot of his thesis he prepared a long manuscript on the origin of bird migration completed by Meise after Mayr had left for New Guinea (see p. 48).

Mayr's notebooks of 1926–1927 in Berlin also document his early interest in zoogeography. He not only planned a detailed analysis of the "Geographical derivation of the birds of Germany" but also a book on "Zoogeography" or "Zoogeography of land animals" and made entries on the arrangement of subject matter (history, zoogeographical regions, geological development, Middle American landbridge, barriers to dispersal, colonization of oceanic islands, laws of extinction, significance of climatic fluctuations for the development of faunas, etc.). These ideas were triggered by Richard Hesse's textbook, *Ecological Zoogeography* (1924), a title which Mayr considered rather misleading, since it was a textbook of geographical ecology rather than of zoogeography. He wanted to present the dispersal and range expansion of birds and other animals and to enlarge upon the general portions of his serin paper (1926e). But this early plan was never realized.

## Dynamic Faunal Approach

Researchers of the late 19th century like Schmarda, Sclater, Wallace, Reichenow, and many others had subdivided the world into zoogeographical regions, provinces, and ever smaller units with no agreement ever reached as to their delimitation and hierarchical classification. In contrast to this descriptive and static method, several zoologists of the early 20th century had initiated a new dynamic faunal approach to causal zoogeography, e.g., E.R. Dunn (1922, 1931) for the reptile fauna of North America, E. Stresemann (1926, 1933, 1939) for birds of Europe and the Malay Archipelago, E. Lönnberg (1926, 1927, 1929) for those of North America and Africa, and B. Stegmann (1938) for the bird fauna of the Palearctic Region. They found that attempts to analyze faunas in terms of their respective elements with different histories fruitfully supplement the paleontological approach, particularly for groups with an inadequate fossil record. G.G. Simpson (1940, 1943, 1947) became the leader of this new dynamic approach, particularly as far as mammals were concerned, and Mayr for birds. Avian distribution patterns in the Indo-Australian archipelago are consistent with the hypothesis of transoceanic dispersal, as Mayr documented in many of his papers published since 1931 (e.g., on the birds of Rennell Island [1931e] and especially those of Polynesia [1940i] and of the Sunda Islands [1944e]).

From the references 1931e and 1942e, 1942h, it is shown that Mayr emphasized that no fauna can be fully understood until it is segregated into its components ("elements") and until one has succeeded in explaining the separate histories of each of these different species groups (Mayr 1965q). The dispersal ability of each species and the geological histories of the areas involved need to be studied in detail to determine the zoogeographical history of a group of animals as a dynamic and continuing process. This endeavor implies the proper evaluation of (a) the relative age of the various taxa depending on the degree of their respective differentiation, (b) the determination of the dispersal capacity of the various taxa, and (c) the distribution of related taxa. Some bird species disperse readily across water gaps and others do not. A constraint limiting the range expansion of readily dispersing species is the occurrence of competing close relatives beyond the range limits. In these cases ecological competition prevents further range expansion, as shown by the mosaic distribution patterns of the members of superspecies on continents and in island regions. The basic zoogeographical processes underlying Mayr's views are summarized in Table 4.1.

Range disjunctions of populations or taxa may be primary or secondary (Hofsten 1916) depending on whether they originated actively (through jump dispersal across a preexisting barrier) or passively (through a disruption of a previously continuous range). A variety of geological processes led to such range disjunctions (vicariance) and subsequent rejoining of populations. Climatic-vegetational cycles on the continents of India, Africa and Australia split the ranges of forest birds in two or more widely separated portions leading to the differentiation of isolated populations (1942e: 231, 1950b); Pleistocene sea-level fluctuations caused the separation

**Table 4.1.** Concepts underlying Ernst Mayr's zoogeographical views

| | Zoogeography | | Geographical (allopatric) speciation |
|---|---|---|---|
| Causes of disjunction | Pattern | Process | |
| Colonizing ability of animals across barriers | Primary disjunction | Jump dispersal | Peripatric (founder model) |
| Climatic-vegetational cycles, sea-level fluctuations, tectonic uplift/subsidence, continental drift | Secondary disjunction | Vicariance | Dichopatric (dumbbell model) |

and rejoining of populations in the Philippines and Solomon Islands (Delacour and Mayr 1946k, Mayr and Diamond 2001g: 7–9); tectonic uplift in southern Central America led to faunal interchange between North and South American faunas (1946h, 1964c); continental drift caused the separation of North American and European faunas during the early Tertiary (Mayr 1990b). Until the concept of plate tectonics was accepted during the 1960s, zoogeographers like Mayr, Darlington, Simpson and others, did not consider continental drift to explain extant distribution patterns in birds. This is understandable, because the evolution of most extant groups of birds and mammals took place during the Tertiary after Gondwana had split into the series of southern continents.

Speciation models reflecting the processes of jump dispersal and vicariance are the founder model (peripatric speciation) and the dumbbell model (dichopatric speciation), respectively. Modern authors occasionally call the leading zoogeographers of the 1930s to 1950s "dispersalists," thereby overlooking the fact that their theoretical framework included various types of vicariance which they proposed had an effect on faunal differentiation (Table 4.1).

## North and South America

Most of the North American families and subfamilies of birds are either Old World in origin, South American in origin, or members of an autochthonous North American element. This latter endemic element developed during the period of geographical isolation of North America during the Tertiary, when the southern half of North America (from about Honduras to the present Canadian border) had a tropical-subtropical climate. During that time, most of northern Middle America was a southern peninsula of North America. The tropical North American element includes the Mimidae, Vireonidae, Parulidae, Troglodytidae, nine-primaried oscines, and Momotidae. Mayr (1946h, 1964c) identified the following components (elements) in the North American bird fauna: Unanalyzed (oceanic, shore and freshwater birds, hawks, eagles, and others), Holarctic (Pan-Boreal), Pan-American, Pan-Tropical (trogons, barbets, parrots), South American, North American, Old World (Eurasian). The North American element comprises up to 50%, or even more, of the North American bird fauna in all habitats except the arctic. Therefore the North American bird fauna cannot be included in a "Neotropical" or in a "Holarctic" region. The South American element increases gradually southward. No boundary line separating North American and South American bird faunas can be drawn in Panama, because they were mingled in Pliocene time. Many more Palearctic birds immigrated into North America than in opposite direction (e.g., the wren colonized Europe). Several North American groups had secondary radiations in South America (Parulidae, Thraupidae) and in the Old World (Emberizinae). The suboscines with 10 families are clearly of South American origin, and only the Tyrannidae and the hummingbirds had secondary radiations in North America. Faunal interchange between North and South America occurred mainly toward the end of Tertiary, when the last gap between Panama and Colombia was closed. The number of recent South American elements is much smaller in the arid habitats of Central America than in the tropical rainforest which was massively invaded by South American species. Therefore, the arid habitats of Central America reflect the composition of the Tertiary North American bird fauna more accurately than the humid tropical habitats.

The North American continent was connected with Europe (via Greenland) in the early Tertiary and has had intermittent connections with Asia across the Bering Strait bridge (Mayr 1964c). It was separated from South America by several water gaps and its fauna evolved in isolation during the first half of the Tertiary, when the southern half north to 38–40° latitude was humid and tropical ("tropical North America"). Indigenous North American bird families which here evolved include the wrens (Troglodytidae), mockingbirds (Mimidae), vireos (Vireonidae), wood-warblers (Parulidae) and buntings or American sparrows (Emberizidae).

In an article on the "Age of the distribution pattern of the gene arrangements in *Drosophila pseudoobscura*" Mayr (1945i) discussed the long-distance dispersal ability of these flies which occur in western North America and Middle America. They may be assumed to have jumped across the lowlands of the Isthmus of Tehuantepec in Mexico by aerial transport in fairly recent geological time. It is not so much the factor of transport that matters, but rather the ability to get established and to survive because of competition and other ecological factors. A clue to the possibly rather early origin of the gene arrangements (as opposed to their current distribution patterns) is that they are found both in *Drosophila pseudoobscura* and *D. persimilis* and may have been present already in the ancestor of these species during the late Tertiary.

Mayr (1964c,s) discussed the Neotropical Region and its bird fauna and analyzed, with W. H. Phelps, Jr., the avifauna of the impressive table mountains (tepuis) of southern Venezuela and the border region of Brazil and Guyana which they named collectively Pantepui (Mayr and Phelps 1955f, 1967c). 96 bird species are subtropical elements of this region's montane avifauna (of which 29 are endemic). Among the 48 species of long-distance colonists at least 24 presumably came from the Andes to the west, 19 from the coastal cordilleras of northern Venezuela, and five from more distant areas. In this Pantepui region there is a completely even gradation from endemic genera to species that have not even begun to develop endemic subspecies. Fewer than one third (29 species) of the subtropical bird fauna are endemic species. These facts provide conclusive evidence for the continuity and long duration of the colonization of the Tepui Mountains, which have been open to colonization for millions of years. Many of the older endemics probably are extinct and have been replaced by younger immigrants during a process of faunal turnover. In contrast to the bird fauna, the flora of Pantepui shows relationships primarily with that of the highlands of southeastern Brazil or even with Africa. Only 11 percent of the 459 plant genera known from the summit of Pantepui are related to Andean plant genera (Steyermarck 1979). What characterizes all local bird faunas is their composite nature. They are composed of colonists from different source areas.

## Polynesia

The symposium volume on "The origin and evolution of Pacific island biota" edited by A. Keast and S.E. Miller (1996) is dedicated to "Ernst Mayr: Modern pioneer of Pacific biogeography" who began studies of the evolution of Pacific island birds in the early 1930s. His writings established the basic features of Pacific ornithogeography and continue to inspire modern researchers in this region and elsewhere. As mentioned above, Mayr (1933j, 1940i) clearly discussed what became later known as the equilibrium theory of insular biogeography (MacArthur and Wilson 1963, 1967).

The native land birds of Polynesia comprise 469 species and subspecies (185 non-Passeres and 284 Passeres, songbirds). The fauna is characterized by its poverty in families and genera and by the high degree of speciation within certain genera (Mayr 1933j, 1940i), e.g., the white-eyes (*Zosterops*) with 33 species and subspecies, the *Aplonis* starlings (31), the *Halcyon* kingfishers (28), reedwarblers *Acrocephalus* (25), whistlers *Pachycephala* (25), fruit pigeons *Ptilinopus* (23), fantail flycatchers *Rhipidura* (18), etc. Except for *Acrocephalus,* all these most speciose Polynesian genera are derived from the west that is from the Papuan Region (New Guinea). Some fairly recent immigrants from the Palearctic Region reached western Micronesia (*Acrocephalus* penetrated even eastern Polynesia) and some Australian elements invaded southern Melanesia. Steadman (2006) published a modern review of the Polynesian avifauna based on recent surveys on many islands and a more complete knowledge of fossil species that became extinct in this region as a result of human impact. Such paleontological data were unavailable to Mayr in the 1930s.

The bird fauna of the Galapagos Islands is derived from South America, and half of that of the Hawaiian Islands came from North America (7 of 14 immigrants), only two immigrants are Polynesian (Mayr 1943a), but the Polynesian element prevails among the extinct bird fauna. Thus in terms of its living birds Hawaii is in the Nearctic Region. On the other hand, the Polynesian elements prevail in Hawaiian plants, insects, arachnids, and mollusks. Recent molecular studies of Hawaiian birds confirmed Mayr's conclusions (Fleischer and McIntosh 2001). Mayr (1933j, 1940i) included under "Polynesia" the following areas: Micronesia, southern Melanesia (including the Santa Cruz Islands, New Hebrides (Vanuatu), Loyalty Islands and New Caledonia), central Polynesia (including Fiji, Tonga, Samoa, and others), and eastern Polynesia (Cook, Society, Marquesas Islands, and others). Some island groups inhabited by Melanesian people (e.g., Vanuatu and Fiji) carry a Polynesian bird fauna.

Zoogeographically, all the Pacific islands (except New Caledonia and New Zealand) are "oceanic," i.e., they received their faunas across the sea and not by way of a land connection, in contrast to "continental" islands. This is also true for most of the avifauna of New Caledonia. However, the oldest endemic bird there, the kagu (*Rhynochetos jubatus*), probably is a relict of the bird fauna of the early Tertiary "continental" period of this island (which, like New Zealand, originally formed part of Gondwana land).

Mayr's (1940i) zoogeographic analysis of the central Polynesian bird fauna, especially of Fiji, Tonga, and Samoa, is methodologically of interest, because he applied here the hypothetico-deductive method. He proposed sequentially three different theories and then tested them against the data at hand: (1) Fiji, Tonga and Samoa are remnants of a large land mass; (2) Samoa is an oceanic island, whereas Fiji is continental, having received a considerable part of its fauna by way of a land bridge; (3) both Fiji and Samoa are zoogeographically oceanic islands which derived their faunas across the sea. Only the last theory stood up to the test.

Several general principles of Pacific zoogeography (Mayr 1933j, 1940i) are applicable to other island regions as well. These principles may be summarized as follows:

(1) Species numbers decrease with increasing distance from the source area.
(2) Zoogeographically, islands are either "oceanic" or "continental." The latter have a relatively complete and balanced faunal representation because of the broad former contact with a rich fauna, whereas oceanic islands lack many groups. They are inhabited by a selection of faunal elements particularly suited to long-distance dispersal.
(3) The oldest islands have, all other conditions being equal, the richest faunas.
(4) The largest (and therefore ecologically most variable) islands have the richest faunas; size and elevation appear more important than geological age.
(5) Subspeciation and speciation are characteristic features of islands and island groups. Because of frequent island hopping and the frequency of small, isolated populations, these processes are relatively rapid in islands. The small groups of individuals that founded new island populations carry only a fraction of the species' total gene pool. This is the basis of Mayr's founder principle (1942e: 237).
(6) The smaller the island population, the greater is the risk of extinction; an unexpectedly high rate of faunal turnover takes place on islands (Mayr 1965p).
(7) Successfully dispersing and colonizing bird species show a tendency to be social, that is (a) to travel in small flocks, (b) to be ecologically flexible with respect to food and habitat, (c) to be able to discover unoccupied habitat, and (d) to move up into the air where they might be caught by hurricanes (Mayr 1965n).

The open sea is like a sieve, Mayr explained, which lets pass only certain specially adapted elements: White-eyes (*Zosterops*), starlings (*Aplonis),* trillers *(Lalage*), honey-eaters (*Meliphagidae*), fruit pigeons (*Ptilinopus*), and certain thrushes form small flocks. They are among the most successful colonizers of highly isolated islands, whereas solitary birds like woodpeckers are easily stopped by barriers. When a small flock is carried to a far-distant island, both sexes will presumably be represented and a founder population can be established at once. The requirements for dispersal (barrier crossing) and colonization (becoming established) explain the relative uniformity of the animal life of Polynesia. Of the 62 bird families in New Guinea only 23 (37%) have reached Fiji and only 16 families form the major portion of the bird fauna of Melanesia and Polynesia. David Lack (1976) postulated that rather than decrease in immigration rate with distance, "ecological impoverishment" explains the decrease in species numbers on remote islands. This is very doubtful because birds are able to utilize broad resource types.

Colonizing organisms may affect the members of the indigenous fauna either as competitors/enemies (e.g., rats) or as vectors of disease organisms. Island birds are particularly vulnerable: 80 percent of all birds that became extinct in historical times were island birds, even though the number of island species is less than 10 percent of all bird species. A comparatively rapid faunal turnover is taking place on islands independent of the arrival of man. Mayr (1940i, 1965n, 1967f) established this important evolutionary phenomenon and formulated the simple rule that the percentage of endemic island species increases with the size of the island at a double logarithmic rate (Mayr 1965p). The smaller the island the more rapid the turnover which is due to the extinction of the native element and its continuous replacement by new immigrants. Mayr (1942e: 225) stated: "There is little doubt that [...] well-isolated islands are evolutionary traps, which in due time kill one species after another that settles on them." Apparently the genetic composition of island populations becomes so uniform that any change of environmental conditions including the introduction of a disease may be fatal.

In a report as chairman of a zoogeography committee of the Seventh Pacific Science Congress, Mayr (1954j) refuted once more previous attempts to build landbridges in the Pacific which is an old ocean basin. Volcanic eruptions of the Tertiary never formed extensive land masses, but merely stepping stones passable only for organisms capable of transoceanic dispersal.

All Pacific archipelagoes are "effectively accessible" to dispersing birds. In most of these islands empty niches for birds were filled more rapidly by colonists from outside than by evolution of endemic taxa in situ (Diamond 1977). However, the Marquesas, Society and Fiji Islands are sufficiently remote for birds to have permitted speciation within the archipelago at least in some groups (the pigeons *Ptilinopus mercierii* and *P. dupetithouarsii* on the Marquesas, the kingfishers *Halcyon tuta* and *H. venerata* on Tahiti (Society Islands), and the "flycatchers" *Mayrornis versicolor* and *M. lessoni* on Fiji). Only the Galapagos and Hawaii archipelagoes are so remote and immigration has been so low that intra-archipelagal speciation became dominant ("adaptive radiation" of the Geospizidae and the Drepaniidae, respectively).

## Northern Melanesia

During the 1970s, E. Mayr and J. Diamond investigated jointly the biogeography, ecology and evolution of the island birds of northern Melanesia (Bismarck and Solomon Islands) regarding the origin of the montane avifauna (Mayr and Diamond 1976f), the species-area relation (Diamond and Mayr 1976a), and the species-distance relation (Diamond, Gilpin and Mayr 1976d). In these studies more than half of the work and results are J. Diamond's. Most montane species have lowland populations on some islands. For example, the Island Thrush (*Turdus poliocephalus*) and the "flycatcher" *Rhipidura [spilodera]* are montane on some islands but inhabit the lowlands of other islands in the same archipelago. These niche shifts can often be correlated with major differences in the biological environment of competing species in the lowlands and mountains. Most montane birds either dispersed by direct "jumping" from mountain to mountain or they evolved from a relative in the lowlands. Almost all of the "great speciators," with at least five distinct subspecies or allospecies in the Solomons, are rather common short-distance colonists. Correlated with their high inter-island immigration rates, long-distance colonists show no geographical variation throughout the Solomons.

In their large book, *The Birds of Northern Melanesia. Speciation, Ecology, and Biogeography* Mayr and Diamond (2001g)[6] investigated the processes of geographic speciation, as documented in the northern Melanesian avifauna by island populations that have reached various stages of taxonomic differentiation. Overwater dispersal ability of each species and the weak to more conspicuous geographical variation of conspecific populations on different islands are discussed in detail. Speciation is complete when the new taxon developed not only reproductive (genetic) but also ecological isolation and was able to invade the range of its parent population (19 species pairs). These processes can be followed by a quantitative analysis of this rich, yet still "manageable" tropical bird fauna (191 native zoogeographical species, i.e., taxonomically isolated species and superspecies). The authors also analyzed the varying degrees of endemism in the different island groups, the varying ecological strategies of species, their geographical origin (mostly from New Guinea and Australia), dispersal barriers and extinction of species as well as numerous other ecological and biogeographical aspects, including the effect of Pleistocene sea-level changes on the distribution and current differentiation of these island birds (Fig. 4.12).

Their study led Mayr and Diamond (2001g) to conclude: "We cannot recognize any case that could plausibly be interpreted as a stage in sympatric speciation" (p. XXII). All evidence favors the model of allopatric speciation, more specifically peripatric speciation from founder populations: "Peripheral isolates attract attention because the geographically most peripheral taxon in a species or superspecies is usually the most strikingly divergent one. However, in Northern Melanesia it is rare for peripheral isolates to expand into [species rich] central locations and become important evolutionary novelties, because of the problems of upstream colonization and faunal dominance" (p. 287). Therefore, at least in

[6] The basic manuscript was Mayr's, but in the end J. Diamond (*1937) had at least done half. Diamond is a professor of physiology at the University of California at Los Angeles and the author of several award-winning books. He has been elected to the National Academy of Sciences, the American Academy of Arts and Sciences, and the American Philosophical Society. He is also the recipient of a MacArthur Foundation fellowship and was awarded the 1999 U.S. National Medal of Science and the 2000 Lewis Thomas Prize of the Rockefeller University, New York. His father, L. K. Diamond, had also collaborated with Ernst Mayr in a study on human blood groups in 1956 (p. 281).

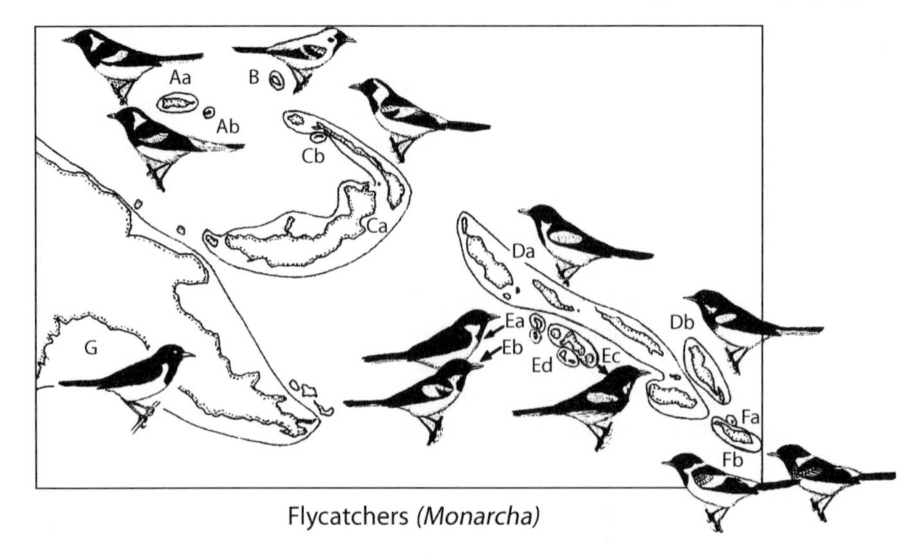


**Fig. 4.12.** Distribution of the monarch flycatcher superspecies *Monarcha [manadensis]*. A *M. infelix*, B *M. mencke*, C *M. verticalis*, D *M. barbatus*, E *M. browni*, F *M. viduus*, G *M. manadensis* in northern Melanesia. Plumage color is basically black and white. This superspecies is one of the extremes among "great speciators," with six allospecies, 11 megasubspecies, and two weak subspecies in Northern Melanesia, plus four allospecies outside Northern Melanesia. The superspecies thus comprises forms of a wide range of distinctness, from weak subspecies to very distinct allospecies, with many borderline forms about which it is difficult to decide whether to classify them as allospecies or subspecies. All Northern Melanesian taxa are allopatric, but speciation has produced sympatric related species on New Guinea and several other Papuan and Lesser Sundan islands. Adapted from Mayr and Diamond (2001, map 41 and plate 3) which see for identification of subspecies

this region and bird fauna, the general evolutionary significance of peripheral isolates is relatively low.

Although molecular sequence data are needed to confirm the genealogical relationships of the bird taxa studied, such information is unlikely to change the major conclusions reached. This massive handbook again demonstrated the overriding importance of time and barriers to dispersal for the speciation of island birds. It "is an impressive synthesis, unique in scope, and inspiring by the challenges it sets forth. It represents the fulfillment of a research program" (Grant 2002a: 1881). Hopefully, similarly comprehensive analyses of geographical differentiation and speciation will be published for other groups like insects, land snails or plants. The model system of birds should be compared with other groups of organisms to see whether the generalizations made are true also for other organisms or whether they have their own regularities and laws. The book on the bird fauna of northern Melanesia summarized Mayr's latest thinking in the field of biogeography.

## New Guinea

New Guinea forms one zoogeographical unit with Australia, as shown by the lack of placental mammals (except for bats and some rodents) and the wealth of marsupials. The bird fauna of New Guinea resembles the mammalian fauna in the strong prevalence of groups which it has in common with Australia (Mayr 1954k). Papuan or Australo-Papuan families and subfamilies include those with the highest number of New Guinea species, like the Meliphagidae (61), Malurinae (24), and Pachycephalinae (24). At the level of genera the Australo-Papuan elements also far outweigh the Indomalayan (Asiatic) elements, the other major faunal group represented in New Guinea; and there is a high degree of striking endemism in the Papuan Region. The Moluccan Islands are inhabited by an impoverished Papuan bird fauna with a considerable admixture of Asiatic elements.

In contrast to these relationships of the faunas, the flora of the tropical belt from Malaya through New Guinea to the Solomon Islands and Polynesia forms one phytogeographical unit (Malesia) which differs strikingly from the flora of temperate and warm-dry Australia. The reasons for this difference between the zoogeographic and phytogeographic situation are according to Mayr (l.c.): Dispersal between Malaya and New Guinea demanded island hopping, but birds, in spite of their mobility, are easily stopped by ocean barriers. On the other hand plants disperse without difficulties across water gaps if the land area beyond the water barrier is located within the same climatic zone. The repeated and sometimes long lasting connections between New Guinea and Australia permitted a free faunal interchange. The nearest relatives of many rainforest birds of New Guinea inhabit brush savannas and semideserts of Australia. Such ecological shifts are much more difficult for plants (and certain invertebrates). Their establishment after dispersal is more closely dependent on climatic and edaphic factors than that of birds. These facts explain the striking differences between the floras of New Guinea and Australia in contrast to the respective bird faunas.

Nothing was known about climatic-vegetational changes in the tropical lowlands of New Guinea and corresponding faunal movements during recent geological periods, when Ernst Mayr traveled in these regions. Botanists had suggested that the grass savannas around Lake Sentani near Hollandia (Jayapura) in northern New Guinea (Fig. 2.8) had originated through deforestation by man. Mayr was immediately doubtful of such an interpretation because he found several endemic subspecies of birds to inhabit these savannas (e. g. *Lanius schach stresemanni*, *Saxicola caprata aethiops*, etc.) indicating a very old age and a natural origin of these plant formations. They had been merely enlarged through later rainforest clearing. In his expedition report Mayr (1930f: 25) stated with respect to the grasslands around Lake Sentani:

"The many indigenous subspecies prove that the grassland must be of very old origin. Nowadays the natives burn the grass regularly, and the forest is going back every year, but I am convinced (contrary to the opinion of botanists) that this grassland here is a very old one. It is isolated now more or less from the grassland patches of Eastern New Guinea, but I think that in former geological periods the

Pied Bushchat *(Saxicola caprata)*

**Fig. 4.13.** Distribution of the Pied Bushchat (*Saxicola caprata*) in the Malay Archipelago and in New Guinea from where this open-country species of Asian origins has reached only the Bismarck islands of New Britain and New Ireland and the nearby recently defaunated volcanic islands of Long and Uatom. Like most other open-country colonists of Northern Melanesia, it has scarcely differentiated: the Bismarck populations still belong to the same subspecies as the New Guinea source population. From Mayr and Diamond (2001, map 31) and bird sketches from MacKinnon and Phillipps (1993, plate 71)

steppe had a much wider distribution in New Guinea than now" and in a letter to Erwin Stresemann dated 25 April 1939 discussing the distribution pattern of the Pied Bushchat (*Saxicola caprata*; Fig. 4.13) Mayr wrote: "I believe that at some time in the past there was a drier zone along the north coast of New Guinea which was the immigration route of this form. The three species which I discovered in the Arfak Mountains: *Megalurus timoriensis, Acrocephalus arundinaceus* and *Lonchura vana* are an indication of this line of immigration. There are quite a number of the savanna species which have not reached Australia. I am fairly convinced that these species reached eastern New Guinea via the Moluccas and western New Guinea. These species are *Saxicola caprata, Merops philippinus,* and *Lanius schach*" (see Haffer 1997b: 509). *Saxicola caprata* probably did not have to fly across the ocean from the Philippines to eastern New Guinea, as tentatively indicated on Fig. 4.13, but was able to hop from one island to the next through the Moluccas and western New Guinea during a cool-dry climatic period of the recent geological past, when these regions were partially covered with open nonforest vegetation.

Similarly, the fact that the grasslands of the island of Guadalcanal, one of the Solomon Islands, are the home of two endemic subspecies of birds (a finch and

**Table 4.2.** Degree of endemism in montane species of Passeres on three New Guinea mountains (after Mayr 1940c); two percent values under Arfak slightly changed

| | Arfak | Cyclops | Saruwaget |
|---|---|---|---|
| Not endemic | 36 = 40.4 % | 21 = 72.4 % | 59 = 68.6 % |
| Endemic subspecies | 46 = 51.7 % | 8 = 27.6 % | 23 = 26.7 % |
| Endemic semispecies | 5 = 5.6 % | 0 = 0 % | 3 = 3.5 % |
| Endemic full species | 2 = 2.2 % | 0 = 0 % | 1 = 1.2 % |
| Total species | 89 = 100 % | 29 = 100 % | 86 = 100 % |

a button quail) indicated to Mayr (1943h) that this plant formation is older than human colonization.

Geographical isolation of the faunas inhabiting the three New Guinea mountain ranges which Mayr explored in 1928–1929 led to the differentiation of the respective taxa of birds as illustrated in Table 4.2. There are several endemic species and numerous endemic subspecies. Even most of those species which are listed as undifferentiated are slightly different on each mountain, only these differences are below the "taxonomic threshold" of subspecies (Mayr 1940c).

In the introduction to his *List of New Guinea Birds* Mayr (1941f) explained: "In 1930, on my return from New Guinea, I planned to write an ornithogeography of the Papuan region. It soon became apparent to me that no such work was possible without a reliable list of the birds of the New Guinea region and that it would be my first task to prepare such a list" (p. V). At the annual meeting of the American Ornithologists' Union in Pittsburgh (1936) he presented a zoogeographical analysis entitled "The birds of the New Guinea region," illustrated with lantern slides, and in 1969 (e, p. 11) he stated "a more detailed study [of bird speciation in New Guinea] by Mayr is in preparation," but the manuscript was never completed. It should be noted, however, that several aspects of the zoogeography and speciation of New Guinea birds are treated in Mayr's book of 1942(e).

Meanwhile, some data on the changing distribution of rainforest and more open vegetation in the lowlands of New Guinea and the Malay Archipelago during the last several million years have become known (Hope 1996, Morley 2000). According to Pratt (1982, Beehler et al. 1986) these changes caused temporary separation and differentiation (i.e., subspeciation and speciation) of animal populations in ecological refugia.

## Australia

Mayr's zoogeographic analysis of the Australian avifauna (1944e, 1944k), still valid, revealed endemic families, genera, species, and subspecies. Most immigrants came from Asia and the region of the Malay Archipelago across ocean barriers (see also

1972c, 1990b). Thus most of the Australian fauna developed through "continued single origin colonization," that is the steady and continuous colonization from the northwest (1965q). Corresponding to their differentiation and relative age, Mayr distinguished five major layers of colonists between which there are, of course, no clear cut discontinuities (1944k):

(a) Strongly endemic families and subfamilies, about 15,
(b) Less isolated indigenous families and subfamilies, about 8,
(c) Endemic genera of non-endemic families, about 30,
(d) Endemic species of non-endemic genera, about 60,
(e) Recently immigrated species, not differentiated at all, or only subspecifically, about 40.

The oldest layer comprising, besides emu and cassowary, the Megapodiidae, *Anseranas, Pedionomus* and perhaps *Stictonetta,* probably represents Gondwana elements, less than 3% of the bird fauna (Mayr 1990b). Australia split from Antarctica about 50 million years ago. At that time Australia must have had a fairly rich Gondwana bird fauna, but owing to circumstances that are not yet fully explained, this older fauna became almost completely extinct and was replaced during the Tertiary by immigrants from Asia. The ancestors of the Corvida immigrated into Australia ca. 30 million years ago and radiated into the characteristic families of Australian songbirds.

An island archipelago probably existed between eastern Asia and Australia during the Mesozoic at least since the early Cretaceous and during the entire Cenozoic (Tertiary-Quaternary). This archipelago probably included small drifting continental plates (split off northern Gondwanaland) which today form portions of Burma, Malaya, Borneo, Sumatra, Java and western Sulawesi; portions of what later became Timor and New Guinea were located off northern Australia (Audley-Charles 1987). The islands of this early Southeast Asian archipelago probably served as stepping stones for Asian immigrants into Australia as well as an area of intensive speciation and faunal differentiation during the Cretaceous and Tertiary. Many animal groups originated here which today form part of the rich fauna of the Malay Archipelago and the New Guinea region. The last faunal exchanges between Asia, the Malay Archipelago and Australia took place during several periods of low sea-level stand during the Pleistocene, when large portions of the shelf regions of the world became dry. As an example, the distance Timor-Australia during periods of lowered sea-level was only 50 miles instead of 300 miles today. This greatly facilitated the exchange of certain members of the bird faunas, mostly species of open savanna vegetation that was widespread on these shelf regions during cold (glacial) periods of lowered sea-level (Mayr 1944e, 1944k).

There is no question about the Asian origin of nearly all of the ancestors of Australian birds, as Mayr's analysis had revealed. Recent authors speculate that the ancestors of songbirds (Oscines) which reached Australia from Asia, also had originated in Gondwana land. How they may have arrived in the northern hemisphere (via contact of "Greater India" with Asia during the Eocene?) remains open (Cracraft 2002).

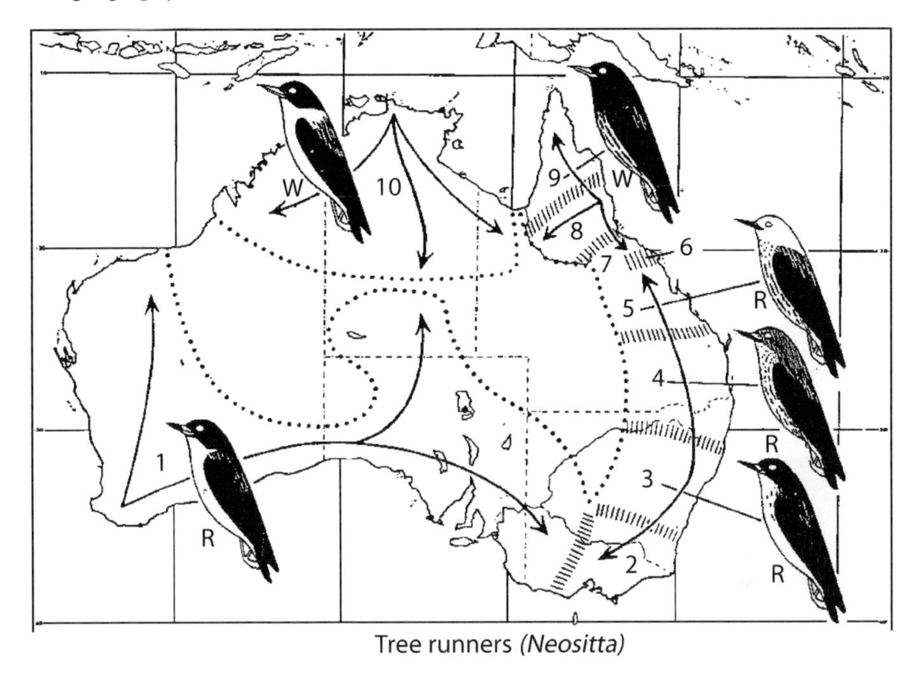


**Fig. 4.14.** Secondary hybrid belts among tree runners (*Neositta*) of Australia. The *arrows* indicate expansion from Pleistocene aridity refuges. Wherever former isolates have met, they have formed hybrid belts (indicated by hatching). R, subspecies with red wing bar; W, subspecies with white wing bar. From Mayr (1963b: 373)

As to the differentiation of Australian birds at low taxonomic levels Mayr (1950b) conceived a theory of speciation and subspeciation in tree-runners (*Neositta*): During a Pleistocene or post-Pleistocene arid period (or several periods) a previously widespread ancestral species had been separated into a number of moist refuges around the periphery of the continent, where these isolated populations acquired the diagnostic characters of the taxa of this genus now recognized (Fig. 4.14). After the return of more humid conditions the previous isolates spread out from the refuges together with the expanding vegetation zones, hybridizing in the areas where they came in contact with one another. Keast (1961) and Short et al. (1983) strongly supported this interpretation. It was the first detailed presentation of the refuge theory of Pleistocene speciation for birds of the southern hemisphere. Zoogeographers had proposed similar models for the differentiation of European birds and other animals which were assumed to have survived in humid Mediterranean refuges during the generally dry glacial periods.

## Malay Archipelago

A schematic eastern delimitation of the Indomalayan Region follows the border of the Pleistocene Sunda land from the straits between Bali and Lombok in the south, between Borneo and Celebes (Sulawesi) and between Palawan and the Philippines in the north. This is "Wallace's line" of most zoogeographers (Fig. 4.15). A corresponding western limit of the Australo-Papuan Region follows the western border of the Sahul Shelf which includes the islands of Aru, Misol and Waigeu ("Lydekker's line"). There is a third line where western and eastern faunal elements of mammals and birds reach similar proportions of faunal balance (50 : 50%), "Weber's line" west of Tenimber, Buru and the northern Moluccas. Weber's line separates the islands with predominantly Indomalayan fauna in the west from islands with a predominantly Papuan fauna in the east. Mayr (1944d) documented that the shelf margins of both the Sunda land and Sahul land represent conspicuous breaks in a general faunal transition zone between the Indomalayan and Australo-

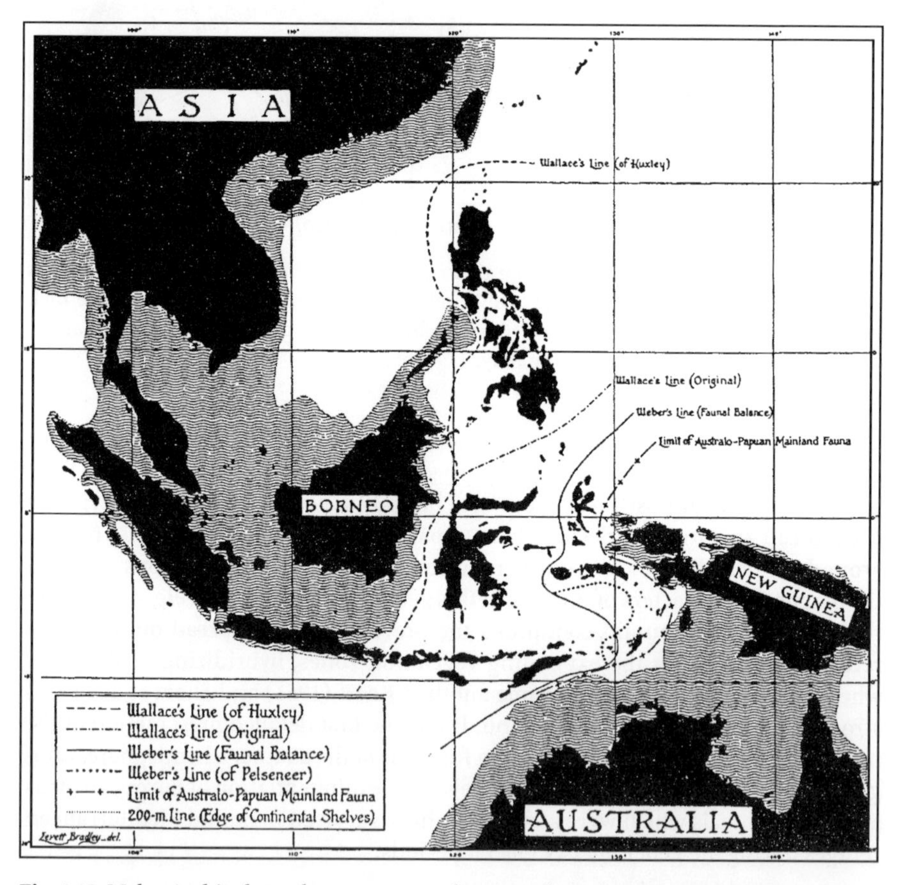


**Fig. 4.15.** Malay Archipelago, the contact zone between the Indo-Malayan and the Australo-Papuan faunas. The shaded area in the west is the Asian (Sunda) shelf, and in the east lies the Australian (Sahul) shelf. The area between the two shelves, never connected by a land bridge, is referred to as Wallacea. The real border (line of balance) between the Asian and the Australian faunas is Weber's Line (from Mayr 1944d)

Papuan Regions (see also Keast 1983; Clode and O'Brien 2001). Generally speaking, this island region is inhabited by such a faunal mixture of western and eastern provenance and so heterogeneous that according to Mayr (l.c.), it should not be considered as a separate zoogeographical unit "Wallacea." All four areas that are supposed to make up "Wallacea," the Lesser Sunda Islands, the Celebes region, the Philippines, and the Moluccas, are more different from each other and each of them is more similar to an area outside Wallacea than to make sense whatsoever of the recognition of "Wallacea." The western elements of reptiles, butterflies and plants penetrated farther to the east than those of birds and mammals. Plant geographers even consider New Guinea part of the Malayan (Malesian) Region.

The islands of Timor and Sumba are inhabited by a predominantly Indomalayan bird fauna containing numerous eastern elements from the Papuan Region (Mayr 1944e) all of which reached these islands by dispersal across ocean barriers. Like Wallace (1869) Mayr found no support for any of the numerous land bridges in this region that other zoogeographers, e.g., Rensch (1936), had proposed to explain the immigration of birds and other animals. Four phenomena confirm the interpretation of dispersal across ocean barriers in the Malay Archipelago and contradict the assumption of land bridges (Mayr, l.c.):

(1) Faunal relationships are independent of submarine contours (below the 200-m line) but are closely correlated with the distances of the islands from each other (providing size and ecology are comparable);
(2) Faunal affinities exist even between geologically unrelated islands;
(3) The small percentage of endemic species on the islands of the Sunda arc is explained by the superior dispersal faculties of the birds of the Lesser Sunda Islands which crossed easily the water gaps to the north and south;
(4) The spread of birds inhabiting high mountains in the Malay Archipelago can be understood only by assuming great dispersal powers, because their montane habitats have never been directly connected, not even during the coldest periods of the Pleistocene.

This important zoogeographical analysis (which Mayr considers one of his best papers in this field) established the "oceanic" nature of the eastern islands of the Malay Archipelago strongly supporting Stresemann's (1939) conclusions. When the latter congratulated Mayr, in early 1946, on his zoogeographical studies of the bird faunas of the Malay Archipelago ("You succeeded in making zoogeography a fascinating subject again"), Mayr refused "credit for having started anything new. It was your zoogeographical analysis in the *Birds of Celebes* (1939) which made me think and got me interested in analyzing the birds of Timor (1944e) in a somewhat similar fashion" (21 March 1946). This is another instance of the important influence of Stresemann's methods and thinking on Mayr's systematic and zoogeographic work (see also Glaubrecht 2002; Vuilleumier 2005b:66–67).

About Rensch's book, *History of the Sunda Island Arc* (1936) Mayr had written to Stresemann: "I thoroughly disagree with most of Rensch's conclusions. He compares the faunas of two islands and then draws conclusions as to their possible connection on the basis of the percentages of the animals common to both. In this, he pays no attention whatsoever (except in the general introduction) to the fact, that these islands might be totally different in their ecology: If you compare a savanna-covered island with a completely wooded island, you may get only a slight agreement of the faunas, although the geological history of the two might easily be the same. On the other hand, Rensch makes very little allowance for the distribution across the open sea. My work on Polynesian islands and even such islands, like Rennell and Biak , has convinced me that the open sea is not always as much of a barrier as commonly supposed. As soon as two islands are fully settled such as Batanta and Salawati, there is only very little exchange of individuals, but as long as one island is full of open ecological niches, it will permit the settling down of immigrants. This point has been largely disregarded by Rensch. I believe you cannot make such definite statements as he does in many cases, neither do I believe that all of the island connections have existed as postulated by him" (25 April 1939).

The distribution of birds in the Philippines is influenced by three main factors (Delacour and Mayr 1946k): (1) Climate and vegetation, (2) former land connection between islands, (3) proximity to neighboring areas. These factors jointly have contributed to the special bird faunas of the various islands and island groups.

# 5 Biological Species and Speciation—Mayr's First Synthesis

## A Modern Unified Theory of Evolution

Ernst Mayr's early interests in evolution and genetics (pp. 26–29, 45) led to his decisive contributions to the modern synthesis of the late 1930s and 1940s, when a largely unified evolutionary theory emerged. Biological evolution comprises two components–adaptive development of populations during long geological periods (anagenesis, evolution as such) and multiplication of species during relatively short periods (speciation, cladogenesis). Evolution as such and the theory of common descent were accepted by biologists within a few years of the publication of the *Origin* in 1859 during the *first Darwinian revolution*. A synthesis of modern genetics and evolution as such was accomplished when, by 1932, the mathematical geneticists R. A. Fisher, J. B. S. Haldane and S. Wright had convincingly shown that small mutations and natural selection play the main roles in the gradual process of adaptive evolution of populations through time, thereby solving one of the two major problems of evolutionary biology, the problem of anagenesis, a historical accomplishment which Mayr (1999k, 2004a) called the "*Fisherian synthesis.*" During the so-called *Evolutionary Synthesis* of the period 1937–1950 (the *second Darwinian revolution*) the other main problems of evolutionary biology were solved or generally accepted (gradualism, speciation, and natural selection) and the processes of speciation were combined with those of adaptive evolution (Mayr 1993a). None of the mathematical geneticists had discussed the phenomenon of speciation or did so only superficially. In general, the period of the modern synthesis, when a new unified theory of evolution originated, saw a synthesis (1) between the thinking in three major biological disciplines–genetics, systematics and paleontology, (2) between an experimental-reductionist approach (genetics) and an observational-holistic approach (naturalists-systematists) and (3) between an anglophone tradition with an emphasis on mathematics and adaptation and a continental European tradition with an emphasis on populations, species, and higher taxa. The three genetical aspects (then not new insights) that were firmly and universally adopted during the evolutionary synthesis were (1) that inheritance is hard, there is no inheritance of acquired characters, (2) that inheritance is particulate, that is, the genetic contributions of the parents do not blend but remain separate, to be differently recombined in future generations, and (3) that most mutations are very small and evolution therefore is gradual. The evolutionary synthesis also led to a refutation of the three anti-Darwinian paradigms (a) the typological-saltational, (b) the teleological-orthogenetic and (c) the transformationist-lamarckian theories.

Based on his background as a naturalist-systematist in Russia during the 1920s, Dobzhansky (1937) produced a first synthesis between the views of the naturalists-systematists and the geneticists. The other "architects" of the evolutionary synthesis widened the path which Dobzhansky had blazed–Ernst Mayr (1942e, species and speciation), Julian Huxley (1942, general evolution), George Gaylord Simpson (1944, paleontology), Bernhard Rensch (1947, macroevolution), and Ledyard Stebbins (1950, botany).[1] Because most mutations are very small, Th. Dobzhansky (1937) and E. Mayr (1942e) were able to demonstrate that there was no conflict between the results of the population geneticists and those of the systematists who had discussed gradual differentiation of populations and geographical speciation for a long time (although, until the early 1930s, based largely on the assumption of the Lamarckian inheritance of acquired characters). The original Darwinian paradigm of variation and selection was confirmed during the evolutionary synthesis. Between 1937 and 1950, the "architects" combined in synthetic publications the results of their own research with those of population genetics. The process of this unification of evolutionary biology is referred to as the evolutionary synthesis and the product as the synthetic evolutionary theory (Synthetic Darwinism, Junker 2004). Mayr's specific contributions to the evolutionary synthesis were his analyses of the nature of biological species and of the origin of organic diversity (speciation) making "the species problem" a central concern of evolutionary biology. Speciation and other processes in evolution are not simply a matter of genes but of populations and of species. His 1942 volume explained a large part of evolutionary theory well-known to naturalists-systematists but not to geneticists, particularly species and speciation and the role of geography in the evolution of populations and species. He discussed populations at various intermediate stages between variously differentiated subspecies and variously differentiated biological species in line with gradual Darwinian change. Mayr's work demonstrated the importance of taxonomic research for evolutionary theory.

At an international conference in Princeton, New Jersey, in January 1947 there was general agreement among the participating geneticists and naturalists-systemists on the nature of species, the gradualness of evolution, the importance of natural selection, and the populational aspect of the gradual origin of species. A synthesis indeed had taken place, but it goes without saying that, by present standards, this synthesis was still incomplete. It did not include molecular evolution, comparative genomics, evolutionary developmental biology or phylogenetics, i.e., the actual, detailed history of life on Earth (Wilkins 2007). Some differences that remained at that time included the problem of the target of selective demands which, for the population geneticists, continued to be the gene, whereas the naturalists-systematists insisted, as had Darwin, that it was the individual organism as a whole. The individual either survives or it does not; it either reproduces successfully or it

[1] It should be noted that the "Evolutionary Synthesis" was an international research program to which also several other European biologists contributed, especially E. Baur, N. Timoféeff-Ressovsky, and W. Zimmermann in Germany, S. Chetverikov and N. Dubinin in the Soviet Union, G. Teissier in France, and A. Buzzati-Traverso in Italy (Mayr 1999a; Reif et al. 2000; Junker 2004); see p. 357.

does not. The gene is never isolated and can never be selected by itself, as Mayr (e.g., 1959f) argued in his attack on "beanbag" genetics showing that it is individuals that really count. In general, there are 2 or 3 possible targets of selection. One is the gametes which are directly selected. The next one is the individual and the third one are certain types of social groups, those consisting of cooperating individuals (e.g., in early humans).

In general, a unification of biology "was not an objective in the minds of any of the architects of the synthesis during the 1930–1940 period. They were busy enough straightening out their own differences and refuting the anti-Darwinians to have time for such a far-reaching objective. It wasn't until the 1950s that most of the previous difficulties had been resolved that one could begin to think seriously about the role of evolutionary biology in the whole of biology and about the capacity of evolutionary biology to achieve a unification of the previously badly splintered biology" (Mayr 1993a: 33; see also Smocovitis 1997: 202).

## Collaboration with Th. Dobzhansky

Early in the 1930s, Mayr had given up his Lamarckian in favor of Darwinian selectionist views (as had independently E. Stresemann and B. Rensch in Berlin) convinced by the publications of geneticists and long conversations with James Chapin, explorer of the Congo rainforest and his colleague at the AMNH (p. 120). The first volume of Chapin's *Birds of the Belgian Congo* (1932) was the best work on the ecology, behavior, and biogeography of tropical birds of that time. F. M. Chapman, the chairman of the Bird Department, still believed in direct environmental influences and in saltation, as did Mayr's colleagues R. C. Murphy, J. T. Zimmer, G. K. Noble, and others. After Chapman's retirement in 1942, Ernst Mayr became the dominant force within the Department of Ornithology showing new theoretical and conceptual ways to explain the development of species and other evolutionary processes (Lanyon 1995).

Mayr's taxonomic studies of the 1930s were written for taxonomists and did not reveal the fact that their author was collecting data for a comprehensive evolutionary analysis of geographic variation and speciation in birds (and other animals). What he intended is mentioned in his second letter to Theodosius Dobzhansky (Pasadena, California) dated 25 November 1935:

> *Dear Dr. Dobzhansky,*
> *Many thanks for your kind letter of November 12th and for your reprint. I will send you in the next days a few of my taxonomic papers, although I doubt if they mean much to you. I have restricted myself in the past to purely descriptive species and genus revisions and am waiting for the completion of these detailed taxonomic studies before I want to draw any conclusions. I am mainly working with insular birds, which do not lend themselves as easily to the study of gradual differences of populations as do continental birds. In fact, the majority of the subspecific characters of these island birds are those of the type, but could have easily originated as conspicuous gene mutations. These birds have,*

*however, one advantage. They show clearly that Goldschmidt's contention that subspecies are not the building material of new species is wrong. In fact, about 60% of these subspecies are considered good species by the majority of living ornithologists.*
*There is one genetical phenomenon which seems to be of particular interest to the taxonomist. This is what one of the Russians, I believe Severtsov, has called 'homologous series of mutations.' It is the fact that the common and most conspicuous mutations in the genus* Drosophila *(or any other genus) occur in every species of the genus, while the characters that separate the various taxonomic units are much more stable and much harder to define than "white" or "sooty." If you have a* D. melanogaster *which has defect mutations in regard to its eye, to its wing and to its abdomen, a qualifying biologist will still tell you without hesitation that this is* D. melanogaster.
*The same is true for birds. The real differential characters between species seem to be of a different kind than typical gene mutations. I am well aware that most of these statements are mere "hunches" and are hard to substantiate, but whatever experiments have been made with geographical races or species seem to indicate that many characters of geographical races do not inherit in Mendelian fashion. I would like to refer to the latest* Peromyscus *studies of the Michigan school. We get the same thing with our wild populations in continental birds. There is a perfect blending from one population to the next. Birds have one great advantage, i.e., that their taxonomy is better known than that of any other group of animals. They also have one great disadvantage, which is that they are hard to keep in captivity and that their reproduction is very slow as compared with insects. I believe that the task of the bird student will be to make suggestions to the geneticist interested in evolution, but it will be up to the experimental entomologist to prove or disprove these suggestions.*
*I am very willing to help you in getting taxonomic papers that might be of interest to you. The majority of taxonomic papers, however, is purely descriptive and would mean nothing to you. Rensch has compiled up to 1929 most of the literature which has a bearing on this subject. Since then a good many scattered contributions were made, but it is hard to say what would be significant and what not. Why do you not send me specific questions? They are more easily answered.*
*Yours sincerely,* (Ernst Mayr, 1935)

This letter again documents Mayr's and many other zoologists' belief that two different types of characters are involved in the differentiation of populations: (1) conspicuous but unimportant Mendelian characters and (2) blending characters important for geographical divergence. After reading Dobzhansky's (1933) paper on "Geographic variation in lady-beetles" Mayr had exclaimed "Here is finally a geneticist who understands us taxonomists!" (Mayr 1980n: 419) and had written him a "fan letter" on 7 November 1935 emphasizing the need for an integration of genetic and taxonomic research, as he had tried in his early paper on the snowfinches (Mayr 1927f; see here p. 45). Dobzhansky answered him on 12 November 1935:

*Dear Dr. Mayr:*
*Many thanks for your kind letter of Nov. 7th, which is so highly flattering to me, and to which I certainly want to reply.*
*The need for a reconciliation of the views of taxonomists and geneticists I feel very keenly, but it seems to me that all what is to be reconciled are just the viewpoints, since I do not perceive any contradictions between the facts secured in the respective fields. Of course, this is a big "just". So far geneticists appear to think that they need not pay any attention to what taxonomists are doing, and vice versa. To my mind this is the root of the trouble. Probably no less than 75% of geneticists still believe that there is nothing in particular to be gained from studies on the races of wild animals as compared with races in bottles. You and myself will probably have no disagreement as to the absurdity of this view.*
*On the other hand, I also fail to see the difference between the "qualitative" and "quantitative." May be I am blind, but the only distinction which is of consequence is that the former are easy to study, and the latter are notoriously difficult. Of course, there are geneticists which agree with you (e.g., Castle). Can you, however, name anyone who got anywhere in particular with this distinction?*
*The fly I am working with (*Drosophila pseudoobscura*) is as favorable a material for studies on geographic variation as its more widely known relative is unfavorable (*Dr. melanogaster*). If nothing unexpected happens, I hope to be able to furnish you in the future with some more information on the subject. In turn, I feel the deficiency of my knowledge of the studies in the same field by taxonomists; our library here has no taxonomic literature at all. Hence, I would like to conclude this my discourse with a request to you to include my name in your mailing list. I think you have some of my older reprints, and in a few days I shall send you another bunch.*
*Sincerely yours, Th. Dobzhansky.*

Both Dobzhansky and Mayr followed a continental European tradition in evolutionary research and systematics studying geographical changes of populations leading to speciation and macroevolution, i.e., the "horizontal" (geographical) dimension of evolution. Other representatives of this tradition were Plate, Stresemann, Rensch, Stegmann, Mertens, Reinig, Wettstein, Baur, Philipchenko, Timoféeff-Ressovsky, and, prior to 1933, Goldschmidt. This list includes not only evolutionary systematists but also several continental geneticists (see also Harwood 1993: 129–137). On the other hand, an Anglophone tradition emphasized the study of adaptive genetic change in populations through time, i.e., the "vertical" dimension of evolution, indicated by the names of T.H. Morgan, H.J. Muller, S. Wright, R.A. Fisher and J.B.S. Haldane. Of course this distinction is not always clear-cut. Several scientists in North America and Britain also studied problems of diversity and the "horizontal" aspects of evolution like F. Sumner, L.R. Dice, D.S. Jordan, J. Grinnell, E.B. Poulton, E.B. Ford, and Karl Jordan (Mayr 1992i, 1993a).

When Dobzhansky visited New York a year later, in October–November 1936, to lecture on *Genetics and the Origin of Species* Mayr naturally attended these talks which for him were like an "intellectual honeymoon," learning an enormous amount of evolutionary genetics and being ever more enthusiastic about speciation. Dobzhansky's book with the same title as his lectures appeared in late 1937[2] and Mayr delved into it immediately. He appreciated Dobzhansky's emphasis on the importance of morphological taxonomic work and the still imperfect understanding of many aspects of evolution. Dobzhansky pointed out that the majority of mutations are very small and frequently beneficial (rather than large and lethal as most German taxonomists still believed). Geographical variation and its adaptive nature are discussed at length, including Mayr's information on the Golden Whistler (*Pachycephala pectoralis*). Dobzhansky did mention the difference between processes of speciation (called "cladogenesis" later) and changes in a single gene pool over time ("anagenesis"), but–despite the title of the book–there is no separate chapter on the origin of new species. In the treatment of isolating mechanisms (a most useful term coined by Dobzhansky [1935, 1937]) he defined species as reproductive communities. Among the isolating mechanisms he recognized extrinsic ones (like geographical barriers) and intrinsic properties of organisms. Mayr (1942e) later provided an improved classification and strong arguments in favor of geographical speciation which Dobzhansky had treated rather ambiguously. Mayr disagreed strongly with Dobzhansky's definitions of subspecies (races) and species as "stages in the evolutionary process." Rather, Mayr (1942e) argued, they are populations or sets of populations.

They exchanged many letters until, in late 1939, Dobzhansky joined the Department of Zoology at Columbia University, New York. Ever since they had established contact in 1935, they were striving fervently for a synthesis of genetic and taxonomic data and research without realizing how close they were to that goal. Through their discussions, presentations and publications they actively forged the evolutionary synthesis[3]. In December 1939, Mayr spoke on "Speciation phenomena in birds" stating in the introduction:

"Evolution is a very complicated and many-sided process. Every single branch of biology contributes its share of new ideas and new evidence, but no single discipline can hope to find all the answers or is justified to make sweeping generalizations that are based only on the evidence of its particular restricted field. This is true for cytology and genetics, for ecology and biogeography, for paleontology and taxonomy. All these branches must cooperate. [...] It is obvious that the taxonomist will

---

[2] On the spine of Dobzhansky's book (1937) appears an artistic diagram of a cell dividing which, however, has the chromosomes in the division figure backward. Ernst Mayr delighted in pointing out this error to Dobzhansky who, so far, had not noticed it: "The bird taxonomist telling the working cytologist that the cell division pictured on the spine of his book is ridicuously in error" (Provine 1994: 111).

[3] Regarding Mayr's and Dobzhansky's roles in the development of evolutionary studies in the United States from 1936–1947, in particular the foundation of the Society for the Study of Evolution and its journal *Evolution*, see Jepsen (1949), Cain (1993, 1994) and Smocovitis (1994a, 1994b).

not find out very much about the origin of new genetic characters nor about their transmission from one generation to the next. On the other hand, the taxonomist will be able to give answers to certain questions which are not attainable by the geneticist since speciation is not a purely genetic process" (Mayr 1940c: 249) and in a lecture next year he said: "We seem to be at the beginning of a new phase of the study of evolution, that is the study of the origin of the discontinuities between species. This is a typical border line field which must be studied jointly by the naturalist and ecologist, by the geneticist and taxonomist" (Mayr 1941i: 139).

Mayr had been thoroughly influenced by the ideas of Stresemann and Rensch at the Berlin Museum of Natural History who fully endorsed the biological species concept and the importance of geographical variation and barriers for the process of speciation, and he assumed that any taxonomist with some knowledge of the literature would also adopt this concept and interpretation. However, several experienced workers published critical statements like Goldschmidt (1933,1935) who denied geographical speciation, Robson and Richards (1936) who minimized it and the paleontologists (Osborn, Beurlen and Schindewolf) who ignored it. "I suddenly realized that it was important to present massive documentation in favor of geographic speciation so that at least this particular uncertainty could be eliminated from the panorama of evolutionary controversies" (Mayr 1980n: 420).

At the American Association for the Advancement of Science (AAAS) meeting in Columbus, Ohio in December 1939, the American Society of Naturalists and the Genetics Society of America sponsored a joint symposium on "Speciation." Th. Dobzhansky as organizer invited Mayr to participate as a speaker which was a decisive step into his evolutionary career. It was the first time that he generalized on aspects of geographic variation and speciation in birds using many of his examples from the Pacific islands and New Guinea (Mayr 1940c).

The story of his presentation on 28 December 1939 is spiced with a slight bit of malice (Mayr 1997d: 179):

"The speaker scheduled just before me was the famous geneticist Sewall Wright, a brilliant mathematician, but not an orator, to put it mildly. We lectured in the largest hall at Ohio State, seating several thousand people. The stage from which we spoke was huge; it was also used for concerts and for theatrical performances. There was a lectern with a fixed microphone at the very front of the platform, and a series of blackboards some way at the back of it. Wright started his lecture in the front of the microphone, but became totally inaudible when he soon went back to the blackboards, on which he wrote a long series of mathematical formulae. Once in a while he became aware that he probably could not be heard by the audience, and would walk about half way toward the microphone, but of course was still quite inaudible. Soon he happily returned to the blackboard and his mathematics. This went on for nearly an hour.

Most of the huge audience had come to hear the famous Sewall Wright, and after his lecture there was a great exodus. Fortunately, those truly interested in speciation stayed. When I came onto the platform, having learned from Sewall Wright what *not* to do, I stayed right next to the microphone and I showed a set of beautiful color slides of the geographic speciation pattern of the birds collected by the Whitney Expedition. This was before the days of Kodachrome, but we had an artist at the AMNH who did a beautiful job coloring lantern slides. It seems that my lecture was such a contrast to Sewall Wright's preceding one that everybody thought I had given a splendid lecture."

This was one of those historical accidents or chance events which shaped Mayr's career in evolutionary biology (p. 4). He liked to say jokingly that he basically owes it to Sewall Wright.

## The Jesup Lectures on Evolution[4]

Within half an hour after Mayr's lecture in Columbus, Ohio, the geneticist Leslie C. Dunn (1893–1974) of the Department of Zoology, Columbia University (New York) approached him asking whether he would like to be the Jesup Lecturer at Columbia University on the "new systematics" in the near future. "I was scared stiff by the challenge to give these prestigious lectures, but what could I do but accept?" (Mayr in retrospect 1997d: 179). During the next two months they considered the preparation of a book on modern taxonomy (or "new systematics") and of a lecture series on animal and plant systematics by Ernst Mayr and Professor Edgar Anderson, geneticist at the Missouri Botanical Garden, St. Louis. This is documented by the following letters:[5]

*Dear Mayr, March 8, 1940*

*I talked last year with several people about future plans for the Columbia Biological Series and we agreed that a book on the new taxonomy or on taxonomy as interpreted by modern workers would be very valuable and timely. Your mention of Anderson the other afternoon suggests that you might not be averse to undertaking such a book jointly with a man like Anderson who is interested in taxonomy as a botanist. One scheme we discussed was whether you and he might jointly undertake the Jessup [sic] Lectures provided we could recapture the fund which was withdrawn after Dobzhansky's and Northrop's lectures. I have no authority to discuss the lectureship, but if you were interested in such a possibility I could consult my colleagues here and the Administration and see what the prospects might be.*

*Sincerely yours, L. C. Dunn.*

[4] Named after Morris K. Jesup (1830–1908), a New York philanthropist and one-time president of the AMNH. "Jesup Lectures" were held for a few years after 1905 and revived by L. C. Dunn in 1937 (Cain 2001).

[5] The letters and documents quoted in this section are in the Mayr Papers, Ernst Mayr Library, Museum of Comparative Zoology, and Pusey Library, Harvard University Archives, Cambridge, Massachusetts (HUGFP 14.7, Box 1, Folder 37 and Box 2, Folders 71 and 74).

*Dear Professor Dunn, March 20, 1940*

*Your suggestion that I should co-operate with a botanist in the preparation of a series of lectures on modern taxonomy appeals to me very much. I have given the matter a good deal of thought and believe that a good job can be done on this subject, although I feel that somehow or other the subject should be tied up with the question of speciation. One might, for example, choose a title like "taxonomy and the origin of species" to make it a companion of Dobzhansky's work. There is no need for a purely technical treatise of the methods of taxonomy because there are several books of that sort available. What is needed is something that would stimulate the taxonomist to try to go beyond the merely technical treatment of his material and, on the other hand, something that would give the general biologist an idea what really can be gotten out of a good modern taxonomic work.*

*I realize that it is still too early to go into details, but I have drawn up a plan of some of the possible chapters of such a book. There is no question that some of this arrangement will be found impractical when it comes to the actual working out of the material, but it will help to crystallize our ideas if I put down this preliminary plan.*

*Sincerely yours, E. Mayr*

*Dear Dr. Mayr, April, 3, 1940*

*I return your outline with a few comments. I think it should serve as the basis for a good book, which the Press ought to be glad to get. (We must still keep book and lectures separate since the latter are very uncertain.) The books for our Biological series ought to be read by all kinds of biologists, therefore ought to avoid the technicalities of special fields as much as possible. This should not be difficult with taxonomy which speaks a lingua franca (often of yesterday) of biology. This would only be a problem in parts like your section 2. Since addressed to so varied an audience the treatment ought to be elementary but not popular; and provocative rather than conclusive. For myself I should like to see discussed somewhere the question whether principles of taxonomy exist; if so whether they are of really general application, or how they are limited.*

*It seems to me that the common grounds on which biologists often meet are 1) historical, 2) philosophical, including the sharing of systems of reasoning and of words, so these are not to be avoided in a discussion for a general biological audience. But Edgar Anderson or Dobzhansky or others would be much better critics of your plan than I.*

*It seems to me the next step might be to sound out Anderson to see whether he's interested.*

*Yours ever, L. C. Dunn.*

Funds for the Jesup Lectures were obtained during the summer of 1940 and Professor Anderson was contacted in September. Toward the end of the year the latter had agreed to participate in the program and between December 1940 and February 1941, Anderson and Mayr exchanged a series of 13 long and detailed letters. It is obvious from these letters that they both agreed on many basic aspects such as the importance of the population (or museum series) rather than the individual, the problems of delimiting species and genera, hybridization and introgression as well as many other topics in the "new systematics."

Since he had not done any genetic work himself, Mayr encouraged Anderson to "take over that part of the program. You may be as heterodox as you want to be, since I am not so sure that all evolution proceeds exactly as the *Drosophila* School thinks. [...] It seems odd that we have never met before and are now joint contributors to a lecture series. I am looking forward with pleasure to undertake this work in collaboration with you" (December 19, 1940). Anderson wrote: "When I read over your last letter and when I went through much of this first chapter I had a curious sensation which I have never had before. It was as though I were reading something which I had written and forgotten about, or as though there were another me, working elsewhere and independently on the same problems. [...] As usual I quite agree with you as to facts and have profited by the clarity of your analysis" (January 25, 1941). Mayr considered the audience to be expected: "I was at Dobby's [Dobzhansky's] lecture 4 years ago and the audience consisted of graduate students and professors, mostly from Columbia but a few from other institutions of learning in the neighborhood. This time I am sure there will be a few of the younger members of the museum staff and, if you invite them, a few of the men from the Botanical Museum. The total audience is not likely to exceed 50 and may drop down to 30 or 25. I am always surprised how few listeners there are to a lecture that has a certain technical quality. I may be all wrong about these estimates but this is as close as I can guess it. At any rate the audience would be fairly high class and you would not have to try to be 'popular.' I am going to try to illustrate some of the more interesting cases by actually demonstrating specimens which can be done very well in the case of birds. I also have some lantern slides to illustrate some of my discussions" (January 28, 1941).

Arrangements for the lectures were completed in early 1941 and an announcement appeared in the *New York Times* on Wednesday, February 19:

"Columbia names two as Jesup Lecturers.

Drs. Ernst Mayr and Edgar Anderson here next month.

Dr. Ernst Mayr, associate curator of the Whitney-Rothschild collection of the American Museum of Natural History, and Dr. Edgar Anderson, Professor of Botany at Washington University, St. Louis, and geneticist of the Missouri Botanical Garden, have been named Jesup Lecturers in the Department of Zoology at Columbia University, it was announced yesterday.

Endowed by the late Morris K. Jesup, a president of the American Museum of Natural History and of the New York Chamber of Commerce, who died in 1908, the lectures were established in 1905 by the university and the museum 'to present in popular form the results of recent scientific discoveries.' 'Speciation and Evolution' will be the subject of this year's lectures, which will be given Tuesday and Thursday each week from March 4 to March 27 in Schermerhorn Hall [...]."

The invitation card distributed by the Department of Zoology (Columbia University) mentioned eight lectures on "Systematics and the Origin of Species" to be given on 4, 6, 11, 13, 18, 20, 25, and 27 March, four by E. Anderson and four by E. Mayr in Room 601 at Schermerhorn Hall, Columbia University, at five o'clock. The card summarized the contents of the lectures as follows:

Professor Anderson will discuss: Taxonomy the art versus taxonomy the science. The effect of taxonomy the art on taxonomy the science. The nature of taxonomic work and its efficiency as a scientific technique. Species as a phylogenetic unit. Genetical and biochemical evidence for species. The morphological nature of species differences. Regional variations in speciation. Internal variations in speciation. Phylogenetic patterns. The facts of genetics with regard to species hybrids. The taxonomic consequences of hybridization. The genetics of species hybrids.

Dr. Mayr will discuss: Principles and methods of systematics. The old and the new systematics. Systematics and genetics. Individual and geographical variation. Taxonomic characters and their variation. What is a species and how does it originate? The biology (ecology) of speciation. The higher categories and the rules of classification.

Mayr proposed: "I have now arranged my lecture number 1 definitely in such a manner that it can only follow your lecture number 1. In other words, I am scheduled for March 6th and you for March 4th, I hope that this will be in line with your plans. As far as the six other lectures are concerned, I think we can arrange the sequence after you have arrived here in New York" (February 18, 1941). The lectures took place as scheduled. But more than a year thereafter, Anderson confessed (see also Kleinman 1993):

*Dear Mayr: June 12, 1942*

*I was glad to hear from you and to learn how the book was coming on. After I heard from you in December I worked for a time on my manuscript hoping to produce a complementary volume. However, as you may have heard I am now studying Corn very intensively and I am always more interested in what I am doing than in what I have done. The manuscript grows very slowly and the times being what they are, I have put the whole project away for the present at least.*

*This summer I am going to be working in the East at the University of Virginia Experimental Farm at Blandy which is not far from Winchester.*

*The prehistoric Corn which Dr. Wissler was kind enough to let me study has proved to be extraordinarily interesting and I have a paper in the press describing it in some detail.*

*Sincerely yours, Edgar Anderson*

## *Systematics and the Origin of Species*—Evolutionary Biology

Because E. Anderson had declined, L.C. Dunn and the publisher suggested that Mayr expand his contribution so that it could be issued as a separate volume. Thus originated *Systematics and the Origin of Species* (1942e; Fig. 5.1), one of the founding documents of the new synthetic theory of evolution. This was another one of those "historical accidents" furthering Mayr's career (p. 4) which, in this case, provided him with the opportunity to write a more complete manuscript than what would have resulted from his lectures alone. The title of the book remained the same as the joint lectures, although the contents now referred primarily to animal species and speciation. One wonders how it was possible that Mayr as a museum taxonomist having worked exclusively on bird collections during the preceding years was able to prepare this wide-ranging survey in such a short time. The answer: Since his arrival in New York in 1931, Mayr had spent every Tuesday afternoon at the magnificent library of the AMNH going through the newly received

SYSTEMATICS AND

THE ORIGIN OF SPECIES

*FROM THE VIEWPOINT OF A ZOOLOGIST*

By ERNST MAYR

THE AMERICAN MUSEUM OF NATURAL HISTORY

COLUMBIA UNIVERSITY PRESS

*NEW YORK*

**Fig. 5.1.** Title page of E. Mayr's book (1942e) on species and speciation that became a cornerstone of the evolutionary synthesis

journals. Afterwards he took notes on articles dealing with species and speciation or with problems of systematics and evolution. He had also participated in the seminars at the zoology department of Columbia University (L. C. Dunn) since about 1932 (see p. 132) and had been in close communication with Dobzhansky since 1935. When he received the request, it was "simply" a matter of organizing this material and writing (pers. comm.).[6] The manuscript was finished in early 1942 and L. C. Dunn delivered it to the Press in March of that year. The book appeared toward the end of 1942; A. H. Miller received his review copy on December 15. Mayr's Preface began:

"During the past 50 years animal taxonomy has undergone a revolution as fundamental as that which occurred in genetics after the discovery of Mendel's laws [...], the change from the static species concept of Linnaeus to the dynamic species concept of the modern systematist."

The book was not intended to show that the data of the systematists are consistent with the newly developed principles of genetics which in fact they are. Mayr felt that he had little to add to what Dobzhansky (1937; and S. Wright through Dobzhansky) had said on genetics. As I mentioned above, the real objective of the volume was to explain a whole set of phenomena well known to systematists (naturalists) but not to geneticists, particularly species and speciation and the role of geography in the evolution of species and populations. It should be noted that certain topics treated in this book were later modified by Mayr himself. The book was a continuation of the thinking expressed in Dobzhansky's book (1937) and dealt with many aspects of species and speciation which the latter had neglected or discussed only briefly. Mayr's volume was also written in response to Goldschmidt's (1940) ideas on saltational speciation through systemic macromutations:

"Even though personally I got along very well with Goldschmidt, I was thoroughly furious at his book [1940], and much of my first draft of *Systematics and the Origin of Species* was written in angry reaction to Goldschmidt's total neglect of such overwhelming and convincing evidence" for the concept of geographic speciation (1980n, p. 421). "Goldschmidt [1940] confuses the sympatric with the allopatric gap [between species] and this is the reason why he denies speciation through geographical variation. He shows very clearly what all of us, of course, have always known that sympatric gaps are bridgeless gaps because otherwise we would have hybrid flocks. He then proceeds to state that all gaps between species are bridgeless even those between allopatric species. This conclusion is not only not justified but actually contradicted by his own material and quotations" (Mayr in a letter to E. Anderson, 21 January 1941). In 1952, when both Mayr and Goldschmidt were teaching courses at the University of Washington (p. 260), Mayr once asked him how a new "hopeful monster" would react to the other (normal) members of the population. He was quite nonplused and finally said "Well, I never thought of it that way." And Mayr (1982d: 381) later: "There are literally scores of

[6] Mayr also registered the publications in various other fields like general ornithology, anthropology, genetics, behavior, and paleontology to broaden his general knowledge. "My reading in those days was quite enormous, because I was equally interested in many animal groups and almost all aspects of biology."

cases in the history of science where a pioneer in posing a problem arrived at the wrong solution but where opposition to this solution led to the right solution," e.g., Goldschmidt-Mayr or Lyell-Darwin.

On Richard Goldschmidt (1878–1958) Mayr reported: "So far as I remember, I never met Goldschmidt when I was a student at the University of Berlin. Presumably, during that time (in 1925–1926) when I took courses at Dahlem he was absent in Japan. When I first met him in the States, I don't know. I do know that in the 1930s he visited me several times at the American Museum of Natural History, and that on those occasions I showed him all the marvelous examples of geographic variation as displayed by the South Sea island birds, particularly those of the Solomon Islands. Obviously, I did not convince him because he refers to this only in a parenthetical footnote in his 1940 volume on *The Material Basis of Evolution.*

Our relationship was quite cordial in spite of our scientific disagreement, and this is illustrated by the following anecdote. Probably in 1936 Goldschmidt visited me on a Saturday in the museum, and he was so much interested in my demonstrations that it came close to my lunchtime. He accepted mine and Gretel's invitation to a simple lunch (lentils and frankfurters), came to my apartment in northern Manhattan, and we had an altogether delightful time together.

The next thing was that I went up to New Haven together with my friend Herman Spieth to attend Goldschmidt's Silliman Lectures. Of course, I did not approve at all of his saltationist origin of species through hopeful monsters, but it was interesting to hear him present his ideas. As we were talking together after one of his lectures, Herman asked him about the origin of birds from reptiles. He clearly, unequivocally said, 'Well, the first bird hatched out of a reptilian egg.' He was that much of a saltationist.

This experience, and particularly his rather cavalier treatment of my work in his 1940 volume, made me rather angry. I considered it almost unscientific to suppress this splendid evidence for geographic speciation. And, as I have said on other occasions, part of my *Systematics and the Origin of Species* was written in a mood of anger over Goldschmidt's behavior.

He was a typical Geheimrat—as the modern sociologists call it, a mandarin. He was fully aware of his position in life and had little patience with ignorance and poor manners. Interestingly, the fact that we disagreed scientifically did not disturb our relationship. I am told that he was quite upset when coming to the University of California at Berkeley about the secondary role he played there. Having been the director of a Kaiser Wilhelm Institute, practically being the Pope of German biology, it was a terrible let down for him to be only one professor in a large department.

He had a peculiar tendency always to come out with unorthodox biological theories. This was true not only for speciation, but also for sex determination, for gene action, and I don't know what other subjects. Invariably almost everybody opposed him, and as history shows, he was almost always completely wrong. However, he himself was fully convinced that he was right, and he would say, 'Well, this will probably not be adopted in my lifetime, but I am perfectly sure that eventually it will be shown that I am right.' There is a very good biography of Goldschmidt by Curt Stern [National Academy of Sciences USA, Biographical Memoirs 39 (1967): 141–192]. On the other hand, Goldschmidt's autobiographical reminiscences *Portraits from Memory* (1958), are not reliable. Those German biologists who were his friends are usually praised too highly, and those who were his enemies, particularly if they were known to have been anti-Semitic, were presented usually as being very poor zoologists also. This is definitely in some cases misleading as, for instance, in the case of Ludwig Plate" (1862–1937).

As mentioned above, Mayr's specific contribution to the evolutionary synthesis was the analysis of the origin of organic diversity (speciation), i.e., the causes of divergence and discontinuity: "To fight for the inclusion of speciation, the origin of diversity, was my major task and contribution to the evolutionary synthesis." Species and speciation formed the center of his research. In contrast to his emphasis on the geographic ("horizontal") component of evolution, earlier evolutionary biologists and geneticists had studied almost exclusively adaptive ("vertical") changes within populations along individual phyletic lineages. The reductionist approach of the geneticists concentrated on the genotypic level of organisms and considered evolution as changes in the genetic composition of populations, whereas the naturalists (systematists and paleontologists) considered the phenotype, the entire organism, as the target of selection and defined evolution as descent with modification (anagenesis) *and* multiplication of species (cladogenesis). The Evolutionary Synthesis in North America led to a more general acceptance of natural selection as a mechanism of evolutionary change and eliminated non-Darwinian theories like neo-Lamarckism and orthogenesis. Cooperation between systematics, genetics and paleontology was established through the major publications of Dobzhansky, Mayr, Huxley, Simpson, Stebbins, and several authors in Europe (e.g., Rensch). Mayr (1980f: 1) summarized their conclusions: "(1) Gradual evolution can be explained in terms of small genetic changes ('mutations') and recombination, (2) the ordering of this genetic variation by natural selection, and (3) the observed evolutionary phenomena, particularly macroevolutionary processes and speciation, can be explained in a manner that is consistent with the known genetic mechanisms."

Among the architects of the synthetic theory of evolution Mayr and Rensch had conducted extensive research in avian systematics and zoogeography (as had their teacher Erwin Stresemann) and Huxley had studied the behavior of birds for several years (Junker 2003). One may ask the question whether birds were particularly suited for such a generalizing approach. The answer is probably "yes": Birds demonstrate geographical variation better than most other animal groups. More importantly, there was so much information about birds available, far more than for any other group of organisms. Nevertheless, Alden H. Miller in California, with virtually the same facts at his disposition, wrote a very formalistic monograph on "Speciation in the avian genus *Junco*" (1941) and missed many if not most of Mayr's general questions, even though his basic ideas were similar (see Mayr's review of Miller's work [1942d]). A broad synthetical overview on the relevant data from many different biological disciplines was required, and this was supplied by Mayr (1942e).

Reviewers of *Systematics and the Origin of Species* (see 1942e for references) were enthusiastic about the wide scope of the book, the clear and logical presentation of the topics of the "new systematics" and geographical speciation illustrated by the author's own taxonomic work and the findings of modern genetics and other fields of science. The reviewers emphasized the book's general biological-evolutionary significance when stating, e.g.,

"Dr. Mayr's lucid and stimulating book [...] should be read and understood by every individual who would be called a systematist. The subject is approached from the broad viewpoint of a general biologist who not only understands the principles of genetics, ecology, morphology, physiology, and geographical distribution, but who is able to apply these to problems of systematics" (Ownbey);

"This work is of great value in bringing about a better understanding of the contributions of systematics to evolutionary theory and in facilitating the synthesis of the results of different lines of investigation into a unified and logical picture" (NN);

"No systematist can afford to overlook any point that Dr. Mayr raises. The author has presented a scientific synthesis" (Zirkle);

"A broad groundwork of sound generalization is laid for the construction of a modern philosophy of speciation; systematics and other branches of zoology are coming to a rapprochement" (Hubbs), a "synthetic exposition of modern research in zoology" (Jepsen), "an attempt at a general synthesis," although limited to animals (Sturtevant). "Very helpful to American students of speciation will be the data and interpretations of European systematists; Mayr's familiarity with the far-flung systematic literature is amazing" (Hubbs); similarly also Emerson, Schmidt, and Ripley.

Critical comments concern the claim that speciation is almost exclusively geographic (allopatric), a less detailed treatment of aquatic animals, and the lack of examples from plants. Therefore Mayr entitled his follow-up work *Animal Species and Evolution* (1963b).

Lack, Miller, Ripley, Griscom, and Kramer noted with pride the advanced state of their special field of interest and the leading role ornithologists have played in developing the methods of "new systematics." Their application of trinominal nomenclature led to the study of geographical variation and later to the concept of geographical speciation. Besides the book pointed the way toward more intensive utilization of museum material for general studies.

Mayr himself wrote in retrospect (in 1982) in his autobiographical notes: "When rereading my 1942 book 40 years later I cannot conceal how impressed I am by the maturity of much of my discussion. I open up many important evolutionary problems and the solutions I propose are by and large the ones that are still considered valid today. In view of the fact that I had rather slight contacts with other evolutionists at that period I was at first puzzled as to how I could have reached such maturity. I then recollected that after all I had been a working taxonomist for

15 years and that I had read voluminously. Even though I did not maintain a 'Notebook on Transmutation of Species,' as Darwin did, I nevertheless had kept massive excerpts from the literature always with comments of my own. Unfortunately all this has now disappeared; I probably threw it out when I moved from New York to Cambridge.

What is also evident to a careful student is that in various major issues I tended to diverge from both Simpson and Dobzhansky, as much as all three of us stood on the common ground of the Evolutionary Synthesis. My emphasis on populations and horizontal evolution is altogether absent in Simpson. My emphasis on the essential completion of the building-up of isolating mechanisms during geographical isolation is not shared by Dobzhansky. And there are various other differences. Even though I still express one or two reservations concerning the power of natural selection, I nevertheless was a far more consistent selectionist than Dobzhansky, who had been strongly influenced in his thinking by Sewall Wright."

## The New Systematics

E. Stresemann and B. Rensch, Mayr's mentors at the Museum of Natural History of Berlin, pursued their work as biologists rather than as species cataloguers. Systematics had become in their hands truly a branch of evolutionary biology. Other scientific staff members in Berlin were also excellent. Three of the entomologists more or less acknowledged and practiced new systematics: W. A. Ramme, worked on orthopterans and introduced Mayr to the concept of sibling species; E. M. Hering, with butterflies and mining insects, and H. Bischoff, who prepared a major book on the biology of hymenoptera. None of them was just an alpha taxonomist. When Mayr came to the United States, he was appalled at the typological spirit dominating taxonomy. Each species was thought to have an essential nature, and species were defined by their degree of difference from one another (Mayr 1999k: XIX). He therefore decided to include in his book (1942e) a complete presentation of the European approach, one whose spirit Julian Huxley (1940) had captured in the phrase "new systematics" (although that book itself contained very little of this new approach). Mayr's book (1942e) consists of two parts, (1) an introduction to the new systematics and (2) an analysis of species and speciation. Providing the first comprehensive treatment, after Rensch (1929, 1934), of modern systematics at the species level in English, *Systematics and the Origin of Species* became highly popular among taxonomists. Whereas the English-American literature made not much of a contribution to the evolutionary synthesis, central European and also Scandinavian systematics contributed massively to the synthesis by stress on population thinking, the biological nature of species, the study of geographical variation, and the clear understanding of geographical speciation in a wide variety of animals (besides birds also in mammals, reptiles, amphibians, fish as well as invertebrates). One of Mayr's major achievements was to summarize and synthesize this scattered European literature which, for linguistic reasons, had remained unknown internationally. Among a total of 450 references, the bibliography includes 174 European, mainly German, titles. This integration of North American and European systematic research in clear and straightforward presentation, the interpretation of many facts in the light of modern systematics and population genetics soon extinguished neo-Lamarckian and typological views in contemporary biology. Allopatric speciation, population thinking, i.e., an emphasis on the "horizontal" (geographical) component of evolution, and the demolition of the typological species concept were the general topics treated in detail. Additional aspects were geographically variable polymorphism, clinal geographical variation, population structure of species, sibling species, the biological species concept, and monotypic and polytypic species taxa (it should be noted that the distinction between category, concept, and taxon was not yet established at that time). Mayr promoted an adaptationist view of subspecies and species differences, but stressed "the point that not all geographic variation is adaptive" (1942e: 86). His volume includes a summary of his own 12 years of intensive taxonomic research in the spirit of new systematics, particularly in the island regions of Indo-Australasia.

In his first chapter Mayr contrasted old and new systematics like this:

"Old"—the species and purely morphological views occupy a central position; the individual, not the population, is the basic unit.

"New"—subdivisions of the species, subspecies, populations, individual and geographic variation are studied in detail; the population or the "series" of the museum worker has become the basic taxonomic unit. The morphological species definition has been replaced by a biological one which takes ecological, geographical, genetic, behavioral, physiological, biochemical, karyological and other factors into consideration. The new systematist approaches his material more as a biologist and less as a museum cataloguer, he attempts to formulate generalizations and syntheses. He inquires into the nature and the origin of the taxonomic units with which he works.

"The ornithologist knows his material so well that he can do what the geneticist also does that is to pick out one particular character and study its fate under the influence of geographical variation, and in the phylogenetic series. In the less worked groups the taxonomist is forced to consider his taxonomic units as complete entities and all he does is to arrange them in the most natural order. I think I shall add that among the differences between the old and the new systematics" (Ernst Mayr to Edgar Anderson, when preparing his Jesup Lectures, 28 January 1941; HUGFP 14.7, Box 2).

Mayr also referred to the procedures of the species systematist (collecting and analyses of the material, naming of populations and the selection of a type specimen) and the rules of nomenclature. When, in early 1946, he sent copies of his publications during war time to Erwin Stresemann in Berlin, he stated:

"A book came out here last year entitled *The Reader over your Shoulder* [by R. Graves and A. Hodge]. It was a treatise on the technique of writing and on the good usage of English. It emphasized that you should always write as if somebody was looking over your shoulder reading what you put down. In a somewhat different way you were always the reader over my shoulder. Time after time, when writing my *Systematics and the Origin of Species* (1942) or my Timor report (1944e), it was you with whom I really discussed my problems while I put them down on paper. Perhaps you can read this between the lines of some of the passages" (24 January 1946).

Stresemann praised Mayr's *Systematics and the Origin of Species* as "a synthesis of taxonomic, genetic, and biological ways of viewing evolution, ... [which] will long remain a reliable guide for systematists working in the complicated labyrinth of phenomena through which his predecessors had tried vainly to find their way during the past 150 years" (1951: 281, 1975: 277–278).

The new systematics had a significant impact on the development of population genetics, and genetics in turn profoundly influenced the ideas of new systematists. Mayr (1948c) reviewed this progress in the field of systematics for the benefit of geneticists (polytypic species and trinominal nomenclature, population thinking, biological species concept). A majority of scientists now agreed on the following seven statements (Mayr 1948c): (1) Animal species taxa have reality in nature and are well delimited except in borderline cases; (2) The species concept is defined biologically, using reproductive isolation as a criterion; (3) Species are composed of distinct populations (subspecies or geographical races); (4) All characters (morphological, ecological, physiological) are subject to geographical variation; (5) Geographical separation results in genetic differentiation which varies among species; (6) Isolating mechanisms will inhibit the interbreeding of the daughter populations when they come in secondary contact; (7) Except in borderline cases, "bridgeless gaps" in morphological and other phenotypical features separate sympatric species. These considerations changed the entire field of animal taxonomy (Huxley 1940, 1942; Mayr 1942e). A revolution had occurred—a change from the static species concept of Linnaeus to the dynamic species concept of modern systematics.

In the years since 1942 many North American systematists have told Mayr that his book had entirely changed their professional lives. They had been conventional taxonomists, had described new species and made generic revisions, but had never risen above the essentially descriptive level. It was Mayr's book that showed them the immense potential of systematics and demonstrated to them that systematics at the species level occupied an important realm that was inaccessible either to the geneticist dealing with the gene level or the paleontologist dealing with phyletic lines and higher taxa. The volume was a real revelation to the traditional taxonomists. One of them was Ralph Chermock, an assistant professor at the University of Alabama and E.O. Wilson's teacher in 1947. As the latter wrote in his autobiography:

"The prophets of the Chermock circle were the architects of the Modern Synthesis of evolutionary theory. [...] The sacred text of the Chermock circle was Ernst Mayr's 1942 work *Systematics and the Origin of Species*. Mayr was the curator of birds at the American Museum, but his training had been in Germany, a source of added cachet. The revolution in systematics and biogeography that Mayr promulgated was spreading world-wide but especially in England and the United States. [...] The Modern Synthesis reconciled the originally differing world-views of the geneticists and naturalists. [...] The naturalists were given a hunting license, and for the Chermock circle Mayr's *Systematics and the Origin of Species*, following upon Dobzhansky's book, was the hunter's vade mecum. From Mayr we learned how to define species as biological units. With the help of his written word we pondered the exceptions to be expected and the processes by which races evolved into species. We acquired a clearer, more logical way to think about classification by using the phylogenetic method" (Wilson 1994: 110–112; see also Wilson 1998: 4).

Working on a revision of his successful 1942 book, Mayr soon decided to separate methodological aspects of systematic zoology from the science of species and evolutionary biology. In this way originated his textbook, *Methods and Principles of Systematic Zoology* (1953a, with Linsley and Usinger) which covered methodological aspects (p. 321) and his masterly monograph on *Animal Species and Evolution* (1963b) treating the species problem and aspects of evolutionary biology.

In his article, "Trends in avian systematics" Mayr (1959e) emphasized two main topics: (1) Population systematics and (2) macrotaxonomy. Subspecies are unsuitable for describing the population structure of a species. This can be done in terms of three major population phenomena: (a) Geographical isolates which are particularly common near the periphery of the species range and which represent important units of evolution; (b) The population continuum, a continuous series of populations which make up the main body of the species, and (3) Zones of secondary intergradation between a former geographical isolate (which did not yet attain full species rank) and the main body of the species. To make such an analysis of all species of a family, and to record the relative frequency of the three elements, is one of the steps leading to "comparative systematics." Recording the position of belts of secondary hybridization permits the reconstruction of formerly existing barriers and of the location of "refuges" caused by drought conditions in low latitudes during the Pleistocene. With respect to macrotaxonomy, Mayr proposed to utilize various new sources of information like promising new character complexes of behavior and biochemistry to determine the phylogenetic relationships between genera, families and orders.

Behavioral characters are completely equivalent to morphological characters (Mayr 1958g). If there is a conflict between the evidence provided by morphological characters and that of behavior, taxonomists are increasingly inclined to give greater weight to the ethological evidence, e.g., in the case of grasshoppers, in certain swallows (p. 327), ducks (p. 327), and finches (p. 329). Based on behavioral characteristics the weaver finches (Estrildidae) and weaverbirds (Ploceidae) are not as closely related as previously thought. Movements often precede special morphological features that make them particularly conspicuous. Comparative studies of behavior permit statements about trends in the evolution of behavior, as shown, e.g., by the work of Whitman on pigeons, Tinbergen on gulls, Heinroth and Lorenz on ducks, Meyerriecks on herons, Hinde on finches, Morris on weaver finches. Other groups studied in this way include grasshoppers, *Drosophila*, hymenopterans, and cichlid fishes.

In 1964(n), Mayr again summarized the then "new systematics" mentioning that certain aspects can be traced back to some authors of the mid-19th century who collected "series" of specimens (population samples). A more general change in the working methods of systematists occurred during the 1930s instigated in Europe by Rensch (1929, 1934). "New systematics" is a viewpoint, an attitude toward taxonomic work. The emphasis is on non-morphological characteristics derived from behavior and ecology, bioacoustics, biochemistry, individual and geographical variation, weighting of characters and computer analysis. Basic taxonomic description and cataloguing is still necessary. Genetic research had shown that the phenotype is the product of the total genotype: Many genes shape a character (polygeny), and a single gene contributes to the expression of many characters (pleiotropy). This had an impact on the thinking of systematists. Other improvements were, e.g., the recognition of sibling species which seem nearly identical morphologically, the biological definition of the species concept, the new evaluation of many geographical isolates previously considered to be species as subspecies of polytypic species. The taxonomist has to (1) identify and define populations, subspecies and species; (2) assemble the units into aggregates, groups, taxa, and (3) assign such taxa to taxonomic categories (ranking).

The 1930s and 1940s saw a preoccupation with the species and subspecies, that is microtaxonomy, whereas in the 1950s and 1960s (Simpson, Cain, Rensch) the discussion shifted to higher taxa, that is macrotaxonomy. Mayr (1971g) returned to methods and strategies of taxonomic research, some of which he had reviewed in his textbook (1969b). With this article he intended to stimulate curatorial practices and taxonomic publications. Owing to major ecological projects and rescue operations in areas where the habitat is being destroyed, the burden on the taxonomists has increased, although the number of positions for taxonomists has not increased anywhere near as much. This requires streamlining of taxonomic operations. Mayr felt that at least half of a curator's time should be devoted to research and inspired by projects not restricted to the study of local faunas and floras. Advice is here also given on illustrations, the preparation of a manuscript for publication, the cataloguing of collections and the (questionable) maintenance of filing and cross filing systems. "Even though a classical branch of biology, taxonomy does not need to be old-fashioned."

## The Biological Species Concept and Species Taxa

In this section I trace the historical development of Mayr's views on the theoretical species concept and on species taxa which were central to his research. It has become evident in recent decades that there are only two theoretical species concepts:

(1) The typological or essentialistic species concept postulates that a species consists of similar individuals sharing the same "typological essence," is separated from other species by a sharp discontinuity, remains constant through space and time (i.e., species limits are fixed), and variation within a species is severely

limited. (Regarding Mayr's conceptualization of typological *versus* population thinking since the late 1940s see p. 350.)[7]

(2) The biological species concept states that a species is a group of interbreeding natural populations that is reproductively (genetically) isolated from other such groups because of physiological and/or behavioral barriers; the limits of biological species are open.

Other "species concepts" proposed are those of the "vertical" (historical) cladistic "species" (W. Hennig) and of the paleontological or evolutionary "species" (G. G. Simpson) which, however, refer to portions of phyletic lineages rather than species (see p. 213). The "chronospecies" of paleontologists, i.e., artificially delimited subdivisions of phyletic lineages, are not species in the sense of biological species. Mayr (1942e: 154) stated: "The 'species' of the paleontologist is not necessarily always the same as the 'species' of the student of living faunae." The recognition concept is a different formulation of the biological species concept (Mayr 1988h). Under each of these as well as the typological and biological species concepts mentioned above authors delimit narrow, intermediate or wide species taxa, depending on whether they place the taxonomic species limits at relatively low, intermediate or rather high levels of differentiation among the geographically representative populations, respectively. Narrow species limits emphasize differences, wide species limits emphasize similarities among the various geographically representative taxa. The so-called *phylogenetic species concepts* are instructions of how to delimit particularly narrow species taxa, but not theoretical species concepts. Hey (2006) reviewed the long-standing discussion on species concepts under the title "On the failure of modern species concepts."

In their comprehensive treatise of speciation Coyne and Orr (2004) basically adopted Mayr's biological species concept, because *the* problem of speciation is "the origin of discrete groups of organisms living together in nature. [...] It seems undeniable that nearly all recent progress on speciation has resulted from adopting some version of the biological species concept" (pp. 6–7).The conceptual reasoning as to species limits in molecular population studies is also based on the biological species concept.

While a student in Berlin under Erwin Stresemann, Ernst Mayr had become a representative of the Seebohm-Hartert "school" of European systematic ornithology which had originated in the late 19th century and, during the course of 100 years, widely influenced ornithology through the work of several well-known scientists: Seebohm-Hartert-Hellmayr-Stresemann-Rensch, and later Ernst Mayr himself (Fig. 5.2). The views of these ornithologists as well as those of E. B. Poulton, K. Jordan, and Ludwig Plate on biological species, subspecies and speciation

[7] This was the species concept of numerous 19th century biologists prior to Darwin (1859) regardless whether they believed in the origin of species through divine intervention (i.e., creation) or through autochthonous (spontaneous) generation from organic matter during periods of special environmental conditions. The views of Linnaeus himself and of several other early naturalists on species were less essentialistic than often stated (Mayr 1982d: 259; Winsor 2006).

around 1900 eventually influenced, through Ernst Mayr, the entire fields of systematic zoology and evolutionary biology.

The concepts of the naturalists of the Seebohm-Hartert "school" (Haffer 1994a) are based on the Darwinian idea of a common descent of all species and on the belief that forms (taxa) exist in extant faunas that represent intermediate stages between subspecies and species. Species are reproductive communities and speciation occurs through the differentiation of relatively small and geographically isolated populations, as discussed, e.g., by Stresemann (1927–1934: 644–645 and in earlier publications; see Haffer et al. 2000: 195–198). The study of geographical variation of populations in widely distributed subspecies and species permit an interpretation of the concept of biological species and the delimitation of rather broad species taxa.

Mayr's views on the biological species concept and the process of geographic speciation developed from the publications of Stresemann (1919a,b, 1920, 1927–1934) and Bernhard Rensch (1929, 1934). Mayr himself (1999a: 23) pointed out the decisive influence of Stresemann stating "Virtually everything in Mayr's 1942 book was somewhat based on Stresemann's earlier publications." The principles of the biological species concept and the importance of geographic speciation in evolution had been established by naturalists during the late 19th century (p. 230), but it was largely due to Mayr that this information came to the attention of the geneticists. Mayr promoted the acceptance of this species concept with its emphasis on populations and on reproductive isolation. Although not originally his own it was the support he gave it and its concise formulation that led to its rapid subsequent adoption by the majority of zoologists. Stresemann had stated:

"Forms which, under natural conditions, pair successfully through generations, represent together a species regardless of the morphological differences, [...] whereas all forms that under natural conditions maintain themselves side by side without intergradation, are specifically distinct" (1920: 151–152) and "forms of the rank of species have diverged from each other physiologically to such an extent that they can come together again [after the removal of a geographical barrier] without intergradation [...]. Morphological divergence is thus independent of physiological divergence"(1919b: 64, 66). "Sexual affinity and sexual aversion, respectively, under *natural* conditions is considered as the test for the [specific or subspecific] relationship of two forms" (1920: 151; transl.).

Stresemann knew that species can be very different from or very similar to each other, i.e., the outer appearance is not a measure of species status. In the case of insular (allopatric) distribution of related forms their status as subspecies or species should be inferred from several auxiliary criteria (p. 46).

Stresemann always supported the interpretation of the origin of species from relatively small and geographically isolated populations (geographical or allopatric speciation), as developed during the 19th century by Moriz Wagner and John T. Gulick. In his handbook, *Aves*, Stresemann stated (1931: 633–634) that the rate of differentiation of an isolated population depends on "(1) the strength of the factor initiating the change, (2) the number of generations per unit of time [...], (3) the size of the population; the smaller the number of individuals, the greater

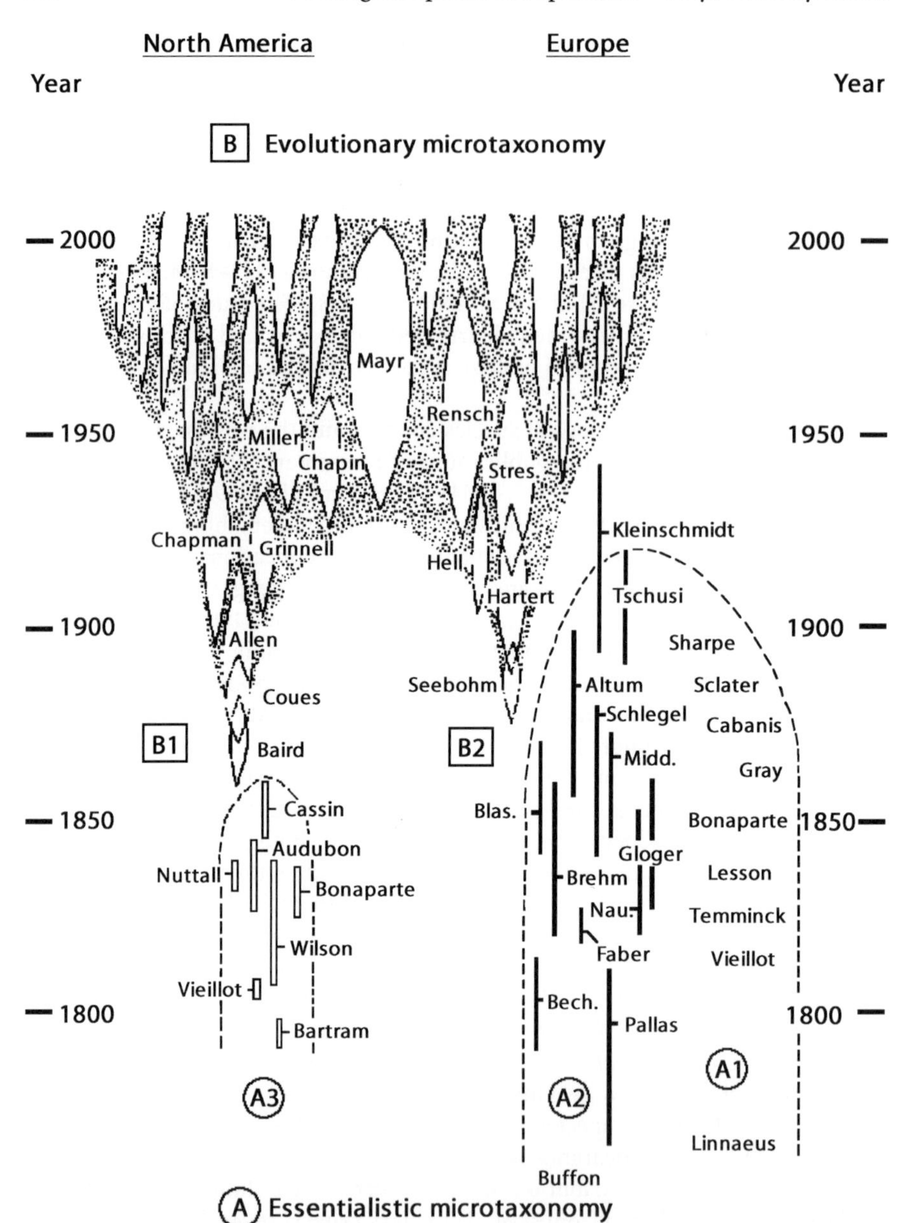


**Fig. 5.2.** Research traditions ("schools") of systematic ornithology during the 19th and 20th centuries. A Typological (essentialistic) microtaxonomy; A1 Linnaeus "school," A2 Pallas-Schlegel "school," A3 Wilson "school;" B Evolutionary microtaxonomy; B1 Baird-Coues "school," B2 Seebohm-Hartert "school." The main publishing periods of major ornithologists are indicated symbolically. Most presently active representatives of evolutionary microtaxonomy are indicated anonymously. Abbreviations: Bech.–Bechstein, Blas.–J. H. Blasius, Midd.–Middendorff, Nau.–Naumann, Stres.–Stresemann. From Haffer (1997a)

the chances for a mutation to prevail." He reemphasized (1931: 644) that speciation occurs only through spatial separation of populations (either through jump dispersal or disruption of the continuous range of the parental species) and pointed out the particularly favorable conditions for speciation on island archipelagoes like the Hawaii and Galapagos Islands. Speciation occurs in isolated populations through small mutations and natural selection. During the 1920s he had studied polymorphism in birds and had emphasized repeatedly that conspicuous morphs ("mutations") have nothing to do with the origin of new species. Regarding ecological segregation in speciating populations, he concluded (1939: 360): "The correspondence of two forms may have been reduced to a correspondence of only their ecological requirements with, at the same time, divergent differentiation of their sexual activities. Two forms at that stage of the differentiation compete with each other for space and where they meet by range expansion, they abut sharply against each other without forming hybrids. [...] Examples of such situations are probably much more common than currently known" (Haffer et al. 2000). He concluded that hybridization, after the removal of a geographical barrier between incipient species invariably leads to a secondary intergradation of these populations, and not to speciation.

However, Stresemann never prepared a general treatment of species and speciation, mainly because his main interests during the late 1920s had shifted to functional anatomy and physiology[8] and because he may have considered B. Rensch's book of 1929 as having covered the subject sufficiently well.

Bernhard Rensch (1928) introduced the concept of Artenkreis (superspecies Mayr 1931). In his book, *The Principle of Polytypic Species and the Problem of Speciation* (1929) Rensch showed that in many groups of animals numerous geographical taxa may be combined as subspecies of polytypic species. He also discussed numerous borderline cases between subspecies and species which document that in the majority of cases, species originate from geographically isolated populations. Conspicuous phenomena of geographical variation are described in Bergmann's, Allen's, and Gloger's Rules. At that time, he interpreted these rules by a direct influence of the environment. His book was the first manifesto of "the new systematics" and Rensch the first "new systematist." In further publications he explained the geographic principle, the geographical replacement of conspecific taxa, individual and geographical variation, superspecies, and speciation in what he called *Instructions for zoological-systematic Studies* (1934). He also demonstrated the adaptive nature of some subspecific differences in birds. No equivalent title for Rensch's booklet of 1934 existed in the English literature until Mayr (1942e) published his volume which, in the first chapters, includes an introduction to taxonomic procedures. On June 6, 1941 Mayr wrote to Stresemann: "I am presently busy preparing my book manuscript on *Systematics and the Origin of Species*. One cannot deal with this topic without noticing all the time, how much the solution or at least the clear exposition of these problems owes to our friend Rensch."

[8] "My real interests have shifted [from taxonomy and zoogeography] to very different fields, in particular functional anatomy and physiology" (letter to Mayr dated 14 October 1929 and see p. 141, footnote).

In fact, Mayr (1942e) cited the publications of Rensch more frequently than those of Stresemann and wrote in retrospect: "My own work [on geographic variation and speciation] was a continuation of the work of Rensch" (Mayr 1976m: 119) and "he probably had more influence on my thinking than anyone else; I greatly admired his 1929 book" (Mayr 1980n: 416).

Until the early 1950s, systematists used the term "species" without distinction between (1) a theoretical concept, (2) a category in the taxonomic hierarchy, and (3) a particular species taxon, including Mayr, as he himself has said repeatedly.[9] Modern definitions read (Mayr 1969b: 4, 5, 26):

(1) The theoretical *species concept* of general biology (the "nondimensional" biological species concept): "Species are groups of interbreeding natural populations that are reproductively isolated from other such groups."
(2) "The *species category* in the taxonomic hierarchy is a class the members of which are the species taxa."
(3) The *species taxon* is a taxonomic group that is considered to constitute a particular species, e.g., the Robin (*Turdus migratorius*). Under the same biological species concept authors may delimit species taxa more broadly ("lumpers") or more narrowly ("splitters") and systematists delimiting species taxa similarly, may adhere to different theoretical species concepts, i.e., no relationship exists between the species concept adopted by a particular systematist and the recognition of narrowly or more broadly delimited species taxa (Haffer 1992, p. 118, Table 2).

In Mayr's early publications on species and speciation these different meanings are usually, but not always, obvious from the context in which he used the term "species." Terminological confusion concerned mainly the theoretical species concept and the species taxon, such as Mayr's use of the non-dimensional species concept (the actual concept) and the multi-dimensional species concept (which is actually the species taxon; see below). "There is an undeniable tension between these two aspects of the word species and from 1942 until the present time, I have never ceased to struggle with this problem" (Mayr 1992i: 9). Mayr also failed to distinguish completely the species concept and the species category in taxonomy, as he used the definition for the species concept also for the category (Bock 1995b).

In his first theoretical paper Mayr (1940c) approached "Speciation phenomena in birds" as a taxonomist. Considering numerous species taxa of the Australasian bird fauna and how new species probably originated he singled out for "species in the making" strongly differentiated peripherally isolated populations with ranges detached from the main species range on an island, on a mountain or beyond some other geographical or ecological barrier. Such forms are geographical representatives of their near relatives and show the common origin still very clearly. They are

[9] The word taxon "recently coined" by botanists was already mentioned by Mayr et al. (1953a: 36, 323).

differentiated at the level of subspecies or species, in some cases even as monotypic genera. His examples are sufficient to show that geographical variation and isolation of populations lead to the formation of new species taxa (geographical or allopatric speciation). In view of his taxonomic approach the species definition he proposed was that of a polytypic species taxon:

"A species consists of a group of populations which replace each other geographically or ecologically and of which the neighboring ones intergrade or hybridize wherever they are in contact or which are potentially capable of doing so (with one or more of the populations) in those cases where contact is prevented by geographical or ecological barriers" (1940c: 256).

Hence, interpretation was required of all the factual information to decide whether allopatric populations are "potentially capable" of interbreeding, that is, whether to name them subspecies or species. The differences among representative island populations were not only quantitative and continuous but often qualitative and discrete as observed among congeneric species. For this reason he replaced the typological concept by a concept of species taxa as aggregates of geographically variable populations. In this article he did not (yet) discuss the general species concept based on reproductive isolation between populations but he did so during the following year.

The central topics of Mayr's Jesup lectures in March 1941 were the biological species concept and the origin of new species taxa. Only four months later, on 31 July 1941, he spoke at the Marine Biological Laboratory (Woods Hole, Cape Cod, Massachusetts), on his main research topic, "The origin of gaps between species" (Mayr 1941i): Evolution is a continuous process, he stated, but the units produced by evolution are discontinuous, a problem that the geneticists had left open. Two different classes of gaps between species need to be distinguished:

(a) Absolute "bridgeless" gaps between sympatric species at a particular locality. The main phenomenon is reproductive isolation between species. These have objective reality, e.g., the five similar species of thrushes of the genus *Hylocichla* of which up to three inhabit the same woods in northeastern North America without the slightest intergradation, and
(b) Relative (gradational) gaps between allopatric representatives. In widespread species taxa geographically differentiated continuous populations interbreed and merge into each other. Toward the periphery of the species range there are often representative forms which are geographically separated by a barrier. In such situations the taxonomist must use his judgment to infer the taxonomic status of these latter forms. Obviously, such species taxa cannot be delimited in an objective way. Geographical variation coupled with geographical separation completely blurs the borderline.

Geographically separated populations are of the greatest importance for speciation. Mutation, difference of selective factors and differences of random gene loss will, through time, produce an increasing divergence. Eventually, isolated populations will become distinct species and may come into contact with one another without interbreeding, provided reproductive (genetic) isolation is complete (allopatric speciation). Occasionally it happens that a geographical barrier breaks down before that is the case. The result is extensive hybridization in the zone of contact. Sympatric speciation "is of common occurrence in certain animal groups with very specific ecological requirements, but almost completely absent in all other animal groups" (Mayr 1941i: 142; see below, p. 223).

Summarizing, Mayr (1941i) distinguished clearly reproductively isolated populations of different species taxa at a particular locality and geographically representative interbreeding populations belonging to the same species taxon. From these considerations he derived in this article his definition of a polytypic species taxon (1940c). The theoretical species concept with respect to the absolute gap between sympatric species taxa was presented in his book, *Systematics and the origin of species* (1942e: 120) based on his Jesup lectures and his lecture at the Marine Biological Laboratory in 1941. Here he gave definitions for both the theoretical species concept ("Species are groups of actually or potentially interbreeding natural populations, which are reproductively isolated from other such groups") and for the polytypic species taxon (Mayr 1940c, 1941i and quoted above). "The gaps between sympatric species are absolute, otherwise they would not be good species; the gaps between allopatric species are often gradual and relative, as they should be, on the basis of the principle of geographic speciation" (1942e: 149). Criteria to infer the subspecies or species status of allopatric populations are here suggested as follows: "We must study other polytypic species of the same genus or of related genera and find out how different the subspecies can be that are connected by intermediates, and, vice versa, how similar good sympatric species can be. This scale of differences is then used as a yardstick in the doubtful situations" (1942e: 166–167).

Mayr (1942e: 148; 1943e, 1946l, 1955e, 1963b) continued to illustrate the absolute gaps between sympatric species by the five thrushes of the genus *Hylocichla* in northeastern North America several of which may occur at the same locality.[10] They are very similar, but completely separated from one another by biological discontinuities. However, he emphasized that reproductive isolation in nature does not necessarily mean sterility. Ability to cross in captivity is not a decisive test.

Mayr introduced the term "sibling species" for morphologically very similar populations that were nevertheless reproductively isolated and discussed this phenomenon repeatedly. The striking discontinuity between local populations of animals and plants, the "nondimensional" species (Mayr 1946l: 273), had already impressed J. Ray and C. Linnaeus during the 17th and 18th centuries, respectively, and instigated the development of the biological species concept. Also some native people recognize as such the same natural units that are called species by the biologist:

"Some 25 years ago, when I was in the mountains of New Guinea I was all alone with a tribe of very primitive Papuans, who were excellent hunters. I sent them

[10] Four of these spotted thrushes are currently included in the neotropical genus *Catharus* and only the Wood Thrush (*Hylocichla mustelina*) remains in this genus.

out every morning with their guns, and for every specimen that they brought back I asked, 'What do you call this one?' I recorded the scientific name in one column and the native name in another. Finally, when I had everything in the area, and when I compared the list of scientific names and the list of native names, there were 137 native names for 138 species. There were just two little greenish bush warblers for which they had only a single name. At the time, I took this for granted because as a naturalist I always believed in species, but whenever I read statements by armchair biologists who deny the existence of species, I always marvel at the remarkable coincidence that the scientist and the native in New Guinea should by pure accident have an imagination that is so closely similar that they assign the mountain birds of New Guinea to the same number of species" (Mayr 1956g: 5; see also here p. 58, Mayr 1943e, 1949m, 1963b: 17, and Diamond 1966).

In later publications Mayr usually quoted the definition of the theoretical biological species concept (the "nondimensional species"), sometimes without the qualifying "actually or potentially" in front of "interbreeding," because it is irrelevant for species status whether the isolating mechanisms are challenged at a given moment (e.g., Mayr 1949j, 1953b, 1969b, 1982d: 273).

Mayr (1949f: 290) considered an emended species definition: In case speciation occurs without a complete separation of the populations an extended definition may become advisable: "Species are groups of actually or potentially interbreeding natural populations that are either completely reproductively isolated from other such groups or whose genetic differentiation (owing to mutation, selection, etc.) outweighs an actual or potential gene interchange with other such groups." He added "But there are also some serious objections to such an emendation." After ecological factors had been emphasized by various workers Mayr (1951l: 92) defined the species as "an aggregate of interbreeding natural populations which are not only reproductively isolated from other such aggregates but also ecologically specialized sufficiently so as not to compete with other such species" or more formally: "A species is a reproductive community of populations (reproductively isolated from others) that occupies a specific niche in nature" (1982d: 273). The reason why he added the qualifying clause "that occupies a specific niche in nature" was that "it seemed to me that no population has completed the process of speciation until it is able to coexist with its nearest relatives. [...] Time will show whether this additional qualification is useful or confusing" (1987e, p. 214).

If the ecological factor would be raised to a criterion for species status, such a definition would reduce to subspecies status most closely related paraspecies which are reproductively isolated but exclude each other because of ecological competition. For this reason such a definition (with the above qualifying clause) never caught on and Mayr (1992a: 222) himself returned to this simple wording: "A species is an interbreeding community of populations that is reproductively isolated from other such communities." Local representative populations of species have particular niches, but not a species as a whole.

In Mayr's early publications of the 1940s and 1950s, there is no terminological distinction between the species concept and the species taxon. In a lecture on "The species as a systematic and as a biological problem" Mayr (1956g: 7) introduced for the polytypic species a new "multidimensional species concept" (see also Mayr 1957f: 16, 1963b: 19). When he had clarified for himself the distinction between the theoretical species concept and the species taxon during the early 1960s, "it became evident that the polytypic species is merely a special kind of species taxon but does not require any change in the concept of the biological species category" (Mayr 1982d: 290). Therefore in his writings of the late 1960s and following decades he no longer mentioned a "multidimensional species concept" but stated: "The species concept is meaningful only in the nondimensional situation: multidimensional considerations are important in the delimitation of species taxa but not in the development of the conceptual yardstick" (1982d: 272; see also Hey (2006: 448) for a summary of Mayr's changing categorization of species concepts).

Beurton (2002) presented a detailed historical review of Mayr's struggle with the biological species concept during more than 50 years. It would have been easier for the reader to follow Beurton's arguments if he had clarified from the start the early terminological confusion regarding (a) the biological species concept per se and (b) the polytypic species taxon and if he had applied Mayr's later clarification to his early writings (rather than presenting the early confusion again in detail without applying Mayr's later clarification). Mayr (1940c) discussed species taxa and that is what his definition referred to (see above). In his Jesup lectures he talked about the general "species problem" referring both to the absolute gaps between sympatric species and to the relative gaps between allopatric taxa, as is obvious from Mayr (1941i), an article Beurton (l.c.) did not consult. Mayr (2002a: 100) himself commented: "It is a basic weakness of Beurton's account that the confusion between species concept and species taxon is not clarified. [...] Those authors who fail to make the distinction, provide a confused and misleading analysis." Mayr (1942e) then presented the distinction between the biological species concept (absolute gaps) and the polytypic species taxon (relative gaps), as in 1941.

The theoretical concept and definition of biological species (Mayr 1942e and later) found entrance into all textbooks of biology. Bock (1986, 1992a, 1994, 1995b, 2004b) emphasized that genetic isolation was meant when Dobzhansky (1937) and Mayr (1942e) spoke of the "common gene pool" and the "harmonious genotype of a species." Mayr (1968i: 164) stated: "Possession of a shared genetic program is the common tie uniting individuals derived from the gene pool of a given species." Therefore Bock (l.c.) emended the definition of biological species to read: "A species is a group of actually or potentially interbreeding populations of organisms which are *genetically* isolated in nature from other such groups." Bock's emendation appears useful also in view of the discovery in recent years that several geographically representative taxa, especially of insects, hybridize freely along the contact zone for lack of premating isolating mechanisms, but such hybrids are infertile due to fully developed postmating isolating mechanisms. In addition, Bock (l.c.) made a distinction between the sets of isolating mechanisms for genetic isolation and reproductive isolation which Mayr failed to do. Some bird species which meet along "zones of overlap and hybridization" (Short 1969) may also represent taxa which are genetically isolated but not fully isolated reproductively. These biospecies would be considered conspecific taxa under Paterson's

(1985) "recognition concept" of species. Evolutionists need to distinguish between genetic isolation, reproductive isolation, and ecological isolation of species (Bock 1979, 1986, 2004d). Genetic isolation must be complete before the neospecies establish secondary contact in order not to hybridize, but reproductive and ecological isolation may be perfected in sympatry through natural selection and mutual selective demands.

The "horizontal" biological species concept refers to genetically separated groups of populations which live during a particular time plane like the Present or any time plane of the geological past (Fig. 5.3). Under this concept, species have, strictly speaking, no origin, age or duration. They represent horizontal "cross sections" of vertical phyletic species lineages in the time dimension (Bock 1979, 1986, 1992a; Szalay and Bock 1991). A phyletic lineage is the continuum of a species in time and documents its history; it does not participate in the development of species (populations do). The "phyletic lineage" and the "species" should be kept separate conceptually.

Mayr's (1931b, 1942e) consistent application of the *superspecies* concept in his systematic work was of general significance because it permitted the combination of several closely related geographical representatives under one superspecific name without reducing these allo- or paraspecies to subspecies status. Such 'superlumping' had been advocated by several workers (e.g., Kleinschmidt and

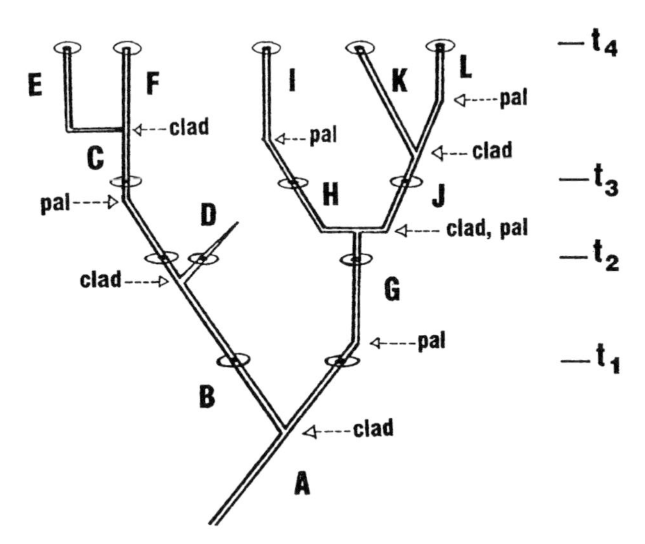


**Fig. 5.3.** Several imaginary phyletic lineages illustrate "species" limits under the cladistic concept (clad.) and the paleontological concept (pal.). Schematic representation. Groups of populations representing the various lineages at particular time levels ($t_1$–$t_4$) are different biological species (*oval circles*). Vertical scale-geological time; horizontal scale–morphological and other biological changes. A–L represent paleontological "species," except C–F, which together are one paleontological "species" but represent 2 cladistic "species." The current time level is $t_4$

Stresemann) during the 1920s, even though many of such allopatric forms actually were too distinct to be treated as subspecies.

The units of a regional fauna are *zoogeographical species* (Bock 1956, Mayr and Short 1970f) composed of two elements, (a) taxonomically isolated species (isospecies; that are not members of superspecies) and (b) superspecies (Amadon 1966; Bock and Farrand 1980; Amadon and Short 1992). Much earlier Mayr had already emphasized that one must count zoogeographical species rather than individual species in a particular region of the world, in order to avoid misleading implications. In his letter to Stresemann dated 29 September 1936 he wrote: "A renewed count of the birds of New Guinea resulted in 513 breeding zoogeographical species (reckoning [the superspecies] *Astrapia* and *Parotia* as one species each)." However, it was not until 1970 that Mayr (l.c.) introduced the term "zoogeographical species" formally (actually already used in Bock 1956). He himself credited it later to B. Rensch (Mayr 1989k: 156) without a specific reference. However, Rensch (1934) had used the term *species geographicum* only in the sense of polytypic species (not in the sense of superspecies plus taxonomically isolated species).

In their analysis of zoogeographic species Mayr and Short (1970f) studied the avian species taxa of a well-defined region (North America) to answer the question of how often the application of the biological species concept leads to difficulties, controversies, or ambiguities presenting in full their data on which their conclusions were based. Most of the tabulation and accompanying text was written by L. Short and evaluated in discussions with Mayr. They showed that invariably the application of the biological species concept helped to clarify difficult situations. In a number of peripheral isolates one could not be sure whether or not they had already reached species level, but this created as much difficulty for a morphological as for the biological species concept. Only in Mexican populations of towhees, *Pipilo*, the same two forms behave in one area as two biospecies and in another area as one because of extensive hybridization. Several other cases of "interspecific hybridization between largely allopatric members of the same superspecies" are not clear-cut either, especially those where there are large zones of geographical range overlap and extensive hybridization.

The theoretical concept of biological species is nondimensional and refers most clearly to genetic-reproductively isolated populations at a particular locality (Mayr 1941i, 1942e, 1946l, 1953b, 1963b). Because a fully differentiated biospecies represents a genetic unit, a reproductive unit *and* an ecological unit, Mayr (1951l: 92, 1982d: 273) specified as one aspect of biological species a specific niche in nature, i.e., ecological isolation permitting sympatry with competitors. This theoretical notion of biological species must be distinguished from the multidimensional species taxon. If several differentiated groups of populations are in contact and intergrade, they belong to the same species taxon (Fig. 5.4). Allopatric, i.e., geographically separated, representative taxa are assigned subspecies or species status on the basis of inference (Mayr 1948c, Mayr et al. 1953a: 104, 1969b: 197, Mayr and Ashlock 1991i: 104–105). Auxiliary criteria used for that purpose include:

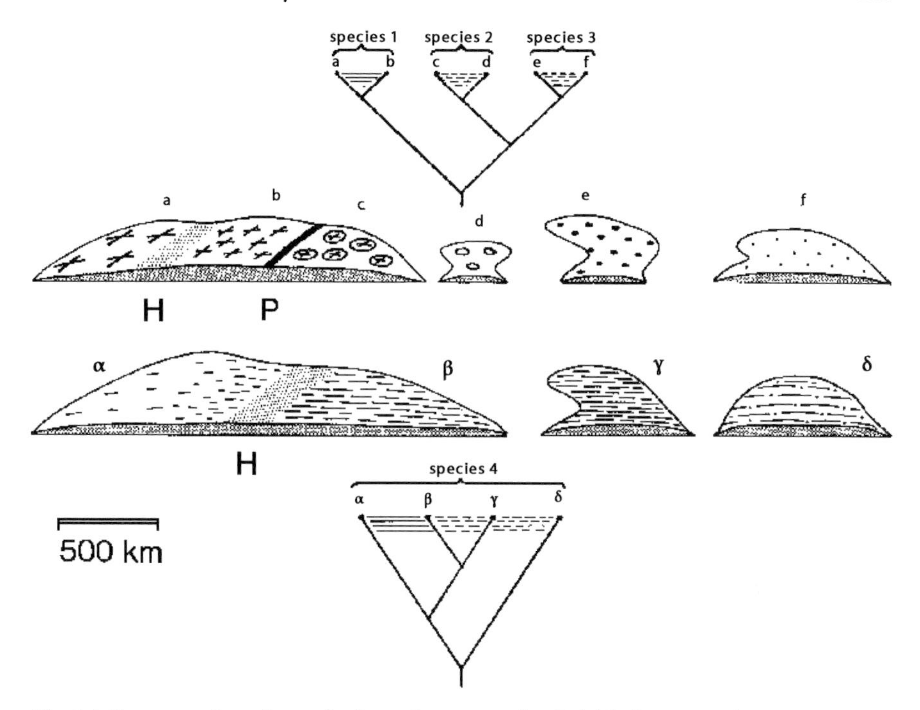


**Fig. 5.4.** Two sets of species and subspecies taxa each of which forms one large continental unit (*left*) and several geographically isolated populations (*right*) with their hypothetical cladograms. Schematic representation. H hybrid zone between subspecies, P parapatric contact zone between species (geographic exclusion without or nearly without hybridization). In the cladograms hatching indicates known intergradation (hybridization), dashes indicate presumed hybridization. In all areas the respective sympatric populations (taxa) of these two sets are specifically distinct with respect to each other (biological species). In the upper set, forms a and b hybridize and together represent species 1 which does not hybridize with parapatric species 2 (consisting of subspecies c and d). The status of the island populations d–f as species or subspecies (i.e., their hybridization or non-hybridization if they were in contact) is judged on the basis of inference (see text). Because forms alpha and beta of the lower assemblage hybridize freely where they meet, forms gamma and delta are also assumed to hybridize if they would establish contact. All four taxa are considered subspecies of one polytypic species. Species 1 to 4 are all monophyletic

(1) degree of difference between sympatric species, (2) degree of difference between intergrading subspecies within widespread species, (3) degree of difference between hybridizing populations in related species. Occasional criticisms of the biological species concept refer mostly to its practical application in delimiting species taxa rather than to the theoretical notion of biological species itself. The highlights of Mayr's views on the Biological Species Concept, as expressed again in several recent articles (1988h, 1996e, 2000e), may be summarized as follows: A biological species is a reproductively (genetically) cohesive assemblage of populations, an assemblage of well-balanced harmonious genotypes. The devices which

maintain the integrity of species, protecting the harmonious gene pools, are isolating mechanisms, a term proposed by Dobzhansky (1935, 1937: 228–258) who, unfortunately, did not distinguish between external geographical/ecological barriers operating in the allopatric phase of speciation and intrinsic mechanisms. Mayr (1963b: 89–109) stressed the enormous diversity of these devices, e.g., sterility genes, chromosomal incompatibilities, ecological exclusion, behavioral properties especially in higher animals. Species usually occupy ecological niches that are sufficiently different to permit the sympatric occurrence of these species. However, member species of a superspecies exclude one another allo- or parapatrically because of ecological competition. The Biological Species Concept is not applicable to asexual organisms (agamospecies) which form clones, not interbreeding populations, as defined by the biologist. Their gene pools do not require protection by isolating mechanisms, because they usually are not subject to genetic recombination. A list of Mayr's publications dealing with the species problem totals 94.

## Allopatric Speciation

Evolutionists understood since the "Fisherian synthesis" (p. 183) how and why populations of species change over time, yet species are distinct–they do not hybridize with or blend into one another. How can one separate species arise from another (parental) species or one species split into two? Some geneticists ("saltationists" like W. Bateson, H. de Vries, R. Goldschmidt) believed that new species arise instantaneously by large mutations, sudden steps in which either a single character or a whole set of characters together become changed. By contrast, the naturalists were "gradualists" and believed that speciation is a populational process, a gradual accumulation of small changes often by natural selection. To solve the apparent contradiction between sympatric species of a local fauna separated by bridgeless gaps on one hand and the idea of gradual speciation on the other hand Mayr presented many intermediate "borderline cases" of geographic variation from his work on the birds of Oceania pointing to a gradual origin of new species from isolated (peripheral) groups of populations of an ancestral species (1940c, 1942e, 1951l, 1963b). The differences he observed among representative island populations were not only quantitative and continuous but often qualitative and discrete as observed among congeneric species. For these reasons he replaced the typological concept of species by a concept of species taxa as aggregates of geographically variable populations.

"What I did, basing my conclusions on a long tradition of European systematics, was to introduce the horizontal (geographical) dimension, and show that the process of geographic speciation is the method by which a gradual evolution of new species is possible, in spite of the gaps in the non-dimensional situation" (Mayr 1992i: 7).

As he stated in this quote and in other historical reviews (p. 230), the principles of the biological species concept and of allopatric speciation had long been established by the naturalists (systematists), but it was Mayr's clear and synthesizing discussion which convinced the geneticists and zoologists generally of the biological species concept and of the common occurrence of allopatric speciation: Either a previously continuous species range is split into two or more parts (through sea-level changes or climatic-vegetational changes) or a small isolated founder population is established by dispersal of a few individuals of a parental species across a barrier. Each isolated population then evolves independently, gradually diverging from one another. When sufficient genetic differences have accumulated, the parental and daughter populations may come in contact without hybridizing–a new (daughter) species has originated. Despite the striking gaps between the species of a local flora and fauna, the gradual evolution of new species was no longer a puzzle.

There is no case in birds or mammals that would require sympatric speciation (without geographical separation of populations) and this may also be true for many, perhaps the majority of insect families (butterflies, Carabidae, Tenebrionidae, etc.). To appreciate the effectiveness of allopatric speciation one only needs to compare the high number of endemic species in island archipelagoes with the low number of such species in a continental region of comparable size.

Mayr summarized the various stages of the differentiation process during allopatric speciation in his "dumbbell" model (1940c: 274–275, 1942e: 160) as follows [with some additions]; see Fig. 5.5:

Stage 1: A uniform species with a large range; followed by
Process 1: Differentiation into subspecies; resulting in
Stage 2: A geographically variable species with a more or less continuous array of similar subspecies (2a all subspecies are slight, 2b some are pronounced); followed by
Process 2: (a) Isolating action of geographic barriers between some of the populations; also (b) development of isolating mechanisms in the isolated and differentiating subspecies; resulting in
Stage 3: A geographically variable species with many subspecies completely isolated and some of them morphologically as different as good species [note that Stage 3 does not require Process 2b, only 2a]; followed by
Process 3 (often in connection with 2): Development of intrinsic isolating mechanisms for genetical isolation, as well as the development of reproductive isolation and ecological differentiation, resulting in
Stage 4: Expansion of range of such isolated populations into the territory of the representative forms; resulting in either
Stage 5a: Noncrossing, that is, new species with restricted range or
Stage 5b: Interbreeding, that is, the establishment of a hybrid zone (zone of secondary intergradation).

During the course of time, isolated populations differentiate as members of the following microtaxonomic categories: local population-subspecies-species-superspecies-species group with numerous cases of intermediate differentiation

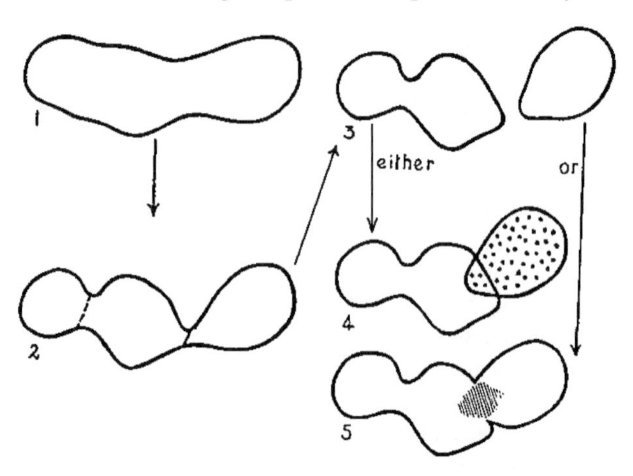


**Fig. 5.5.** Stages of geographical (allopatric) speciation (from Mayr 1942e, Fig. 16)

found which were difficult to classify (see Table 4.2 on p. 177). However, such "intermediate" forms illustrated for Mayr the process of speciation graphically. Strictly speaking it is not the subspecies but the geographically isolated population which may reach species status. However, little was known about the genetic basis of the speciation process itself which both Dobzhansky (1937) and Mayr (1942e) felt certain occurred as a continuation of microevolution. In a superspecies the member species are not ecologically compatible, for which reason the respective species populations, if their ranges abut, compete along the contact zones and exclude each other geographically without (or only rarely) hybridizing. The closely related members of a species group may be sympatric in parts of their ranges. In ring species the overlapping populations are so strongly differentiated that they no longer hybridize and would be considered species if their direct intergradation through the "ring" of subspecies were to be interrupted (an example are the Asian greenish warblers *Phylloscopus trochiloides*, Irwin et al. 2005). According to the "Wallace-Dobzhansky model" isolating mechanisms of neospecies originate due to selection forces in sympatry, whereas according to the "Darwin-Muller-Mayr model" isolating mechanisms arise as incidental, pleiotropic by-products of divergence in allopatry (but see also under sympatric speciation, p. 223). David Lack (1944, 1949, 1971) added important data on ecological aspects of the speciation process, as acknowledged by Mayr (1982d: 274).

Besides the "dumbbell" model of geographic speciation, Mayr (1942e) developed an alternative speciation model: a new species originates from jump dispersal of a few individuals forming a small strongly isolated peripheral founder population (founder principle, 1942e: 237; 1954c; 1963b: 539, 554; peripatric speciation, 1982l). This principle relates the reduced variability of small founder populations not to accidental gene loss, but to the fact that the entire population was started by a single pair, a few individuals or even by one fertilized female which contained only a fraction of the gene pool of the parent population. These "founders" carried with them only a very small proportion of the genetic variability of the parent population. This idea goes back to an early formulation by Rensch (1939: 184) who, however, did not appreciate its importance. He emphasized (transl.) that "the gene pool of a few individuals [establishing a new island population] never corresponds to the total gene pool of the numerous individuals of the parent population. In such cases *the newly developing island race is not characterized by an increase but by a decrease of genes*" (italics in the original).

In a later article on "Change of genetic environment and evolution" (which he considered as one of his most important publications) Mayr (1954c) discussed the genetic processes during speciation of peripherally isolated founder populations. His main idea was that in the limited gene pool the selective value of the genes would be different from the original condition in the parent population particularly owing to inbreeding and a rise in homozygosity. Later the population will enlarge and gradually build up its depleted genetic variability. Rapid evolution in founder populations may lead to the development of strongly differentiated bizarre peripheral isolates (p. 158). Such forms, however, have low evolutionary success, when subsequently exposed to competition with mainland stocks. Mayr did not claim, however, that every founder population speciates, that every genetic change in a founder population is a genetic revolution or that speciation occurs only in founder populations. On the other hand, he was already fully aware of the consequences of his theory pointing out that geographical isolation and the small size of speciating founder populations may explain the phenomenon of lack of documentation of speciation in the fossil record.

As he stated, "Many paleontologists have postulated various kinds of typostrophic 'saltations' in order to explain the absence of crucial steps from the fossil record. If these changes have taken place in small peripheral isolated populations, it would explain why they are not found by paleontologists" (1954c: 210).

Many years later the paleontologists Eldredge and Gould (1972) based their theory of "punctuated equilibrium" on Mayr's observations. Although they did refer to his relevant publications, this intellectual debt got almost lost in later years. In developing his model of genetic revolutions during the speciation process Mayr (1954c) applied the "new genetics" of Dobzhansky, Wallace and others who like Wright emphasized the interaction of genes and their varying selective values depending on their "genetic environment" or "genetic background." Mayr (l.c.) introduced these latter terms which were widely appreciated and applied in discussions of the integration of the gene pool or of coadaptation among genes (Williams 1966: 59). Mayr got the idea of "genetic reorganization" (or "genetic revolution") in peripherally isolated populations when visiting Naples, Italy during the summer of 1951 and lectured on it already in Oxford during September of that year. At that time, E. B. Ford asked him to contribute a paper on this topic to a festschrift for Julian Huxley which only appeared 3 years later. Until then Mayr was "mortally afraid that someone else would get ahead of me."

In retrospect Mayr commented on his idea of "genetic revolutions" during speciation in an interview as follows: "I still think that this is an important idea. The key thing is that the smaller the population, the more important the chance factor becomes because it no longer involves a very strong selection for just one kind of thing. By pure chance a lot of genes and gene combinations simply drop out and so in small populations, improbable combinations are quite frequent and they may be the starting point of something very interesting which in a large population would never happen. Among the classical population geneticists, the only one who saw that was Sewall Wright. Fisher and Haldane never saw it and they would say that small and large populations are one and the same thing. In fact, Fisher said something that was completely and utterly wrong, he said 'the larger a population, the faster it will evolve.' We now know that the truth is exactly the opposite. The larger a population, the more inert it becomes because it is difficult for any change to penetrate through a large population. That is why Wright had his shifting balance theory by which a little piece of a population went outside, in the spirit of my 1954 paper. Then as a unit, it goes back inside again and spreads into the large population. The critics of Wright, in particular J. Coyne [et al. 1997], claim that this whole thing of Wright's is quite impossible, it just can't happen for many reasons. But you have to give Wright credit, that at least he saw there was a genuine problem there, which Fisher and Haldane didn't see" (see Wilkins 2002: 966).[11]

Current evidence provides little support for founder effect speciation and it appears that selection is more important for speciation than genetic drift, which also plays little part in morphological evolution (Coyne and Orr 2004). One recent case study of founder effects in silvereyes (*Zosterops lateralis*) on Pacific islands appears to be inconclusive (Grant 2002b).

In his review of different patterns and theories of speciation, Mayr (1982l) refuted the theory of stasipatric speciation and discussed the weakness of the theories of parapatric and sympatric speciation. Chromosomal changes and geographical isolation occur simultaneously. Similarly, Mayr and O'Hara (1986a) refuted an earlier claim that parapatric speciation had occurred in West African birds.

Mayr summarized the various speciation models in the following table (1987a: 311–312): New species originate:

(A) Through a speciation event

  (a) Instantaneous (e.g., polyploidy, stabilized hybrid)[12]
  (b) Very rapid (peripatric speciation, conceivably sympatric speciation)

[11] Provine (2005) criticized (1) Mayr's concept of genetic revolution as being "devoid of genetic content" and (2) his usage of terms like gene pool and homeostasis of gene pools as being "biological nonsense." However, Futuyma (2006) rejected these criticisms pointing out that (ad 1) "given what Mayr had learned from his population geneticist colleagues, his hypothesis [...] had as much genetic content, I believe, as Wright's shifting balance theory" and (ad 2) that "Mayr was taught genetic homeostasis in one of the major schools of evolutionary genetics of his day." Moreover, unlike Darwin, Mayr has shown "that species of sexually reproducing organisms are real and that they exist by virtue of reproductive isolation rather than of phenotypic distinctiveness" and often originate in allopatry.

[12] New evidence indicates that hybrid speciation occurs not only in plants but also in animals and is more frequent than previously thought (O. Seehausen, Trends in Ecology and Evolution 19: 198–207, 2004 and J. Mallet, Nature 446: 279–283, 2007).

(B) Without a speciation event (parental species transformed)
  (c) Dichopatric speciation
    (split by a geographic barrier with gradual divergence)
  (d) Gradual phyletic transformation of a single lineage.

The crucial process in allopatric speciation is geographic isolation (not selection). This statement only refers to the geographic position and separation of the speciating populations. It still remains completely unknown as to what happens genetically during speciation. Possibly rather different genetic mechanisms are involved in the speciation of different kinds of organisms and under different circumstances. Of course, the speciating populations remain well adapted through natural selection which is always present.

Some authors have suggested that Sewall Wright's writings have influenced Mayr's thinking and that in particular, his model of the "adaptive landscape" had given Mayr the idea of geographic speciation.[13] Nothing could be further from the truth. Geographic speciation was discussed among naturalists since the 19th century (p. 230) and was standard thinking in Erwin Stresemann's department in Berlin; Mayr described geographical speciation already in 1927(f) in his paper on the Rosy-Finches (*Leucosticte*). Moreover, Wright's model of the "adaptive landscape" was an abstract model, not one of a three-dimensional landscape; it has very little to do with speciation, but refers mainly to "the evolution of the species as a whole" (quote in Mayr 1992i: 11). As Mayr (1999k: XXIV) pointed out, S. Wright was never particularly interested in problems of speciation and, although he published reviews of all the major works on evolution, the only one he omitted was Mayr's *Systematics and the Origin of Species* (1942e).

Species originate through "splitting" and "budding" (Fig. 5.6). "Budding" occurs when, e.g., a derivative population of a widespread mainland species reached species status on a nearby island. This speciation event had no effect on the parental biospecies (no. 3, Fig. 5.6) on the mainland from which neospecies 4 has budded off. The mainland species is real in the sense that it represents a biological unit characterized by close genetic-reproductive and ecological relations among its component subspecies taxa. The cladistic analyses schematically illustrated in Fig. 5.6 (if feasible at that intraspecific level) yield relevant phylogenetic ("vertical") and biogeographical data on the origin and relationships of the various groups of taxa. Mayr (e.g., 2000e: 164) feels that most new species originate by budding, splitting is less common.

One may ask the question why Mayr, in 1942, preferred the dichopatric model of speciation (the "dumbbell" model), although the island populations of the Polynesian and Melanesian birds he studied in detail certainly had originated through jump dispersal (and peripatric speciation). When he studied a particular aberrant peripheral island population, the question foremost in his mind at that

[13] Thus M. Ruse (1999: 118) wrote: "Mayr, who in later years put distance between himself and Wright, based his original picture of the evolutionary process on the hypothesis" (the shifting balance hypothesis of S. Wright).

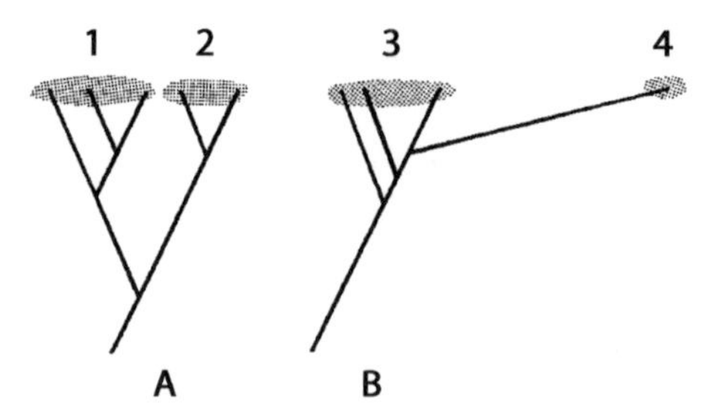


**Fig. 5.6.** Speciation through splitting (A) and budding (B) resulting in monophyletic biospecies 1 and 2 (consisting of 3 and 2 subspecies, respectively) and paraphyletic biospecies 3 (three subspecies). Species 4 which budded off from species 3 is monotypic and may demonstrate its species status by reinvading the ranges of some or all subspecies of species 3. Shading indicates genetic cohesion and intergradation of subspecies along contact zones. From Haffer (1992)

time was a taxonomic one, whether this form should be ranked as a subspecies, species or as a monotypic genus. The evolutionary questions why and how these aberrant forms originated on isolated islands were not yet foremost in his mind.

Another question is to what extent his fieldwork in New Guinea and the Solomon Islands influenced Mayr's ideas as developed and promoted in his 1942 volume and in later publications. He gave the following answer in an interview:

"The thing is that the influence [of my fieldwork] was only limited because I already had most of the basic ideas. They had been developed further in recent years by Stresemann and several other workers in this area including in Russia. Naturally, everything I had done and observed in New Guinea and the Solomon Islands fitted exactly on these basic ideas of the European workers, but they never had developed all the ideas in these areas: What are the possibilities? What are the methods to be used? What are the mistakes you might make?" (Bock and Lein 2005, video CD-ROM).

Mayr's 1942 book had an enormous influence on the development of evolutionary studies worldwide, but various general questions still remain open: What is the percentage of species of major taxonomic units (e.g., birds or mammals) that originated through peripatric ("budding") and dichopatric speciation ("splitting"), respectively? To what extent did peripatric speciation take place on continents? How many of the taxonomic species currently recognized are, in a cladistic sense, monophyletic and how many are paraphyletic?[14] And how important is sympatric speciation?

[14] Monophyletic species contain all descendants of a common ancestor and paraphyletic species contain some, but not all descendants of a common ancestor. Coyne and Orr (2004: 472) conclude: "Because there is no unitary *genetic* history at the population level, it is almost impossible to recognize true paraphyly among closely related taxa using genetically based phylogenies."

## Sympatric Speciation

Mayr stated repeatedly that geographic speciation prevails among animals but never ruled out that non-geographic speciation may occur.

"Isolating mechanisms are, for example, distinct and nonoverlapping breeding seasons, or in the cases of parasitic or monophagous species strict host specificity with the mating taking place on the host. Enough such cases have been described to make me believe that sympatric speciation is of common occurrence in certain animal groups with very specific ecological requirements, but almost completely absent in all other animal groups" (1941i: 142).

In his 1942 book he concluded "that bona fide evidence for sympatric speciation is very scanty indeed. [...] There is some indirect evidence for the importance of 'host races' for speciation:" Monophagous and oligophagous genera of butterflies and moths are much richer in species than the polyphagous ones (*Nepticula*, 140 Palearctic species; *Lithocolletis*, 100 sp., *Coleophora*, 140 sp.). "Certainty as to the relative importance of sympatric speciation in animal evolution cannot be expected until a much greater body of facts is available than at present" (p. 209, 215). In 1947(e) Mayr pointed out that the existence of species-rich genera of host specialists by no means proves sympatric speciation. It may merely mean that they have more niches available than generalists. Genetic considerations made it also increasingly difficult to visualize the occurrence of sympatric speciation:

"The realization of the genetically highly complex nature of the isolating mechanisms and, more broadly speaking, of the intricate integration of the total gene pool of a population, makes it exceedingly difficult to conceive of a mechanism that would permit the building up of genetic isolating mechanisms within a physically undivided gene pool" (1959a: 226).

In his 1963(b) book he carefully evaluated the available evidence and concluded: "Host races [of parasitic insects] constitute the only case indicating the possible occurrence of incipient speciation" (p. 460) but assumed that complete stabilization on a new host cannot occur without geographical isolation. "The possibility is not yet entirely ruled out that forms with exceedingly specialized ecological requirements may diverge genetically without benefit of geographical isolation" (p. 480; see also 1976m: 144; 1982d: 605; 1988e: 376). Mayr did not rule out sympatric speciation categorically but merely pointed out that those cases mentioned in the literature did not prove it. Sympatric speciation requires the simultaneous acquisition of mate preference and niche preference, something Mayr considered improbable in 1942. However, he accepted it when it was demonstrated in cichlid fishes (see Mayr 1984e, 1999k: XXX; 2001f: 100, 180; see also Barluenga et al. 2006; Pennisi 2006; Meyer 2007). The females of these fishes have a definite preference for a particular part of the environment, the pelagic or the benthic part, and simultaneously for the particular males that occur in this particular environmental niche. This joint preference could quickly produce a new sympatric species. That both of these aspects could be selected at the same time is something that had not occurred to Mayr, but it has been proven. However, in many other groups speciation is exclusively allopatric, e.g., mammals, birds, butterflies (except in some highly host specific forms). Mayr (2004a: 108) summarized his latest views stating: "After 1942 allopatric speciation was more or less victorious for some 25 years, but then so many well-analyzed cases of sympatric speciation were found, particularly among fishes and insects, that there is no longer any doubt about the frequency of sympatric speciation."

In recent years, several authors have claimed that if natural selection drives speciation, then gene flow will not affect diversification in areas (like, e.g., Amazonia) that are much larger than the per-generation dispersal range of individuals. In view of sufficient genetic variation and ecological diversity in extensive tropical forest regions, species may originate through "isolation by distance." Actual splitting of populations, as required under the model of allopatric speciation, supposedly is unnecessary and splitting of populations in space may be the result rather than the cause of genetic differentiation during "adaptive speciation" (Schilthuizen 2001; Knapp and Mallet 2003; Tautz 2003; Dieckmann et al. 2004; critical discussion by Gavrilets 2005). Allopatric speciation is well supported and uncontroversial, whereas speciation in the face of gene flow is less well supported and more controversial. Comparative work suggests that it is far less frequent than allopatric speciation (Coyne and Orr 2004: 7).

Work on speciation over the past several decades mostly focused on the geography, ecology, and timing of speciation, in the tradition of Mayr's 1942 book. Molecular studies confirmed that Pleistocene events caused substantial intraspecific differentiation and the origin of several closely related pairs of North American bird species. Many other birds split from extant relatives at earlier times (Lovette 2005). Topics such as the role of sexual selection and the frequency of sympatric speciation are now also being addressed in several avian systems (Edwards et al. 2005). Rapid speciation may be frequent in birds in view of the importance of prezygotic isolating mechanisms including song, whereas the slow development of intrinsic postzygotic isolation will facilitate continuing hybridization. In recent years the search for speciation genes has been accelerated which eventually may permit understanding the genetics of avian speciation processes. Such genes seem to have very rapid rates of adaptive amino acid replacement (Orr 2005).

## The Species Problem in Mayr's Empirical Research

Mayr's theoretical conclusions on the biological species concept and the geographical (allopatric) origin of species were based on the results of numerous case studies which comprised not only birds but also snakes, fossil man, and marine organisms as well as many other examples reported in the literature.

In his early paper on snow finches (1927f), he discussed geographical variation and geographic speciation in a group of finches (p. 45). His studies in Pacific island birds convinced Th. Dobzhansky, when Mayr showed him his material in New York (p. 133). In August and December 1939, Mayr presented many examples of speciation from isolated populations among the birds of Oceania, first at the 6th Pacific Science Congress held in California emphasizing zoogeographical aspects (not published until 1941k) and, second, at a symposium on speciation organized by Dobzhansky (Mayr 1940c).

Mayr (1954e) asked the question whether speciation in marine organisms differed from speciation in land animals (birds, mammals, and butterflies). At the time he was working with West Indian *Cerion* snails (see p. 331) and therefore selected the West Indian shallow water echinoids for study. Each genus consists of one or several groups of allopatric species, some of which should be considered as subspecies. Allopatric populations show every grade of distinctness, ranging from slight difference through subspecific to specific rank. This pattern corresponds to that of land animals. This is also suggested for deep sea organisms (crinoids) and pelagic organisms (Scyphomedusae). Dispersal barriers which caused speciation in warm water echinoids are the eastern Pacific, the Isthmus of Panama and the cold waters of southwestern Africa. There was no evidence suggesting any other mode of speciation in sexually reproducing marine organisms than geographic speciation. Palumbi and Lessios (2005) largely confirmed Mayr's interpretation of geographic speciation in sea-urchins using DNA sequence data. They also showed that the origin of reproductive isolation depends not only on genome-wide steady accumulation of substitutions but also on the rate of evolution of gamete recognition proteins.

Sibling species were found among the colorful birds of southeastern Asia (p. 161) and the polymorphic Californian king snake proved to be one biological species only (p. 226). Particularly interesting cases of geographical speciation in birds include the Black Bulbul (*Hypsipetes madagascariensis*; p. 160), trillers (*Lalage*), fantails (*Rhipidura*), Reef Heron (*Demigretta sacra*), drongos (Dicruridae, p. 158), *Halcyon* kingfishers, Australian tree runners (*Neositta*; p. 179), wheatears (*Oenanthe*, p. 226), paradise magpies (*Astrapia*, p. 155), and honeyeaters (*Melidectes*). When experimentation with birds turned out as logistically too difficult, Mayr studied isolating mechanisms in *Drosophila* (p. 228). He also applied the principles of new systematics to paleanthropology (p. 330). Numerous additional examples of geographical variation and speciation from peripheral isolates may be found in his major books on species and evolution (1942e, 1963b, Mayr and Diamond 2001g). The number of generic and family revisions, faunistic papers or check-list entries requiring decisions on the species status of geographically isolated populations totals about 60.

Mayr (1946a) estimated the number of bird species in the world at 8,616 species and Bock and Farrand (1980) at 9,021. This increase in the number of species taxa was caused in part by the discovery of genuine new species and in part by reinterpreting the rank of numerous allopatric taxa as species rather than subspecies. In a study of a local flora, Mayr (1992a) tested the validity of the biological species concept for plants concluding that the great majority of taxa (93.5%) form biological species; only 6.44% pose difficulties by apomixis and by autopolyploidy; Whittemore (1993) published a critical discussion of these results from a botanical point of view.

## The Nature of Species in Bisexual Animals

Sexuality leads to genetic recombination and integration of natural populations into species. Most species taxa are polytypic, that is they are composed of lesser units, subspecies and local populations, which deviate more or less from one another. Species are, through their geographical extension, multidimensional systems. The subspecies is not a biological unit like the local population and the species, but on continents often an artificially delimited taxonomic unit. Island populations are more clearly defined but their subspecies status must be inferred. Taxonomists study clinal variation, especially in species with extensive ranges on continents, which is correlated with gradients in selective factors of the environment. Peripheral or otherwise separated forms of a group of relatives are often the most distinct, e.g., in birds *Junco vulcani* (Miller 1941), *Ptilinopus huttoni* (Ripley and Birckhead 1942), *Dicrurus megarhynchus* and four other taxa of drongos (Mayr and Vaurie 1948d), *Dicaeum tristrami* (Mayr and Amadon 1947d) and many others. Many of these are borderline cases between subspecies and species and indicate the course of speciation.

Most morphological characters and numerous physiological species differences are neutral with respect to the maintenance of reproductive isolation, as they are merely accidental by-products of the genetic divergence of the populations during geographic speciation (Mayr 1948c, 1949f). Other characters promote the coexistence of species and have selective value, like different ecological requirements which also reduce or prevent the meeting of potential mates functioning as sexual isolating mechanisms (Dobzhansky 1937). Behavioral barriers like courtship differences in ducks and *Drosophila* function similarly. Mayr (1948c) emphasized that isolating mechanisms are not simple lock and key mechanisms, because often several different factors and mechanisms are involved. Natural selection will improve imperfect premating isolating mechanisms between populations when they come in secondary contact.

Difficulties with the biological species concept may arise, when morphological variants occur within a single population which may be due to simple Mendelian inheritance (e.g., in the Californian king snake, Mayr 1944g) or to balanced polymorphism (e.g., the wheatear genus *Oenanthe*, Mayr and Stresemann 1950f). Another problem is the interpretation of the taxonomic status of allopatric populations. Since isolating mechanisms are usually correlated with a certain amount of morphological differences rather typical for a given genus, the taxonomist uses this evidence to work out a yardstick which can be applied to allopatric populations (pp. 214–215). Mayr's opinion was that it is preferable to treat doubtful cases as subspecies, because allopatry indicates inability of coexistence (given the fact of dispersal) and closest relationship.

## Natural and Sexual Selection

As a young systematist Mayr, like many of his colleagues including Stresemann and Rensch, thought that the differences between animal species, e.g., *Drosophila melanogaster* and *D. simulans*, were surely something entirely different from the conspicuous mutations (white-eye, yellow body, crumpled wings, etc.) in these species. To them environmentally induced continuous geographical variation was far more important in evolution than such mutants observed by geneticists in the laboratory and not relevant to the mutation-selection theory (Mayr 1992i: 2, 23). They believed that the evolutionary significance of natural selection was rather limited explaining, e.g., the similarity (mimicry) in egg color between the brood-parasitic European cuckoo and its host species. Since the early 1930s Mayr was persuaded through modern genetic publications and conversations with his colleague J. Chapin at the AMNH that indeed natural selection was the crucial factor explaining adaptation and the differentiation of populations, particularly since the modern geneticists had shown that natural selection works on minute mutations and recombinations as the raw material of evolution. Interest of evolutionists in the study of sexual selection increased only during the 1970s (Mayr 1972g; see below, p. 279); this factor was not mentioned in the textbooks and monographs of the 1930s and 1940s, including Mayr's book of 1942(e).

## Ecological Factors

Rensch (1928, 1929, 1934) and Stresemann (1939: 360) had considered closely related bird species which exclude each other geographically on continents and in island groups because of competition owing to equal or similar ecological requirements. These observations were the basis for Rensch's (l. c.) *Artenkreis* concept (*superspecies* of Mayr 1931b). In the section on "The biology of speciation" Mayr (1942e) discussed the influence of ecological and behavioral factors. Speciation may happen rather rapidly or extremely slowly in different groups. The geographical ranges of species may be small or very extensive and geographical barriers may be very effective for some species but ineffective for others, because their dispersal capacity varies greatly. Speciation is balanced by extinction which is frequent on small islands. Where isolating mechanisms between species fail, hybridization occurs.

Stresemann (1943) discussed the evolutionary role of ecological differences among local populations of birds giving many examples. Lack (1944, 1947, 1949) based his discussion on the ecological differences between species as an indispensable prerequisite for coexistence and Mayr (1944n) was so enthusiastic about Lack's (1944) first article on these topics that he immediately reviewed it in an American ornithological journal: "Lack makes the very important point that reproductive isolation alone is not enough for two species to coexist. They must have also developed certain ecological differences—dissimilar habitat or food preferences, for example—that prevent competition with each other. In many cases there is considerable overlap, but it is never complete" (Mayr 1944n).

Ecological competition may prevent two species from invading each other's ranges, as shown by many members of superspecies which exclude each other geographically along sharply defined contact zones. Lack (1947) showed how competition determined which ecological niche various species of *Geospiza* occupy on a given Galapagos Island and how this affects the size of the bill. It was Lack who brought ecology into the evolutionary synthesis, Mayr (1947e) remarked in his article on "Ecological factors in speciation."

As to the origin of higher taxa Mayr (1942e: 298) concluded that this process "is nothing but an extrapolation of speciation. All the processes and phenomena of macroevolution and of the origin of the higher [taxa] can be traced back to intraspecific variation, even though the first steps of such processes are usually very minute."

## Isolating Mechanisms in *Drosophila*

Mating behavior is a very important isolating mechanism. Mayr wanted to know how such species specificity was obtained. In a letter to R. C. Murphy, chairman of the bird department at the AMNH, dated 18 March 1943, he explained:

"I have come to the conclusion in connection with my work on the origin of species that a study of the factors which control the mating of animals is one of the most badly needed research jobs in this field. There are certain factors, the so-called isolating mechanisms, which prevent the mating of individuals which do not belong to the same species. I'm planning to conduct a number of experiments [...] through which I want to determine how far the isolating mechanisms are inborn and to what extent they are conditioned during the early life of the individual" (cited from Cain 2002b).

As a first step he was going to work on *Drosophila* which would "make the problems clearer." Mayr spent mid-June to mid-July 1943 in Cold Spring Harbor (Dobzhansky was in Brazil that year) and hoped to start experiments with birds a few months later in the AMNH. A room on the top floor, the necessary cages and individuals of three species of estrildid finches (*Lonchura*) were acquired. Alexander Seidel, the artist for Mayr's books on the *Birds of the Southwest Pacific* and the *Birds of the Philippines* (p. 150), was hired to run the bird colony. However, the logistics of this project and several problems, e.g., how to keep the species separated from each other, were overwhelming in view of other museum tasks and his long commuting trips each day.[15] Mayr restricted research on isolating mechanisms to *Drosophila*. His collaboration with Dobzhansky in Cold Spring Harbor from 1944 until 1946, working at the same table, gave him another chance to broaden his knowledge of evolution and evolutionary genetics. Mayr and his family continued to spend 1–2 summer months there until 1952 (except in 1951 when they were in Europe). He participated in seminars and discussions and interacted with many other geneticists and scientists who were developing molecular biology at that time. These annual stays at Cold Spring Harbor were invaluable for his future career (pp. 243–250).

In a first set of experiments Dobzhansky and Mayr (1944j) tested geographical strains of neotropical *Drosophila willistoni*. The methodology was developed by

[15] Such experiments were later carried out successfully by Klaus Immelmann (1935–1987) and his students in Bielefeld, Germany (e.g., Immelmann et al. 1978).

Dobzhansky, the experiments were mostly conducted by Mayr. The method used was multiple choice: one kind of males was placed in a container with two kinds of suitably marked females, and it was tested whether and to what extent insemination was random. The authors decided for male choice, although they realized that in animals, female choice prevails (but is more difficult to test).

Innate species recognition is the rule in most of the lower vertebrates and invertebrates, might be influenced by conditioning. Mayr and Dobzhansky's (1945d) findings with *Drosophila pseudoobscura* and *D. persimilis* indicated that males inseminate a higher percentage of females of their own than of the other species and that the degree of preference can be altered slightly through conditioning, by previous association with the respective females. Since there is little doubt that males' preference of their own species is mainly controlled by genetic factors, Mayr (1946d) tested the hybrids in regard to their position on the mating preference scale: Males of *D. persimilis* inseminate more hybrids than females of their own species. Males of *D. pseudoobscura* show a slight preference for their own as compared to hybrid females. Mayr (1946g) also analyzed their courtship behavior (wing vibration, circling the female), described copulation and discussed the role of sense organs in species discrimination. In multiple choice experiments mating is not random, but highly discriminative; conspecific matings are much more frequent than heterogamic matings.

What Mayr labeled his "best experimental work" concerns the role of antennae in the mating behavior of female *Drosophila* (Mayr 1950c). After many difficulties, he mastered the operation to remove the antennae from these very small flies. The readiness for mating was then drastically lowered. However, since the antennae are not the only organ by which the females receive stimuli from males, even antennae-less females will eventually copulate. The removal of the antennae also decreased the females' ability to discriminate between males of the sibling species *D. pseudoobscura* and *D. persimilis*. In retrospect it must be noted that these experiments were indecisive because, in nature, females choose their mates. Dobzhansky had persuaded Mayr to base his experiments on male choice for technical reasons.

"Dobzhansky was an amazingly hard worker. Part of our work was to determine what percentage of the females had been inseminated. This meant anaesthetizing these females (killing them) and then dissecting their abdomen to see whether there was sperm in the genital tract. Each examination took a certain amount of time, as did collecting the material, killing the females, spreading them out on a glass slide, etc. Dobzhansky succeeded to do 250 such dissections in a single day. I, who sat opposite him and tried to work just as hard and as fast, never got above 180 females. The only consolation I had was that I said to myself that my examination was more thorough than Dobzhansky's, and I sometimes discovered spermatozoa in females which Dobzhansky had recorded as uninseminated. This illustrates one of his weaknesses. Dobzhansky was a very impatient person, and the painstaking analysis of something was very difficult for him."

## Brief History of the Biological Species Concept

The history of the biological species concept (BSC) goes back to the early 19th century (Mayr 1955e, 1957f, 1959a, 1963b: 482–488, 1980f: 33, 1982d: 270–272; Grant 1994; Glaubrecht 2004; Haffer 1992, 2006). The first authors who used a biological "shared essence" (fertility of matings) for a species definition were John Ray (1627–1705) in 1686 and Georges Buffon (1707–1788) in 1749. However, their language indicates that species, although real entities, are essentialistically constant and invariable. Buffon's species concept was widely adopted in Europe during the late 18th and early 19th centuries including Carl Illiger (1775–1813) in Germany whose "Thoughts on the concepts species and genus in natural history" Mayr (1968i) translated into English. He cited (1942e: 156; 1963b: 483) the amazingly perceptive remarks on species and allopatric speciation by Leopold von Buch (1774–1853) in 1819 and 1825 about which Mayr may have learned from the articles by Robert Mertens (1894–1975) in 1928 or several earlier authors who had also referred to L. von Buch's theory. This theory deeply influenced Charles Darwin who had adopted a biological species concept in his notebooks during the 1830s, but later gave it up. Major contributions to the formulation of the biological species concept and the theory of allopatric speciation during the 19th century were then made by H. W. Bates, A. R. Wallace, B. Walsh, H. Seebohm and later during the early 20th century by Karl Jordan, Edward Poulton, Ludwig Plate, Erwin Stresemann and B. Rensch (Fig. 5.7).

Dobzhansky (1935, 1937, 1940) discussed genetic differences between every discrete group of individuals and the development of isolating mechanisms through natural selection. He proposed the most useful term "isolating mechanism" between species, but he included not only intrinsic but also extrinsic geographic barriers which, of course, are something entirely different (Mayr 1942e). According to Dobzhansky species may be conceived "statically" and "dynamically." Statically, "a species is a group of individuals fully fertile inter se, but barred from interbreeding with other similar groups by its physiological properties (producing either incompatibility of parents, or sterility of hybrids, or both)" or, in other words, "discrete non-interbreeding groups of organisms" (1935: 353). His "dynamic conception" of species was less satisfactory, because species are populations, not stages in a process: "Dynamically, the species represents that stage of evolutionary divergence, at which the once actually or potentially interbreeding array of forms becomes segregated into two or more separate arrays which are physiologically incapable of interbreeding" (1935: 354). Dobzhansky's endorsement of the BSC undoubtedly contributed to its increasing popularity.

Neither Dobzhansky nor any earlier Russian scientist were the originators of the BSC, as Krementsov (1994) wrongly implied. He traced the roots of Dobzhansky's ideas backward to several Russian entomologists. While this is probably correct, Krementsov (l. c.) conveyed the impression that these were the first to develop the BSC. Some of his remarks read as follows: "The idea of biological species attracted little attention in the west and was not much discussed" (p. 36), "Western entomologists were generally less interested in discussing these ideas

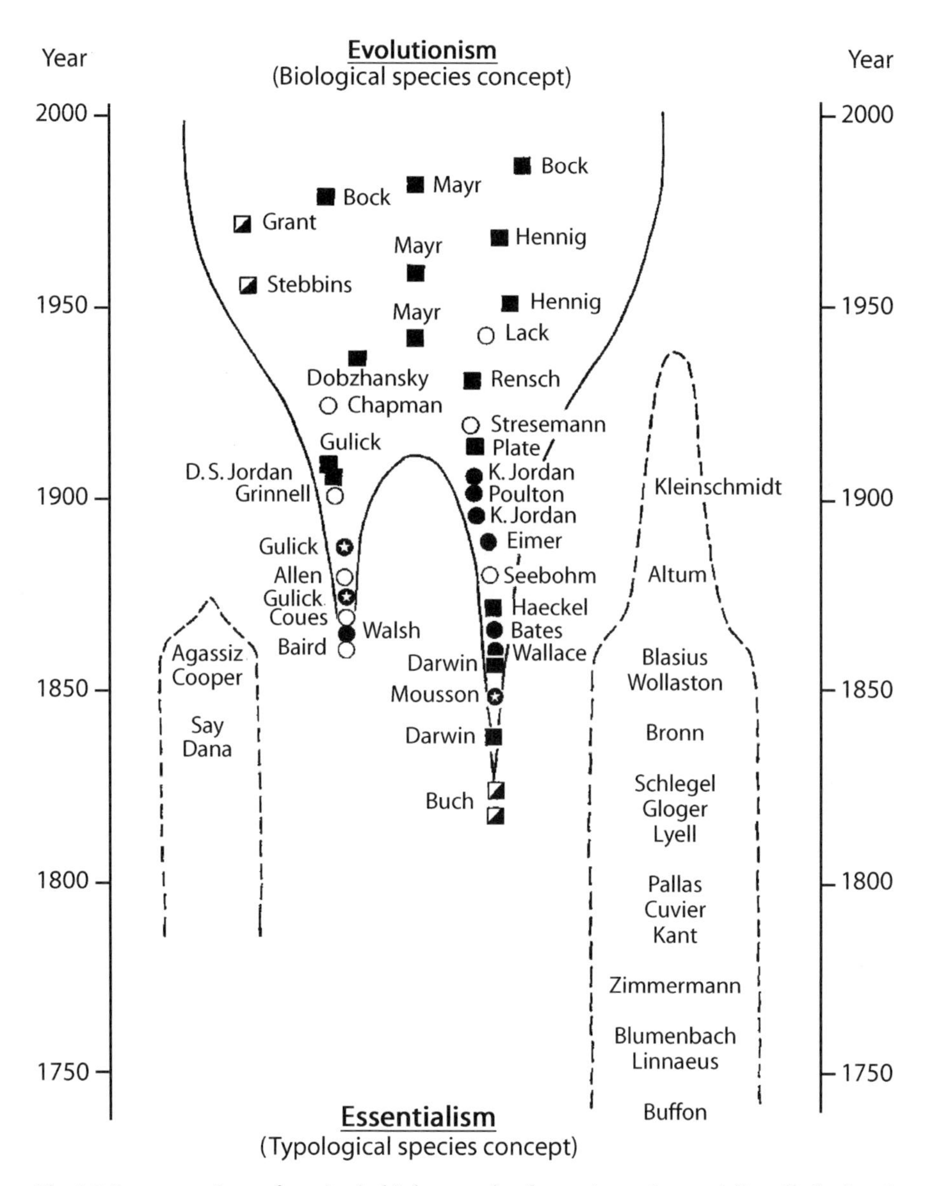


**Fig. 5.7.** Interpretations of species in biology under the notions of essentialism (*below*) and evolutionism (*above*), from Haffer 2006.
The names of selected authors are indicated at their respective time levels. *Solid squares* refer to publications on animals in general, *half-filled squares* to those on plants, *solid circles* to insects, *circled stars* to mollusks and *open circles* to birds.

[of geographical isolation as a causal factor of speciation] than were their Russian colleagues" (p. 38). "In my opinion, the essence of the biological species concept was clearly formulated in the theoretical papers of leading Russian entomologists in the prerevolutionary decades." However, they were not the first to develop the BSC, nor were E. B. Poulton or K. Jordan whose work Mayr (l.c.) discussed in detail.

Building on the work of Stresemann, Rensch, Dobzhansky, and several earlier authors mentioned above, Mayr (1942e, 1963b, 1970e) combined systematic, genetic and ecological aspects, analyzed the speciation process and thus established the theoretical BSC in all its ramifications. Through his contributions, the BSC became one of the central tenets of the modern synthetic theory of evolution. Although he was not the originator of the BSC, he demonstrated its validity more convincingly than anyone else before and proposed a superior and concise definition which has been widely adopted.

## Community Architect

During the 1940s Ernst Mayr became increasingly active in founding specialty groups beyond ornithology (where he had initiated intellectual changes first in the Linnaean Society of New York and then in the American Ornithologists' Union; see pp. 109 and 122–124), as an aggressive reformer and administrator in committees and societies, and active in modernization of research infrastructure. Whereas some of his colleagues like Simpson and Dobzhansky were willing to be presidents of societies but avoided administrative burden, Mayr saw himself repeatedly at the center of such enterprises. He became an organizer first within the "Committee of Common Problems of Genetics, Paleontology, and Systematics," later within the Society for the Study of Evolution, the Society of Systematic Zoology and again in the American Ornithologists' Union. He strove to produce formal networks of communication, i.e., professional journals and approved programs at scientific meetings, for he wanted more than merely exchanging information. His interests were synthetic and constructive; he aimed at theory-building and integration. Joe Cain (1994 ff.) studied the progress of the evolutionary synthesis during the 1930s and 1940s emphasizing professional and intellectual infrastructure, i.e., journals, societies, committees, workshops, meetings, communication networks, funding, etc. In conjunction with these analyses he has traced Ernst Mayr's efforts to include systematics and the work of "museum men" into the evolutionary synthesis. He mentioned the short-lived Society for the Study of Speciation (1939–1941) and Mayr's attempts at improving its effectiveness (Cain 2000a). He also acknowledged Mayr's leadership role in the "Committee of Common Problems of Genetics, Paleontology, and Systematics," 1942–1949 (Cain 2002b) and finally discussed his 3-year term as founding editor of the new society's journal *Evolution*, 1947–1949 (Cain 2000b). The following account is based on these detailed reports (see also Cain 2004 and Smocovitis 1994a,b).

## The Society for the Study of Speciation, 1939–1941

During the meeting of the American Association of the Advancement of Sciences (AAAS) in Columbus, Ohio in late December 1939 (see p. 189) Th. Dobzhansky, J. Huxley, and A. Emerson founded the Society for the Study of Speciation (SSS), designed to simplify communication among widely scattered workers in this field (Cain 2000a). The objective was to create an informal information service or information network, not a journal-issuing society, focusing on the dynamics of the origin of species. Emerson was recruited as secretary and principal agent for the organization of the group and publication of a booklet about twice a year to present bibliographical citations and notes on original work from various members and laboratories as well as new ideas. The first and only bulletin (29 p.) was distributed to 374 individuals not before March 1941. Emerson could not spend enough (or any) time on this project. Moreover, he did not receive sufficient suitable material for future bulletins since members delayed work or shifted to other projects because of the United States' entry into the war. Emerson put the SSS and the bulletin on hold until a later year.

The objective of the SSS was close to the center of Ernst Mayr's scientific interests. He was very eager to see it succeed and closely followed its development. However, he disliked the delay and the quality of the first bulletin, which finally arrived when he had just concluded his Jesup lectures on "Systematics and the origin of species." Although not on the executive committee of the SSS but with Dobzhansky's knowledge, Mayr took the initiative to replace A. Emerson as secretary and approached in this regard his friend A. G. Richards, an entomologist at the University of Pennsylvania. Mayr also made several suggestions to Emerson himself how to improve the contents of future bulletins, e.g., different persons should follow speciation work regarding freshwater organisms, marine organisms, island birds, etc. and the bibliography on speciation should be combined with *Biological Abstracts*. Emerson replied that "your comments on our recent mimeographed material were about the best we received in the way of constructive comment," but because of his limited time he was unable to be the active editor and secretary Mayr envisioned. Due to Emerson's "lack of initiative" the SSS expired.

## The "Committee on Common Problems of Genetics, Paleontology, and Systematics," 1942–1949

This Committee was founded upon a suggestion of Walter Bucher (1889–1965), Professor of Geology at Columbia University, New York, and chair of the National Research Council's Division of Geology and Geography (serving 1940–1943). Eleven scientists, including G. G. Simpson, Dobzhansky and Mayr, met in the Zoology Department of Columbia University on 17 October 1942 to discuss future research in the borderland between genetics and paleontology. The final plan envisioned a Western Group of 10 scientists centered in Berkeley that would emphasize the cooperation between an equal number of geneticists and paleobotanists; and an

Eastern Group of 20 scientists that would be centered in New York City and emphasize genetics, zoology and paleozoology. This proposal was formally accepted on 6 February 1943 and the "Committee on Common Problems of Genetics and Paleontology" (CCP) officially launched during that month.

At the same time the two leaders of the Eastern Group left the country; the paleontologist G.G. Simpson had joined the army in North Africa in December 1942 (returning in late 1944) and the geneticist Th. Dobzhansky went to Brazil in January 1943 to study natural populations of *Drosophila* in the tropics under a government program to promote inter-American cooperation (returning in the spring of 1944). Before his departure, Dobzhansky had asked his friend Ernst Mayr to replace him as chairman of the genetics section of the Eastern Group during his absence and to create the genetics part of the summer program and to plan the group's summer 1943 meeting (in consultation with W. Bucher and the paleontologist G. Jepsen, G.G. Simpson's replacement). Ernst Mayr took over this task with great enthusiasm and quickly moved into a leadership role of the CCP in the New York region using the opportunity for programmatic reforms. Again, we might say, an example of his unshakeable self-confidence.

By that time, Mayr was well known among geneticists many he had met personally. The 2-day summer 1943 meeting of the Eastern Group in the American Museum of Natural History was a success. The themes were genetic variation, discontinuity, rates of evolution, and evolutionary trends. Eight paleontologists, seven geneticists and a number of invited guests attended. Within the session on "discontinuity" Mayr and Curt Stern lectured on isolating mechanisms and the differences between sympatric and geographic speciation. Probably it was Mayr who argued that papers for a later symposium of the entire CCP should: "set forth for workers in the two fields the precise nature of the best data now available [...] that bear on the major problems of evolution. Each section [of the Committee] is to select specific topics that represent the most instructive and complete cases and lead to the theoretically most significant conclusions [...] Each subject is to be documented by illustrations and tables as full as possible and presented in such a manner that it may be fully comprehended and critically evaluated by the workers in the other field who are not familiar with the details of terminology and techniques used. In each case the theoretical implications of the results are to be set forth explicitly for those not sufficiently acquainted with the lines of reasoning peculiar to the two fields of science" ("Report of meetings," p. 9; cited from Cain 2002b: 300).

Because of the war, preparations for a joint symposium of the Western and Eastern Groups including the identification of suitable case studies for discussion had to be postponed from 1943 to a later year. However, Mayr proposed to begin "letter exchanges." He announced in January 1944 that the Eastern Group would "start a discussion by correspondence of topics which fall within the field of interest of the Committee." He volunteered to serve as "the central agency" for this communication. During the following months, members corresponded actively sending him copies of their letters which he collected and, once sufficient material had accumulated, distributed (often with his own comments) in mimeographed form to all the members as a "Bulletin." The first number was distributed in May 1944. This led to further discussion among other members or with the original correspondents and Mayr edited three additional "Bulletins." Gretel Mayr contributed also to this effort by typing mimeographed stencils. Otherwise there was, during World War II, virtually no interaction between the Eastern and Western Groups.

According to Cain (2002b: 302), Ernst Mayr was an aggressive editor, prompting queries, recruiting additional materials, instigating interaction. Under wartime conditions he was leading a highly successful "synthetic" and integrative program. Arguing that systematics was "the vital link" between paleontology and genetics, Mayr proposed and was granted a change in the name of the CCP in April 1944, adding systematics: *Committee on Common Problems of Genetics, Paleontology, and Systematics*. At the same time he began thinking and talking to colleagues, in particular Dobzhansky (back from Brazil in spring 1944), about a Society for the Study of Evolution with its own journal, i.e., projects along the lines that he as a member of the defunct Society for the Study of Speciation had envisioned earlier. Because, by rule, Committees under the auspices of the National Research Council had to be temporary, Mayr was planning for the future of the CCP after the war had ended.

When G. G. Simpson returned to New York in the autumn of 1944, he added, as chairman of the Committee, a preface to Mayr's fourth Bulletin praising the progress made and stating: "From the whole series of letters in the Bulletin there has emerged concrete evidence that a field common to the disciplines of genetics, paleontology, and systematics does really exist and this field is beginning to be clearly defined." He also agreed with Mayr's and Dobzhansky's ideas about founding a Society for the Study of Evolution and its journal. Specific planning was somewhat delayed until December 1945 and led to the formal launching of both the Society and the journal *Evolution* in June 1946 (Fig. 5.8).The next meeting took place during the December 1946 AAAS meetings in Boston. It was voted that the purpose of the society would be "the promotion of the study of organic evolution and the integration of the various fields of biology."

The main sporadic activity of the CCP under Simpson and Dobzhansky during 1945 and 1946 was the planning and organizing of the postponed major symposium in Princeton. Mayr was a member of the steering committee and G. Jepsen headed the local organization. The conference on "Genetics, Paleontology and Evolution" in January 1947 was culmination of work of the CCP and at the same time signaled the beginning of the new evolution society (Jepsen 1949). Due to Mayr's insistence David Lack from Britain was invited to bring ecology into the synthesis. But, as Mayr revealed in his autobiographical notes, Dobzhansky's coworker on *Drosophila*, J. T. Patterson of Texas, had not been invited (due to a confusion with Bryan Patterson of Chicago). Dobzhansky, enraged, was about to leave Princeton. Several of his friends managed to persuade him to stay, but he refused to contribute to the conference volume (1949) and to be listed as one of its editors.

The conference in 1947 was a confirmation of the evolutionary synthesis (or rather a synthesis because, by modern standards, it was far from complete). The books by Dobzhansky (1937), Mayr (1942), Huxley (1942), and Simpson (1944) had

**Fig. 5.8.** Ernst Mayr as cofounder of the Society for the Study of Evolution, 1946 (AMNH Library photographic collection, negative no. 122774)

all appeared several years earlier. The geneticists had realized that there was a large field of study in biodiversity and its origin, and the naturalists had realized that the evolutionary ideas of the early Mendelians based on large mutations (saltations) had been most misleading. Therefore, by 1947, there were no more arguments because both sides, the geneticists and the naturalists (systematists), understood that there was no conflict any longer between their thinking. As to Mayr's personal interactions with other North American "architects" of the evolutionary synthesis, his close relations with Dobzhansky have been detailed above (pp. 133, 185ff.) and his non-interaction with G. G. Simpson in New York during the 1930s–1940s was mentioned on p. 120. During the 1920s Bernhard Rensch had been Mayr's colleague at the Museum of Natural History in Berlin whose writings influenced Mayr as much as or even more than those of his teacher Erwin Stresemann (pp. 207–208). As to his personal interactions with Julian Huxley (1887–1975), Sewall Wright (1889–1988), J. B. S. Haldane (1892–1964) and Ronald A. Fisher (1890–1962), Mayr said in an interview (Wilkins 2002):

"I knew Julian Huxley for a very long time as an ornithologist and we had met at international congresses and he was an enthusiastic outdoor birdwatcher so we got along just fine. Now Huxley was a really good friend; we visited each other's houses and our wives were friends. My only criticism of him was that he was so full of ideas and plans that there was very little cohesion between the things that he did. His book, '*Evolution. The Modern Synthesis*' (1942) was very good in detail but was chaotic in its composition. I now know also why he used to come and visit me in New York. He would sit across the table from me. He would put out a pad and say 'Now Ernst, what's the latest in evolutionary biology?' I would tell him and he would eagerly scribble and then I think he worked it out in a little more detail, fed it to his secretary, she put it together and there was his book! I'm exaggerating but you get the spirit. And so there really was very little in his 1942 book that was his original ideas. And no one ever quotes him as the originator of this or that idea. He is known mostly for coining the term 'the modern synthesis' and he had that term before there was a modern synthesis."

"The story that I was greatly influenced by Sewall Wright is mostly the concoction of Michael Ruse [e. g. 1999: 118]. Actually I was not influenced very much, if at all, by Sewall Wright. He was a mathematician and looked at everything from that point of view and it just didn't make sense to me. I used to sit down next to him at Cold Spring Harbor when there were no other dinner companions and I'd try to get a conversation going and I never was successful." (See also p. 220). Mayr (1959) considered Wright as one of the mathematical geneticists whose theories referred to the same "beanbag genetics" as those of R. A. Fisher and J. B. S. Haldane who calculated independent effects of individual genes without taking into consideration genic interactions and factors other than random drift. Later Mayr admitted that his early views about Sewall Wright's theoretical contributions were not quite correct (p. 274).

"I never knew that J. B. S. Haldane was such a good outdoors man until one day when I visited him in Calcutta he took me to Orissa and we stayed overnight at a government guest house. This was right next to the fields, to a little native village and to really untouched Indian nature. I was an ardent birdwatcher at that time, this was in 1960 when I was about 56 years old. In the morning at about 5 AM I was out there and there were the most marvelous birds, the natives came out of their huts, it was a brilliant morning and I was just absolutely inebriated by this beautiful landscape. I came back to the guesthouse at 7 AM for breakfast and I was still full of enthusiasm and I held forth on how wonderful it was and all that. Suddenly Haldane interrupted me rudely saying 'Ernst, why didn't you take me along?' I said 'Well, I didn't know you'd be interested' and he said 'Of course.' So the next morning the arrangement was that he would knock at my door at 5 AM and he surely did and I took him out and we saw all these wonderful things again the next morning and he was just as enthusiastic as I was. We came home and we had had the most wonderful outdoor excursion and ever since that time I realized that in addition to all his mathematics and physiology, how much he was basically a friend of Nature.

I had a number of interactions with Ronald Fisher. I was editor of the journal *Evolution* and he and E. B. Ford submitted a joint paper in which they criticized a paper by Sewall Wright that I had published the issue before. In his paper, Wright showed (or thought he showed) that a particular change in a polymorphic pattern of one of their butterflies that they worked with was not a systemic selective change but could as well be due simply to random drift. Fisher and Ford were infuriated by this paper and they sent a rebuttal which not only used language unsuitable for a scientific journal but also didn't in any way answer Wright's criticism. They simply reiterated that it was systemic selection. Well, I read it as editor, and I said to myself 'I cannot publish this as it is.' I gave it to two other readers and they both agreed it would have to be changed, first of all the language had to be cleaned up, and secondly, they had to come to grips with the actual point of Wright's criticism. So with both of these things I returned the manuscript to Fisher together with the reviewers' letters and I got a very curt reply back from him saying that since obviously the editor of *Evolution* refuses to publish any paper that doesn't conform to his ideas, they didn't want to embarrass him and they were herewith withdrawing their paper.

Now there is a continuation to this story. A year or two later, one day at the luncheon table at the American Museum, Simpson said to me 'Oh, Sir Ronald came to see me this morning and when we had discussed what we wanted to discuss, I was going to bring him up to your office but at that moment Sir Ronald realized or remembered that he had an important appointment in downtown New York and left in a great hurry.' About another year later, I was in Cambridge, England, visiting Bill Thorpe. He was a very friendly person and as soon as I was there he said now we have to make a schedule for you, whom you have to see and all that and he mentioned Sir Ronald. I said 'Oh, don't bother. Sir Ronald doesn't want to see me' and Thorpe being such a kind person said 'Of course he would, naturally we must see him.' Well, I said 'OK, but it's your responsibility.' So he called his secretary and Sir Ronald wasn't there but she said that he'd be back at 11 o'clock. And so we timed it so that Thorpe herded me to Fisher's villa at 11 o'clock. Sir Ronald himself opened the door and he stepped outside just between me and Thorpe and in such a way that he turned his back to me and was facing Thorpe and started to talk to Thorpe and I was totally cut out. At this point, I thought well I can play that game too and I rotated around, facing him again from the other side. Then he had to give up and we went inside and, after that everything was reasonably civil."

## Founding Editor of the Journal *Evolution*

Again, Ernst Mayr was willing to contribute his share in the administration of the new evolution society and to take over the job of founding editor of its journal *Evolution*. He implemented the plan that he and his collaborators Dobzhansky, Simpson, Huxley, Emerson, and others developed. He had gathered initial experience in editing and publishing the Proceedings and Transactions of the Linnaean

Society of New York (see p. 110). His proposals for the title of the journal and for the editorial policy were approved without objection. In late 1946 the funding of the journal *Evolution* had been secured and the first manuscripts were in hand. Mayr fully realized the opportunities for reform and recommended that all papers should deal with evolutionary factors and forces, i.e., with the process of evolution. A logistic problem ensued, when many printers had no spare capacities, as they worked through large wartime backlogs. Eventually Lancaster Press (in Lancaster, Pennsylvania) was able to accommodate Mayr's schedule. The first double issue of *Evolution* (vol. 1, numbers 1–2) was distributed to 746 subscribers in July 1947.

The new Evolution Society, Mayr wrote to Stresemann, "is a joint affair of geneticists, taxonomists, and paleontologists, somewhat equivalent to the circle that Timoféeff-Ressovsky had gathered [in Germany during the 1930s and early 1940s]. Huxley, Dobzhansky, and myself are the prime movers of this new venture. The taxonomists and naturalists of this country, as well as of England, are beginning to realize that they can contribute a great deal to the study of evolution, and we are trying to focus these efforts and break down the borderlines between these fields" (24 January 1946) and "It is important to emphasize the evolutionary angle as a counterbalance against the assertions of the physicists and chemists who see nothing in this particular branch of research. Science in this country is literally swimming in money but it has to be atomic science or medical research, otherwise no money at all is available. People like Dobzhansky, Simpson, and myself try to counteract this trend but it is very difficult" (25 January 1949).

For his 3-year term Mayr solicited manuscripts on specific topics to balance the coverage of various fields and corresponded with the authors. He took his job very seriously making numerous suggestions on the contents and presentation of nearly every manuscript, and provided additional references to the literature. As a systematist, he also envisioned the journal as an outlet for publications on the evolutionary byproducts of "museum men," i.e., morphologists, biogeographers and ecologists, whether neontologists or paleontologists, zoologists or botanists. He continuously attempted to balance such topics against the prevalence of genetical articles on *Drosophila*, with partial success only because not all promised contributions from the fields of anthropology, paleontology, botany, and taxonomy were submitted despite Mayr's active solicitations.

"I find it difficult to get a stock of good manuscript every three months. The forthcoming issue has two or three rather weak contributions in it, but the June issue again will have some excellent papers. However, like every good editor, I must do an almost incredible amount of letter writing to get contributions from the right kind of people. The promises I now have are sufficient to fill two full volumes, but unfortunately one can't print a journal with promises. My manuscript drawer is empty each time an issue goes to press. This means that I have to do some rather hectic editing of the last-minute manuscripts before they go to the printer" (to Stresemann on March 9, 1948).

As mentioned above, Mayr also intended to bring the systematists into the community of evolution workers. Discussions of evolutionary topics by systematists, if published at all, were often buried in the introductions of taxonomic revisions or even in the species accounts of faunal papers and thus inaccessible to workers outside their specialty. However, in the field of evolutionary systematics Mayr had also considerable difficulties to obtain suitable manuscripts and set out to recruit selected manuscripts himself with partial success. He wanted discussions of general biological aspects of systematics, explanation and analysis of evolution as illustrated in particular groups of animals or plants. However, he was unwilling to risk the journal's reputation and rejected papers which supported processes like orthogenesis or macromutation that could not be justified by current genetic knowledge. He did confess though (1997g) that he published in the first 2 years several papers that he would not have accepted if he would have had a larger supply of manuscripts. In 1949, Mayr introduced a new feature, a set of critical "Comments on recent evolutionary literature," and continued publishing such reviews until 1952, when the eighth installment in this series appeared. It included a total of 163 of his reviews of papers ranging from paleontology and ornithology to genetics and behavior. His principal objective was to counteract the narrowness of many evolutionists and to make known to American readers papers from the international literature.

Although asked to serve another term as editor, Mayr stepped down in 1949. His health had suffered while preparing the "Biology of Birds" exhibition (see p. 129 and 302). His successor as editor was the paleontologist E. Colbert. Mayr was elected president of the evolution society. In subsequent years, he took his reform efforts to the Society of Systematic Zoology (resulting in a sharp increase in the number of papers devoted to evolutionary systematics that appeared in its journal *Systematic Zoology;* Hull 1988; Cain 2000b: 257), to Harvard's MCZ, to the National Academy of Sciences and to the National Science Foundation (see below).

Mayr summarized his activities during the late 1940s in a letter to W. Meise in Berlin:

"You are quite correct when you say that I seem to be leading a busy life. My main job is, of course, the curating of our collections and the daily correspondence. Most time-consuming, however, is perhaps the editing of '*Evolution*' and the enormous correspondence connected with that job. During the last year I have also helped in editing a volume of a symposium on evolution [Jepsen et al. 1949]. For a time there was a great deal of correspondence with several European ornithologists about literature and reprints. It took a lot of my time even though Gretel handled most of the work. In addition, there are always two or three volunteers working in the Department who require much of my time. Last spring I taught at the University of Minnesota and next spring I plan to teach at Columbia University. Fortunately I have a very good secretary who takes care of most of my routine matters. Also, I have a Dictaphone and can dictate my correspondence in the evenings so that I can devote myself to museum matters when I am in the museum. My correspondence is really quite enormous. The other day, after a three-day absence from the museum, I found no less than fifty-three pieces of mail on my desk. The unfortunate result is that I don't have nearly as much time for ornithological research as I would like.

Also, I ought to revise my *Systematics and the Origin of Species* and that also is not proceeding too rapidly. Years ago I signed a contract to write a biology of birds and that also is not making any progress. I just thought that you might be interested to collaborate with me on that job and I could translate your part into English. What do you think about this plan?" (17 November 1949; see p. 129).

# 6 Life in North America during World War II

## New York and Tenafly, New Jersey

When World War II broke out in August 1939, the British ornithologist David Lack (1910–1973) who was traveling in the United States happened to be a houseguest with Ernst and Gretel Mayr in Tenafly, New Jersey. They had long conversations and reflected on the certainty that the war would cause destruction and misery in many parts of Europe. Their friendship became deeper and stronger through this shared experience. During the first 2 years, communication with families and friends in Germany was maintained via Switzerland and Japan. When the United States entered the war in late 1941 exchange of letters was, of course, no longer possible and the question loomed large: Would the American Museum be able to keep the German at his work or would they be forced to dismiss him? At that time Mayr was sitting feverishly over his major work, *Systematics and the Origin of Species* (1942e), often far into the night. "Later we found out," Gretel Mayr wrote in her reminiscences, "that this aroused suspicion with some neighbors who denounced Ernst to the police suggesting that he was up to some clandestine, hostile activity. I went to a secretarial school in Englewood and learned shorthand and typing so that in an emergency I could get a job."

Mayr was allowed to continue his work at the museum (no doubt through the influence of Dr. Sanford) and all of his colleagues continued to be friendly and supportive, but he had to give up his car. Also, no "enemy alien" was permitted to possess a shortwave radio, a camera or a gun. These he gave to friends for safekeeping. Some neighbors looked at the Mayrs suspiciously, but others remained friendly. Someone suggested that he change his name to "Ernest Mayer," but he refused. Although many Germans were arrested and sent to internment camps, Ernst and Gretel Mayr remained protected because they had American-born children. However, in August 1942 FBI agents came and searched their house for two hours (particularly for a copy of Hitler's book *Mein Kampf*, this the Mayrs had never owned or even read). The parents were taken to the local courthouse where they were legally arrested, photographed and fingerprinted. They were later released on parole, in the charge of the Tenafly police department and two sponsors, Professor Franz Schrader (Columbia University) who lived in Tenafly and Professor Bill Glenn (New York University), their neighbor friend. The Mayrs had to contact the police and the two sponsors once a week and were not allowed to leave Tenafly without permission. Ernst automatically got a permit to go to New York, his place of work, but Gretel had to write to the justice department in Trenton, New Jersey, whenever she wanted to leave the state of New Jersey. However, visits to the dentist or doctor in New York were easily granted and she could then stay all day, go shopping and visit friends. Uncle George and Aunt Helen Simon (p. 103) who lived in Bernardsville, had an apartment in New York and the Mayrs could occasionally see them there. In 1943 they obtained permission to spend their summers in Cold Spring Harbor (see below). After the initial anxiety, life was not unbearable during the war. Of course, they were worrying about their families and friends in Germany. Occasionally it was possible to send and receive a message of 20 words via the Red Cross.

Several months after their "arrest" at the Tenafly police department their case came up for a hearing before a board in Newark which had been set up to deal with enemy aliens living in New Jersey. The prosecutor's only aim was to prove that the Mayrs were guilty and dangerous. They were interrogated separately, but afterwards it was decided that they could remain free on parole. In addition to the weekly visits to the Tenafly police department, they now had to go once a month also to Ellis Island, off the southern tip of Manhattan, to report to the Department of Immigration and Naturalization. This involved a lot of traveling and took nearly a whole day. In spite of these inconveniences the Mayrs were not unhappy realizing how lucky they were to be far away from the events of the war.

## Cold Spring Harbor Laboratory (1943–1952)

A "Biological Laboratory" had been founded on the shores of Long Island Sound in 1890 followed by a "Station for Experimental Evolution" in 1904 (renamed the "Department of Genetics" in 1921). Both were under the direction of M. Demerec during the 1940s and 1950s (Watson 1991). During the 1940s, the Biological Laboratory sponsored the work of summer researchers, while the Department of Genetics supported by the Carnegie Institution had a number of permanent staff members. Several European experimentalists who had fled their homelands gathered at Cold Spring Harbor (CSH). By the mid-1940s these scientists had forged to a large part a new discipline in the biological sciences–"molecular biology." Among them were the physicist Max Delbrück and the microbiologist Salvador Luria who conducted joint experiments on bacteria-eating viruses (bacteriophages). For most of the year they had teaching duties at the universities of Nashville and Indiana, respectively, but took advantage of the opportunity afforded by the summer research program at CSH to continue their own research. From 1945 onwards they taught for many years at CSH their highly popular "Phage course."

When Mayr's attempt at studying experimentally the nature of isolating mechanisms in birds had failed (p. 228), he complained about this to Dobzhansky who answered "Why don't you do this kind of work with *Drosophila*? I shall spend next summer at Cold Spring Harbor. Let us then do some joint papers. I contribute the *Drosophila* technique and you set up the experiments asking the right kind of questions." Hence, beginning with 1943 and continuing until 1952, Mayr and his family spent their summers in Cold Spring Harbor (CSH), Long Island, one hour away from New York City. Dobzhansky left for Brazil in early 1943 on short notice. Therefore, cooperation began in 1944 continuing until 1946. The Mayrs usually spent about two months in Cold Spring Harbor each year. He would commute to his office in New York, but at least one month was vacation. The results of Mayr's experimental work on isolating mechanisms in *Drosophila* during these years have been discussed above (p. 228). Natural history observations which Mayr made in the outdoors included a song sparrow that went every day to an ant nest to ant himself (Mayr 1948f); yucca moth larvae which remained in the yucca "fruit" until there was a heavy rain. Then at once they bored themselves free, dropped to the soft, wet ground and dug themselves in to pupate. Other observations concerned the species of insects found in CSH mushrooms and photic signaling of fireflies. Mayr was also a frequent participant of the annual Cold Spring Harbor Symposia (1946, 1947, 1950, 1955, 1957, 1959) and presented papers at two of them: 1950 on the *Origin and Evolution of Man* (where he spoke on "Taxonomic categories in fossil hominids," 1951g) and 1959 on *Genetics and Twentieth Century Darwinism* (his talk entitled "Where are we?," 1959f).

For Mayr these summer weeks meant an incredibly important development. It was there that he acquired his extensive knowledge of advanced genetics and molecular biology. He had numerous long conversations particularly with Bruce Wallace during the summer of 1950 who taught Mayr the new genetics of "gene pools," "in which the selective value of a gene depended on the genetic environment rather than on having a constant value. The gene pool had a cohesion, or exhibited homeostasis" (Provine 2004: 1044). "I not only became a close friend of Max Delbrück and other leading molecular biologists but also of many foreign visitors, particularly the 'French contingent' (Monod, Lwoff, Jacob, Ephrussi) as well as others," like Hershey, Dulbecco, Spassky, and Wallace.

"I had perhaps even more contact with local biologists and saw a great deal of Barbara McClintock who was rather aloof but quite delighted when a visitor dropped in once in a while. Dobzhansky's student Bruce Wallace did very interesting work in population genetics but, perhaps more importantly, was an even stronger holist than Dobzhansky. Jim Watson was there for several summers and became a close friend of my daughter Christa. Indeed I met a large percentage of the more prominent biologists of the period in Cold Spring Harbor. I never would have had this opportunity at the American Museum. I have often been asked where I had acquired my extensive knowledge of molecular biology but the answer is quite simple: at Cold Spring Harbor.

Whenever new discoveries were made at Cold Spring Harbor or elsewhere, they very frequently became the subject of heated arguments. I remember when Avery discovered that nucleic acids were the crucial genetic material, Cold Spring Harbor was split into two camps. The phage group under the leadership of Delbrück did not believe it nor did Alfred Mirsky and several others, while another group consisting of Rollin Hotchkiss, Ernst Caspari, Bruce Wallace, and myself accepted it rather quickly."

Mayr described the atmosphere in CSH in his letters to Stresemann:

"Tomorrow I shall be leaving for Cold Spring Harbor Biological Laboratory where I usually spend part of the summer. Gretel and the children get some swimming, while I watch what the male of *Drosophila* species A does to the female of species B. However, even I get all the swimming I want. A whole group of geneticists spend the summer there and it is a most stimulating atmosphere" (12 July 1946). "During the first week I take a holiday and spend the mornings and afternoons at the beach. Last night Mrs. Delbrück arrived to be followed by her husband within two days; they will remain until 11th July. The von Koenigswalds will arrive this afternoon. Michael White, the British cytologist, who intends to settle in Texas, is here and I understand that Prof. Willier will appear soon; also Ernst Caspari, a student of Kühn. As you see, we live in a very stimulating 'intellectual environment.' I am learning immensely, in particular in fields where I know nothing. During the last months new discoveries in the field of bacterial genetics have been made almost daily. One knows now roughly how many genes a bacteriophage has!" (30 June 1947; transl.).

Most of the scientists came to CSH with their families and therefore their wives and children became friends as well. The international atmosphere, in spite of the war and postwar times, was a great relief for all of them. As "enemy aliens" Mayr and many others owned no car and, of course, they had no close relatives in the States. These circumstances and the facilities given at CSH were the reasons for so many foreign scientists gathering there. But of course numerous North American colleagues came also to pursue their research and to participate in the workshops and seminars.

The physical and social environments of the Biological Laboratories in Cold Spring Harbor were as congenial as the intellectual environment, and it is not surprising that the Mayr family all loved being there. The campus during the 1940s and 1950s included two lovely old houses built in the 1830s as residences for families working in the whaling and textile industries owned by the Jones family. They had been renovated into numerous apartments for summer rental and the Mayrs lived in several different apartments in Hooper House and Williams House over the years. The campus clustered around a lawn, which sloped toward the bay from Long Island Sound. There was a tennis court, and play equipment for children near the edge of the water stream, which flowed into the bay. It was an excellent environment for birdwatching and blue herons, green herons and swans were permanent residents. Families made use of a dock near the head of the bay for swimming and boating.

The community building, Blackford Hall, was a general gathering place where people congregated for meals, mail collection and socializing (Fig. 6.1). There was a large room opposite the dining room which was used for frequent seminars and other less serious performances. People often hung out at Blackford Hall in the evening playing music, playing ping-pong, discussing science and testing their theories with their peers.

Children had an unusual amount of freedom in the safety of the Bio Lab campus. They would often play in groups of wide ranging ages which offered some

**Fig. 6.1.** Ernst Mayr on Blackford Porch, Cold Spring Harbor Laboratory, 1950 (Photograph courtesy of E. Mayr)

supervision from the other children. Adults were also usually nearby. Children could entertain themselves at the water or in the woodlands. Even young children attended some afternoon seminars and listened to the adults talk science. As adolescents, the Mayr girls both worked as lab assistants in the Bio Lab, washing Petri dishes and sterilizing equipment. After the Mayr family had moved to Cambridge in 1953, their younger daughter spent two summers in CSH during college vacations working at the Bio Lab on her own.

Ernst Mayr's reminiscences of five friends from the days at CSH follow below: A. Buzzati-Traverso, E. Caspari, M. Delbrück, B. Wallace, and C. Stern:

(1) *Adriano Buzzati-Traverso* (1913–1983): "Adriano was my closest Italian friend. We met in Cold Spring Harbor and we were equally interested in modern population genetics. Adriano, however, in contrast to many others, tended to study the genetics of wild populations and was very much interested in scientists who worked on freshwater organisms or like myself on birds.

In 1951, he invited me to give a series of lectures in Pavia, which were subsequently published. I was put up at the Collegio Ghislieri, a renaissance palace, very elegant, but not heated after the 1st of April, not heated. I nearly froze to death. They put an electric heater on the podium where I lectured. The Po Valley can be extremely cold, even in April. Adriano was always full of plans of how to bring Italy up to the international level of research. He wanted to have his own institute in Naples funded by UNESCO but attached to the University of Rome. The genetics establishment at Rome didn't want to have this separate institute, but Adriano was afraid that if it was under the director of the genetics department at Rome he could not carry out his plans. This ruined the last years of his life. He died rather unexpectedly and rather young of liver cancer in 1983.

Adriano was always full of plans and full of ideas. It was always fun to talk with him. As a person he was very simpático."

(2) *Ernst Caspari* (1909–1988): "One of the greatest pleasures of the summers that Gretel and I spent at Cold Spring Harbor was that we met there just about every summer also the Casparis. He was one of Alfred Kühn's most prominent students and some historians believe that he should be given credit for discovering the one-gene-one-enzyme relationship. Being Jewish, he was driven from Germany, and eventually found a haven in Rochester. His original research never reached again the heights of his success in Germany. Caspari was a broadly interested and widely read biologist, intelligent and critical, and was particularly good in writing review articles. His wife Hansi was also German Jewish, and Gretel and I found that we were very close in our thinking in most matters, so we were in each other's company most of the time. Caspari, in contrast to the phage group, was one of those who were at once convinced by Avery's demonstration that DNA was the genetic material and he quickly convinced me also. Whenever I had problems in experimental zoology or physiological genetics, I always discussed them with Caspari. He was an inveterate smoker and eventually he developed cancer of the bladder and succumbed to it after a long struggle."

(3) *Max Delbrück* (1906–1981): "As with Caspari, it was Cold Spring Harbor which brought Gretel and me together with Max and Manny Delbrück. We saw them there virtually every summer in the 1940's and up to 1950. As everyone knows, Max was a charismatic figure, almost always surrounded by a group of his admirers. Although trained as a physicist, he had become interested in biology through Niels Bohr in Copenhagen who always thought that there were some undiscovered physical laws lurking in living organisms. This is well described by Lily Kay (1985) in a paper on Max Delbrück. In Max' lab no experiment was ever made without having been thoroughly analyzed in an almost Socratic manner. If the experiment would not lead to clear-cut results it was not to be done. Whenever anybody attended his seminar he could expect afterwards a severe dissection by Delbrück. I remember a paper presented by Alex Novikoff who argued against extreme reductionism and in favor of higher levels of integration. Alas, Novikoff was too far ahead of his time and Delbrück massacred him unmercifully.

Gretel and I spent many evenings with the Delbrück's quite often playing bridge, which I normally don't do. And I still remember one evening when Max and I bid a grand slam and made it. I knew this would never happen to me again, and I didn't play much more bridge after that occasion. After supper in the evening we always sat on a wooden railing along Bungtown Road talking about things, particularly talking shop. This is where I learned much of my molecular biology, because I was one of those fortunate people who, so-to-speak, were present at the birth of molecular biology in Cold Spring Harbor. This is where I met Ephrussi, Monod, Lwoff, and all the other greats of molecular biology. One evening when I sat on that railing Delbrück sat next to me and started asking me questions about evolution. I was most happy to give him all the answers I could. The next evening when I sat there again Max sat down next to me and continued the questioning. After a while I said to him, 'Max, why do you want to know so much about evolution?' He answered, 'Because I plan to give a course on it next winter.' I said, 'But Max, how can you give a course on evolution, you know nothing about it!' to which he answered, 'That is precisely why I want to give this course.' I think this conversation sheds a good deal of light on his enterprising spirit.

Whenever we went west we visited the Delbrück's. On three occasions we were invited to come along on one of their famous desert trips. They were quite unforgettable. One usually drove as far as one could into the desert, usually continued walking for a certain distance, then laid down in one's sleeping bag under the starry sky. Eventually Max was diagnosed to have multiple myeloma. The treatment at that time was not yet very well established, and an experiment, very much encouraged by Max himself to try interleukin was a disaster. As a result he succumbed to the disease rather quickly. However, he was still alive when we found out that Gretel had exactly the same disease. He wrote her a remarkably humorous letter saying that it was a wonderful disease having all sorts of advantages, the greatest one being that doctors couldn't operate on you. I forgot what the other advantages were that he had somehow or other invented. We stayed on good terms with Manny and have visited her since, but there is no doubt that Max's death has left quite a void in the circle of our friends."

(4) *Bruce Wallace* (*1920): "He was one of Dobzhansky's best students. It so happened that I was on his PhD. committee and I have been in contact with him ever since. If I remember correctly, I published his dissertation, or at least the gist of it, in *Evolution*.

Wallace's first job was in Cold Spring Harbor where he did population genetic work inspired by Dobzhansky. He was at Cold Spring Harbor during many of the years when I was there in the 1940s and early 1950s, and there was no one in Cold Spring Harbor with whom I talked as much as with Wallace. Like Michael Lerner, he belonged to the relatively small group of population geneticists who were particularly interested in the interaction of genes rather than as the mathematicians, only in additive factors. As far as my general attitude toward the genetics of evolution is concerned, I was probably more influenced by Wallace than I was by Dobzhansky. Dobzhansky was not nearly as holistic as Wallace, perhaps because Dobzhansky was more like Sewall Wright, who was equally only mildly holistic. The manuscript of my 1954 paper was read and criticized by Wallace, and in fact one of the figures in that paper was inspired by Wallace's suggestion.

Wallace was a superb experimenter. He knew how to set up the crucial experiments, and didn't mind the immense amount of work some of them involved. What is unfortunate is that he was very poor in expressing himself in lectures and equally poor at presenting his ideas in print. After one of his lectures you were always wondering what he had really proved. One of the consequences of the opaqueness of his writing is that his work did not have nearly as much impact as I think it should have had."

(5) *Curt Stern* (1902–1981): Mayr knew the geneticist *Curt Stern* from the latter's lectures at Columbia University and then they met at Cold Spring Harbor: "Another German refugee who was a good friend of Gretel's and mine was Curt Stern. He, together with Tracy Sonneborn, was the clearest lecturer I ever encountered. He was able to make the most difficult genetic interactions perfectly simple. His special field, morphogenetics, is very difficult, and he was perhaps the only specialist in his time. Earlier he had made some decisive discoveries which facilitated his finding a position after he had to flee from Germany. He was always grateful that he had managed to bring his parents to America, something neither Ernst nor Hansi Caspari had been able to do.

Stern had a tremendous sense of justice. He had what in German is called 'Zivilcourage.' During the war when Germans were viciously attacked in the American newspapers and even in *Science*, Stern pointed out that such accusations were not true for all Germans but that the majority of the Germans, he was quite sure, did not share the Nazi ideas. To publish something like that in the middle of the war required great courage and indeed, Stern encountered a good deal of hostility.

I first encountered his name when I got my PhD in Berlin at age 21 with the predicate *summa cum laude*. At that time, I was told that the only person in recent years who had achieved the same distinction was a person with the name of Curt Stern. I did not actually encounter him until many years later in the United States when he was lecturing at Columbia University. I believe Stern originally was a student of Max Hartmann, but later felt much closer in his interests to Goldschmidt. Stern always stood up for the right things, but did it in such a pleasant and modest way that we all admired him. Having been trained in the Dahlem embryological-cytological school, he had only an incomplete understanding of evolution. I once attended a lecture of his, the ultimate aim of which was to demonstrate the possibility of sympatric speciation. He did so because owing to his Dahlem upbringing he felt that most new species originated by saltation. The memoir about Goldschmidt which he prepared for the National Academy [*Biographical Memoirs*, vol. 39 (1967): 141–192] is quite admirable in bringing out both the great achievements of Goldschmidt without minimizing his faults. I always regretted that he was so far away, first in Rochester, then in Berkeley, so that I had only few chances of direct interaction with him."

In 1950, Mayr joined the Laboratory's board of directors, serving until 1958. Ten years later, James D. Watson became director of the Cold Spring Harbor

Laboratory and, through his fund-raising skills, inspiration and the ability to attract able researchers, raised that institution to even greater reputation as a center of advanced studies. In 1953 J. Watson and F. Crick had discovered the structure of the genetic material and, in 1956, Watson "arrived [at Harvard] with a conviction that biology must be transformed into a science directed at molecules and cells and rewritten in the language of physics and chemistry [...]. He treated most members of the Department of Biology with a revolutionary's fervent disrespect" (Wilson 1994: 219).

"Not surprisingly," Mayr told me, "Watson made quite a few enemies. One of them was young E. O. Wilson who attempted to join Watson's group. But Watson brushed him off rudely. This offended Wilson deeply, who was always friendly and obliging. Now he hated Watson. This is why in his biography [1994: 219] he calls Watson the 'Caligula of biology.' This reveals his anger but not a particular knowledge of the classical literature. To the best of my knowledge, Watson never indulged in any of the vices attributed to Caligula.[1]

Contrary to much gossip Watson was not unfavorable to organismic biology. After all he had been an ardent birdwatcher in his younger years and had supported in Cold Spring Harbor natural history courses for the children of the scientists. When the three rooms in the Blackford dining hall were given names, Watson was responsible for having one named the ERNST MAYR room. This, I am sure, was to recognize the contribution I had made in the 1940s to the general intellectual atmosphere at Cold Spring Harbor. Even though in most years I did not run a lab, it is at CSH where I did much of my scientific writing in those years."

Blackford Hall was renovated and enlarged in 1991–1992. Its new reading room was named for Efraim Racker (1913–1991), noted biochemist, molecular biologist, and artist, and the two new upstairs dining rooms were named for molecular biologist and geneticist Rollin Hotchkiss (*1911) and for Ernst Mayr. The dedication ceremony of these rooms took place on October 4, 1992. The entire Blackford dedication in early 1993 was part of a celebration honoring the 40th anniversary of the discovery of the structure of DNA. The Racker Reading Room and both upstairs dining rooms, Hotchkiss and Mayr, are adorned with portrait sketches of scientists working at CSH, drawn or painted by E. Racker, such as those of Ernst Mayr, M. Delbrück, A. Hershey, R. Hotchkiss, J. Monod, B. Wallace, J. Watson, and several others. All these facts document the friendly relations within the CSH community over more than half a century.

## Citizenship

Trying to open avenues for Mayr to return to Germany, Erwin Stresemann recommended him strongly as the Director of the Zoological Museum in Dresden in a letter written on 28 May 1940 (Haffer 1997b, pp. 932–933). But this was in the

[1] Caligula (12–41 A.D.) was Roman emperor from 37 to 41 A.D., in succession to Tiberius. Accounts of his reign by ancient historians are biased against him. He displayed wild, despotic caprice, pretense to divinity and is accused of torture, blasphemy and murder.

middle of World War II, and Mayr did not know of Stresemann's recommendation. In hindsight, of course, this is what saved Mayr's life. In Germany, he would have been drafted into the army with little chance of surviving the war. Legally, Ernst and Gretel Mayr remained "enemy aliens" in the United States for the duration of the war, but thanks to the broadmindedness of his colleagues, he never had any difficulties in pursuing his ornithological work at the AMNH, in arranging meetings of the "Committee on Common Problems of Genetics, Paleontology and Systematics" or publishing articles and books. When F. M. Chapman heard that Jean Delacour was coming to the Department of Ornithology (AMNH) in 1940, he was afraid that they (a Frenchman and a German) "would kill each other." Instead, they behaved like old friends and started writing a joint paper within a few weeks after Delacour's arrival in New York.

For more than 10 years, Ernst and Gretel Mayr suffered a most unpleasant treatment by the Immigration Service, as repeatedly referred to in Mayr's letters to Stresemann:

"I have to be grateful that I was not deported because of my relief work for the German ornithologists" (27 March 1950). "I had hoped to tell you personally [at the IOC in Uppsala] through what *hell* we have gone these past 8 years! With letters being no doubt read all the time I have to refrain to write about it. There is no sense in going abroad until I have my citizenship. Here I am protected by my American born children, abroad I am vogelfrei [outlawed]. Actually all my American friends and acquaintances, and particularly the Jewish ones, have been standing by us in the most wonderful manner. It is merely the *Beamten* [civil servants] who act like that!" (28 April 1950).

The delays by the U.S. Immigration Service to issue his passport prevented Mayr from attending the first postwar International Ornithological Congress in Sweden (June 1950). He was afraid that if he left the States without a valid U.S. passport, he would not be allowed to reenter the country. Mayr wrote on the issue of citizenship in his autobiographical notes:

"The *Bulletin of the American Academy of Science* Vol. 44, no. 4 (January 1992) has a very perceptive essay on the meaning of citizenship and the treatment of immigrants or foreigners. It brings back to my mind my situation when I came to the States in the early 1930s. In America, nationalism in the European sense of ethnocentricity is hardly developed at all, and one is proud of the country being a melting pot. A third generation immigrant is usually English-speaking and hardly distinguishable from any other American. In Europe each country has a somewhat different attitude toward foreigners or immigrants. Germany is almost at one extreme; in the 1920s there was still the memory of the unification of Germany in 1871, the end of a long period of extreme separatism, of literally hundreds of little states. The result was an almost extreme nationalism and emphasis on the ethnic aspect. According to the above mentioned article, this is true even today. Even a third generation Turkish immigrant is considered a Turk.

This caused a real dilemma for a German, particularly an upper-class German, who in the 1920s or 30s immigrated to America. The Americans expected him to cast off his past citizenship like a shirt and become at once an American. The

German was willing to become a law-abiding American citizen but he felt, as the heir of a very different culture, something that could not be cast overboard at once. The tension between the two attitudes is obvious. It is also well illustrated by most of the upper-class Jews who were driven by Hitler from Germany to America. In spite of the horrible treatment they had received, a large proportion of them still felt as Germans and as aliens in the American culture.

The attitude my wife and I had about going slow in petitioning for American citizenship, particularly in view of the uncertainty of my position at the American Museum [p. 96], was very much resented by the officials of the Naturalization Office. They considered it glaring evidence of hostility toward America."

Ernst and Gretel Mayr first applied for US citizenship in 1939, but during the waiting period, war broke out in Europe, and the Department of Immigration and Naturalization stopped granting US citizenship to Germans. After the end of WW II in 1945, the Mayrs reactivated their application for US citizenship and their application was rejected. They submitted a new application, which was again rejected owing to the anti-German hostility of the Immigration office in New York. Eventually, they hired an attorney, and with his help and with fifty (!) signed affidavits in hand supporting the Mayrs from colleagues, neighbors and friends, the courts finally granted the Mayrs US citizenship in December 1950.

# Part III
# Professor of Zoology at Harvard University

# 7 The Harvard Years (1953–2005)

In the fall of 1952, while in Seattle as a visiting professor at the University of Washington, "I received a telephone call one day from Alfred Romer, then director of the Museum of Comparative Zoology, when he asked me whether I was interested in being appointed an Alexander Agassiz Professor at Harvard University. To say that I would be interested would be the understatement of the week! I had long wanted to have a teaching position, but there were really only three institutions in the whole United States that would be suitable for my particular classification: Harvard, Michigan, and Berkeley. In early 1953, Romer passed through New York and we met at Grand Central Station where he told me the conditions under which I would be working at Harvard. I agreed with everything and I soon received the notice that I had been appointed."

Upon my enquiry, Ernst Mayr answered (5 January 1994):

"No, my shift from New York to Harvard was not a flight. I was in a way quite happy at the American Museum, but the position at Harvard was so infinitely better in every possible way that I could not have rejected it. To begin with, I wanted to live in a small town. The daily commuting from New Jersey to New York (one hour each way) was a great sacrifice which I made so that my children could be raised in a small country town. Also, I had no opportunity to have PhD students staying at the AMNH.[1] Finally, the total intellectual environment was bound to be far more stimulating at Harvard than at the AMNH, and this certainly proved to be true.

Actually, there were no ill feelings whatsoever over my leaving, because everybody understood that this was an offer I could not possibly decline. The director, Dr. Parr, tried very hard to make me stay, and offered all possible incentives, but he simply couldn't match what Harvard offered. However, the AMNH insisted that I stay connected to the bird department as an honorary curator. Later on I was even elected a trustee of the AMNH and I served as such for two full terms. However, as I got older, traveling down to New York for these more or less social meetings of trustees was getting to be too much and I therefore finally resigned. The bird department still considers me to be part of them and requests that I annually send my list of publications which they include in their annual report as the publications of a member of their department. All this shows how warm our relations have remained."

[1] Most of the zoology professors at Columbia University did not take evolution, systematics and natural history very seriously and would not permit Mayr to have PhD students. When word reached Columbia University about the Harvard offer, he received a telephone call giving him permission to have students, etc. as of that day, but then it was too late.

The move to Cambridge meant to Mayr, most of all, "exchanging New York for a pleasant and stimulating university town which perhaps has more of a European character than any other town in the United States. I will be able to walk to work, instead of standing for an hour in a crowded train or bus. I still plan to spend several months every year at the American Museum, although my official association will be with the Museum of Comparative Zoology" (to Erwin Stresemann, Berlin, on 25 March 1953). Mayr enjoyed "the contact with outstanding scholars in all fields, and I find the place scientifically stimulating. You realize, of course, that the existence (in the same area) of Harvard Med[ical] School, Mass[achusetts] Gen[eral] Hospital, M. I. T., and various branches of the Harvard Biology Department provide an almost endless opportunity for scientific stimulation" (to J. Lederberg, Madison, Wisconsin, on 28 February 1957).

Mayr took with him to Harvard University the large empirical database on geographical diversification of animal populations he had assembled during his work at the AMNH and began his career as evolutionary biologist and later historian and philosopher of biology. A revised edition of *Systematics and the Origin of Species* (1942e) had been under way since the late 1940s separating subject matter on the principles of "new systematics" from evolutionary aspects. The former were included in his textbook (with E.G., Linsley and R.L. Usinger) on *Methods and Principles of Systematic Zoology* (1953a), whereas the latter formed the central theme of his synthesis on *Animal Species and Evolution* (1963b). Continued work on these topics led to the publication of thoroughly revised editions of these books in 1969 (*Principles of Systematic Zoology*, again revised, with P. Ashlock as coauthor, in 1991) and in 1970 (*Populations, Species, and Evolution*), respectively. Although the subtitle of this latter work states that it is an "Abridgement" of *Animal Species and Evolution*, actually several chapters, particularly Chapter 10, were thoroughly revised, and certain phenomena such as gene flow and the role of chromosomes in speciation and evolution were dealt with in a novel fashion.

With his move to Harvard, Mayr did not lose contact with ornithology. He became editor as well as contributing author for more than half of the 16-volume *Check-list of Birds of the World* which was completed in 1987, a detailed catalogue of all species and subspecies of birds of the world (Bock 1990). No comparable list exists for any other group of animals. Mayr was President of the American Ornithologists' Union (1957–1959) as well as President of the XIIIth International Ornithological Congress (Ithaca, USA, 1962) and he summarized repeatedly the progress of ornithology and its relation to general biology (1963r, 1980d, 1983h, 1984a, 1988d, 1989c,k) emphasizing that throughout the history of biology, ornithology has played a leading role in making new discoveries and in developing new concepts.

Since Mayr's biological interests and national participation had greatly broadened, the percentage of straight ornithological publications decreased since 1953 and those on evolutionary biology greatly increased (Fig. 4.2). His activities now also comprised committees of the National Academy of Sciences, governing boards in the National Science Foundation, discussions on evolutionary biology (Fig. 7.1), biogeography, book reviewing, presidency of the 13th International Ornithological

**Fig. 7.1.** At Harvard University (*left* to *right*) Alfred Romer, Ernst Mayr and Julian Huxley, 14 November 1959 (Photograph courtesy of E. Mayr)

Congress (Ithaca 1962, Fig. 7.2), and the Commission of Zoological Nomenclature (1954–1976). As a member of the Biology Council (National Research Council, 1950s) he wrote a report on "Preserved materials and museum collections" to bring out the importance of museum collections and systematics generally, including a plea for better support. With the upsurge of molecular biology he was on continuous vigilance to prevent that all financial resources and new positions would be given to this new field. Again and again he emphasized the important contributions of systematics to the conceptual framework of biology such as population thinking, evolutionary biology, the basis of ecology and behavioral biology (besides its descriptive and service functions of taxonomy).

He wrote to Erwin Stresemann: "Did you see the Karl Jordan festschrift yet? I am working toward his election, despite his age, as an overseas member of the National Academy of Sciences (USA). Perhaps I'll succeed! I am anxious that systematics as such is honored" (11 March 1956; transl.) and "This again has been an exceedingly busy autumn, primarily because I have to serve on so many government committees as Mr. Systematicus. It is unfortunately true and realized by everybody that our field loses out whenever I am unable to represent it on high-level policy committees. I cannot do this forever but as we few in German [systematic] ornithology there is no one in this country who has quite the same authority on committees of the National Science Foundation or National Academy of Sciences as I do. Within the current 3-week period I have to be in Washington,

**Fig. 7.2.** Ernst Mayr as President of the International Ornithological Congress, Ithaca (USA), June 1962. Members of the Executive Committee of the I.O.C. Index: 1 A. H. Miller, 2 L. von Haartman, 3 E. Mayr, 4 D. Lack, 5 C. Sibley, 6 R. Falla, 7 D. Serventy, 8 R.-D. Etchécopar, 9 H. Lloyd, 10 E. Stresemann, 11 W. Thorpe, 12 S. Ali, 13 G. H. Lowery, 14 B. Biswas, 15 C. Lindahl, 16 F. Gudmundsson, 17 A. Schifferli, 18 Loke Van Tho, 19 D. S. Ripley, 20 P. Scott, 21 W. Meise, 22 R. Kuhk, 23 G. von Rokitansky, 24 J. Macdonald, 25 K. Voous, 26 J. Fisher, 27 R. Peterson, 28 G. Mountfort, 29 A. Wetmore, 30 G. Rudebeck, 31 R. Drost, 32 Sir Landsborough Thomson, 33 E. Schüz, 34 A. Rand, 35 J. Dorst, 36 Y. Yamashina, 37 G. Niethammer, 38 K. Bauer (Archives Museum of Natural History Berlin, Orn. 240, 1)

Philadelphia and New Haven for a total of 11 days and 6 different agencies or committees. What a life!" (9 November 1966; transl.).

His emphasis of systematics and organismic biology cannot be construed, as some authors have done, as an opposition to the rise of molecular biology as such. To the contrary, he was enthusiastic about it!

Although Mayr had been interested in the history of science since the time he wrote his dissertation, the activities before and during the Darwin centennial (1959) and his own participation in the celebrations stimulated and boosted his historical interests considerably. The number of titles on the history and philosophy of biology increased annually, especially after his "retirement" in 1975, and in 1982 (d), he published his masterly synthesis on *The Growth of Biological Thought* followed by *Toward a New Philosophy of Biology* (1988e), *One Long Argument. Charles Darwin and the Genesis of Modern Evolutionary Thought* (1991g), *This is Biology. The Science of the Living World* (1997b), *What Evolution is* (2001f), and *What makes Biology Unique? Considerations on the Autonomy of a Scientific Discipline* (2004a). Generally speaking, these publications deal mostly with the history and structure of biological theories and ideas, in particular with Darwin's theses of evolution, with Darwin's predecessors, his successors and opponents and with the development of the synthetic theory of evolution (1937–1950). In his philosophical writings he emphasized the autonomy of biology, in particular its independence from physics.

## Teaching and PhD Students

At high school, Ernst Mayr suffered from stage fright. He overcame this condition when, in the United States, he presented papers at ornithological meetings and was not nervous at all before his first major public lecture at the AAAS meeting in Columbus, Ohio on December 29, 1939.

When preparing his Jesup Lectures in early 1941 he wrote: "I have never taught in a class-room and I give only one or two lectures a year, which means that lectures for me are always somewhat of an ordeal. I have to go by lecture notes and I prefer to have a manuscript near me on which I can fall back on when I feel that I have lost my trail" (Ernst Mayr to Edgar Anderson, 28 January 1941; HUGFP 14.7, Box 2).

Later he developed a teaching routine along with a course on "Evolution and speciation" he held as a visiting professor at the University of Minnesota in 1949:

"This was 7 years after my book, I was editor at that time of the journal *Evolution*, and I had all the manuscripts of all the latest work that was going on. I remember that I had outlines of the first fifteen or twenty lectures, and I brought the first five with me on the podium as I gave the first class. By the end of the class, I had used up the first three outlines for the first three lectures, and I was just panicking that I would run out of material. I went to Professor Minnich, the chairman of the department, and said 'Look here, I'm very inexperienced, that's what happened to me. What should I do?' He said 'Well, the first thing you do, of course, is that everything you say, you repeat about three times,' and the second thing was, he said

'Always have in your folder a table with columns and lines and all that and if you think you've run out of material, you say, 'Well, this is all very well illustrated by a table, which I will now put on the blackboard,' ' and 'if you make the right kind of a table, you'll use up fifteen minutes and keep the class fully busy all the time.' Well, I think I used it once. But Professor Minnich helped this young inexperienced first-time teaching professor through a crisis."

Shortly afterwards Mayr became an adjunct professor at Columbia University, New York (1950) and taught a course each winter. He related these developments in several letters to E. Stresemann (Berlin):

"The great news here is that I have been offered a guest professorship at the University of Minnesota. I am to give a course on evolution and speciation" (7 February 1949). "Since usually about 20–25 professors of zoology, paleontology, botany, entomology, etc. also listen to the great authority from New York, I have to prepare myself well. My book [*Systematics and the Origin of Species*, 1942e] is well known here and I am expected to lecture on the literature that appeared since 1942" (20 May 1949; transl.). "My visit to Minnesota was very stimulating and has induced me to consider whether or not I should take up teaching more seriously. Life in a university town would certainly be more pleasant than in New York and far less strenuous than the daily commuting. However, it would break Dr. Sanford's heart if I should leave here, and I don't expect at the present time to make any changes, even though I had two additional offers" (15 June 1949). "Perhaps I shall make a compromise and do some teaching at Columbia. [...] In most places they do everything to discourage young taxonomists rather than the opposite. This has been one of the reasons why I have been tempted to go into teaching. I feel that it is very necessary to provide some counterbalance against the strictly physiological, bio-chemical trend in our zoology departments" (8 August 1949).

Mayr renewed such a teaching experience with a course on species and speciation as a visiting professor at the University of Washington, Seattle, in the fall semester of 1952. This was an exciting event, particularly because Richard Goldschmidt (1878–1958) was teaching there at the same time. Mayr gave his class on Monday, Wednesday, Friday, and Goldschmidt on Tuesday and Thursday. What amused the faculty and rather puzzled the students was how often Goldschmidt and Mayr disagreed in their interpretations. Their personal relations were very cordial, Mrs. Goldschmidt coming from the same little town in Bavaria (Kempten) as Mayr. They met in private several times, and Mrs. Goldschmidt always prepared some Kempten specialty for dinner.

When Mayr arrived at Harvard University in 1953, he learned that evolution had not been taught there in many years. He was to change this situation within a short time. His main graduate courses were entitled "Principles of Evolutionary Biology" (with G. G. Simpson in some years, with S. J. Gould in others) and "Methods and Principles of Systematic Biology," which he offered alternately every other year. He was also occasionally guest lecturer on "Evolution and speciation" for more general courses organized by a colleague (e.g., on "Biology of the vertebrates" and "Biogeography of animals"). In addition he offered more specialized research courses for graduate students of the MCZ, for example, in the spring term of 1955 on "Systematics and evolution," together with a laboratory. There, each student had to select a genus of animals or plants and study it thoroughly with particular emphasis on some systematic or evolutionary problem that turned up during the study. To his surprise the students did very well and nearly everyone produced something worth publishing.

The most interesting course in evolutionary biology Mayr gave at Harvard was a seminar in which a student's essay was copied and distributed among the (some 8 to 12) participants, each of whom had to ask the author of the essay three critical questions. The whole seminar of two hours duration at least, consisted of the defense of his thesis by the essayist against the onslaught of critical questions by the other members of the class. This exercise became so famous that one had to move into a larger room with the members of the seminar sitting in the middle at a large table and chairs around for those who wanted to listen to the discussions. Only the members were permitted to ask questions. For years Mayr also held a private seminar at the Biology Laboratory in the evening and at the noon hour in his office and later in his house. He invited about 10 students interested in special problems or controversies. Gretel Mayr made a dessert usually something particularly delicious of the German cuisine. The participants still remember these "dessert seminars" 30–40 years later. They were discontinued when Mayr was absent on a visiting professorship and then not taken up again. These latter seminars illustrate Mayr's concern for his students and Gretel's interest in them working very hard on these special desserts.

Ernst Mayr's PhD students (in chronological order):

1955. J. Allen Keast: Bird speciation on the Australian continent (Bull. Mus. Comp. Zool. 123: 305–495, 1961). Professor emeritus, Queens University, Kingston, Canada. He published on Canadian freshwater fish ecosystems as well as on Australasian birds and ecological biogeography.

1958 Andrew J. Meyerriecks: Comparative breeding behavior of some North American herons (Publ. Nuttall Ornith. Club no. 2, 1960). Results of additional fieldwork were published in the accounts of the *Handbook of North American Birds* (R. S. Palmer, ed., vol. 1, 1962). Professor emeritus at University of Southern Florida, Tampa.

1959 Walter J. Bock: The palatine process of the premaxilla in the Passeres: a study of variation, function, evolution and taxonomic value of a single character throughout an avian order (Bull. Mus. Comp. Zool. 122: 361–488). Professor of Evolutionary Biology at Columbia University, New York City; published many papers on evolutionary morphology, particularly of birds, the theory of classification, and the biological species concept. Permanent Secretary of the International Ornithological Committee (1986–1998), and President of the 23rd International Ornithological Congress (1998–2002).

1960. Carl W. Helms: Activity patterns, energetics, and migration in birds: a study in ecology and physiology of the annual cycle. PhD dissertation, Harvard University. Professor emeritus at Clemson University, Clemson, South Carolina. He did additional studies on the physiology of bird migration.

1960. Terrell H. Hamilton: Studies on the hormonal control of urogenital differentiation in the chick embryo. PhD dissertation, Harvard University (co-supervised with Frederick L. Hisaw). He worked on adaptive trends in the geographical variation of polytypic species of birds and later on the actions of several endocrine systems. Formerly Professor at the University of Texas at Austin.

1961. W. John Smith: Behavior homologies and classification in the avian family Tyrannidae. PhD dissertation, Harvard University (Communications and relationships in the genus *Tyrannus*, Publ. Nuttall Orn. Club no. 6, 1966). Professor emeritus at University of Pennsylvania, Philadelphia.

1962. Robert Barth: Comparative and experimental studies on mating behavior in cockroaches. PhD dissertation, Harvard University. Professor at University of Texas at Austin.

1964. Ira Rubinoff: Morphological comparisons of shore fishes separated by the Isthmus of Panama. PhD dissertation, Harvard University. Research scientist at the Smithsonian Tropical Research Institute (STRI) in Panama.

1964. Richard W. Thorington: The biology of rodent tails: a study of form and function. PhD dissertation, Harvard University. Curator, at the National Museum of Natural History, Smithsonian Institution, Washington, D.C.; published on the taxonomy and behavior of mammals.

1965. Roderick A. Suthers: Acoustic orientation by fish-catching bats. PhD dissertation, Harvard University (co-supervised with Donald R. Griffin). Professor at the University of Indiana, Bloomington. Later he did pioneering research on the role of the syrinx in avian sound production.

1967. Francois Vuilleumier: Speciation in high Andean birds. PhD dissertation, Harvard University. Curator emeritus, at the American Museum of Natural History, New York City. Published on the ecology and zoogeography of Andean and southern South American birds.

1968. Susan T. Smith: Communication and other social behavior in *Parus carolinensis* (Publ. Nuttall Ornith. Club no. 11, 1972). Chose a political career.

1970. John Alcock: Discrimination learning and observational learning by birds. PhD dissertation, Harvard University. Professor at Arizona State University at Tempe. Published on the behavior of birds, insects, and other animals.

1970. Robert E. Jenkins, Jr.: Ecology of the three species of saltators in Costa Rica with special reference to their frugivorous diet. PhD dissertation, Harvard University. Active in conservation at Biodiversity Institute, Northville, N.Y.

1971. George L. Hunt: The reproductive success of Herring Gulls (*Larus argentatus*) in relation to man's activities. PhD dissertation, Harvard University. Professor at University of California, Irvine. Studied the biology, ecology, and distribution of marine birds, particularly in the Gulf of Alaska.

1973. M. Ross Lein: The biological significance of some communication patterns of wood warblers (Parulidae). PhD dissertation, Harvard University. Professor at University of Calgary, Calgary, Canada and Secretary of the American Ornithologists' Union.

This list of 16 graduate students is increased to 17 by William Coleman who did his thesis on G. Cuvier in the Department of the History of Science, but exclusively under Mayr's guidance.[2] Nearly all of them obtained very respectable positions at major universities or museums and most published regularly. About half of them worked on animal behavior, very few on systematics, speciation, and biogeography, although most studied evolutionary problems. A majority worked on birds (twelve), two on mammals, one each on fish and insects. This diversity reflects Mayr's range of interests and his supervisory style (Lein 2005) encouraging students to follow their own interests. As mentioned above, two of the graduate students studying birds stayed with the Mayrs at their farm for several summers (W. J. Smith and M. R. Lein).

Charles Fleming (1916–1987) in New Zealand was never actually Mayr's student, but considered himself as such. His entire work as an ornithologist and biogeographer was an application of Mayr's principles to New Zealand birds and other animals. David Lack (1910–1973) was also influenced by Mayr as far as evolutionary ideas are concerned. However, bringing ecology into the evolutionary synthesis was his own contribution. Therefore, Mayr insisted that Lack be invited to the Princeton Conference in 1947 where he lectured on ecological aspects of evolution.

Quite a number of people received informal training and advice during prolonged periods at the Department of Ornithology (AMNH) or, after 1953, at the Museum of Comparative Zoology (Harvard University). They were either employed by the museum or on fellowships: Cardine Bogert (mid-1930s), Dean Amadon (late 1930s–early 1940s), Hugh Birckhead (late 1930s), Eleanor Stickney (late 1930s), Dillon Ripley (early 1940s), Charles Vaurie (early 1940s), Martin Moynihan (mid-1940s), Biswamoy Biswas (late 1940s), Daniel Marien (late 1940s), Thomas Gilliard (late 1940s), Kate Jennings (early 1950s), Robert Wolk (early 1950s), A. J. Cain (early 1950s), Pillai (mid-1950s). Post-docs at Harvard University include Michael Ghiselin (mid-1960s), Gerd von Wahlert (late 1950s), L. von Salvini-Plawen (early 1970s) and Thomas Junker (mid-1990s).

In later years, Mayr was visiting professor or lecturer at Cornell University, University of California (at San Diego, Riverside, Davis) and at the Collège de France in Paris.

## Director of the Museum of Comparative Zoology[3] (1961–1970)

When, at the time that Alfred S. Romer was to retire as director of the MCZ, Dr. Nathan M. Pusey, President of Harvard University, offered Ernst Mayr the directorship of the museum, he at first declined, because he was so immersed in his research that he did not want to be bothered with administration. However, in the end, Pusey simply told him that on July 1, 1961, he would be the new director. Mayr

[2] Other students and close associates in the field of the history of science include R. W. Burkhardt, Jr., F. J. Sulloway, M. Winsor, S. J. Gould, F. B. Churchill, and G. E. Allen.

[3] Today the Harvard Museum of Natural History.

served with distinction until June 30, 1970, when Alfred W. Crompton succeeded him.

For every year of his directorship, Mayr published a detailed annual report summarizing new developments and the respective status of the institution, the turnover of staff, teaching and facilities, discussing future plans and the research and publications of staff members and visiting scientists. Other topics mentioned include expeditions and travel, new exhibitions, increase of the collections and of the main library as well as exchanges with and loans to other museums. Marjorie Sturm, Mayr's administrative assistant, recalled: "The annual report was his baby. He wanted that every year and he really put his heart and soul into the annual reports; they were down to the last publication of each professor or curator. A description of what each department had done for that year and he worked on it."

There were five to six Alexander Agassiz Professors at the MCZ and 12 additional research zoologists on the curatorial staff as well as a varying number of administrative personnel. 30–50 graduate students were working in the museum or under direct guidance of staff members.

From the start of his directorship Mayr pursued vigorously two major objectives: (1) Construction of a laboratory wing or an "experimental wing" of the MCZ and (2) Acquisition of a tract of relatively undisturbed land to establish a field research station fairly close to Harvard University. He pointed out that the modern museum naturalist increasingly studies the diversity of living nature in all of its aspects which he investigates in the museum, in the laboratory, and in the field. Behavior, competition, distribution, niche occupation, population structure, environmental physiology, and all aspects of evolution, genetics and ecology are the concern of the systematist in his study of biodiversity. As a naturalist Mayr appreciated the study of live animals and the ten summers at Cold Spring Harbor had taught him the necessity to be acquainted with adjacent fields of research. The new "experimental wing" should be devoted to the integration between classical taxonomy and modern evolutionary biology. It would provide laboratory facilities for the researchers and house aquaria, insectaria and aviaries for maintenance of live animals. Before the end of his term as director Mayr had raised the funds but by the time the actual construction of the building began in 1970, his successor had taken over. He invited Mayr to give the opening speech on "Museums and biological laboratories" when the new wing was inaugurated (Mayr 1973k). Henceforth, the presence of specialists in behavior, population biology, and biochemical evolution enriched the intellectual atmosphere of the museum.

Mayr had acquired for the MCZ about 700 acres of lovely woodlands, the Estabrook Woods in Concord, about 35 min driving from Harvard, and, in 1966, established headquarters with field laboratories nearby at Bedford, Massachusetts. Estabrook Woods comprise mixed woodland with streams, wooded swamps, and ponds. Mayr's further plan to raise the endowment for a professorship in Environmental and Conservation Biology was not pursued by his successor and, to his great regret, nothing was done with respect to this final step in the creation of a Department of Environmental and Conservation Biology at the MCZ (or to the establishment of an ongoing field program at Estabrook Woods).

In his last Annual Report (for 1968–69) Mayr gave an overall view of the years of his directorship which have been a period of transition in many respects. He pointed out that in his opinion, the MCZ had changed in three major ways: (1) The staff had become more professional, resulting in higher scientific standards; (2) Teaching had received greater emphasis, and (3) Research had been increasingly directed toward the study of living animals, although straight taxonomic work (monographs, revisions, description of new species) remained basic requirements. On this occasion, he also acknowledged the work of Marjorie Sturm who took a large portion of administrative burden off his shoulders (see her reminiscences of Ernst Mayr as the Director on page 289).

While living in New York, Mayr corresponded with numerous ornithologists in the United States and in many other countries of the world. From the 1940s onward, he communicated increasingly also with other zoologists and geneticists with an interest in evolution and speciation. All incoming letters and carbon copies of his answers are preserved in Harvard Archives (Pusey Library, Papers of Ernst Mayr) and listed in a computerized inventory. These letters document Mayr's public and professional life at the AMNH and at Harvard's MCZ. Some correspondence and other historical material is also filed in the archives of the AMNH, Department of Ornithology, and of the MCZ, Ernst Mayr Library. I reviewed in some detail the above inventory of his correspondence for the 1960s, while he was the director of the MCZ, and the contents of selected letters. In the second half of this decade he exchanged numerous letters with Philip Handler and with W. C. Steere in conjunction with the work on several committees during the preparation of the "Handler Report" on the status of biology among the sciences (p. 312). Genetical problems were discussed in many letters with E. B. Ford, I. M. Lerner, J. B. S. Haldane, R. Lewontin and M. J. D. White, evolution and adaptation with G. G. Simpson, V. E. Grant, and C. H. Waddington. Of course, close communication continued during the 1960s with Mayr's fellow ornithologists Amadon, Deignan, Delacour, Eisenmann, Friedmann, Gilliard, Grenewalt, Lack, Miller, Moreau, Phelps, Ripley, Salomonsen, Selander, Serventy, Sibley, and Vaurie. German colleagues with whom he exchanged many letters included his friend Erwin Stresemann, and also Niethammer, Immelmann, Rensch, Stein, and Sick, the latter in Brazil since 1939. Numerous letters were exchanged with his former or current PhD students W. Bock, W. Coleman, C. W. Helms, T. H. Hamilton, A. Keast, A. J. Meyerriecks, M. Moynihan, I. Rubinoff, W. J. Smith, and R. MacArthur (although the latter was not one of his graduate students). Corresponding post-doc fellows during those years included A. J. Cain, M. T. Ghiselin, and G. von Wahlert. The letters frequently included detailed technical discussions and left Mayr's office on a daily basis despite his busy schedule. Even at an age of over 90 years he regularly exchanged letters with numerous people, often only one or two letters per year but with many colleagues he corresponded on a monthly basis; a total of 230 names were on his list. A careful study of this correspondence would be very interesting.

# 8 Evolutionary Biology—Mayr's Second Synthesis

The Princeton Conference in January 1947 showed that a consensus had been achieved among geneticists, paleontologists and systematists and that evolutionary biology as an independent biological discipline had been established (Mayr 1997g). As an active community architect, Mayr was at the center of activities for forming a Society for the Study of Evolution in 1946 and establishing its journal *Evolution*, for which he served as its founding editor (1947–1949; see pp. 238–240). He contributed enormously to the development of the new field of evolutionary biology during the following decades, in particular through his publications on variation and population thinking, species concepts, the ontological status of species, the dual nature of evolution (phyletic evolution [anagenesis] and speciation [cladogenesis]), the unity of the genotype, and accident versus design in evolution, as reviewed by Bock (1994a). Additional contributions concern discussions of Darwin and Darwinism and of many philosophical aspects of evolutionary biology like causality, teleology, essentialism, the advance of science and the autonomy of biology. His important papers on evolutionary biology have been reprinted in two volumes of essays (1976m, 1988e). Most of Mayr's work in evolutionary biology refers to species-level changes rather than macroevolution, although he also published some important work on this latter subject. He did not publish a textbook on evolution, although his books of 1963(b) and 1970(e) come close and he published many articles on general aspects of evolution and a modern exposition of evolution (2001f). "Most biologists would probably name Mayr as the emblematic, even iconic, figure of 20th century evolutionary biology. The centrality of his role is attributable both to the multiplicity of ways in which he participated in the creation of modern evolutionary biology and the effectiveness with which he played these parts" (Wilkins 2007). I summarize below first his general overviews of evolutionary biology or portions of it followed by a discussion of a series of particular topics which he dealt with in separate articles.

## Overviews of Evolutionary Biology

Mayr periodically analyzed and synthesized from diverse sources an enormous amount of information on evolutionary biology which he presented as advanced contributions to fellow biologists or as fairly elementary expositions for knowledgeable readers. In these general articles he is a missionary or crusader against the ignorance of the modern evolutionary synthesis, the distortion of biology by molecular biologists, and the neglect of major personalities in the development of evolutionism. As a clear and convincing lecturer, he was a participant of most symposia in this field during the 1960s and 1970s. When the National Academy of Sciences in Washington celebrated its centennial in 1964, the president invited Mayr (1964k) to present a lecture on the essentials of evolution in such a manner that nonbiologists would understand what the current interpretation and open problems were. He emphasized evolution's early "invention" of sex, i.e., the ability of different organisms to exchange genetic information with each other. In this way, ever-new combinations of genes can appear and be tested by the environment in every generation. Therefore Nature organized the well-integrated, coadapted, harmonious genotypes into biological species each of them protected by isolating mechanisms which prevent the interbreeding between individuals of different species and the formation of less viable organisms. New species may originate through polyploidy (in plants) or through geographical speciation (in plants and animals) leading also to ecological specialization. Mayr developed here for the first time his concept of closed and open behavior programs in animals (see below) and ended his presentation with an explanation of natural selection as a two-step process, (a) the production of new genetic diversity (randomness predominates) and (b) the selection of superior individuals (controlled by environmental demands).

In his contribution to a symposium on chemical evolution, Mayr (1964r) described the contrast between the chemical unity of all organisms and the vast diversity of life on earth. During much of evolutionary time the same basic chemical constituents or classes of chemical constituents have been utilized with a great deal of modification of macromolecules. Evolutionary changes have been gradual, controlled by natural selection, there was no threshold where a sudden jump was made involving biochemical revolutions. "No case is known to me in which a change in body chemistry initiated a new evolutionary trend. Invariably it was a change in habits or habitat which created a selection pressure in favor of chemical adjustment" (p. 1233). He concluded this lecture stating that "a healthy future for biology can be guaranteed only by a joint analytical and systems approach."

In 1967(b), Mayr challenged a rather strange mathematical interpretation of evolution and emphasized the difference between biological and physical phenomena, especially the uniqueness and great variability of biological processes in contrast to physical processes. Speciation may be rapid or slow, evolutionary rates may be high or low; different genes will be favored depending on population size. Small founder populations of great evolutionary importance are usually not preserved in the fossil record. Mayr's main objective here was to emphasize that the simplistic and unrealistic assumptions in the equations of several mathematicians and physicists were inadequate for a study of evolution. They had assumed that all the individuals in a species are identical and a mutation has to spread through all the individuals in the population, and this must be followed by another mutation, and so on. On the basis of these unrealistic assumptions they had concluded that the earth was not old enough to have permitted the enormous diversification and adaptations of all organisms.

The report edited by Philip Handler on the status of the life sciences (p. 312) was published under the title *Biology and the Future of Man* in 1970. Mayr had contributed the chapter on "The diversity of life" (with several coauthors but mostly written by him; Mayr 1970h). While physicists deal with only a limited number of elementary particles, biologists estimate that five to ten million (or more) different species of organisms exist today, each of which represents a different genetic system, and each member of these species is also genetically unique. The knowledge of each species and a classification of related species are of biological and practical importance (e.g., in biological control). Emphasizing the organismic aspects of biology Mayr (1970h: 30) felt that "one day there must emerge a new philosophy of science, based largely on the findings of biology rather than those of physics." Man is part of the evolutionary stream and continues to evolve. Evolutionary biology provides the tools for a scientific approach to a study of the future of man. Useful and noxious species of animals and plants must be studied in detail for pest control and potential future use, respectively. An inventory of the world's plants and animals has progressed unevenly and is far from complete. The taxonomist should continue to devote part of his research time to the description of new species and to developing classifications which are essential in comparative biology and represent important information retrieval systems.

At several symposia Mayr presented even two talks. For example, at the First International Congress of Systematic and Evolutionary Biology in Boulder, Colorado, he spoke on "The unity of the genotype" (1975i) on 9 August 1973 and gave an address entitled "The challenge of diversity" (1974c) at the All-Congress Dinner on the following evening. In this latter presentation he again emphasized the importance of the study of organic diversity which, until fairly recently, had been almost totally ignored by geneticists who concentrated on evolutionary change as such, primarily on adaptation. Speciation, the origin of higher taxa and the entire field of evolutionary innovation can only be studied through research on organic diversity.

Overviews of neo-Darwinian evolutionary theory and clarifications of misunderstandings which Mayr published in Germany (where some outdated thoughts on mutation and selection still prevailed) include his articles on "Selection and evolutionary trends" (1965g), "Basic thoughts of evolutionary biology" (1969i), and "How far have the basic problems of evolution been solved?" (1975h). Several physicists who attended the latter lecture tried to demonstrate that there was no difference between biological and physical (cosmic) evolution. Since that time Mayr devoted more efforts to describe the differences between the concepts and phenomena of the biological and the physical sciences. He published summaries of "Evolution" (1978f), of "The current status of the problem of evolution" (1982k), of "Natural selection" (1985e), and "An overview of current evolutionary biology" (1991e) describing the theories of saltational, transformational (Lamarckian), and variational evolution (Darwinism). The latter is contingent on variation and selection and does not strive toward "perfection" or any other goal. Evolution is opportunistic, hence unpredictable and consists of (a) the origin of diversity (cladogenesis) and (b) the origin of adaptations (anagenesis). The discussion of natural selection is facilitated by a distinction between (a) "selection of" and (b) "selection for." In almost all cases the target of selection are individuals, either single cell organisms or complex whole organisms, as in higher animals and plants, or certain social groups of cooperating animals or early humans (see also 1986l and 1990r). The question of "selection for" refers to what kind of traits might be favored by selection.

**Anti-reductionism.** Mayr's anti-reductionist (holistic) evolutionary attitude of an "organismic naturalist" has been obvious since his attack on "beanbag genetics" during the 1950s (see below) and since his repeated statements that the target of natural selection are the individuals of a population, rather than genes (as maintained by the mathematical population geneticists). Despite his clear anti-reductionist statements in his publications Mayr followed inexplicably and "unthinkingly" (see below) the reductionist definition of evolution as given by the population geneticists (evolution is "a change of gene frequencies in populations") and stated in his publications

"It [organic evolution] refers to a change in genetic properties from generation to generation, owing to differential reproduction" (1959a, p. 8) and "Evolution means change. To be more precise, it means the replacement of genetic factors by others" (1963b, p. 168, quoted from Beurton, 1995).

During the 1970s, Mayr realized that the gene frequency definition as given by the population geneticists is unacceptable, as it refers to a result of evolution, an indirect by-product of the superior success of certain individuals. He also realized that his previous acceptance of the reductionist definition has been a mistake and contradictory to his own anti-reductionist attitude as expressed in his publications. This realization did not mean a change in his theoretical attitude in the sense that he only now became an anti-reductionist (as Beurton 1995 assumed), rather it meant a correction of a mistaken definition as seen from his unchanged anti-reductionist standpoint, a mistake he only now appreciated (Mayr 1969d: 126; 1976m: 307; 1977f: 45; 1980e: 12). From then on Mayr labeled the earlier reductionist (genetic) definition as "absurd" (1984h: 126) and defined evolution as "a change in adaptation *and* in the diversity of populations" or similar wording. He explained the correction of his definition of evolution as follows:

"With the rise of population genetics, more and more frequently it was claimed that the gene was the object of selection. [...] This reductionist interpretation had many consequences. It led, for instance, to a definition of evolution as 'a change of gene frequencies in populations,' but although I myself accepted this definition unthinkingly for 35 yr or so, I finally concluded that it was singularly misleading. This definition focuses on the entirely wrong hierarchical level. Evolution is something that happens to whole organisms and populations. Evolution consists of changes in the interaction of individuals (or their propensity to interact) and is usually referred to as changes in adaptation, even though some of the structural and behavioral changes may have been the result of stochastic processes. What is worse, the gene frequency definition of evolution completely ignores the fact that changes in organic diversity (speciation and adaptive radiation) are an equally important component of evolution. The changes in gene frequencies are the result of evolution, not the reverse" (1986e: 235–236). And again: "Even before the synthesis, I, like most naturalists, was a holist. Evolution for me concerned the whole organism, and the organism as a whole was the target of selection. This was, of course, the Darwinian tradition. I admit that during the synthesis, I used the standard formula of the geneticists that 'evolution is a change in gene frequencies,' even though it was actually incompatible with my holistic thinking. But I did not appreciate this contradiction until many years later (Mayr 1977f)" (2004a: 125).

**Post-synthesis developments.** The task of evolutionary biology after the synthesis of the 1940s was to convert the coarse-grained theory of evolution into a fine-grained, more realistic one. For example, more attention was paid to stochastic processes and pluralistic solutions (Mayr 1963b). It was increasingly understood that recombination rather than mutation directly supplies the material for natural selection. Molecular biologists discovered the fact that with minor exceptions, there is only a single genetic code for all organisms, from the simplest prokaryotes up. The one-way transfer of information from nucleic acids to the proteins (the "Central Dogma") explained why an inheritance of acquired characters is impossible. Other post-Synthesis developments include the recognition that all speciation is simultaneously a genetic and a populational phenomenon; that selection is probabilistic, not deterministic; that many genetic changes at the molecular level are probably neutral or near-neutral (i.e., non-evolutionary); that the history of life has been drastically affected by major extinctions that may have fallen largely at random (Mayr 1988e: 525–554; 1991g: 141–164).

These discussions among evolutionary biologists have all taken place within the framework of Darwinism. However, several authors construed them as indications that the Evolutionary Synthesis is incomplete or even needs to be replaced by a new theory of evolution, a false claim that has been refuted effectively. Mayr also repeatedly defended or clarified the modern concepts of evolutionary biology in newspaper articles in order to reach the general public (1972a, 1984h,i, 1991d). In his article on "The triumph of the Evolutionary Synthesis" (1984h) he criticized the misconceptions of several young authors who implied "the end of the synthetic theory of evolution and the rise of a new evolutionary theory." Mayr ended this review stating that "none of the recent developments contradicts or refutes the basic Darwinian paradigm in any way whatsoever." During the mid-1990s he discussed with fellow biologists "basic problems and misconceptions of evolution" (1994p) and published a summary of the modern evolutionary theory (1996c). "There are still so many people who deny evolution. You have to explain to these people how overwhelming the evidence for evolution has become," he said, when, in the early 1990s, he started writing his manuscript of a new book, *What Evolution is* (2001f). In this book which appeared when he was 97 years old, he gives a clear and concise exposition of modern evolutionary biology for all interested persons, the best summary of uncontested evolutionary thought. After writing for his fellow scientists during most of his life, he here spoke to the general public and, in particular, to those persons who want to know more about evolution, to people who accept evolution but are skeptical about the Darwinian explanation for it, and to "creationists" to find out more about what they are up against (p. XIII). The material is clearly partitioned into three sections: (a) The existing evidence for evolution, (b) the nature of the (adaptive) changes within a population (anagenesis) and (c) the origin and meaning of biodiversity (cladogenesis). Mayr here also included a summary of many recent results of molecular biology and speciation research. The virtues of this book are clear and concise discussions of, among others, the following topics: Darwin's five major evolutionary theses, natural selection as a process of elimination, limits to the effectiveness of natural selection, adaptedness as an *a posteriori* result of elimination, biological species, allopatric and sympatric speciation, refutation of neutral evolution, altruism, a novel ecological interpretation of geographical aspects of human evolution, and evolution as a gradual populational process, whether at a relatively slow or relatively fast rate.

"In determining what qualifies as an adaptation," Mayr emphasized (p. 149), "it is the here and now that counts. It is irrelevant for the classification of a trait as an adaptation whether it had the adaptive quality from the very beginning, like the external skeleton of the arthropod, or acquired it by a change of function, like the swimming paddle of a dolphin or a *Daphnia*." Mayr here objects to the opinion of some evolutionists who classified the latter alternative not as an adaptation but as an "exaptation."

As to progress in evolution, Mayr favored Richard Dawkin's thesis which emphasizes its adaptationist nature: Progress is "a tendency of lineages to improve cumulatively their adaptive fit to their particular way of life, by increasing the number of features which combine together in adaptive complexes" (p. 215). Two appendices of the book are especially interesting and deal with (1) criticisms that have been made of evolutionary theory and (2) answers to frequently asked questions about evolution. Examples of new findings which are discussed and which affect Mayr's previous interpretations are sympatric speciation in cichlid fishes and in host-specific insects as well as the origin of eyes in 40 branches of the evolutionary tree. Salvini-Plawen and Mayr (1977d) had considered these to be independent convergent developments. Molecular biology has now shown that this is not entirely correct, because a regulatory master gene (Pax 6) seems to control the development of eyes in the most diverse branches of the evolutionary tree. Mayr also answered in this book of 2001 the question whether intelligent life exists elsewhere in the universe and said: "For all practical purposes, man is alone" (p. 263). He considered the chance of successful communication with another civilization in outer space an improbability of astronomical dimensions (p. 374).

"This book turns out to be an admirable, in fact masterful, summation of the details of evolutionary principles. There is probably no other contemporary professional evolutionist who could articulate them as well as Mayr does" (H. L. Carson, BioEssays 25: 91, 2002). Mayr was also very pleased over S. Stanley's evaluation of his book in a private letter: "Dear Ernst, You may remember that I mentioned I was going to use your *What Evolution Is* book in my course on Darwin and Darwinism. I did so at the start of the present term, and it worked beautifully. I had previously been frustrated by the lack of anything like it. Your book is wonderfully logical and lucid–also highly original in the way it lays things out" (April 7, 2004).

Several significant stages in the modification of the evolutionary paradigm since the late 19th century are listed in Table 8.1. These seemingly discontinuous periods are, of course, broadly transitional. For example, the role of diversity in evolution was stressed by the naturalists from Darwin on, but was almost totally ignored by reductionistic geneticists into the 1930s. Mayr feels that none of the new discoveries of molecular biology have required an essential revision of Darwinism. New findings on sympatric speciation in certain freshwater fishes and host-specific insects indicate that this mode of speciation is probably more frequent than previously thought. A wide variety of evolutionary phenomena has become

**Table 8.1.** Significant stages in the modification of Darwinism (from Mayr 1991g: 144)

| Date | Stage | Modification |
|---|---|---|
| 1883; 1886 | Weismann's neo-Darwinism | End of soft inheritance; diploidy and genetic recombination recognized |
| 1900 | Mendelism | Genetic constancy accepted and blending inheritance rejected |
| 1918–1933 | Fisherism | Evolution considered to be a matter of gene frequencies and the force of even small selection pressures |
| 1936–1947 | Evolutionary synthesis | Population thinking emphasized; interest in the evolution of diversity, geographic speciation, variable evolutionary rates |
| 1947–1970 | Post-synthesis | Individual increasingly seen as target of selection; a more holistic approach; increased recognition of chance and constraints |
| 1954–1972 | Punctuated equilibria | Importance of speciational evolution |
| 1969–1980 | Rediscovery of sexual selection | Importance of reproductive success for selection |

increasingly obvious and multiple solutions to almost any evolutionary challenge are apparent, e.g., the greatly varying age of species and varying morphological differences between species and subspecies, the varying amount of gene flow within species, varying occurrence of parapatry, and other phenomena. The new frontiers of evolutionary biology, according to Mayr, are the molecular structure and cohesion of the genotype and the role of development. However, the basic Darwinian principles are more firmly established than ever.

Kutschera and Niklas (2004) dedicated an overview of "The modern theory of evolution: an expanded synthesis" to Ernst Mayr on the occasion of his 100th birthday. They emphasized the validity of the Darwinian paradigm and discussed post-synthesis developments like the study of endosymbiosis, eukaryotic cell evolution, phenotypic plasticity, and experimental bacterial evolution describing the modern "evolution" of evolutionary theory which "remains as vibrant and robust today as it ever was."

## Integrated Gene Complexes versus "Beanbag Genetics"

Based on his conversations with B. Wallace in Cold Spring Harbor during the summer of 1950 Mayr began to think in terms of different selective values of genes in a founder population as compared with the parent population and spoke of "genetic background" and the importance of the "genetic environment" in evolution (p. 219). In 1952(f) he pointed out in a review (p. 250) that "experimental genetics until recently has been essentially single-locus-genetics. The student of phylogenies, however, deals with integrated gene complexes." In several later articles he advanced his theory of the importance of integrated gene complexes. His summary for the 1955 Cold Spring Harbor (CSH) Symposium entitled "Integration of genotypes: synthesis" (Mayr 1955h) was his first concerted attack on "beanbag genetics," a term not yet used here, but throughout the text he emphasized that phenotypes and their evolution can only very incompletely be described in terms of the genetic analysis of simple Mendelian characters and simple frequencies of genes in natural populations but rather concern "entire genetic complexes." Further aspects stressed were (1) the "uniqueness which is responsible for the numerous types of indeterminacy found in living organisms and their evolution" (p. 331), (2) the multiplicity of answers which natural selection may find to an environmental challenge ("multiple pathways of evolution," Bock 1959), and (3) the differences between closed laboratory populations and natural populations which are open to gene flow. Although true introgression (i.e., gene exchange among species) is rare in animals, intraspecific gene flow among populations is widespread leading to the comparative uniformity of species over wide areas in comparison to the striking deviations of isolated, usually peripherally, populations (1955h, p. 332).

"It has often been stated that there will be no evolutionary advance in populations that are too small because their genetic situation is too precarious [Sewall

Wright's claim]. It is necessary to stress that owing to dispersal the same may well be equally true for populations that are too large and with too much gene exchange throughout the species" (p. 332). Mayr here repeated one of his main contentions of his 1954(c) paper by saying that "population size [...] becomes important for its effect on the selective properties of genes" (p. 333). The same gene may be of high selective value in large or panmictic populations but of low selective value in small or inbreeding populations. Mayr's final conclusion was that "population genetics can no longer operate with the simplified concepts it started out with" (p. 333).

In his celebrated keynote introduction "Where are we?" to the 1959 CSH Symposium, Mayr (1959f) stated that population genetics had two sources, (a) mathematical genetics of R. A. Fisher, J. B. S. Haldane, and Sewall Wright who had established the power of natural selection, and (b) population systematics of Sumner, Chetverikov [mentioned in the reprint 1976m], Timoféeff-Ressovsky, and Th. Dobzhansky. Certain simplifying assumptions such as an absolute selective value for each given gene were necessary to advance population genetics during its early stages of development but should now be replaced by a more realistic approach. "The emphasis in early population genetics was on the frequency of genes and on the control of this frequency by mutation, selection, and random events. Evolutionary change was essentially presented as an input or output of genes, as the adding of certain beans to a beanbag and the withdrawing of others. This period of 'beanbag genetics' was a necessary step in the development of our thinking" (p. 2). Evolution was defined as a change in gene frequencies, the replacement of one allele by another; thus it was treated as a purely additive phenomenon. Only very few authors stressed interaction among genes, the integration of the genotype.

"Rereading this whole paragraph in 1984 I still insist that it was essentially legitimate and did not justify the intense hostility which it engendered in Wright as reflected in several of his papers published in the 1960s, 70s, and 80s. Of course, Wright's writings were not identical with those of Fisher and Haldane, who were far more single-gene-reductionists, and Wright far more than they, considered the genotype or gene pool as a whole, yet when one looks at his equations and graphs there is no denying that he thought largely in terms of (1) closed populations, (2) genes with constant selective values, and (3) a more or less independent selective fate of individual alleles" (see also 1982d: 556).[1] Coyne and Orr (2004: 3) concluded that "Wright, who accepted both the biological species concept and allopatric speciation, nevertheless largely ignored the topic."

Mayr continued in this article (1959f) stating that the evolutionary synthesis of the 1940s has not solved all the problems. How much there still remains to be done he discussed under the headings Natural Selection, Gene and Chromosome, Population, The Species, Peripheral Populations, and others. And again in the summary: "In spite of the almost universal acceptance of the synthetic theory of evolution, we are still far from fully understanding almost any of the more

[1] Provine (1986: 482–484) discussed the differences and similarities of their views in some detail concluding "that the ideas of Mayr and Wright on the mechanisms of speciation are mostly complementary rather than contradictory."

specific problems of evolution" (p. 13). The great value of molecular biology should not lead us to neglect the importance of evolutionary biology, which is of such great importance for the understanding of man's environment and the future of mankind.

Mayr's attack on "beanbag genetics" during the 1950s demonstrated his anti-reductionist attitude. He continued to call attention to the frequency of epistatic interactions among genes and to the general cohesion of the genotype. When discussing changing selective values of genes in different gene combinations, he has called this concept, somewhat jokingly, the relativity theory of genes. He further developed his theme of the unity of the genotype in Chapters 10 of his 1963(b) and 1970(e) books calling also attention to this concept in his presentation at the First International Congress of Systematic and Evolutionary Biology in Boulder, Colorado in 1973 (Mayr 1975i). Here he stated that "free variability is found only in a limited portion of the genotype. Most genes are tied together into balanced complexes that resist change" (p. 377). Genes are not the units of evolution nor are they, as such, the targets of natural selection. Rather, genes are tied together into balanced adaptive complexes, the integrity of which is favored by natural selection. This cohesion of the genotype permits the explanation of many unusual phenomena of speciation and macroevolution, e.g., the evolutionary inertia or stagnation of certain evolutionary lines, the sudden flowering or "explosive evolution" of certain previously long-stagnant evolutionary lines, and the conservative nature of the *Baupläne* of the major animal types.

## *Animal Species and Evolution* (1963b)

The manuscript for his course on "Evolution and speciation" at the University of Minnesota in 1949 (p. 260) was the very first draft for a new book which eventually appeared in 1963, a thorough and authoritative, masterly summation of the biogeography and natural history of species and speciation, a magnificent synthesis of population genetics, variation, adaptation, the origin of new species, and adaptive radiation. The evidence for allopatric speciation is summarized in detail. The unifying theme of Mayr's discussions is the impact of population thinking (as opposed to typological thinking) on the present understanding of the species problem and evolutionary processes. An attempt is made to counteract the modern trend to forget all about the pioneers. In a number of areas the development of a concept is traced backward to its earliest beginnings, e.g., geographical variation, polytypic species, and speciation. Reviewers praised the excellent stimulating presentation of the vast and highly complex mass of information in this book which, as its predecessor of 1942, refers mainly to sexually reproducing animals, especially birds on whose geographical variation and speciation so much work had been done. However, Mayr draws examples from many other groups of animals as well. His ideas on the gene pool and on the effect of separating a population into two or more portions or establishing a small founder population are here expanded into a general theory of the unity of the genotype and the integrated nature of the gene pool of a population which is held together in large populations by gene flow. Reproductive isolation of a species is a protective device against the breaking up of its well-integrated coadapted gene system. Small founder populations may "pass from one well-integrated and stable condition, through a highly unstable period to another period of balanced integration," a genetic revolution (p. 538). Reproductive isolating mechanisms here treated brilliantly are believed to arise as an incidental by-product of genetic divergence in isolated populations and are defined as follows: "Biological properties of individuals which prevent the interbreeding of populations that are actually or potentially sympatric" (p. 91). Mayr's reasons why sympatric speciation is very unlikely, especially in birds and other higher animals, are discussed in detail ("in not a single case is the sympatric model superior to an explanation of the same natural phenomenon through geographic speciation," p. 476). The biological species concept, sibling species, secondary contact zones and hybridization, avoidance of competition through occupation of different portions of a habitat, isolating mechanisms and their occasional breakdown are further topics convincingly and forcefully presented. Mass extinction of species "is always correlated with a major environmental upheaval" (p. 620). Numerous studies have since substantiated this suggestion. The final chapters treat (a) the evidence that the process of evolution above the species level are a continuation of those responsible for the origin of subspecies and species, i.e., macroevolution is a continuation of microevolution, and (b) the human polytypic biological species and man's evolution. Several generations of graduate students have used this volume as their major textbook which, like Mayr's earlier book of 1942, remains a "classic."

Leslie C. Dunn (1964) of the Zoology Department at Columbia University, New York, who, over 20 years earlier, had invited Mayr to give the Jesup Lectures praised this volume of 1963 as of "capital importance which everyone interested in the history of modern science should know about. [...] I know of no better exposition of the integration which is occurring in biology through the joining of systematics, paleontology, population genetics, and ecology in elucidating the processes of speciation. There is only one term which describes the field which combines all these–it is evolutionary biology, and this book is the evidence that Ernst Mayr has become a master of it."

The main topic of the 1942 book had been the origin of the discontinuity between species, whereas in 1963 adaptive changes within populations were also fully discussed in addition to the above topic. Gould (2002: 540–541) asked Mayr about this difference of treatment. The latter "denied any personal augmentation of adaptationist preferences through the intervening years" and explained this difference in a private letter to Gould dated December 20, 1991 as follows:

"Dear Steve,

I gave considerable thought to your question how my 1963 book differed from the 1942 one, and why adaptation was so much more featured in the later volume. I think I now have the answer.

Remember that I consider evolution by and large to consist of two processes: 1) the maintenance and improvement of adaptedness, and 2) the origin and development of diversity.

Since (2) was so almost totally ignored by the pre-Synthesis geneticists, I focused in 1942 on (2). By the 1950s the study of diversity had been fully admitted to evolutionary biology, owing to the efforts of Dobzhansky, myself, Rensch and Stebbins, and in my 1963 book I could devote a good deal of attention to (1). This was rather easy because, as you know, I used to be a Lamarckian. And Lamarckians are adaptationists. Hence, it is not that from 1942 to 1963 I had become an adaptationist, rather I reconciled in 1963 my adaptationist inclination with the Darwinian mechanism."

In one respect, Mayr had indeed become a stronger selectionist after 1942: Several cases considered as evidence for neutral polymorphism in 1942 were reinterpreted in the 1963 book as selectively balanced polymorphism because of the absence of large fluctuations (Mayr and Stresemann 1950f, Mayr 1988e: 528). However, generally speaking, more and more authors of the post-Synthesis period pointed out the existence of stochastic processes and constraints to selection preventing the achievement of "perfection."

Mayr published a new abridged version of the above book of 1963 under the title *Populations, Species, and Evolution* (1970e). Despite its subtitle *An Abridgement of Animal Species and Evolution* it is largely rewritten with many new ideas and discussions, but remains one of the least read of Mayr's books among evolutionists (Bock 1994a), largely because it was presented as an abridged version. Especially the chapters dealing with species concepts, genetic variation, and the genetics of speciation contain much new material. Mayr now deleted the ambiguous expression "actually or potentially" from his new species definition : "Species are groups of interbreeding natural populations that are reproductively isolated from other such groups" (p. 12).

## Behavior and Evolution

Throughout his life Ernst Mayr was interested in the behavior of animals beginning with his intimate observations of bird behavior during his youth (p. 24). In later years he published occasionally brief notes on his behavioral observations of certain bird species (1941l, 1948f,g) and was particularly interested in behavioral isolating mechanisms between closely related species. An outcome of this interest were his experimental studies on the behavior of two species of *Drosophila* flies (p. 228). Also about half of his graduate students chose behavioral topics for their thesis work (9 of 16 PhD students, p. 262). During the 1960s and 1970s, Mayr repeatedly discussed the relations between behavior and evolution in his text books (1963b, 1970e) and in a series of articles (1958g, 1971c, 1974j, 1976c, 1977c). While behavioral characters are quite important at the species level, Mayr found that relatively few higher taxa possess diagnostic behavioral features. The first paper mentioned above (1958g) also lists cases where analogous behaviors were acquired by unrelated groups and others where a study of behavior led to an improvement of classification. This paper shows also that behavior is often the pacemaker in evolution, a behavior pattern appearing prior to the origin of more specialized, facilitating structures. In 1974(j) Mayr discussed his concept of closed and open behavioral programs in more detail than previously as well as their respective roles in evolution. Most organisms have a fixed (closed) program for certain "purely instinctive" activities. Other behaviors are steered by previous experience (open programs), particularly in species which have a long period of parental care, when sufficient time for learning exists in the offspring. However, an open program is by no means a tabula rasa and certain types of information are more easily incorporated than others. Mayr concluded that display activities which are in part permitting the recognition of conspecific individuals must be based on closed programs. There is no opportunity for learning, and any errors in such behavior are usually strongly selected against. On the other hand, non-communicated behavior deals largely with the exploration of the environment, and flexibility of this type of behavior (open programs) should be of selective advantage. Later suggestions that sexual selection may play an important role in the evolutionary change of behavioral isolating mechanisms indicated that some programs for courtship behavior and pair formation are less closed than Mayr had expected such as those of, e.g., the strikingly different geographical representatives of several groups of birds of paradise (*Parotia, Astrapia*, Fig. 4.6). Since these are relatively recent species the change probably occurred rather rapidly without having been greatly impeded by any selection for maintenance of the ancestral species recognition marks. P. Ekman (1998: 386) commented favorably on Mayr's distinction between open and closed programs in his Afterword to the reprint edition of C. Darwin's *The Expression of the Emotions in Man and Animals.*

## Particular Topics of Evolutionary Biology

***Parasites.*** At a symposium on the co-evolution of parasites and their vertebrate hosts in Neuchatel, Switzerland in 1957, many cases of "host transfer" were discussed which prevent a complete parallelism between the classifications of parasites and vertebrate hosts. Also, the rate of morphological change in different lines of parasites is unequal (Mayr 1958b).

***Evolutionary novelties.*** At the Darwin celebration organized by the University of Chicago Mayr (1959d) discussed the emergence of evolutionary novelties which neither Rensch nor Simpson in their recent books dealing with macroevolution had treated in detail. Mayr stressed the importance of a change in function and the fact that evolutionary novelties are in most cases acquired gradually. During the new wave of saltationism in the 1970s, this paper which established gradual evolution so well was almost completely ignored.

***Chance and necessity.*** At another Darwin Centenary Symposium in Australia, Mayr chose as the subject of his contribution the eternal problem of chance or necessity in evolution, or as he called it, "Accident and Design, the Paradox of Evolution" (1962b). He again emphasized here that natural selection is a two-step process, the first step consisting of the production of genetic variability and dominated by chance (mutation and recombination), and the second step, the differential survival and reproduction of the new generation, determined by natural selection. Other aspects of adaptation and selection were also dealt with. His final conclusion was: "The solution of Darwin's paradox is that natural selection itself turns accident into design" (p. 14). He was aware, of course, that natural selection is always constrained by the potential of the existing genotype, i.e., by the preexisting general morphological construction of an animal or plant.

During the course of evolution the overall amount of variation produced varied and "periods of high phenotypic variation (as in the Pre-Cambrian, Cambrian) have alternated at other geological periods with periods of relative stasis. As natural selection is active at all periods, it would seem that evolutionary change depends more on the availability of variation than that of selection. [...] What environmental factors favor high variability? What others favor stability?" are challenging questions, as Mayr (2005) emphasized in his last publication, a Foreword to a book, *Variation. A central concept in biology.*

***Adaptation.*** The topic of adaptation is treated in many of Mayr's publications from the 1950s onward and has been occasionally the subject of separate articles. At the International Ornithological Congress in Helsinki 1958 he introduced a symposium on adaptive evolution stressing "that not every aspect of the phenotype is necessarily adaptive. What is being selected is the total phenotype and through it the total genotype [...] The phenotype is almost always a compromise between several opposing selection pressures" (1960e, pp. 496–497). The word adaptation refers to the condition of being adapted. To counter claims in the literature that a major part of evolution is "Non-Darwinian," effected by neutral genes and neutral characters, Mayr (1982f) discussed several concepts of evolutionary biology, particularly adaptation and selection, and concluded that none of the new insights necessitate any essential correction of the basic neo-Darwinian theory. Since the individual as a whole is the target of selection, many neutral or even slightly deleterious genes may be carried in a population as "hitch hikers" of favorable gene combinations. Selection never succeeds in achieving perfection because of chance factors and several constraints of which Mayr (l.c.) listed no less than seven. In his defense of the "adaptationist program" (which attempts to determine what selective advantages have contributed to the shaping of the phenotype), Mayr (1983a) concluded that "its heuristic power justifies its continued adoption under appropriate safeguards."

***Sexual selection.*** Darwin's most useful thesis of sexual selection was hardly ever mentioned during the first two-thirds of the 20th century, the age of genetics, and none of the architects of the evolutionary synthesis discussed the presumed significance of sexual selection as an evolutionary factor. Some reasons for this neglect are: (1) The assumption that the gene rather than the individual is the target of selection virtually wiped out the distinction between the two kinds of selection, (2) during the evolutionary synthesis the emphasis was on speciation and those aspects of behavior that served as isolating mechanisms. This drew attention away from the behavioral interactions of conspecifics which is where sexual selection takes place (Mayr 1992f). Mayr (1963b: 199–201) called attention to this neglect and Ghiselin (1969) dealt with sexual selection in more detail. Revival of general interest in sexual selection occurred during the centenary of the publication of Darwin's *Descent of Man* in 1971 and led, in 1972, to the publication of a book edited by B. Campbell, to which Mayr (1972g) contributed an essay on sexual and natural selection. Here he discussed female choice, females' sense for "beauty," struggle among males and features which do not contribute to the fitness of the species but merely to the reproductive success of the possessor of these characters.

***Speciational evolution.*** The theory of "punctuated equilibria" (Eldredge and Gould 1972) was based on Mayr's views of peripatric speciation in founder populations (1954c, 1982l), but claimed that all species enter pronounced stasis after completion of the speciation process. In other words, evolutionary change occurs exclusively during speciation. These claims of the prevalence of total stasis and of the impossibility of evolutionary change without speciation, are clearly invalid. Mayr (1982p, 1989i, 1999g) clarified the false claims and misunderstandings associated with "punctuationism" and, like other authors, admitted its stimulating impact on evolutionary discussions and the conduct of fruitful empirical research. Speciational evolution is fully consistent with Darwinism. The discussion of punctuationism "finally brought general recognition to the insight of those who had come from taxonomy (Poulton, Rensch, Mayr) and had consistently stressed that the lavish production of diversity is the most important component of evolution" (1989i, p. 156).

***Haeckel's "biogenetic law."*** Based on the observations of earlier authors Haeckel, during the 1860s, became enthusiastic about the phylogenetic information content of the embryological development of animals and formulated the slogan: "Ontogeny is the short and rapid recapitulation of phylogeny." However, he failed to provide a convincing causal explanation assuming that phylogeny worked like a physiological process. Mayr (1994l) reviewed the history of the "biogenetic law" and showed how often proximate (physiological or genetical) and ultimate or evolutionary causations were confounded in attempts at explaining recapitulation. He emphasized that this phenomenon must be explained not only physiologically but also in terms of ultimate causations, i.e., recapitulated features must have some function to be preserved by natural selection. It became obvious in recent decades that recapitulated ancestral features were necessary stages in the developmental process; each stage has a causative influence on the next following stage. The inducing capacities of the surrounding embryonic tissues serve as "organizers" and form a somatic program. The gill arch system of the human embryo was preserved by natural selection because it was needed. It serves as the somatic program for the further development of the neck region in the embryo. This is the basis of an ultimate explanation for recapitulation, which also permits an explanation of why recapitulation is so irregular. Some characters are recapitulated, others are not, because some ontogenetic developments require a somatic program, others do not. The reason for this difference, however, is not yet clear.

Ontogeny recapitulates—with minor or major deviations—the ontogenetic (but *not* the adult) stages of the ancestors.

***Human blood groups and schizophrenia.*** While still in New York Mayr argued repeatedly with Dobzhansky as to the meaning of blood group polymorphism in man. The latter was convinced of the neutrality of the differences, while Mayr seeing how many cases of polymorphism had been discovered in recent years to be balanced polymorphism, was convinced that blood group frequencies were controlled by selective compromises. Dr. Louis K. Diamond made all the blood group data of the Children's' Hospital in Boston available. However, the frequencies of the A, B, O blood groups of the so-called "blue babies" who had gone there for heart surgery, were almost exactly those of the Boston population. Mayr's former associate on the Whitney South Sea Expedition, Dr. Hannibal Hamlin, who had become a neurosurgeon, made the blood group data of the Neurology Division of the Massachusetts General Hospital available. Most of the hard work of this research was done by Mayr's wife Gretel, who copied all the blood group data from thousands of patients' files in several Boston and New York hospitals. There was one malignancy adenoma of the pituitary gland which strikingly deviated from the Boston population and from other brain tumor (a considerable excess of blood group 0 was found). Other medical studies in the meantime also clearly established a correlation between certain blood group frequencies and certain pathologies. The idea that the blood groups were a neutral polymorphism was thereby refuted.

Mayr has also been interested for a long time in the genetic basis of schizophrenia and, with J. Huxley and two medical colleagues, developed a model of balanced polymorphism for it (1964o). Schizophrenia probably depends primarily on a major dominant gene, Sc, with about 25 percent penetrance, that is, is manifested in only 25 percent of the carriers of this gene. The authors concluded that the Sc gene may be in morphic balance conferring certain selective advantages to compensate for its obvious disadvantages.

"For many years I have been trying to get the psychiatric community to take a greater interest in the problem, but without success. The arguments of course are always the same, it is a syndrome that is difficult to diagnose, it is highly variable and no one knows whether it is one or several diseases, it has a large environmental component, etc. To me these defeatist arguments do not make any sense. I agree with Popper that one cannot advance in science unless one proposes models that can be falsified. Sclater's genetic model for the inheritance of schizophrenia, as well as the Mayr-Huxley model of morphism, are models that can be disproven. This is entirely independent of the question whether or not schizophrenics can be chemically identified and whether or not such chemical components are cause or effect. [...] If the paper in Nature stimulates further research I will be happy even if this research should lead to a falsification of the proposed model" (letter dated February 1, 1965).

In a later status report Mayr (1972b) concluded "that there can no longer be any argument about the existence of genetic factors in schizophrenia" (p. 530).

This encouraged the search for new approaches in the treatment of this disease. Mayr emphasized the need for an adequate consideration of multiple causation, "a process particularly important in the shaping of the phenotype," for which both genotypic and environmental factors are important. The purely environmental theory of schizophrenia has been abandoned.

***Explanations of concepts and definitions of terms.*** The clarity of discussions of theoretical concepts depends on clear definitions of terms. Therefore Mayr included in most of his books detailed glossaries and published an article on the "Origin and history of some terms in systematics and evolutionary biology" (1978c), many of which he had proposed or introduced from foreign languages into the English literature. Examples are: allopatric, cladogram, founder principle, genetic revolution, nondimensional species, phenetics, population thinking, semispecies, sibling species, superspecies, teleonomic and others. He discussed and analyzed repeatedly theoretical concepts like "Darwinism," "adaptation" and "selection." As late as 1997(e), Mayr devoted a discussion to the subject of selection coining the new term "selecton" for the target or object of selection (the gamete, the individual organism, and certain social groups of cooperating animals or early humans). However, there is no "species selection," because the species as an entity does not answer to selection. When one species replaces another one due to competition, then this "species turnover" is due to individual selection discriminating against the individuals of the losing species. The expression "selection for" is used for any aspect of the phenotype (or the phenotype as a whole) that favors survival or reproductive success and which, therefore, will be favored by selection.

# 9 Ernst Mayr—the Man

## Personality and General Views

Mayr was a born naturalist and observer. His parents took their three children on nature walks nearly every weekend and taught them flowers, trees and animals. He felt that his interest in theoretical aspects and basic principles in science was due to his education at a German gymnasium: "They train students to ask critical questions instead of just accumulating knowledge. [...] I have always said that my achievements are due to this heritage of culture of the German gymnasium" (Shermer and Sulloway 2000).

Mayr lost his father when he was not yet 13 years old but he was lucky to meet, through his birding activities, fatherly friends with whom he established close personal relationships. First, during his high school years in Dresden, there was Rudolf Zimmermann (p. 20) and later in Berlin he met Gottfried Schiermann (p. 46) and Erwin Stresemann, the latter certainly the most influential man in his life (pp. 39–42).

Contemplating himself in a lonely field camp in New Guinea in 1929, Mayr listed the following attributes of his character: (1) quick perception, (2) critical judgment, (3) management ability, (4) highly ambitious, (5) always pushing himself and (6) well-developed self-confidence (Mayr Papers, notebook, Staatsbibliothek Berlin). He felt that these characteristics occasionally led to his social isolation. "Ambition is the father of all deeds," he wrote to Stresemann on 13 December 1928. From New Guinea he wrote to his younger brother Hans on 25 February 1929 (transl.):

"One must make the most of one's chances, one must be able to work hard, perform productive work and work on one's own personality; one must have certain basic principles, and definitive objectives [...] and work like hell. The so-called great men were almost without exception great and indefatigable workers. Many of them worked daily 16–20 hours. Quite that much is not possible at our age, but 10–12 hours of work are possible."

In a letter to Dr. Stresemann from New Guinea (20 February 1929), Mayr confessed to be "in some respects a terrible guy, pert, conceited, unpleasantly ambitious"—while Stresemann emphasized in a letter to Hartert: "Mayr is quite self-confident and has his own will. [...] In the field one survives better with a very secure, reckless, decided attitude than with a certain pliancy" (9 December 1928). From these and similar remarks the reader may get the impression that Mayr, at least in certain respects, showed egotistical traits when, generally speaking, the opposite was true. In his profession, he was more altruistic than many of his colleagues, always answering enquiries without delay, reviewing in detail and promptly manuscripts for associates, students and journal editors and supporting students or young colleagues in many other ways. Numerous recipients of his help and kindness have stated their gratefulness in print. Mayr was interactive with students and other interested persons, an excellent adviser, supportive and enthusiastic about the work of others (as his teacher and friend Erwin Stresemann had been).

Mayr also often cooperated with and trained volunteers and young colleagues in the Bird Department of AMNH publishing the results of their joint research in coauthorship with them (p. 120). Despite his extremely busy schedule at the museum he also wrote detailed letters to collectors in the field who were working under his supervision giving advice and encouragement in a truly understanding way. In this respect he was carrying on a tradition because, when he himself was on expedition in New Guinea, Ernst Hartert and especially Erwin Stresemann had sent him long letters stimulating Mayr's interest and ambitions effectively.

In scientific societies like the AOU and Evolution Society Mayr played active roles and worked for their improvement without asking for any other reward (Cain 1994). He was an active member of numerous committees, and a responsible editor. Always willing to work hard, he spent much of his time on these cooperative efforts. After World War II, he and his wife Gretel contributed in a most unselfish manner time, energy, and money to the founding and activities of the American Relief Committee for German ornithologists (p. 137). After Stresemann's *Entwicklung der Ornithologie* had appeared in 1951, it was exclusively due to E. Mayr's continuing efforts that a translator and a publisher were found and an American edition was eventually published (Harvard University Press, 1975). An Epilogue by Mayr updated the history of American ornithology.

Evidently, he saw himself as occasionally more of a "nasty fellow" than he actually was. Also, he never took personally any criticism of his work and maintained friendly relations with his main opponents such as Richard Goldschmidt and John C. Greene. When one of his theories was proven wrong, he did not hesitate to agree and change his opinion. Occasionally Mayr was rather quick in passing judgment on older colleagues, as illustrated by a quote from Stresemann's letter written on his return trip to Germany after a visit to the United States (13 October 1937; transl.):

"I almost jumped out of my shoes when you mentioned to Delacour that you consider Wetmore a dry schoolmaster with a small mind [...] By this you help nobody, but you may harm yourself badly. Bet that Wetmore will hear of it within a short while? Somebody [...] mentioned to Tom Barbour that you called him no more than 'a simple Harvard professor' and that you don't think much of him as a zoologist. This Barbour told me personally, with plenty of bitterness. Why make enemies unnecessarily?" Mayr answered repentantly on 6 November 1937: "I thought I had improved which, however, apparently is not the case! I shall take pains over it in the future! The bad thing about it is that sometimes I cannot remember anything, as in the case of Barbour. I have not the slightest idea to have made any deprecatory remarks about him, but wondered for years why he gave me the cold shoulder."

As a result, when visiting the Museum of Comparative Zoology during the 1930s, Mayr was never invited to Barbour's "Eateria", a kind of a salon at the museum, where he provided catered luncheons to a few favored staff members and visiting guests. Another reason for this treatment probably was that Mayr, as "Sanford's boy", belonged to the competitive camp (see p. 100). Barbour developed a greatly improved opinion after Mayr had destroyed the arguments for land bridges between the islands of the Malay Archipelago (p. 181). Upon reading the above letter in 1995 again, Mayr confirmed his remark on Wetmore, but regarding that on Barbour he commented (pers. comm.):

"I am quite certain now that I never made this claimed remark about T. Barbour. He was a herpetologist, and I could not have judged him at all. I suspect that J. Greenway (who sometimes intrigued) had invented this story and I am sorry that Barbour was taken in by him. I was quite friendly with J. L. Peters already in 1931 and helped him with every later volume of the *Check-list of Birds of the World.* Perhaps this made Greenway jealous? My opinion on T. Barbour at that time was completely different from what I supposedly said."

The Society for the Study of Evolution at first had just 300 members. Mayr as its secretary ordered 1500 copies of the first issues of the journal "*Evolution*", seemingly a vast excess. In due time, however, it turned out that the excess stock gave the society a most welcome additional income when more and more libraries subscribed to the journal and wanted to buy the back volumes. Eventually the whole stock sold out justifying his optimism. A similar example is mentioned on p. 111.

Many times in his life, Mayr felt as the odd man in his environment which, as he mentioned, may well be the reason why he tended to assert himself rather vigorously in his scientific controversies. When he came to Dresden in 1917, he diverged by his Bavarian accent. Later in New York he was the only German at the AMNH and much younger than his colleagues in the Bird Department. Stresemann was well aware of Mayr's peculiarity. He commented on the *Principles of Systematic Zoology*: "I like very much your tolerance, shown in many places, of views and principles with which you basically disagree. Meanwhile you have checked considerably your inborn aggressive drive" (15 April 1969). As a versatile and convincing fighter for his views, Mayr rebelled against the older generation in the AOU during the 1930s, against the discrimination of systematic biology versus molecular biology, and against those philosophers of science who only knew a philosophy of physics and no philosophy of biology.

"He fights for ideas and the advancement of knowledge. 'Fights' is the proper verb here because he is relentless in trying to discover what is confirmable and to discard what is not" (Moore 1994: 10). "Mayr was—and still is—an outgoing man who is very much at home at professional meetings and conferences, where he vigorously and tirelessly defends his views. He makes no distinctions between august experts and hesitant graduate students. He is as willing to spend time setting one straight as the other. The vigor and definiteness with which Mayr addresses all comers is at times intimidating" (Hull 1988: 68). However, I think he was exaggerating or joking when he stated "I'm not dogmatic, I'm simply right!" (Coyne 2005).

Exceptional energy and resolution were personal traits effectively displayed when Mayr worked very hard to pass his PhD examination (June 1926) in time to qualify for an opening for an assistantship at the Zoological Museum in Berlin. The same determination guided him when he went, all by himself, on a one-man expedition to New Guinea, as well as on many other occasions in his life. He certainly had a strong ego.

Sulloway (1996: 191) stated that "rebellious laterborns overcome timidity by systematically cultivating openness to experience" and mentioned as a supposedly good example of such a personality transformation Ernst Mayr who, as the middle of three brothers, had to "fight" both ways and had resented his elder brother's domineering manner. Ernst's childhood nanny described him as "a quiet and dreamy child." Even if this was true, a personality change had certainly taken place already by 1923, when he observed the pair of rare ducks near Dresden and nobody believed him: "That criticism really annoyed me and that's why I wanted to get it really nailed down and establish that by golly, yes, this kid Ernst Mayr did see a Red-crested Pochard" (Mayr 1994m: 330). He trusted his own perception and therefore pushed his way up to Dr. Stresemann at the Museum of Natural History in Berlin. With the same conviction, as a 19-year-old student, he proposed a series of very sensible research topics to Stresemann in 1924 (pp. 27–29). In Berlin Mrs. E. Stresemann soon labeled this student "fresh young Mayr" years before he went to New Guinea. His expedition experience may have completed this personality transformation, as Sulloway (l.c.) implied which, however, was noticeable already several years earlier during the process of his becoming an adult. Mayr (1932e: 97) summarized his experiences in New Guinea as follows:

"Looking back on my first expedition, I value more than the discovery of many specimens and facts new to science, the education that it was for me. The daily fight with unknown difficulties, the need for initiative, the contact with the strange psychology of primitive people, and all the other odds and ends of such an expedition, accomplish a development of character that cannot be had in the routine of civilized life."

In another letter to Dr. Stresemann from the Solomon Islands (dated 28 December 1929) he stated: "In any case, the possibility to obtain a good position at the American Museum of Natural History undoubtedly exists (maybe only on probation). It merely requires the necessary imprudence and diplomacies to reach the goal". At the same time, this strong attitude melted away completely in the presence of a member of the fair sex (pers. comm.).

Mayr had an excellent memory and was able to assemble fragmentary knowledge into new and important ideas. He was driven by an unlimited curiosity: "Darwin was interested in everything. In that respect I am very similar to Darwin. All my life I have been interested in a wide variety of ideas and subjects. But maybe because I grew up in Germany where things always went wrong, and I lost my father at a young age, I grew up to be a realist, maybe even a pessimist."

In his autobiographical notes he continued: "In personal contact with other people I am usually friendly, generous, altruistic, readily joking, so that others referred to me as 'a nice guy.' However, in scientific controversies and in political arguments I seem to be a different person. There I seem to be aggressive, intolerant, stubborn, intent on having the last word on any price. I have often thought about the possible reason for this contrast. I rather suspect what is involved is that through much of my early life I found myself again and again at the bottom of the totem pole and had to fight my way up. It started in my childhood because I was the middle of three brothers. The oldest had all sorts of privileges and so did the youngest. There never was a special privilege for the middle brother even though my mother tried to be as just as possible. Being a later born predestined me to be somewhat rebellious, as Sulloway (1996) has shown.

When I was in school owing to my father's promotions and death I found myself several times a stranger in a new school. When I arrived in Munich in 1914 I spoke the Franconian dialect which was very different from the Munich Bavarian dialect. It took quite some time before I was acculturated in this new environment. In 1917 we moved to Dresden in Saxony and again I was an alien element in my class and it took me some time to become adjusted and make friends. When I arrived at the American Museum in 1931 I was 15 years younger than any of the other scientists there and, furthermore, Chapman, the chairman of the department, always had the peculiar feeling that the 4th floor of the Whitney Wing which contained the Whitney-Rothschild and other Old World collections was not really part of the Bird Department. Finally, when at Harvard I was elected to quite a few committees in Washington I always had to fight for the well-being of taxonomy. There was a widespread feeling among the university biologists that taxonomy was not a real science, meaning an experimental science. I believe that my aggressiveness in arguments is related to this set of experiences in my younger years. Even during the evolutionary synthesis and the ensuing years, being a taxonomist and museum person rather than a geneticist or other kind of university scientist seemed to result in discrimination. At the Chicago Darwin Celebration in 1959 the major attending evolutionists received honorary degrees from the University of Chicago. But no Ernst Mayr. Both Dobzhansky and Simpson were elected to the Royal Society in middle age. Ernst Mayr not until he was well into his 80s, and so it goes. As far as my science is concerned this aggressiveness was an advantage but as far as my personality and my relation to people were concerned it definitely was not."

Mayr was not humble or modest but always ambitious and self-confident. However, he did point out that certain aspects of his early career were due to no more than a series of accidents (p. 3). On the other hand, he defended his scientific views and theories firmly, even if they were controversial like eugenics which Mayr considered the only way to "develop mankind further," although he admitted that in practice, this is not feasible. He honored the theories of fellow scientists but claimed very firmly the merits for himself in those instances where he was the first, e.g., to propose theories of island biogeography (1933j, 1940i) and of rapid speciation from small populations (1954c). Examples of his self-confidence are his undertaking a one-man expedition to New Guinea in 1928, his leadership role in evolutionary studies in the United States during the mid-1940s, his spokesmanship for systematics and evolutionary biology against molecular biology during the 1960s and his leadership roles in the fields of history and philosophy of biology. Yet he also knew his limits, for example when he restricted his book on species and evolution (1963b) to animal species rather than also including plants, as several friends had suggested, and when he gave up on preparing the second volume of his history of biology which was to deal with the history of functional-physiological biology (p. 346). He realized his insufficient familiarity with these subject matters.

It is true that Mayr was not a "media scientist" like his colleagues E. O. Wilson, S. J. Gould, R. Dawkins who often write for the general public. Therefore Mayr was not as widely known outside technical circles as these colleagues are. Most of his publications are directed at his peers and knowledgeable persons including his latest books on *This is Biology* (1997b), *What Evolution is* (2001f) and *What makes Biology unique?* (2004a) which explain in general terms genetic variation, adaptation, speciation, the evolution of man and important aspects of the philosophy of biology.

On several occasions Mayr triggered or participated in general discussions on topics outside his own professional expertise. Some time after his "retirement" he published an editorial in *Science* entitled "Tenure: A sacred cow?" advocating a system of periodic tenure review at academic institutions, perhaps at 10-year intervals (1978b). He knew the difficulties when hiring new staff members. As a museum director he had hired, among others, two scientists who seemed to be very promising but later turned out to be failures. [This may indicate that Mayr was not an especially good judge of people's abilities.] Four discussants of his proposal were against it preferring instead to improve the hiring and tenure granting practices, while two letters were in favor of tenure review. Mayr's answer to these letters was longer than the original editorial itself. In his comment on another editorial in *Science* he agreed "that the role of the court ought to be strengthened," e.g., calling experts with no obligation to either the prosecution or the defense. He felt "that the extreme emphasis on the adversarial approach [in our national system] tends to interfere with the finding of the truth" (1989a). However, all systems no matter how well they are arranged can fail because one has to deal with people. Mayr's contribution to a book, *Education and Democracy* is entitled "Biology, pragmatism, and liberal education." Here he dealt with general concepts which are the basis or scaffolding of our world view, such as democracy, freedom, altruism, competition, progress, and responsibility. He concluded that in an up-to-date liberal education more room in the curriculum should be allotted to the study of the concepts mentioned above and to a more detailed analysis of the concepts that are the basis of our belief in democracy (1997i).

Some persons accomplish more than others during any given period of time and Mayr was one of them:

"For very many years I tried to get up every morning at 4:30, take a short breakfast consisting of tea, cold cereal and yoghurt, and start working at around 5:00. After three hours I had a second breakfast, with my wife, and was in the office around 9:00. Of course, I went to bed at 9 or 9:30 so that I had ample sleep all the time. I did not depart from this schedule in the summer when I was up at the 'farm'. But there, at around 9:30 or 10:00 I would go outdoors and either go on a hike or do one of the many outdoor jobs one has to do at such a place. I published my 1963, 1969, and 1970 books while I was a very busy director of the MCZ. It was my morning schedule that made this possible" (pers. comm.) and "In my own case I have found the hours from 5:30 to 9:00 in the morning particularly productive while others of my colleagues do not seem to hit their full stride until after 10 p.m." (1971g: 428).

The only common trait of 800 successful American scientists identified in a study by a psychologist was an above average eagerness to work. Ernst Mayr fit into this picture. Even an honorable position like the presidency of a scientific society for him was not so much an honor as an obligation to do a job. At an age of 96, he listed on a sheet of paper under the heading "To be done" more than 20 projects including: (1) Last minute revisions for *What Evolution is* (2001f), (2) preparations for two lectures, (3) revision of a manuscript on classification, (4) completion of three manuscripts for a new essay volume (to be published in 2004), (5) review of Darwin's *Origin* "sentence by sentence," etc., etc.

Mrs. Marjorie Sturm, Mayr's administrative assistant while he was the Director of the Museum of Comparative Zoology (1961–1970), knew four consecutive directors. She stated that Mayr was "the most dedicated director I've worked with. He was a shining example of what a director should be. He loved the museum and what it stood for. He was interested in all aspects of the museum. He had a commanding presence. [The previous director] Alfred Romer didn't quite have that presence; he was friendlier maybe is the word, outgoing. Dr. Mayr was a little austere. Both he and his wife were very nice and warm to me. [However,] he was the director, he had his presence. And I don't know that he would have won a popularity poll. There was a distance between him and his staff.

Being with him socially quite a bit, going out to his farm and seeing him out there and with his wife he was, of course, more relaxed. He didn't go around patting people on the back or joking or kidding. He was a very serious man. But he also had a sense of humor and he could see humor in himself as well as in other people. We just had really a very nice relationship.

Dr. Mayr is dedicated to whatever he is doing; he wanted to do an excellent job as director and, as I have said, he had a real love for the museum and what it was doing. The next director [Dr. A. Crompton] was completely different from Dr. Mayr as any two people could be. I think that it pained him to see what was happening to his museum. He feels perhaps that the research has declined a little bit. I do know that he is very sad that there are no more annual reports" (summarized from an interview conducted by Dr. R. Creath, Arizona State University, Tempe, Arizona).

To some people Mayr appeared "cold" or even arrogant, as mentioned by the Indian ornithologist Salím Ali in his autobiography (1985:60): "Ernst Mayr is currently Emeritus Professor of Zoology at Harvard University and undisputably among the topmost biologists of the world today. Unfortunately, in marked contrast to his mentor Stresemann's unassuming modesty, Mayr makes you feel he is not unaware of the fact." This sounds like a rather superficial evaluation of both persons. Hardly any visitor to "The Farm" would have agreed with Ali's impression of Mayr as presumptuous and few, if any, German ornithologists would have described Erwin Stresemann as "unassumingly modest." In general Mayr was friendly and easy going. Every secretary praised how pleasant it was to work for him, because he never blew up but just laughed, when she goofed. As the director of the MCZ, he had no visiting hours, his door was always open and anybody could meet him any time. Many authors sent their manuscripts to him for review and the acknowledgments of numerous relevant papers in the 1940s–1960s often include a "Thank you" note by the author.

Mayr maintained fond memories and a "soft spot" in his heart for the Allgäu region of Bavaria, where he was born, and for the Dresden region where he grew up. This is shown by the choice of his Ex Libris (Fig. 9.1), an Alpine scene near Einödsbach in the Allgäu region drawn for him during the mid-1930s by his friend at the AMNH, the well-known painter Francis Lee Jaques (1887–1969) on the basis of a photograph. Mayr used this Ex Libris extensively during the 1930s and 1940s, but the glue his secretaries had available was of poor quality and many Ex Libris have come off in later years, when the program of identifying his books in this

**Fig. 9.1.** Ernst Mayr's Ex Libris which he used during the 1930s and 1940s

way was discontinued. Pictures of Moritzburg and Dresden adorned the walls of his corner room at The Farm.

Coming from Germany Mayr was able to draw from the extensive European literature and to combine this knowledge with what he gathered in North America. Settling in two different cultures gave him a much broader approach to problems and has been one of the major reasons for his success (Mayr 1994a). The training he had received from Professor Stresemann was clearly significant in establishing the research strategy and analytic methods used (Bock 1994a:281). When he was asked "How would you like to be remembered as a scientist?" Mayr answered: "One of the things that most people don't see me as is that basically I am a modest person. They think that I am a great egotist and pusher of my own ideas. But I am a humble man. I just want to understand nature and make a contribution to the body of knowledge about the natural world. And nothing else. I will add that except for losing my father when I was 12, and my wife a few years ago, I have been an extraordinarily lucky person" (Shermer and Sulloway 2000). Further, Ernst Mayr was *generous* with his time and the attention he gave to students and to colleagues; generous with the money he received for academic prizes which he gave to purposes of education and conservation. He was *tolerant* regarding racial, religious or ethnic differences and he appreciated the *unknown* in science and religion and therefore did not speculate about things that we do not know. His driving *enthusiasm* for his work until the last month of his life was remarkable. Asked in an interview as to what had given him particular pleasure in his ornithological work, he answered: Having been ahead of his competitors like, e.g., when he discovered a sibling species of Asian birds which had been overlooked by several experienced ornithologists (p. 161). He was also very pleased with himself and proud when, after 10 years of hard work, he had finished his *List of New Guinea Birds* in 1941, the basis of all later publications on this avifauna (Bock and Lein 2005; CD-ROM).

An outstanding feature of Mayr's thinking has been, from his student days on, a deep concern for *generalizations*. In his publications he emphasized the "broad picture," i.e., the underlying general concepts and ideas but, at the same time, he was always very accurate with the details. There are hundreds of painstaking descriptions of subspecies with many thousands of measurements. He spent about 10 years studying the birds of New Guinea and did not finalize his book (1941f) until he had seen and compared the types of all questionable taxa and most other forms and consulted all the publications of the preceding 60 years attempting to identify every species however insufficiently described. His tendency to draw general conclusions directed him to philosophy and the theoretical framework of biology. In his scientific work, Mayr usually adopted multi-factor rather than single factor interpretations, as shown by the following quote: "For me phenomena such as territory, clutch size, niche partitioning, or species number on an island were rather the phenotypic compromise between several selection pressures, different for each species and different for each locality" (1973j: 434).

During his professional life Mayr dictated most of his letters and manuscripts, but corresponded with handwritten letters in German with his friends in Germany.

His *handwriting* was fairly neat and clear becoming less legible in his old age. After his retirement, when he had only temporary secretarial help, he also wrote many pages of his manuscripts by hand which were typed later (he never made personally the transition to the typewriter, the laptop or other virtues of the computer age). Mayr had learned in school the German script which he used in his notebooks, letters, and manuscripts until the mid-1920s, but in about 1927 he switched to the regular Latin script. He was a non-technical person, but managed or was involved in numerous practical chores at The Farm.

Mayr had many *informal students*. He influenced D. Amadon and trained the dentist C. Vaurie discussing with him every sentence of all of his papers (at least into the 1950s); he trained B. Biswas from India, and was the first teacher of D. Ripley and M. Moynihan. About four or five historians of science (including Coleman, Adams, Burkhardt, Churchill) consider him as their PhD supervisor, even publishing a Festschrift on the occasion of his 75th birthday. Mayr influenced numerous people in North America and other countries of the world through personal contact, through his publications or through correspondence. As examples I cite expressions of several colleagues to show how they looked up to him. Steven J. Gould said when he completed his *magnum opus* on *The Structure of Evolutionary Theory* (2002): "It is my greatest joy that I have finished this while Ernst [at age 97] is still with us. We have been talking about it for a long time;" Douglas J. Futuyma: "To receive a signed copy of such a work [*Birds of Melanesia*, 2001], from my intellectual hero who has been an inspiration to me since my undergraduate days, is truly rewarding."

Irv De Vore: "I don't think I have ever had the chance to tell you how much I have treasured you, over the years, as the mentor who was instrumental in leading me out of the swamp of Social Anthropology into the more granitic and satisfying precepts of biology. It was all those years ago, at the Wenner Gren Conference when I first met you, Simpson, et al., that I joined you on your predawn ambles across the Austrian hills. Those walks with you, in which I was dazzled by the depth of your understanding of all of nature around, were surely the most formative events in my intellectual life."

One of his characteristics that is often overlooked was Mayr's sense of *humor*, although admittedly his humor was sometimes a bit sharp. In conversation he loved to make his partners laugh by making "cracks." On the occasion of a social dinner a colleague once remarked: "I try to get on Mayr's table, because that is the table where there is the most laughing." One of his roommates at 55 Tiemann Place in Upper Manhattan during the early 1930s remembered that Mayr loved to read the comics in the weekend edition of one of the New York newspapers.

The paleontologist Edwin H. Colbert of the AMNH was quite right, when he called Mayr a man of *strong opinions*. Some of his colleagues even considered him as dogmatic. He was never afraid to express his views directly and he said so in some published statements. He was of the opinion that this is the quickest way to get at the truth. "To take an unequivocal stand, it seems to me, is of greater heuristic value than to evade the issue."

"Because I hate beating around the bush, I have sometimes been called dogmatic. I think this is the wrong epithet for my attitude. A dogmatic person insists on being right, regardless of opposing evidence. This has never been my attitude and, indeed, I pride myself on having changed my mind on frequent occasions. However, it is true that my tactic is to make sweeping categorical statements. [...] My own feeling is that it leads more quickly to the ultimate solution of scientific problems than a cautious sitting on the fence" (1982d: 9). He once told Frank Sulloway about his risk-taking in science. He said that he liked to go out on a limb sometimes because this motivated other scientists to try to prove him wrong by cutting off the limb on which he was standing. If it turned out he was mistaken, he said, then something was learned that would not otherwise have been learned had he taken a more cautious stance. Such a bold willingness to be the target of criticism takes a strong ego to put into action, Sulloway commented.

In Dresden the family attended regularly when some special *music* by Bach or Händel was performed in one of the churches. Later in Tenafly, New Jersey Ernst and Gretel subscribed to some concert series and greatly enjoyed hearing the major artists of the 1930s and 1940s. They also subscribed to the Boston symphony when they had concert series at Harvard's Sanders Theater. While they enjoyed classical music, modern music and poetry did not play any major role in their lives. In 1951 Mayr was deeply moved by a performance of Gluck's *Orpheo and Euridice* in the Scala of Milan under the direction of Wilhelm Furtwängler; the staging, the singing, everything was quite overwhelming. When he needed some relaxation, Mayr walked across Central Park from the American Museum to visit the Whitney Museum on a number of occasions.

Although ambitious and driven in his profession, Mayr easily switched over to his *family* when at home. He was available to his two girls and more easy going with them than sometimes their mother. When needed, he helped the children with their homework and to care for them when one was ill. Of course, he spent long hours in his "office" upstairs, a "sacred" place and essentially "off-limits" to the children while the family lived in Tenafly. They both remember heated discussions on technical subject matter when Dobzhansky and his family came over from Manhattan for dinner.

Ernst Mayr was raised in a family in which *charity* was ranked as one of the foremost virtues. His mother's life philosophy was "Geben ist seliger als nehmen" (to give is more blessed than to receive) and Gretel Mayr, a minister's daughter, was equally committed to charitable giving. According to American tax laws one gets credit in the income tax calculation for charitable donations up to one half of one's income. Gretel and Ernst Mayr always tried to use as much of this allowance as possible. During the 1990s, he annually gave about $25,000–40,000 for social purposes (charity, conservation, science) and he never used for himself the money associated with the prizes he won. With the $100,000 of the Balzan Prize (1983) he established the annual Ernst Mayr grants[1], with the $100,000 of the Japan

[1] The principal aim of the Ernst Mayr travel grants (administered by the MCZ) is to stimulate taxonomic research on neglected, difficult or poorly known taxa, in particular the study of types in different museums.

Prize (1994) he assisted the Nature Conservancy to save a unique desert river (Los Mimbras) in New Mexico. With the Crafoord Prize (1999) he created an endowment for the Ernst Mayr Library (Museum of Comparative Zoology). He also created a number of smaller endowments, such as $50,000 for the Carroll Center for the Blind (one of Gretel Mayr's favorite charities), $30,000 to the Society of Systematic Biology for training fellowships. Forty organizations received annual contributions, mostly $100 to $200.

In his will, Mayr made bequests of $100,000.00 each to the Carroll Center for the Blind and the Planned Parenthood League of Massachusetts. According to his will, all royalties received after death were to be added to principal of the Ernst Mayr Fund of the Ernst Mayr Library at Harvard University.

He took care of his children and grandchildren (also in his will) and thus could afford to make these charitable donations. Since his charitable activities reflect on his character, he felt that he ought to mention them to me, but he did not brag about them and many of them were made anonymously. He gave most of his professional library to Harvard University and a large portion of his personal scientific library to the newly founded (1998) Biohistoricum in Neuburg on the Danube River and even paid the cost of transportation to Germany.

His international outlook led Mayr repeatedly to emphasize the possible advantages if only scientists would be in a position to handle *political problems*: "I still remember that we [Mayr and David Lack in August 1939] wistfully remarked how much better it would be for the world if internationally minded scientists could arbitrate all conflicts between nations, not leaving such important matters to politicians and generals" (1973j: 432). Writing to Stresemann on 14 February 1956, at the height of the "Cold War" period between the Western Allies and the Soviet Union, after he had been visiting Cuba (before the revolution succeeded and it became communistic) and Stresemann just returned from Moscow and Leningrad, he stated: "I have always claimed that the whole world would be a single brotherhood if only scientists had their way" (which may have been a somewhat overoptimistic reaction to the political events of those days).

Both of Ernst Mayr's parents were agnostics but his mother occasionally went to church to listen to sermons of outstanding preachers. Yet, his parents clearly had the purest Protestant *ethics* and would never do anything wrong or unjust. Ethics is the standard by which one behaves toward other humans of one's group and no religion is needed to behave ethically. In the gymnasium Ernst had two hours per week "religion" which actually was a course in the history and dogma of Christianity. At the age of 13 and 14 years he became quite rebellious in high school about miracles and other seemingly improbable events reported. To gather support for his arguments he read Ernst Haeckel's *Welträthsel* (1899).

"However, my mother insisted that I was confirmed, and for the better part of a year I took confirmation classes. The minister was an extreme liberal, and managed to make Christianity palatable to me. I came back to Christianity for a while, but this did not last very long. [...] In my student days my closest friend was a Protestant divinity student. Now, 60 years later, I am still corresponding with this Protestant minister" (letter to J. Greene [1999: 231] dated May 1, 1989).

Mayr would get annoyed occasionally with John Greene because of his "harping" on religion. This was also one of the few major points of disagreement with Th. Dobzhansky.

In later years, Mayr emphasized that he fully supported the ethics of Christianity but not its metaphysics. He did not believe in a Providential God, any divine revelation, or a supernatural realm, but was convinced that a person could be deeply religious in the complete absence of theology. He was disturbed at the strictly anthropocentric teachings of the Bible and the disregard of the environment. Since he was always convinced that *religious beliefs* are a very private matter, he never engaged in any controversies about the existence of God. However, on one occasion he was rather aggressively pressed by an interviewer to express why he did not believe in a personal God and answered: "Do you think any God would permit Hitler to kill 6 million Jews and Stalin to kill 30 million Russians? And permit two world wars which brought nothing but misery or death to millions of people or permit destructive earthquakes and volcanic eruptions that kill ten thousands or hundred thousands of people?" "On the other hand, several famous evolutionists were firm believers in a personal God, such as Dobzhansky and David Lack. Frankly, I have never been able to understand this."

Pope Paul II disapproved any method of birth control despite the ongoing population explosion worldwide, the cause of many problems particularly in the Third World. Three biologists—Ernst Mayr, Jeffrey Baker (University of Puerto Rico) and Paul Ehrlich (Stanford University, California)—protested sharply against this *papal Encyclica* (1968) in a declaration which they posted at the annual meeting of the American Association of the Advancement of Science (AAAS) in late December 1968. The declaration was supported and signed by 2600 scientists attending the meeting. The text read in part: "We declare that we shall no longer be impressed by calls for world peace and sympathy for the poor by a man whose deeds contribute to further war and to make poverty unavoidable." The world had hoped for a long time that "the outdated and inhuman policy of the church toward sexuality would be given up." In his book of 1970 (e) he included a sentence without referring to the papal Encyclica stating "It is most regrettable that the adoption of a healthier way of thinking about the perils of overpopulation is impeded by medieval, and in their effects extremely vicious, church dogmas" (p. 410).

Ernst Mayr (pers. comm.) confirmed the above newspaper report and added that the radio news provider in Boston, apparently a strict Catholic, protested against the biologists' declaration, in particular against their using the expressions "crime" and "criminal" in their comments. He kindly invited Mayr to come to his studio which he did immediately. Mayr confirmed in this interview that large differences of opinion might exist in such valuations. However, in this case there could be no doubt that the Pope's denial of birth control was furthering population explosion leading to much unhappiness and often to increased mortality. They, the scientists, had every right to consider this a crime against mankind. Although Mayr doubted that the radio station would include this renewed attack against the Pope in its program, he returned home quickly and was surprised to hear indeed the interview in full. As a result, he received a lot of "hate mail" among which was a letter of a French colleague at Harvard, a professor of Romance languages.

"Like all young Germans, I grew up with the feeling that Jewish people are 'different.' Jewish schoolboys were just *Jews*. They were not treated badly, but perhaps were somewhat envied, because they had usually good marks. I believe that Jewish classmates had no friends among the non-Jewish boys and vice versa. We had no Jewish relatives, only our cousin Alfred Pusinelli married a Jewish girl (Lotte Eger) which caused some displeasure in the family, although she was an unusually nice girl. Alfred and Lotte later lived in Berlin and I was—strictly platonically—almost in love with her. However, I don't think I would have married a Jewish girl.

Like almost the entire German upper class my family was slightly anti-Semitic. My father, as prosecuting attorney in Würzburg (1908–1913), handled many lawsuits because of wine adulteration, and the adulterators were almost exclusively Jews. This was one of the reasons for his anti-Semitic position. My father made a great distinction between the old German Jewish families that he recognized as German without reservation, and the 'Jews of the East' who had immigrated from eastern Europe, especially Poland and Galicia, during the preceding 50 years. He considered them 'outsiders.' My mother's attitude was very international and I do not remember her ever making anti-Semitic remarks.

In Greifswald, I belonged to a student fraternity, the Gilde, developed from the Youth Movement. It was distinctly 'right wing,' but not Nazi (in 1924–1925). At an annual meeting on Lobeda Castle (Thuringia) I had a public dispute with a leading member of the Gilde (later the Nazi leader, Gauleiter, of Thuringia). Following Günther's racial classification he accepted only blond and blue-eyed people as good Germans. When I asked him whether the poet Johann Wolfgang von Goethe had blue eyes, he was so furious that he proposed I be expelled as 'un-German.' His application was turned down.

There were almost no Jewish people at the University of Greifswald, except the chemist Pummerer and a lady assistant at the Zoological Institute, about whom many students were very enthused. On the other hand, Jews were of great influence in Berlin, whereto I moved in 1925, but ornithology did not interest them. I believe none of Stresemann's PhD students were Jewish, not even Theodor Elsässer. There were several Jewish professors, e.g., the philosopher Dessoir, but they were not considered as such. This was the height of the Weimar Republic and the Nazis were gaining in strength. They proclaimed that Berlin was ruled by Jews. The best hospital in Berlin was the Charité. When I rather naively mentioned that I was interested to do my intern period as a young medical doctor there, I was told with a smile that only Jews had a chance at that institution. Gay [1968] described in his book on the Weimar Republic the Jewish domination of theater, the musical world and the arts. This did not bother us in the biological sciences, but it did those in physics. On the other hand, we were disgusted by the blunt anti-Semitism of the newspaper 'The Stormer' ('Stürmer').

Several years later in New York I lived in the International House and later in an apartment with two other bachelors, some of whom were part-Jewish and for this reason in America (Rosental, Koch-Weser, Lenel). After getting married with

Gretel Simon we had many connections with German DPs [displaced persons] who had been banished by Hitler. Of all befriended 'Americans,' the DPs were those with whom we got along best; not only because they (like us) were more fluent in German than in English, but in particular because they were raised in the same German culture as ourselves. Indeed we felt almost as German as American.

When the Second World War broke out and we were legally 'arrested' as enemy aliens, we could not understand that the FBI and the Department of Naturalization registered us automatically as Nazis (p. 242). Thanks to our two American-born children (who were citizens), we were not sent to an internment camp.

Gretel and I had filed our papers for naturalization in 1939. I do not want to describe here in which scandalous manner we were treated during the 10 years (1940–1950) of waiting for our American passports. German friends of ours who lived in Wisconsin were naturalized immediately. The difference was that all civil servants (without exception) in the New York office were Jewish. One of them stated very bluntly: "As long as we are in this office, no god-damned German will get naturalized." We finally sued the Federal Government and, of course, won our case. We had to produce 50 affidavits proving that we were good potential American citizens. The irony was that 17 of these affidavits had been provided by Jewish friends of ours, who were quite outraged by the way we were treated" (transl.).

Mayr's *friendship* with Erwin Stresemann lasted lifelong (see p. 39). But, during the 1940s and 1950s, it was undoubtedly Th. Dobzhansky who had the greatest influence on his thinking. Both had a similar European scientific background and Mayr learned a great deal on genetics in their conversations. They discussed evolutionary matters by the hour. However, because both lived in New York City, their interaction is not documented by correspondence. Since the 1960s Mayr felt he had more in common in his ideas and ideals with John A. Moore, then also at Columbia University and later at Riverside, California.

Mayr reported: "Not very long after I had arrived in the American Museum of Natural History in 1931, a high school boy from West Virginia came to New York with his mother to meet the great Frank M. Chapman, author of the *Handbook of North American Birds*. Chapman was most gracious, but after a little while he did not know what else to do with this schoolboy. He said: 'Oh, I have a colleague downstairs who is quite active in local birding and I am sure he would like to meet you,' and he sent the boy, John Moore, down to the fourth floor where I had my office. This is how I first met John Moore, and now, more than 60 years later, he is not only my oldest living acquaintance in America, but in the meantime since Dobzhansky's death the person whom I consider my best friend. It so happened that Moore eventually came to study at Columbia University and to take his PhD, and I happened to be the outside examiner on his PhD committee, and I have been friendly with him and his wife, Betty Moore, ever since that time. In mid-life Moore left Columbia and went to the University of California in Riverside. Here Gretel and I visited the Moore's repeatedly and he even arranged for a visiting professorship for me one year at the university in Riverside.

Moore by training is an embryologist, working particularly on the embryology of frogs. However, he is also an excellent naturalist and a lover of the outdoors.

I have several times come to California to admire in his company and under his guidance the flowering of the desert for which California is so famous. Moore, all his life, has fought for the proper recognition of organismic biology, and more particularly zoology. He is most interested in education—he was the main author of the so-called Yellow Version (by far the most successful one) of the BCC textbooks of biology and later on of the book series *Science as a Way of Knowing*. What is perhaps most outstanding about Moore is his wonderful sense of humor. When I am with him we never stop fooling with each other. His wife, Betty, has always been a splendid friend also, and a close friend indeed of Gretel. I am most grateful that I have had the great fortune to acquire such a wonderful friend."

Mayr continued an active correspondence with E. Stresemann and B. Rensch in Germany until they died in 1972 and 1990, respectively. He saw them after World War II repeatedly at congresses, when they visited North America or when he came to Europe (Fig. 9.2). Mayr also followed several invitations to participate in symposia or to lecture at various society meetings in Germany (Fig. 9.3). Several younger people among his PhD students and the scientific community at large became his friends in later years, when he grew older and his earlier friends had passed away. He also maintained very friendly relations with the Department of Ornithology of the AMNH (p. 255) and the 60th anniversary of his arrival in New York was celebrated with a reception (Fig. 9.4).

Late in the 1990s, a somewhat younger resident of the Carleton-Willard retirement home, the physicist Joe Lemon, lost his wife and became depressive.

**Fig. 9.2.** Ernst Mayr (*left*) and Bernhard Rensch in Münster, Germany, 1975 (reproduced from Rensch 1979)

**Fig. 9.3.** Ernst Mayr and German ornithologists at the 100th annual DO-G meeting in Bonn, September 1988; from *right* to *left*: E. Mayr, G. Creutz, J. Haffer, J. Martens, H. Sick, S. Eck, and W. Meise. (Photograph courtesy of H. Classen)

Mayr, who had not known him before, invited him to his table and helped him to overcome his problems. They became close friends and from then on had their walks together almost every afternoon. Mayr would point out flowers, trees and birds, and on one occasion stated: "Nature provides the best entertainment in the world. I wonder why not more people appreciate that." Joe Lemon turned out to be a great help after July 2003 when Mayr suffered a severe illness looking after him daily (besides his daughter Susanne who lives in Bedford). He described Mayr as democratic, social and easy going with rather strong political opinions, mostly liberal. After supper they would read the *New York Times* in the reading room for a little while and discuss issues of interest. Lemon also mentioned that Mayr, driving his car (until 2003), always tried to "beat" the traffic, carefully timing his trips to the museum. Later, Joe kindly drove Mayr to different places including the MCZ and to The Farm on 5 July 2004 for the celebration of Ernst's 100th birthday. He was also present on 3 July 2005 for the scattering of Ernst's ashes (p. 306).

After the *National Socialist ("Nazi") regime* had been in power for only one year Erwin Stresemann sent Mayr a highly perceptive letter about the nature and dangers of the new government:

"The State no longer serves the well-being of the individual, but the duty of the individual is to protect the State and to do what the leaders think is right. [...] The revolution means a catastrophe; and the overwhelming majority of scientists, artists, industrialists, etc. are of the same opinion.[...] The socialists have never been closer to the fulfillment of their maximes; the leveling makes tremendous

**Fig. 9.4.** Ernst Mayr at the AMNH on 13 May 1991, reception on the occasion of the 60th anniversary of his joining the AMNH in January 1931 (AMNH Library photographic collection, negative no. 600653-9)

progress. [...] I cannot help standing in sorrow at the ruins of an incomparably more pleasant past of our class. [...] Above all one cannot get rid of the nervous question: Where will all this lead us?" (12 April 1934; transl.; for full letter see Haffer 1997b: 378).

When Mayr visited Germany a few months later, he witnessed the reality of the Nazi regime and replied:

"It would be nice here at home if the evil politics would not embitter one's life. I now understand everything you wrote and told me. Judging by the endless unhappiness and bitter rage which I observe wherever I talk to someone, riots will soon take place here" [which, however, the harsh measures of the police state prevented] (30 July 1934; transl.).

During the following years Mayr became ever more explicit in his denouncement of the Nazi regime in Germany.

## Health

Ernst Mayr enjoyed good health throughout most of his long adult life. As a young boy, besides all the childhood diseases, he was subject to allergies, e.g., against egg white (which caused the mucus glands in his mouth to swell) and mildew or house dust during vacations which occasionally caused severe asthma attacks. However, after he had reached the age of ten or eleven, he was hardly sick at all. As a youth he was not a great sports figure but active in soccer games, bicycling, skiing, sprinting, and volley ball. He did a lot of hiking and bicycling, mostly in connection with birdwatching. He also did some cross country skiing, when he would take the train up one of the valleys of the Erzgebirge (south of Dresden), ski across the ridge to the next valley, and then take the train from there back to Dresden in the evening. During the last 2 years in the gymnasium he joined the youth division of a local soccer club (Guts-Muts) and played quite a bit of soccer. His love for this game stayed with him until his old age, and he always followed the games of the German soccer leagues, the German soccer championship, and the international soccer results. When I visited him in September 2003, Mayr greeted me saying "We could watch today on television the women's' soccer world championship if you like." We did not do so and thereby missed witnessing the German women winning this championship. Mayr also had a general interest in American baseball, especially the teams in New York.

With his fatherly friend Gottfried Schiermann, he went to see several boxing fights in Berlin. Impressed by his robust health at that time Stresemann mentioned to Hartert that Mayr "has a very tough body as I noticed during joint field trips and practicing gymnastics. Are you able to swing your leg over the back of a chair 75 times?" (letter dated 21 September 1928). But, being excessively acrophobic as a young man, Mayr stayed home when his brothers went on high alpine tours; just crossing a rather steep slope was agony for him. For this reason and much to his regret he had to decline an invitation by Richard Archbold to accompany him, in 1932, on a trip to climb several mountains in the Alps, including the Matterhorn.

During his New Guinea expedition and in the Solomon Islands Mayr had his share of tropical diseases like bacterial dysentery, several malaria attacks (eventually knocked out with the help of Plasmochin, a new experimental drug), and dengue fever aggravated by a botched tooth extraction that had resulted in a jaw abscess. On that occasion he felt most miserable with high fever and much pain on the hot expedition boat. When in Java, he picked up smoking (letter to Stresemann dated 23 February 1928) and smoked cigarettes rather heavily for the next 10 years, but in 1939 broke this habit voluntarily (letter 31 July 1947). He was twice operated. First, shortly after his return from New Guinea in 1930, when his appendix got inflammated (the wound got infected and did not heal for 5 or 6 weeks) and, second, on 13 April 1934 when his left kidney had to be removed. The tumor was not malignant; however, the kidney was located so high that the doctors could not get at it properly, and had to remove it completely. At that time this was still a very dangerous operation. Soon after he had left the hospital Dr. Sanford, who all along had shown an almost fatherly interest in his well-being, took him to his home in New Haven, where he stayed for about a week, and then he placed him in a fishing lodge, the Trout Valley Farm, on the Beaverkill (East Branch of the Delaware River) in the Catskill Mountains in New York State. There Mayr convalesced for another two weeks before returning to his work at the American Museum. His health during the following years was fine except for a sinus problem beginning in 1940 as a result of working during spring time in the very dusty cattail marshes, in connection with his redwing studies.

In 1948, Mayr suffered a stress-related nervous heart condition expressing itself in irregular heart beat together with general weakness. He had to take a complete rest for about a month. In his letter to Erwin Stresemann dated 20 September 1948, he referred to "the opening of Sanford Hall [p. 129], as well as various other exciting events at the museum" together with his work as editor of *Evolution* and the post-War worries about relatives and friends in Germany as possible causes for this health problem. In a later letter to Stresemann he also implicated the difficulties with the U.S. Immigration Service in connection with their applications for citizenship (p. 251): "Perhaps Delacour can tell you [...] what chicaneries I have to go through. No wonder I have a nervous heart!" (27 March 1950). Mayr eventually overcame the condition, but it took more than 5 years. After 1950, he was again quite healthy and, in his ninety's, has been taking only a mild dose of blood pressure medicine and occasional aspirin to combat a beginning arthritis in his left foot but some heart fibrillations developed during the last years of his life. He only started using a hearing aid and a cane when he was 95 years old. His habit of extended daily walks in the afternoon may have been one reason why he enjoyed good health until very late in life.

He visited Germany in October 1997 to give the first "Ernst Mayr lecture" of the Berlin-Brandenburg Academy of Sciences and the Science College of Berlin on the philosophy of biology, and again in June 2001, when his Alma Mater, Humboldt University of Berlin, had invited him to celebrate the 75th anniversary of his PhD examination in June 1926. On that day he spoke only for approximately 15 min about his life, but delivered a public lecture on the autonomy of biology at the Museum of Natural History on 26 June 2001. The following winters he did not spend in Florida but stayed at home. He described his daily routine in a letter to his niece and nephew (Roswitha and Peter Kytzia) in Germany dated 15 January 2002 (transl.):

"At my age of 97, I could become a sluggard and enjoy life. But what good is this for the world? Therefore I work as long as I have the necessary energy and prepare contributions toward our understanding of the world. And the reward is that the daily work is good for one's health. Here I notice that many old people who do not force their brain into work become increasingly brain-indolent and they gradually dull. This does not happen to those who read interesting books, attempt still to fill a position, or occupy themselves otherwise intellectually. I could participate in numerous events here but I have enough to do with my own work.

I set my alarm clock every day at 6:30 and before getting dressed I have a simple breakfast (tea, cereal, grapefruit, yoghurt; no egg or bacon) and then I sit down immediately at my desk and, if we have a weather problem, turn on my television set (weather forecast). I work about 3 hours before I get dressed. I have an enormous correspondence, often 3 or 4 personal letters every day, mostly concerning scientific questions. This costs a lot of time but is very stimulating. At noon I prepare myself a simple lunch, either a bowl of soup and a slice of bread with sausage or cheese, or cottage cheese with applesauce. For dinner I get a full meal at our own restaurant. After dinner I mostly read (with or without the record player turned on) and around 9:30 or 10 o'clock I go to bed. Weather permitting I take a walk in the company of a friend between 3:30 and 4:30. On average once a week I drive to Harvard [where he maintained his office until 2003]. I shall not travel this winter. I have here everything I need and Christa and Susie visit me regularly to make sure that everything is in order.

Now you have some idea how I live. [...] Cordial greetings Your Uncle Ernst."

After his wife Gretel had passed away in August 1990, he managed his household alone but insisted: "No, I don't cook but I prepare my own meals. No fancy sauces or other aspects of the culinary art." He spent a number of winters in Gamboa (Panama) as a visiting scholar of the Smithsonian Tropical Research Institute. Here he kept track of the most recent developments in tropical biology and did a great deal of writing particularly for the books published after 1990. Until 2000 he lived, during the coldest months of the year, in Florida where he was a Distinguished Scholar of Rollins College in Winter Park near Orlando. Recently he stayed at his new home in Bedford, Massachusetts where he continued to be as active as he has always been, publishing more articles and books than most younger colleagues. He enjoyed occasional formal lectures and special seminars in which, from the lecture platform, he answered questions by students and fellow scientists and often enlarged upon the respective subject matter for several minutes. He experimented with such informal sessions for the first time during his 1959–60 visit to Australia. Therefore these (highly popular) presentations were referred to as his "Australian lectures." Such seminars usually lasting several hours became his specialty from which he derived much pleasure for himself irrespective whether he spoke in English in the United States or in German in his former home country. Despite the fact that he lived in North America for over 70 years his German was without any accent due to the fact that he and his wife maintained close connections with their families and many colleagues in Germany. His letters to me began in English or in German and very often switched over to the other language in the middle of a paragraph, but when we met in person we usually communicated in German (unless another English-speaking person was present).

On July 16, 1993 a symposium, organized by John Greene, took place in honor of Ernst Mayr at Brandeis University during the meeting of the International Society for the History, Philosophy and Social Studies of Biology with Mayr participating and commenting on each contribution. He considered it "an enormous privilege to be able to be here, alive and kicking, attending my own memorial meeting." The papers discussing his work were assembled in a Special Issue of the journal *Biology and Philosophy* and published in honor of Mayr's ninety's birthday in July 1994.

In the summer of 1997 Mayr moved to the Carleton-Willard Village, a comfortable retirement home in Bedford, Massachusetts, about 30 km northwest of Cambridge, where he lived in a two-room apartment and received almost daily friends, visitors, and science reporters for interviews. Until the summer of 2003, he was still driving to his office at Harvard at least once each week. After he had recovered from an illness his family then requested him to discontinue using his car. On 5th July 2004 he celebrated his 100th birthday in excellent physical and mental health at the farm surrounded by his immediate family and a few friends. During the weeks preceding and following his birthday he gave numerous interviews to science reporters. Many newspapers and journals carried stories about his life or published variously detailed appreciations of the man and his work. Even the President of the United States sent his birthday congratulations. Numerous authors around the world dedicated articles in technical journals to Mayr on the occasion of his birthday. He himself published an autobiographical sketch on "80 years of watching the evolutionary scenery" (2004b), and his book, *What makes Biology Unique?* (2004a) appeared several weeks later. There have been other scientists who reached an age of 100 years. However, none of them, it seems, presented his or her fans another book on this occasion. The Nuttall Ornithological Club had sponsored on behalf of Ernst Mayr a talk by Peter Grant (Princeton) on 3rd April. He spoke on his long-term studies of Darwin's Finches on the Galapagos Islands. A one-day symposium at Harvard's MCZ celebrated Mayr's 100th birthday on 10th May. Speakers included W. Bock, D. Futuyma, A. Knoll, L. Margulis, A. Meyer, I. Rubinoff, K. Shaw, F. Sulloway, and M. J. West-Eberhard. On 7 June, 2004 a half-day symposium took place in Vienna under the title "Ernst Mayr–Darwin of the 20th century." Speakers were Ilse Jahn, Erhard Oeser, Franz Wuketits, and Gerd Müller. The American Ornithologists' Union organized a symposium during its annual meeting in Quebec in August 2004 with the following speakers: W. Bock on "Ernst Mayr at 100—A life inside and outside of ornithology" (who also showed a video interview with Mayr made on 8 November 2003), M. R. Lein on "Ernst Mayr as a life-long naturalist," F. Vuilleumier on "Ernst Mayr's first and last interests–biogeography of birds," M. LeCroy on "Ernst Mayr as the Whitney-Rothschild Curator at the American Museum of Natural History," and R. Schodde on "Ernst Mayr and southwest Pacific birds: Inspiration for ideas on speciation" (see Ornithological Monograph 58, 2005). On that occasion, the Fellows of the American Ornithologists' Union joined "together to send their hearty congratulations to Ernst Mayr on this milestone and their best wishes for a continuing healthy and active life." After the Secretary of the AOU, his former student M. Ross Lein, had informed him of the passage of this motion, he replied as follows:

"There is no other organization with which, over the years, I have been as closely associated as the American Ornithologists' Union. Nothing therefore could have more delighted me than the greetings and congratulations I received from the AOU on the occasion of my 100th birthday. The society has passed through several periods of difficulties but is now so flourishing and productive that one is justified to be proud to be a member.

I deeply appreciate the resolution passed by the AOU at its recent annual meeting and am looking forward to continuing this happy association!"

Another festschrift was published by a group of philosophers of biology in the Mexican journal *Ludus Vitalis* (edited by F.J. Ayala) with contributions by 17 authors. Also the National Academy of Natural Sciences of the United States sponsored a colloquium on "Systematics and the Origin of Species: On Ernst Mayr's 100th Anniversary" which was held in his honor in Irvine, California December 16–18, 2004. Authors of seventeen papers discussed the current knowledge on speciation (Proc. Natl. Acad. Sci. USA 102, Suppl. 1, 2005).

The Bavarian town of Kempten, where Mayr was born, also remembered its famous son. The mayor congratulated him in a long letter mentioning plans for naming a local school after him and placing a plaque on the wall of the house now standing at the site where he was born on the corner of Kotterner and Scheiben Streets. The plaque was fixed on the wall in 2005 and now reminds passers-by of the Harvard professor from Kempten.

He enjoyed all those birthday honors in good spirits, although the incessant interviews eventually became somewhat bothersome. In late July 2004, he wrote me (trans.): "My head is barely visible above the mountain of birthday cards and letters. Among these are many from people totally unknown to me. And the interviews continue, this week again two (over the telephone). Under these circumstances, of course, I can do none of my own work!" This consisted mainly of the preparation of two manuscripts, first "Darwin sentence by sentence," a detailed commentary on the first edition of Darwin's *Origin of Species* (1859) and, second, on the topic of *Intelligent Design* against which Darwin already published convincing arguments. As he told me, two of his colleagues agreed to complete these manuscripts.

His scientific knowledge did not provide Ernst Mayr with the secret to his long life. "If I knew it, I could become very rich by selling this information!" he said and referred to several of his virtuous habits: He gave up smoking in his late 30s, took daily long walks, ate sensible diet, took vitamins, and kept his mind busy all the time with numerous projects and with new plans. The long periods he spent at his country place in New Hampshire and, in later years, his friendships with younger people may explain in part the riddle of his long life. According to a press report fame alone, independently of wealth, may give a life-extending boost and also may have been a contributing factor.

In the fall of 2004 he was diagnosed with secondary cancerous lesions of his liver (the primary site was never discovered, but Mayr believed it to be a stomach cancer) which developed rather slowly. However, he sensed that he would not live another full year. An intestinal infection was cured with the help of antibiotics in early December, but left him so weakened that he was unable to regain his previous strength. He had been transferred to the full care center of his retirement home in early December and he hoped to return to his apartment still doing physical therapy in late December, when Walter Bock visited him. But he never left the full care center, where he passed away quietly and peacefully on February 3rd,

2005. Memorial services celebrating the life and contributions of Professor Mayr were held at the Carleton-Willard Village on 26 February and in the Memorial Church of Harvard University on 29 April 2005. Speakers at the latter service were the following: Thomas M. Ferrick, Humanist Chaplain of the United Ministry at Harvard and Radcliffe, said he met Ernst Mayr about 15 years ago, when Ernst wanted to discuss in detail what his memorial service at Harvard would be and who would speak. Most or all of the five speakers were chosen on that occasion by E. Mayr himself. In his presentation Ferrick read a "text," a passage from Ernst's book *What Evolution Is* (2001), which is a very strong statement of the "fact of evolution" (the final paragraph of chapter 11 on page 264). Ferrick was followed by James Hanken, director of the MCZ, who mentioned that the Accademia dei Lincei (Rome, Italy) had sent a gold medal honoring Mayr's 100th birthday. He had informed him about it in December but because of Mayr's final illness and the snow storms in January a ceremony could not be arranged before he died. Hanken had the medal with him, showed it to the audience and presented it to the family after the service. The other speakers were Frank Sulloway, Edward O. Wilson, Walter Bock, and Jared Diamond who related Ernst Mayr's influence on their personal careers.

"On 3 July 2005, a pleasant sunny summer day with birds singing and a woodpecker drumming in the background, [Walter Bock] joined Ernst's two daughters and their spouses, his five grandchildren and their spouses, and his ten great-grandchildren to scatter his ashes along a path overlooking the lake at The Farm. The ashes of his wife Gretel had been scattered there in 1990. This is the place that Ernst and Gretel loved the best in their beloved Farm" (Bock 2005).

## The Farm near Wilton, New Hampshire

As soon as they had arrived at Cambridge in 1953 the Mayrs searched for a country place in a hilly area in about 60 miles distance. They found a place for sale near Wilton, New Hampshire region, an abandoned farm on a dirt road with about 160 acres of land bordering Burton Pond and they purchased it, even though the house was in very poor condition. It took a lot of time and work by the family and friends to make it acceptable (Fig. 9.5). However, the previous owner had left them something invaluable, a marvelous old kitchen stove and an equally beautiful old cast-iron Franklin stove both of which, to this day, render excellent service.

The main attraction of this area for the Mayr family was its solitude and wildness. Nearly all of "The Farm," as everybody continued to call it, was covered with unbroken broadleaf woods (oak, beech, four species of birch, maple, ash) and some conifers mixed in. The nearest neighbors were more than half a mile away in one direction and more than two miles in the other direction, when they purchased The Farm. They had Burton Pond virtually to themselves with its beavers, otters, and ducks. About 40 acres of pastures and haying meadows were kept mowed by a neighbor who could use the grass for his animals. The postal address is 310 Collins Road, Lyndeborough.

Rabbits and bobcats, common at first, later disappeared completely. Porcupines had to be decimated to save some pear trees and woodchucks frequently got into the vegetable garden despite various defensive devices. For some time milk snakes (*Lampropeltis*) lived in the house and helped to keep the mouse population down. Even moose, coyotes, and fishers have been observed. The bird boxes around the farm house are occupied by tree swallows, bluebirds or house wrens. Besides many other birds, American Woodcocks give their springtime displays at night.

Scientists from many different countries enjoyed the wildlife as guests, e.g., Konrad Lorenz, Niko Tinbergen, Erwin Stresemann, Karl von Frisch, Julian Huxley, E. B. Ford, Paaovo Suomalainen from Finland, Salím Ali from India and Loke Wan Tho from Singapore. The German ornithologist Hans Löhrl (1960–1961) observed the red-breasted nuthatch in the woods and took one brood with him back home for a comparative study in his aviaries between this species and the Corsican Nuthatch (*Sitta whiteheadi*). Ernst Mayr also studied the flora of this region and showed animals and plants on the farm to his graduate students. One of them, F. Vuilleumier, gave the following account how he (and certainly other visitors as well) was introduced to the American Woodcocks:

They "were displaying in full swing. One bird in particular would display wonderfully and then land in the grass under some trees, and with nearly geographic precision, almost always reached the same spot. Ernst and I crept closer and closer, waiting for the bird to take off, then moving a few feet toward the spot, and then waiting for the bird to land. Again and again we did this, until we arrived to just a few feet away from the landing spot. Sure enough, the woodcock landed where we

**Fig. 9.5a.** "The Farm" near Wilton, New Hampshire 2003 (phot. J. Haffer)

**Fig. 9.5b.** Ernst Mayr at The Farm, 20 April 1958 (courtesy of Mrs. S. Harrison)

had predicted. It saw us, looked sideways at us, not knowing quite what to do next. Neither Ernst nor I moved a muscle, our eyes fixed upon the bird, whose breathing we could see very well. After perhaps one minute, the bird relaxed, walked in the grass, then, presumably as his raging hormones told him to get going, up he went for another display. We left the magic spot on tiptoes while he was in the air, went back to the house and glowed with the memory of that special shared observation" (in Lein 2005: 25).

Ernst and Gretel Mayr usually spent the summer months and most weekends from spring to late fall at the farm. The entire family always gathered there for Thanksgiving and very often for birthday parties and on other occasions. A surprise birthday party for Ernst's 75th birthday was held there with most of his former students attending (see Figure 3 in Bock 1994a). Part of every day at the farm

**Fig. 9.5c.** Celebrating Ernst's 75th birthday at The Farm on 5 July 1979 (courtesy of Mrs. S. Harrison)

was devoted to birdwatching and other natural history studies. One little item published (1976b) was the feeding of birds on egg shells and the long distances they flew to use them. Virtually all the books and articles published since the mid-1950s were written largely there, in particular *Animal Species and Evolution* (1963b) and *The Growth of Biological Thought* (1982d). Every ten days or so Mayr went to Cambridge and filled several cardboard boxes with the necessary new books and reprints leaving the previous lot at home. The windows of a cozy corner room where he worked open to meadows and woods. One wall is decorated with a beautiful etching of the Castle of Moritzburg and another one of a view over Dresden where he spent the formative years of his youth.

As director of the MCZ (1961–1970) he organized each fall a museum party at the Farm, when the apple season was at its height. The Mayrs served coffee and applesauce cake and everybody filled their picnic baskets with as many apples as they wanted before they returned home. Two of the graduate students, John Smith and Ross Lein, did the fieldwork for their PhD dissertations at the farm and lived with the Mayrs for several summers around 1960 and 1970, respectively. The farm was very important in the life of the Mayr family. He spent there long periods each year, even when in his nineties and was convinced that the outdoor work and frequent hikes greatly contributed to the maintenance and improvement of his health into an advanced age. Moreover, the Farm enabled him to concentrate on his writings for long periods unlike any other place.

## Residence in Cambridge, Massachusetts

During the first 6 years of their residence in Cambridge, Massachusetts the Mayrs rented an apartment, at first at 33 Washington Avenue and, from November 1955 onward, next door at 21A Washington Avenue, where they had an extra room and he had a study. In 1959, they moved into their spacious home at 11 Chauncy Street, the home with the yellow front and back doors ("Gretel liked them this way," Ernst said, Fig. 9.6), within a short walking distance of the Museum of Comparative Zoology. Here the Mayrs entertained many visitors from other parts of the United States and from Europe, especially Germany; e.g., J. Huxley, E. B. Ford, J. M. Marshall, E. Stresemann, B. Rensch, K. Lorenz, H. Löhrl and G. Kramer.

The elder daughter, Christa, had finished high school in Tenafly and after 2 years at Swarthmore College, spent a year at the University of Munich, Germany, attending classes in biology. In 1956, she married a German co-student, Gerhart Menzel, who became a nuclear engineer. They continued living in Germany for several years until moving to the United States. Once their three children were in school Christa obtained a bachelor's degree at the University of Hartford (Connecticut) and was hired as an adjunct instructor by the Department of Biology to teach laboratory sections for many years. The younger daughter, Susanne, graduated from Boston University, became a special education teacher at Concord Carlisle High School in Massachusetts. She also married and has two children. Both daughters

**Fig. 9.6.** 11 Chauncy Street, home of the Mayr family in Cambridge, Massachusetts (photograph courtesy of Mrs. S. Harrison)

are retired now. Their families have grown and Ernst Mayr became the patriarch of a large family clan including ten great-grandchildren.

Ernst Mayr encouraged his wife Gretel as much as possible to join him in some scientific projects, but with teen-age daughters in the house and the commuting problem, this was difficult as long as they lived in Tenafly. When they moved to Cambridge in 1953, she accepted some full time volunteer work in Boston and later worked many hours for the blind, reading to them, etc. She also helped recently released patients from mental hospitals to recover. She found this more rewarding than doing clerical work in conjunction with his projects. She did collaborate in two of his studies, (1) an investigation of tail molt in owls (1954g) and (2) in a study of the suspected correlation of blood-group frequency in humans and pituitary adenoma (1956i). For this latter project Gretel visited several hospitals in the Boston area and in New York to gather statistical data.

## Adviser to the National Academy of Sciences and the National Science Foundation

After he had become a member of the National Academy of Sciences in 1954, Mayr was assigned to the Biology Council of the National Research Council and asked to prepare a report on the scientific importance of museum collections (see p. 321, 1955g). In 1956–1957 he served as chairman of an ad hoc Committee for Systematic Biology of the American Institute of Biological Sciences which presented a report (1957i) on the needs of systematic biology and discussed taxonomic research and the role of collections in systematic biology as well as various problems in this field like, e.g., centralization of collections and research, under- and over-developed areas, publication problems, etc., and it made various recommendations to the National Science Foundation. The Biology Council agreed "to study trends of biological research and sociological factors affecting it" (Appel 2000: 125). At that time, Paul Weiss of the Rockefeller Institute was chairman of the Biology Council. He was its intellectual father and, as Mayr remembered:

"He considered it a think-tank charged with generating ideas and concepts in biology. Weiss felt, and he was quite right about it, that most biologists were too much concerned with fact-producing researches and gave too little thought to the underlying concepts in the field of biology and to the organization of biology. This he thought was one of the reasons for the neglect of biology in comparison to the physical sciences.

For me the sessions of the Biology Council—as I remember they took place about once every month—were of the utmost fascination. It was by serving on this body that I became friendly with Ralph Gerard, David Goddard, Harry Harlow (the monkey psychologist), Caryl Haskins, Francis Schmitt, Ed Tatum, the horticulturist Harold Tukey, and the bacteriologist Perry Wilson. The meetings of the Biology Council were very informal except that Weiss was a very dictatorial chairman, something that more amused than annoyed us. To those who did not know Weiss too well he was rather pompous and opinionated, but he actually was a very kind person and very much concerned in the integration of biology. He felt strongly that the old divisions into zoology, botany, and psychology and bio-chemistry, etc., had lost most of their usefulness; and he was in part responsible for the new classification of the subdivisions of biology subsequently adopted by the National Science Foundation."

A conference, "Concepts of Biology" organized by the Biology Council, was held at Lee, Massachusetts, in October 1955. Paul Weiss wrote in the "Foreword" to the conference proceedings (edited by R. W. Gerard): "Concepts are the structural elements of the growing body of knowledge. [...] The following conference report centers on the question of whether present-day biology is paying too little attention to its conceptual maturation and, if so, why." Ernst Mayr's active participation in the discussions during this conference is documented in the condensed and transcribed proceedings (Mayr 1958n). Here he spoke about alpha, beta, and gamma taxonomy (already mentioned in Mayr et al. 1953a: 19), on proximate and ultimate causes in almost any natural phenomenon, and about the increasing significance of time (history) as one moves up the levels of integration. A species, its adaptation, its structure, its behavior are understandable almost entirely in terms of its evolutionary history. He emphasized also that all biological phenomena are basically population phenomena, whereas the older philosophies were all typological, going back to Plato's eidos, the proper thing: "I believe that this gradual displacement of typological thinking by population thinking in the last [19th] century has been one of the most profound intellectual and conceptual revolutions in biology" (see here p. 362). The Biology Council was active between January 1954 and October 1956, when it held its last meeting. It formally suspended operations in June 1957 (Appel 2000: 127).

The 1960s were the period when molecular biology achieved primacy among the biological sciences, with numerous consequences. Two of them were particularly troublesome: The rather low support of organismic biology, particularly compared to the lavish support of the physical sciences, and the relatively low ranking of biological disciplines. But owing to the drastic changes in biology in the preceding 40–50 years nobody any longer had a good grasp what biology really was. For instance, the biochemists located in chemistry departments, were usually extreme Cartesians and not open for organismic biology.

At this point biochemist Philip Handler (1917–1981) of the National Academy of Sciences in Washington and Chairman of the Committee on Research in the Life Sciences, decided to undertake a survey of the status of Biology among the sciences. This Committee was appointed in 1966. It had a very difficult assignment owing to the heterogeneity of life sciences. In order to have complete coverage 22 panels were established, each responsible for a particular field of biology. The original list of panels was developed by cellular-molecular biologists and had four panels for molecular biology but was badly deficient for organismic biology. Mayr fought hard for the addition of new panels dealing with the biology of whole organisms and was in part successful. The creation of a panel on biodiversity was a special victory for him.

Each panel wrote its own report, the one on biodiversity was jointly composed by Ernst Mayr and the botanist William Steere (1907–1988), director of the New York Botanical Garden (Mayr 1970h). The other 4 members made minor contributions. Mayr was one of the 29 members of the whole committee and took part in all of its joint meetings and editorial sessions. The 4 years during which the 526-page report[2] was composed (1967–1970) required a great deal of hard and time consuming work and travel. Mayr scrutinized the reports of all the other panels and eliminated or corrected misleading statements about organismic biology made by the biochemists and other non-biologists.

Philip Handler picked Mayr as his major consultant on organismic biology and it was for these services to the National Academy that Handler nominated him for the National Medal of Science in 1969. This was only one of the committees Mayr served over the years. For instance, he was a member, and later chairman, of the Divisional Committee of the National Science Foundation (NSF), after many years' work on several review panels of the NSF. Although himself a biochemist, Handler recognized the importance of organismic biology and always supported Mayr when it came to doing things for systematics or evolutionary biology. As Mayr recalled: "He was a brilliant person, a wonderful diplomat, and a reliable, honest human being. That he had to die so young from lymphoma was a great loss for American science and for me personally."

## National and International Recognition

Due to the steady stream of his taxonomic publications on South Sea island birds during the 1930s Ernst Mayr became known worldwide among ornithologists. This reputation was reinforced by his books on the birds of the southwestern Pacific and the Philippines. Many societies elected him to honorary membership beginning with the German Ornithological Society in 1941.

The Academy of Natural Sciences in Philadelphia awarded him the Leidy medal in 1946 for his important book, *Systematics and the Origin of Species* (1942e). Recognition in evolutionary circles outside the United States followed, when this book became more widely known after 1945, when World War II had ended. Scandinavian zoologists in particular recognized, at an early stage, Mayr's contribution to the evolutionary synthesis by his clarification of the concepts of species and speciation, i.e., the origin of organic diversity.

During an extended visit to Europe in the summer of 1951, Mayr was invited to and lectured at various universities in Italy, Germany, Switzerland, Denmark and France. He was elected a Honorary Member of the Royal Academy of Sciences, Uppsala (Sweden) in 1955 and, at the celebration of the 250th birthday of Linnaeus in 1957, was awarded his first honorary degree by the University of Uppsala together

[2] *The Life Sciences. The World of Biological Research*. National Academy of Sciences, Washington, D.C. 1970. Philip Handler (ed.) *Biology and the Future of Man*. Oxford, Oxford University Press, 1970; also in *The Environmental Challenge*, New York, Holt, Rinehart and Winston, p. 20–49, 1974.

with a ring he wore ever since (and which he instructed his daughters to present after his death to Walter Bock [see Bock 2005]).

"Although the accent of the symposium to which I had been invited was undoubtedly on botany, zoology was also represented with Rensch, Julian Huxley, myself and perhaps a few others. The award ceremony was a most impressive occasion. Every time a recipient received his degree a canon on the roof of the building gave a salute. The Laurel wreath that was placed on one's head supposedly came from a tree in their greenhouses that had been planted by Linnaeus. After one had thus been decorated one moved over to the King of Sweden and had to bow. I still remember how loosely the wreath was sitting on my head, and how afraid I was it might fall off me if I made a proper bow. As a result my bow was more of a courtesy because I didn't dare to let the wreath fall. The occasion of course was combined with visits to Linnaeus' home and his garden, and all sorts of other interesting features of the surroundings of Uppsala."

Mayr repeatedly lectured in Denmark and in Finland (e.g., in 1957). Official recognition by evolutionary and zoological circles in Germany was somewhat delayed (honorary degree by the University of Munich 1968, Honorary Member of the Zoological Society 1970, member of the Deutsche Akademie der Naturforscher Leopoldina 1972, Mendel Medal 1980, Alexander von Humboldt Stiftung Medal 1989, honorary degrees of the universities of Konstanz 1994 and Berlin 2000). Mayr attributed this delay to the absence in Germany of an interest in species-level evolution and in mechanisms of evolution, which continues even today.

Mayr's reputation grew steadily supported by his well-known publications in the fields of ornithology, evolutionary biology, history and philosophy of biology. During the Darwin year 1959, he was invited to the famous Chicago meeting, the results of which were edited by S. Tax in three volumes. Immediately after the end of that meeting Mayr went via Berkeley to Australia where he gave the major lecture at a Darwin celebration of the Royal Society of Melbourne, which at the same time celebrated its 100th anniversary and honored him by a degree. Yale University awarded him the Verrill Medal in 1966. In 1969 six scientists received the National Medal of Science, the highest U.S. scientific award, one of them Ernst Mayr "for notable contributions to systematics, biogeography, and especially the evolution of animal populations." This medal was created in 1959, when the nation strongly reacted to the Sputnik, Russia's first space flight. Numerous signs of appreciation came to Mayr from universities and scientific societies in the United States, Great Britain, France, Sweden, Germany, Canada, Australia, Austria, and Italy totaling 17 honorary degrees, 35 medals and other special awards and 52 honorary memberships (see Appendix 1 for details).

Mayr achieved distinction in five different fields: (1) *ornithology* and *zoogeography*: Brewster Medal (1965), Honorary degree Oxford (1966), Centennial Medal, AMNH (1969), Coues Medal (1977), Eisenmann Medal (1983), Salvin-Godman Medal (1994), Honorary degree Vienna (1994); (2) *Systematics*: Leidy Medal (1946), Honorary degrees Uppsala (1957), Yale (1959), Amherst (1993), Paris museum (1997), Berlin (2000); (3) *Evolutionary biology*: Honorary degrees from the universities of Melbourne (1959), Munich (1968), Paris (1974), Harvard

(1979), Cambridge (1982), Vermont (1984), Bologna (1995), Wallace-Darwin Medal (1958), Balzan Prize (1983), Darwin Medal (1984), Royal Society (foreign member 1988), Japan Prize (1994), Crafoord Prize (1999). (4) *History of biology*: Sarton Medal (1986), Phi Beta Kappa Prize (1991); (5) *Philosophy of biology*: Honorary Fellow, Center for the Philosophy of Science, Pittsburgh (1993), Honorary degrees from Guelph university (1982), University of Konstanz (1994), and Rollins College (1996), Benjamin Franklin Medal (1995), Ernst Mayr Lectureship, Berlin (1997).

In the 1970s, the French historian of science Jacques Roger came to the United States with a film team and made a documentary of Mayr's ideas by interviewing him both in Cambridge and at his country place in New Hampshire. Altogether, they took enough film for a 5-hour presentation, but of course for the commercial film, which eventually was shown in a good many television stations in various countries, they were able to use only 52 min.

"In the late 1980s, I was elected to the Académie des Sciences, Paris. Apparently my friends in zoology, particularly Bocquet, Teissier, L'Heritier, etc. had been trying to get me elected but the election had to be initiated by a nomination made by one of the Sections of the Academie. In my case, it was the Section of Zoology that should have made the nomination but, believe it or not, that Section even today is dominated by Lamarckians and they naturally refused to nominate a Darwinian. Finally, someone in the Section of Molecular Biology—I have always suspected it was François Jacob but have no definite information—suggested to that section that they should nominate me. Indeed, they agreed, and as soon as I was nominated I was elected. Now I am listed in the membership list under 'Molecular Biology.'"

The Rockefeller University in New York awarded Mayr the Lewis Thomas Prize "Honoring the Scientist as Poet" (1998). The award certificate reads in part: "Dr. Mayr has proved himself an epic chronicler of the evolution of human understanding of the living world and an original and forceful philosophical proponent of the unique intellectual character of the life sciences." The university spelled out the principles governing the selection of an awardee as follows:

"Throughout history, scientists and poets have sought to unveil the secrets of the natural world. Their methods vary: scientists use tools of rational analysis to slake their compelling thirst for knowledge; poets delve below the surface of language, and deliver urgent communiqués from its depths. The Lewis Thomas Prize honors the rare individual who is fluent in the dialects of both realms—and who succeeds in spinning lush literary and philosophical tapestries from the silken threads of scientific and natural phenomena—providing not merely new information but cause for reflection, even revelation."

The Lewis Thomas Prize was established in 1993 by the trustees of The Rockefeller University and presented to Lewis Thomas, its first recipient, that year. Other recipients have been François Jacob (1994), Abraham Pais (1995), Freeman Dyson (1996), Max Perutz (1997), Steven Weinberg (1999), Edward O. Wilson (2000), Oliver Sacks (2001), and Jared Diamond (2002).

Nobel Prizes (Stockholm, Sweden) are awarded only to scientists in the fields of physics, chemistry, and medicine. Entire branches of the natural sciences are excluded, in particular the historical sciences like evolutionary biology dealing with conceptual advances such as the explanations of the origin of organic diversity or of the mechanisms of evolution using the methods of observation and comparison.

As Mayr stated: "If there were a Nobel Prize for biology—which there isn't, because Nobel was an engineer and too ignorant of biology!—if there were such an award, Darwin could never have received it for evolution through natural selection, because that was a concept, not a discovery!"

Partly to remedy the narrow focus of the Nobel Prize, several international prizes have been established and are awarded to scientists at large, including those in fields not considered by the Nobel Committee. These are the Balzan Prize, Japan's International Prize for Biology, and the Crafoord Prize, which are considered equivalents to the Nobel Prize. Ernst Mayr got all three, the "Triple Crown of Biology"!

The Swiss-Italian Balzan Foundation (Milan, Italy) was established in 1956 by Angela Lina Balzan to honor her father, Eugenio Balzan, the former head of Italy's leading newspaper. Annually, up to three researchers, poets, or other persons are each awarded 250,000 Swiss Francs for extraordinary achievements. Previous recipients included Mother Theresa and writer Jorge Louis Borges when, in 1983, this prize was designated for the first time in zoology and awarded to "the greatest living evolutionary biologist" Ernst Mayr (Gould 1984; Markl 1984). Mayr also won the 1994 International Prize for Biology ("Japan Prize") as "the outstanding systematist in the world." The prize is awarded by the Society for the Promotion of Science in Japan and consists of a diploma, a medal and 10 million yen, or about US $100,000. These were presented on 28 November 1994 at a ceremony in Tokyo attended by Emperor Akihito and Empress Michiko.

Mayr's daughter, Mrs. Christa Menzel, and Professor Walter Bock (New York) accompanied him on the trip to Japan, which he enjoyed very much. He wrote a detailed report for his friends from which the following excerpts are taken. They flew first-class and nonstop from New York to Tokyo at the invitation of the awarding society on 21/22 November 1994. After a day's rest, they traveled by a bullet train to Kyoto where a scientific discussion meeting took place and a birdwatching trip to a large freshwater lake was scheduled on the following day. Mayr observed several species here that he had never seen alive before. The party took the train back to Tokyo on the morning of November 27, again with a beautiful view of Fujiyama.

"On Monday the 28th was the ceremony of the presentation of the Prize. Particularly interesting and in part amusing was the formalized ritual of the actual award ceremony. This included being formally introduced to the Ministers and to the Imperial Majesties. The Ministers were the Deputy Prime Minister and the Minister of Education, obviously two very high ranking Japanese politicians. Every step and movement of the ensuing ceremony was carefully prescribed. The Chairman of the rewarding society and the Chairman of the Selection Committee both gave detailed reports why I was chosen for the award. After I had received two scrolls dealing with the award and some envelopes containing information, which I had turned over to Christa, came the next step, which was the delivery of the Imperial gifts. These were really quite fabulous, a beautiful silver vase with a chrysanthemum ornament, and a medal commemorating my receipt of the Prize. This was followed by a message from the Emperor, and subsequently by messages from the two Ministers. This was followed by a further congratulatory message delivered by Professor Peter Raven, a frequent visitor of Japan and director of the Missouri Botanical Garden. The moment had now come where I had to step forward, and deliver a five minute acceptance speech. After the applause it was announced that the Imperial Majesties would be leaving and everybody got up and bowed to them as they left the hall. After them the Ministers left the hall and then I, together with Christa, departed, followed by various of our Japanese associates.

This was followed by a reception in a special room in the basement where little tidbits and alcoholic as well as soft drinks were offered. There we were soon joined by the Ministers and the Imperial Majesties. The Emperor engaged me in a lengthy conversation and the Empress did likewise with Christa. During my conversation with the Emperor, at least six or seven television cameras were directed at us and our conversation was recorded for television and the press. We then were taken by our hosts back to our hotel. These trips in Tokyo, although relatively short in distance, always take a long time by limousine because the traffic jams are unbelievable. There is nothing one encounters in NewYork or Boston that corresponds to the jams in Tokyo streets. In the evening there was a dinner party in my honor where we were sitting at a regular table and could either use chopsticks or European utensils. I preferred the latter. There were some excellent dishes served, and at the end the President of the Society for the Promotion of Science (Fujita) gave a little speech and I followed with a thank you address in which I praised the objectives of the Society.

On the 29th a special symposium was held to accompany the award ceremonies. I gave a 40-minute introductory lecture and Walter Bock was asked to introduce me before I began speaking. I left after the first few lectures, because it would have been too tiring to stay through the entire symposium.

In the evening an event took place which was perhaps the most memorable and unique of everything we experienced. Christa and I were invited to a private dinner in the Imperial Palace which even impressed the Japanese. On previous occasions, I was told, the Emperor had often invited the awardee for tea, but this, apparently, was the first occasion where he had invited the awardee for dinner. Professors Raven and Bock were asked to join us. We were carefully instructed what to do and what not to do, and finally the limousine took us to the Imperial compound. We had to pass through I would think about three different gates, each time with a careful check including showing our passports. Finally, we were taken to the reception room without being asked to take our shoes off. Shortly after we had been seated in the reception room the Imperial Majesties appeared and we had a glass of sherry and quite a lively conversation. After a short time we were asked to move to the adjacent dining room where a table was set for six people. The Emperor and the Empress, myself and Christa, and Prof. Bock as well as Prof. Raven, the botanist from Missouri. The seating was very carefully arranged with the Emperor and the Empress in the middle seat (three people sitting at each of the two long sides of the table, none at the end of the table) and the serving was also carefully arranged. Invariably, I was the first person to be served, Christa the second, Prof. Bock third, Prof. Raven fourth, the Emperor fifth, and the Empress the sixth. I had the honor of sitting to the right of the Empress and the Emperor sat slightly diagonally across from me. The meal was, unfortunately, typically western, but was superb, an unforgettable experience. Surprisingly, we seemed to have no difficulties keeping a lively conversation going. There was no doubt that the Emperor and the Empress enjoyed themselves. She actually spoke better English than he, but the Emperor had a larger grasp of vocabulary. The Emperor is a zoologist and a specialist of a particular group of fishes (Gobiidae). It is perhaps the fact that I was the first zoologist to receive the Japan Prize that made him so liberal with his invitation for dinner. At the end he came up with some reprints of his and very proudly pointed out that in one of them he had cited some of my work. There was no doubt he was fully aware of who I was. After dinner, we finally said thank you and goodbye, but the Emperor and the Empress insisted on taking us to the door, in fact, accompanying us all the way to our limousine. It was an incredible honor to be guided by the Imperial Majesties to our car. As the cars drove away, the Emperor and Empress waved goodbye to us."

On the following day, after an interview and a visit to an art museum, another farewell dinner was scheduled for the evening, which Mayr did not attend, because he felt too tired. On the way to the airport on the 29th of November the group visited the Yamashina Ornithological Institute with which Mayr had had many contacts during the lifetime of its founder, the well-known ornithologist Prince Y. Yamashina. The return flight to New York was again quite comfortable.

On 21 September 1999 Ernst Mayr received (jointly with J. Maynard Smith and George C. Williams) the Crafoord Prize for biology from the Royal Swedish Academy of Sciences, Stockholm, in recognition of "their fundamental contributions to the conceptual development of evolutionary biology." The information leaflet continued:

"Ernst Mayr is a leading figure in the creation of the modern version of the theory of evolution now termed 'the modern synthesis.' Here knowledge gained from genetic, systematic, paleontological and ecological research has been integrated to form a coherent theory of evolution. Mayr has contributed to many research fields in this process but is perhaps best known for refining the biological species concept, thus clarifying what is meant by a species and the conditions under which it may be formed."

The Crafoord Prize consists of a gold medal for each laureate and US $500,000 to be shared. Mayr was accompanied to Stockholm by both of his daughters and their husbands, as well as the Kytzias, his niece, and her husband from Germany.

Eventually Mayr became known as the "Darwin of the 20th century", which alluded to the magnitude of his contributions to evolutionary biology but did not mean that his stature and achievements exactly equaled Darwin's. Mayr separated the phenomena of phyletic evolution and the splitting of lineages, he recognized the "species problem," discussed the role of species in nature and solved the origin of species, the riddle of biodiversity, about which Darwin had been rather confused.

Similarities in Darwin's and Mayr's careers include that both had become young naturalists early in their lives, both had started medical school and later returned to their earlier interests, both had gathered field experience in the tropics during expeditions at the beginning of their careers which deeply influenced their later taxonomic work (on barnacles and birds, respectively), and both arrived at broad biological generalizations in their later theoretical work.

In view of his workload, Mayr also had to decline several honors intended for him: When he was vice president of the International Zoological Congress in London (1958), he was offered the presidency of the next of these congresses in Washington (1962). But, by that time he would be the director of his institution (MCZ) at Harvard and, moreover, he was deeply involved in writing his book, *Animal Species and Evolution* (1963). Other positions that he had to decline included the presidency of the Zoological Society of America, and the presidency of the American Academy of Arts and Sciences of Boston, both in the 1960s, and the directorship of a Max Planck Institute of evolutionary biology in Germany (1959). He explained his decision in a long letter to E. Stresemann:

"I don't think I am a particularly good experimentalist. My strength has always been that of critical integration. I don't know how I would have worked out as head of an institute with a number of assistants, etc. And then there are purely personal reasons. You may not realize it, but I have had an unusual series of duties and commitments for the last 25 years which has made it impossible all these years to do expeditions and extensive foreign travels. I have just now reached the point to consider this and will start with a trip to Australia next winter. The idea to plan and equip an institute makes me shudder" (27 May 1959).

To explain his decision further, he attached a copy of the report of his activities from mid-1958 to mid-1959, commenting that this "gives you a little idea of my numerous ties in this country!" The report listed a series of publications, "works in progress" and "in preparation," and seven lectures he gave at conferences and institutions (American Ornithologists' Union; University of Hawaii; Surgical Research Conference; British Ornithologists' Union; British Broadcasting Corporation, London; American Philosophical Society; Cold Spring Harbor, New York). At the same time, he declined five position offers and 12 offers to participate in various symposia, congresses, panel discussions, etc.

# 10 Systematics and Classification

The systematist is interested in zoological uniqueness, in whole organisms, and systems. He determines the characters of taxa, their variation, and the biological causes for differences or shared characters. Classification, the delimitation, ordering and ranking of taxa, makes the organic diversity accessible to systematics at large and to other biological disciplines. The specific method of the systematist is comparison, not the experiment (Mayr 1969b).

## Diversity

The diversity of the organic world presents the systematist with numerous challenges at different levels of integration (Mayr 1974c). Inventory taking, at whatever level, is the first step and in taxonomy this means the delimitation of species taxa. The origin of diversity refers to the development of new genotypes (through mutation, recombination, etc.), new populations (through geographical variation, reduction of gene flow, etc.), new species and genera. Macroevolution is nothing but an extension of microevolution at the level of subspecies and species and a response to a previously vacant ecological zone. Real advances during the course of evolution were the widespread adoption of sexuality and multicellularity, and in plants vascularity and angiospermy. Each was a major step that provided an entirely new platform for the development of more evolutionary diversity. It is the study of diversity which has slowly undermined essentialism which dominated the study of nature for centuries.

The great contributions made by systematists during the 19th century include population thinking (which developed from the recognition of individuality), the theory of geographic speciation and the biological species concept which were incorporated into biology anonymously step by step and in such a gradual way that systematics did not receive due credit (Mayr 1968b). Again systematists played a decisive role in the development of the synthetic theory of evolution. Systematics is one of the cornerstones of biology. Their practitioners investigated the role of geographical isolation and of isolating mechanisms, the rates and trends of evolution, the emergence of evolutionary novelties, polytypic species, population structure and the never ending variety of solutions found by organisms to cope with similar challenges of the environment. Mayr (l.c.) saw no conflict between molecular biology and organismic biology (including systematics): "Each level of integration poses its own specific problems, requires its own methods and techniques, and develops its own theoretical framework and generalizations."

He felt optimistic about the future of systematics, because environmental biology, behavioral biology and even molecular biology all pay increasing attention to organic diversity. In 1983, when the American Ornithological Society celebrated its hundredth jubilee, molecular biology was at the height of its prestige. In his invitation lecture Mayr (1984a) tried to emphasize that other branches of biology, for instance ornithology, also had contributed to our understanding of the living world, e.g., systematics, evolutionary biology, speciation, evolutionary morphology, biogeography, ecology, population biology, the study of behavior, physiology, migration and conservation. In another article Mayr (1989c) listed specifically all the concepts that ornithology has contributed to biology and showed which particular ideas and processes were first recognized and explained by ornithologists. He also emphasized repeatedly the importance of museum collections and the need for their continuing financial support (1955g).[1] Having been published by the National Research Council this paper had a certain amount of authority and was later frequently quoted when support was asked from the National Science Foundation. A second objective of this report was to recommend more up-to-date methods, since Mayr was quite dissatisfied with the curatorial practices in many museums and herbaria.

## Classification

Ernst Mayr's most general and comprehensive contribution to the fields of systematics and classification is his textbook which appeared in three versions. The first was titled *Methods and Principles of Systematic Zoology* (1953a, with E.G. Linsley and R.L. Usinger); the second version was published under the title *Systematic Zoology* (1969b) by Mayr as sole author and the third version (1991a) by Mayr and P. Ashlock (the second edition of *Systematic Zoology*). While Mayr was working on the manuscript of this textbook around 1950, Linsley and Usinger in California, preparing a similar text, heard of his project. They agreed on coauthorship and the volume was widely adopted. This book, about half of which written by Mayr, dealt comprehensively with the fundamentals of both the theory and practice of taxonomy. He emphasized here for the first time in print the "population concept" ("population thinking," Mayr 1959b):

"The taxonomist is no longer satisfied to possess types and duplicates; he attempts to collect large series at each of many localities throughout the range of a species. Subsequently, he evaluates this material with the methods of population analysis and statistics. This type of study was commenced almost simultaneously by ornithologists, entomologists and malacologists in the second half of the nineteenth century. The new systematics has brought recognition of the true role of taxonomy and placed it at the very heart of modern biology" (1953a, p. 11, 15–16).

---

[1] As to the co-authorship of this report, Mayr explained: "Owing to poor editing Richard Goodwin is listed as co-author but he was actually to prepare a report on living materials, a report he never did."

Part 1 (60 pages) of the first edition of this textbook includes brief discussions of "new systematics," of taxonomic categories and concepts, species and lower categories, classification and the higher categories. While it is the function of the species to denote distinction, it is the function of the genus to denote association. The (subjective) genus is a systematic category including one species or a group of species of presumably common phylogenetic origin, separated by a decided gap from other similar groups. Part 2 (140 pages), the largest portion of the book, introduces the reader into various taxonomic procedures (techniques of collecting, identification and discrimination, quantitative methods of analysis, and presentation of the results). Part 3 (80 pages) comprises treatments of zoological nomenclature, its historical and philosophical basis, the principle of priority, the type method and the names of systematic categories.

The text of the second version (1969b) by Mayr alone is thoroughly reworked and represents virtually a new book. It includes discussions of the species category, polytypic species taxa, the higher categories, and theories of biological classification, methods of zoological classification (identification, analysis of variation, procedures of classifying and taxonomic publications). Part 3 of the book presents the International Code of Zoological Nomenclature. Its entire revised edition of 1964 is accompanied by much more detailed interpretations of the Rules than in the Code itself. Most of the text is updated, incorporating a large part of recent literature. Problems at the species level are treated with frequent references to Mayr's *Animal Species and Evolution* (1963b). The sections on quantitative procedures and methods of illustration as well as the section on zoological nomenclature are more condensed than in the first edition. New material includes recent developments in taxonomic theories of classification on which Mayr (1965b,k, 1968j) had published before: (1) *Phenetics* or *numerical taxonomy* is dismissed because important and unimportant characters are not distinguished and because of the arbitrary levels of phenetic distance on which phenetic categories and the delimitation of taxa are based. (2) *Cladistics*: The grouping of taxa is based on the inferred branching pattern of phylogeny determined through the distribution of derived (apomorphous) and primitive (plesiomorphous) characters. The interpretation of character polarity is critical in this analysis which Mayr praised as a sound way of establishing the phylogenetic (branching) pattern of groups of related species. He criticized, however, cladistic classifications based exclusively on branching or splitting events without taking into consideration different rates of phyletic evolution after splitting of the daughter species (autapomorphies). This is simply not part of the concept of cladistic classification and thus ignored. (3) *Evolutionary classification*: Taxa are delimited according to common ancestry *and* subsequent divergence, classification is based on the dual results of speciation (cladogenesis) *and* phyletic evolution (anagenesis) of organisms. It represents a synthesis or compromise of the best of the two more extremist schools of thought (Fig. 10.1). Established groups should be monophyletic (following the definition of Simpson, i.e., derived from the nearest common ancestor); their rank in classification (genus, subfamily, family) is then assigned on the basis of overall similarity or total information content of weighted

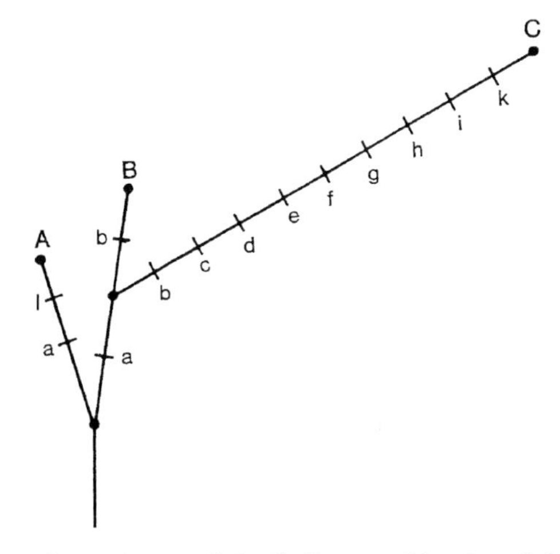


**Fig. 10.1.** Cladogram of taxa A, B, and C. Cladists combine B and C into a single taxon, because B and C share the synapomorph character b. Evolutionary taxonomists separate C from A and B, which they combine, because C differs by many (c through k) autapomorph characters from A and B and shares only one (b) synapomorph character with B (from Mayr 1981a)

morphological or nonmorphological characters as well as between-group character gaps. Evolutionary classification is "horizontal," as it stresses grouping together species in a similar stage of evolution rather than location on the same phyletic line. On the other hand, a cladistic classification is "vertical," because it stresses common descent and tends to unite ancestral and descendent groups of a phyletic line in a single higher taxon ("clade"), separating them from closely related contemporaneous taxa having reached a similar grade of evolutionary change.

After Simpson's (1961) text, the 1969(b) edition is still widely used, even by cladists whose basic philosophy is quite different from Mayr's (and Simpson's) approach. Earlier books were those of Rensch (1934) and Mayr et al. (1953a). To the 1991(a) edition of *Principles of Systematic Zoology* P. Ashlock mainly contributed to the section on macrotaxonomy including the chapter on numerical phenetics. Because his coauthor soon fell seriously ill, most of the work again remained with Mayr. The chapter on nomenclature is now mostly a critical commentary. The volume continues to be an excellent introduction to systematic zoology. It emphasizes that Darwin had actually stressed the importance of phylogeny for classification. Ever since then systematists have shown that genealogies are essential for studying adaptation, coevolution, macroevolution and historical biogeography. Also ecologists, behaviorists, molecular and developmental biologists have successfully used phylogenetic methods in their research.

Mayr (1974g, 1981a, 1986i, 1995b; Mayr and Ashlock 1991a; Mayr and Bock 2002h) published a series of discussions of ordering systems in zoology. The first (1974g) contrasts cladistic analysis with cladistic classification (both introduced by the German entomologist Willi Hennig 1950, 1966 who designated them phylogenetic analysis and phylogenetic classification; the term "cladistic" was coined by Mayr). Taxa are based on the joint possession of derived characters (synapomorphies): "Hennig deserves great credit for having fully developed the principles of cladistic analysis, especially the clear recognition of the importance of synapomorphies for the construction of branching sequences." This step of the analysis is important for the cladist, the evolutionary systematist, and the historical biogeographer. There is no argument over the value of cladistic analysis and the final product, the cladistic diagram or branching pattern for evolutionary studies. Arguments arise over the construction of a classification from the cladogram (Mayr 1974g). Hennig (l.c.) proposed simply to translate the cladogram into a hierarchical classification reflecting only the recency of common descent ("branching point"), whereas evolutionary classification as recommended by Mayr, Simpson and others reflects common descent *and* evolutionary divergence. Mayr's objections to cladistic classifications may be summarized as follows (Mayr and Bock 2002h):

(1) *Relationship* means to the cladist only kinship in a strictly genealogical sense rather than inferred amount of shared genotype or gene content;
(2) *Monophyly* was, since Haeckel, a distinctly "retrospective" term referring to the nearest relatives inferred to have descended from the same stem species. Hennig added the following qualification: "[...] and which includes all species descended from this stem species." Monophyly became a "prospective" criterion and groups which do not include all its descendent species became "paraphyletic." Ashlock (1972) discussed how different Hennigian phylogeny is from Haeckelian phylogeny universally employed by traditional taxonomists. The term monophyly refers to Haeckelian phylogeny and Ashlock introduced the term holophyly for Hennigian phylogeny.
(3) *The neglect of the dual nature of evolutionary change:* All phylogenetic splits have equal weight for the cladist, just as all characters have equal weight for the pheneticist. Autapomorphies or anagenetic invasion of new adaptive zones, the existence of minor or major "grades," and mosaic evolution are all ignored in the construction of a Hennigian ordering system. The cladist acts as if he assumed that all lineages diverge in an equivalent manner and that genealogical distance corresponds to genetic distance. The reptiles, a "paraphyletic" group, represent a well-defined "grade" between the amphibian level and that of the derivates of reptiles: birds and mammals.
(4) *A purely formalistic species definition*: A cladistic species is simply the distance between two branching points on a phylogenetic tree of related species. This is both unbiological and unrealistic. If one "branch" has not changed, it is the stem species which has continued and from which another species has branched (or "budded") off. Also, a simple dichotomy into two daughter species is not the rule, as shown by superspecies often consisting of more than two allospecies.
(5) *The mode of origin of higher taxa*: The evolution of a new higher taxon is correlated with the invasion of new adaptive zones by existing species and has nothing to do with a splitting event (speciation). It is the later process

of anagenetic divergence which leads to ecological changes. It is a misleading formulation to say that higher taxa split.

(6) *A misleading conceptualization of ranking*: Rank is given by cladistic procedures automatically by time of origin, and the same rank must be given to sister groups. This proposition is totally unacceptable and was later given up or modified by cladists. The extremely different evolutionary fates of phyletic lineages derived from the same stem species would lead to highly unbalanced classifications, if the cladistic method of ranking would be applied.

Mayr (1974g, 1980d) further pointed out difficulties (a) of determining the direction of evolutionary character changes, that is the polarity of primitive–derived character sequences, (b) of discriminating between parallelism and convergence (often ignored by cladists), and (c) of mosaic evolution, that is unequal rates of evolution of different characters. He felt that evolutionary classifications are more meaningful biologically, because they pay attention to major adaptive events in evolutionary history which, according to Mayr, are of greater importance for the ranking of taxa than the mere splitting of phyletic lines. Evolutionary classifications permit the greatest number of broad generalizations. Mayr (1981a) proposed a synthesis in the sense of a first ordering of the data as in phenetics followed by an analysis of the hierarchical structure buried in the phenetic clusters based on cladistics and, third, application of evolutionary systematics to produce a mature classification. At that time Mayr (l.c.) thought it would be possible to convert a cladogram into a traditional classification by cutting up a clade into segments of approximately equal degree of difference, that is by converting it into a phylogram. He realized, however, that this cannot be done successfully and that paraphyletic groups cannot be properly indicated (Mayr 1995b). Mayr and Bock (2002h) discussed "Classifications and other ordering systems" from a broad perspective distinguishing between an evolutionary classification and a cladistic cladification. The procedure of classifying is to "make homogeneous classes of similar entities" and the finished product is a "hierarchically ordered system of classes of similar objects" (p. 181). Ordering systems not based on classes, such as sequential listings or cladifications (Mayr 1995b), are not classifications. In W. Hennig's (1966) system all species are assigned to branches (clades or cladons) of the cladogram rather than to classes of similar taxa as in the traditional classification. A clade is holophyletic when it includes the originating stem species and all of its descendants. Hennig's term "paraphyly" does not exist in evolutionary classifications, because the appearance of a new side branch has no retroactive influence on the classification of its ancestors. Even though birds and mammals budded off from the Reptila, this ancestral group remains untouched and as sound an evolutionary taxon as ever, according to Mayr and Bock (l.c.). They emphasized that Hennig (l.c.) was the first taxonomist to have formally introduced the important concept of using apomorphous (derived) characters to determine the branching points and to construct a phylogenetic diagram (cladogram). Cladistic analysis is as useful in evolutionary classification as it is in Hennigian cladification (but see Bock 1992b).

A sound evolutionary classification must be based on a balanced consideration of both cladogenesis (genealogical branching) and anagenesis (similarity or amount of phyletic evolutionary change). Also, in the large majority of cases, a new higher taxon originates not by splitting but by budding (separating) from the parental taxon. An evolutionary classification takes into consideration both ecological adjustment and origin of new biodiversity and it calls attention to the ecological shift in adaptive anagenesis. This is obscured in a Hennigian cladification which places drastically different taxa in a single cladon. However, Mayr did not succeed in convincing cladistic systematists to develop evolutionary classifications. In view of the difficulties to quantify and compare similarity and adaptive changes among taxa, most systematists currently apply the simpler method of cladification based on genealogy or branching. To them, higher taxa are clades—entities resulting from common descent—rather than groups of similar organisms.

A further general aspect of the systematics of a particular group of organisms which Mayr (1965r, 1989g; Mayr and Bock 1994b) mentioned repeatedly is the distinction between a group's phylogeny, classification, and *linear sequence*. Several different, equally valid sequences can be derived from the same two- or three-dimensional phylogeny and classification. At each phylogenetic split the members of one of the sister groups must be listed first and a choice must be made. Three criteria should be applied when arranging taxa into a sequence (Mayr and Ashlock 1991a: 154): (a) Primitive taxa are listed first; (b) derived taxa are listed after the taxon from which they are believed to be derived; (c) every taxon is listed as near as possible to the taxon with which it is believed to be most closely related.

Mayr and Bock (1994b) also distinguished between provisional and standard classifications. The former are proposed for use of specialists and may change rapidly as other workers propose different ones. Standard classifications and standard sequences serve for efficient communication between all biologists interested in a group of organisms. Peters' *Check-list of Birds of the World* (1931–1987) is the standard sequence of avian taxa adopted most widely internationally (Bock 1990). This does not mean, of course, that the classification on which it is based is perfect or that every ornithologist must agree with all of its details. The objective of this standard sequence is to facilitate communication among ornithologists and to assist in rapid information retrieval from collections, faunal lists and other literature. The greatest reasonable stability is essential for the achievement of this objective. Replacement of a standard sequence would be justified only when a new system had become authoritative by being widely supported by avian macrosystematists. Avian systematists may continue their research regardless of what standard sequence the avian biologists use.

Mayr adopted many of the principles summarized above in his publications on the macrosystematics of various groups of organisms, mainly birds but also fossil hominids and bacteria. Many of the methods of taxonomy described by Mayr (1953a, 1969b, 1991h) have been widely adopted, even by cladists.

## Birds

Incorporating the Rothschild Collection into the general collection of the AMNH during the 1930s started detailed systematic studies for the classification of particular families or groups of genera by various ornithologists of this institution (Amadon, Chapin, Delacour, Mayr, Vaurie, Zimmer). Mayr prepared a number of influential papers on swallows (1943f, with Bond), shrikes (1943d), and ducks (1945e, with Delacour). In these macrotaxonomic studies, Mayr took into consideration many details of the birds' ecology, habits, and behavior such as nesting and courtship. With respect to the swallows Sheldon and Winkler (1993: 807) concluded the discussion of their recent biochemical analysis stating "The groups of genera outlined by Mayr and Bond (1943) based on nesting habits and plumage patterns conform to our clades to a remarkable degree." Mayr outlined the text of the paper on Anatidae (Delacour and Mayr 1945e, 1946j) and wrote the more general discussions, whereas Delacour supplied most of the factual detail. They had both agreed that it was unnatural to separate the mergansers (*Mergus*) from the goldeneyes (*Bucephala*), to lump the pochards (*Netta*) with the sea ducks, etc. It was one of the first papers in English-language ornithology (after the comprehensive work of O. Heinroth and K. Lorenz in German) in which life history data were extensively used for reclassification. Both authors believed in large genera, since it is the function of the generic name to express relationship, not distinctness which is expressed by the species name. The broadening of the (polytypic) species necessitated a corresponding adjustment of genus limits. Compared with Peters' *Check-list* (vol. 1, 1931), Delacour and Mayr (1945e) recognized 40 genera in the Anatidae (instead of 62) of which only 22 (instead of 42) were monotypic. They consistently indicated superspecies in the species list. Their macrotaxonomic conclusions were in nearly complete agreement with those reached by K. Lorenz (1941) based on comparative ethological studies of freshwater ducks (Delacour and Mayr were able to comment on this paper in their supplement, 1946j, after receipt of a copy of Lorenz's article which had appeared during World War II). Many publications of later authors have substantiated this early discovery that highly conservative behavior patterns contain historical evolutionary (phylogenetic) information at the level of genera and subfamilies.

The article on the "Classification of Recent Birds" (Mayr and Amadon 1951a) summarized all this work, which gave considerable attention to the Old World families, especially Australasian groups, and separated many of them as distinct families, subfamilies and tribes (Monarchini, Rhipidurini, Pachycephalini, Cinclosomatini; Subfamily Malurinae; family Grallinidae). This was the first classification to emphasize the large and independent radiation of songbirds (Oscines) in the Australasian region comparable to the large radiations of the suboscines in South America and of the nine-primaried oscines in the New World. The authors determined the total number of all bird species as 8,590 compared with 8,616 species listed by Mayr in 1946(a); in later years Mayr's estimate was somewhat higher (9,000 or even near 10,000 species), because it became increasingly clear that a portion of geographical isolates previously interpreted as subspecies of polytypic species possibly have, in fact, reached species status. Mayr and Amadon (1951a) exposed "the abysmal ignorance of the relationships of most songbird families." They stated that habits like, e.g., nest-building behavior are often a better clue to relationship in birds at the level of species and genera than structure.

His global macrotaxonomic knowledge of several songbird families enabled Mayr (1955b) to comment critically on published studies of songbird phylogeny evaluating the methods and principles of the respective authors. He agreed with the separation of the Chaffinch (*Fringilla coelebs*) from the Carduelinae and recommended for it a separate subfamily Fringillinae within the Fringillidae because of this species' cardueline resemblance in plumage coloration and bill structure, in gape color of the nestling and the finely woven nest (see also Mayr 1956e).

James L. Peters' *Check-list of Birds of the World* (16 volumes, 1931–1987, including the index volume) is the most complete survey of all bird taxa that have been described down to the subspecies level giving complete synonymies for all taxa described. It was basically a list of subspecies until Ernst Mayr took over editorship in 1953 (volumes 8–15) and "reintroduced," so to speak, the species as a unit symbolized by a binominal heading of the list of subspecies for each species taxon. The work had been started by J. L. Peters, curator of ornithology at the Museum of Comparative Zoology (Harvard University), in the late 1920s and the first seven volumes had been published, when he died in 1952. Mayr's appointment as an Alexander Agassiz Professor in 1953 initiated his coeditorship of the *Check-list* with J. C. Greenway, curator of ornithology. The overall plan for the continuation of the work was Mayr's, who asked numerous colleagues to participate as authors of particular families. Some of the larger families were treated by several authors (Bock 1990).

Ever since he arrived in New York in January 1931, Mayr had helped Peters with the first volumes of the *Check-list*. Writing to Stresemann on 7 May 1931 he stated (transl.):

"Tomorrow I travel to Boston to work with Peters. His *Check-list* is fine and I admire how well he knows the new literature. I need to correct only very little regarding New Guinea and the Solomons. This is the more surprising since he is so little specialized in this region where Mathews makes work so difficult for every non-specialist. I have the impression that the *Check-list* is an extraordinary achievement. Of course, the American influence is noticeable with respect to the delimitation of genera and species."

Mayr's first task in 1953 after arriving at Harvard was to develop a classification and sequence for the families of passerine birds to be treated in the eight remaining volumes (regarding the difference between a classification and a linear sequence see p. 326). A choice was needed between sequences of the songbird families which ended either with the nine-primaried songbirds (Wetmore 1951) or with the crow family (Mayr and Amadon 1951a). Therefore an independent international committee was appointed at the International Ornithological Congress in Basel (1954) not to decide on classification but to recommend a standard sequence for the oscine families (Mayr 1956d). This "Basel sequence" was intended to provide a standard sequence for communication among ornithologists generally and which Mayr and Greenway could use for planning the remaining volumes of the *Check-list*. The committee voted unanimously for the second alternative, the crows-last sequence. No claim is made in the report of the committee that the adopted sequence is the best possible system or even a final one. At the same time, two different sequences were proposed by Amadon (1957) and by Delacour and Vaurie (1957). As briefly discussed by Mayr (1958e), this diversity of opinion was not very practical. A standard sequence was never agreed upon, although most authors followed the sequence adopted in the *Check-list* during later years. Mayr (1965r) commented on the unfortunate situation approaching "scientific anarchy" that 5 or 6 different sequences of passerine families were in use at that time. Also, within families, some authors had proposed to list genera and species alphabetically like K. Gesner and W. Turner 400 years ago. This seemed inadvisable, because very dissimilar species would often be placed next to each other in lists and museum collections. The search for a natural grouping of species must continue. Mayr (1965l) published the reasoning for several deviations adopted in the sequence of species of the Indo-Australian Zosteropidae to be covered in the next volume of the *Check-list*. Similarly, he (1968c) presented the classification and sequence of genera in the Estrildidae followed in the *Check-list*. This complex family was treated by three authors. Mayr had prepared a preliminary list which was circulated to seven specialists. Their answers to a questionnaire of 19 questions were used to prepare the final list of genera which gives higher weight to behavioral characters than to plumage patterns. I mention these details to illustrate how serious Mayr took the preparation and editing of the *Check-list* volumes. He is the author or coauthor of the catalogues of 25 families, more than any other colleague, and he had the major responsibility for the entire project. He oversaw publication of volumes 8–15 and revised volume 1. One can sympathize with him, when he inscribed the final volume (1986) to Walter Bock: "At last the millstone is off my neck!"

On several occasions Mayr (1959e, 1976i, 1980d) reviewed the status of macrotaxonomy or the classification of birds emphasizing that in contrast to the "science of the species" during the period of "new systematics," little progress had been made in the understanding of the arrangement of species into genera, families, orders and super-orders. He referred repeatedly to Erwin Stresemann's (1959: 277–278) well-known pessimistic statement:

"In view of the continuing absence of trustworthy information on the relationship of the highest categories of birds to each other it becomes strictly a matter of convention how to group them into orders. Science ends where comparative morphology, comparative physiology, comparative ethology have failed us after nearly 200 years of efforts. The rest is silence."

For most of the 28 orders of birds usually recognized there was no certainty which other order is its nearest relative. Almost half of all bird families contain only one or two genera whose nearest relatives are unknown. Moreover, numerous additional genera actually are "incertae sedis" and only very tentatively assigned to one or the other larger family. The symposia and discussion groups at the International Ornithological Congresses at Canberra (1974) and Berlin (1978), which Mayr and Bock, respectively, convened showed a renewed interest in the many open problems in the classification of birds at higher taxonomic levels. The reasons for this interest were the discovery of new characters and the controversies on several new theories of classification (see above).

## Fossil Hominids

Mayr's first papers on fossil hominids were reactions against the extreme taxonomic splitting of previous authors, whereby nearly every fossil find was described as a separate species or even genus (1951g, 1963q). He combined Neanderthal and *Homo sapiens* in one species and also lumped too much among the species of *Australopithecus*. There is now no longer any doubt that there were two sympatric taxa in this genus, one gracile and one robust, both perhaps consisting of several species. Much evidence indicates the existence of several allospecies related to *Australopithecus africanus*. Three factors contributed to the current uncertainty in the proposed interpretations of the history of man: (1) New fossil finds, (2) a more consistent application of geographical thinking to the ordering of hominid taxa, and (3) the appreciation of the importance of climatic changes for the evolution of hominids (1982d, 1997b). Stimulated by theories of S. Stanley and R. Wrangham, Mayr (2001f, 2004a) has recently proposed an entirely new scenario of the whole evolutionary development from chimpanzee to *Homo sapiens*. This model is based on climatic changes in Africa and correlated changes in vegetation leading from rain forest through a stage of tree savanna to the current stage of a widespread bush savanna. New fossils are being discovered all the time and they may ultimately decide how much of the proposed scenario has validity.

In his new proposals Mayr considered not only bones but also our improved understanding of climate, ecology, and behavior. Some australopithecines probably became adapted to the bush savanna when the climate in East Africa became drier. They changed to terrestrial habits and became bipedal, made the first flaked stone tools, may have constructed lances and evolved into *Homo*. Fire was their best defense at camping sites. Selection rewarded especially ingenuity and brain power. This resulted in an increase of brainsize from 450 to 700–900 $cm^3$ in *Homo* (which eventually reached 1,350 $cm^3$ in *Homo sapiens*). The change from *Australopithecus* to *Homo* of the bush savanna, although rapid, was populational and hence gradual. Mayr has decisively contributed to the new interpretation even though some of his historical narratives are bound to be refuted eventually. He stated (1993a): "The evolution of man is permissible within the framework of Darwinism, but it is not an inevitable consequence of it," as suggested by the somewhat finalistic interpretations of Huxley and Dobzhansky.

I include here Mayr's reminiscences of the German anthropologist Franz Weidenreich (1873–1948) whom he met at the American Museum during the war: "I remember the famous anthropologist Weidenreich with a good deal of affection. He was in Germany a well-known embryologist and anatomist who had made particularly distinguished work with respect to the neural crest. When the Nazis came to power he was fired and went to China to work at the medical school in Peking. It was there that he got involved in the description of Peking Man, and did indeed an excellent job not only of describing *Sinanthropus* but also Java Man and the later findings in Java. Like all anatomists, he used a purely typological language and made a new genus and species of nearly every discovery, but he seemed to realize that Peking Man and Java Man were nothing but geographic races of the same species.

When the Japanese attacked China, Weidenreich had to flee again and came to America, where he was located at the American Museum. He loved to drop in on me once in a while to talk German to somebody. Obviously he was rather homesick for Germany. We always had very interesting conversations. Having apparently learned his English rather late in life, he still occasionally mixed the two languages, this sometimes leading to rather funny statements. In the last year of his life he told me that he would prefer not to go back to Woods Hole this summer because the diet there did not suit him 'on Sundays they always serve baked bones.' And I, of course, understood at once what he wanted to say, but it took me a minute or two before I realized how funny his remark had been. Fortunately, he had made some excellent plaster castes of Peking Man so that the tragedy of the loss of the original material is not quite as great as it otherwise would have been. I learned a great deal about the fossil history of man from my talks with Weidenreich."

## Snails

The genus *Cerion* is famous in the evolutionary literature because it poses a challenge to anybody interested in the nature of species. There is a rich literature, and after almost 100 years of controversy, there was a majority explanation among the specialists. However, when Mayr began to study this genus he soon realized how little we really understood its variation. In March–April 1952, he spent several weeks on Bimini Island, an outpost of the Bahamas, to collect *Cerion* snails. Two facts stimulated him to undertake this research: (1) No reproductive isolation had developed among taxa that were strikingly different; seemingly indiscriminate interbreeding took place, where they came into contact. (2) These hybrid populations showed no increases in the coefficients of variation for size and shape characters but were highly variable with respect to sculpture (presence of ribs) and coloration. There was no clear-cut correlation between shell type and habitat. All colonies on Bimini seemed to be conspecific. Mayr planned to go back and therefore marked all individuals on certain carefully measured squares in order to determine their survival and dispersal during the later years. These experiments were not followed up because of Mayr's move to Harvard. In a publication on this project (Mayr and Rosen 1956j), most of the measurements of the material collected and calculations were made by the coauthor.

On the suggestion of Bill Clench at Harvard University, Mayr went to the Banes Peninsula in Cuba, where the richest assemblage of *Cerion* species occurs. Here again he found a narrow hybrid zone between two extremely different taxa, and again in this belt the coefficient of variability did not rise appreciably for dimension and shape. He also found zones of overlap of *Cerion* species not very different in morphology and encountered individuals of certain taxa in strikingly different habitats (exposed sandy beach and on top of a raised coral reef in open woodland). On the other hand, along a stretch of flat exposed sandy beach about 4–5 exceedingly different morphological types occurred in an identical environment. Mayr was unable to go back to Cuba, when Fidel Castro took over the government in 1959. Therefore only a summary of his findings is included in the book, *Animal Species and Evolution* (Mayr 1963b: 398 and Fig. 13-9). S. J. Gould later continued the study of *Cerion* snails and summarized his findings in his recent book (2002).

### Two Empires of Organisms or Three?

The difference between a Hennigian cladification and a Darwinian classification underlies the argument between a group of microbiologists and Ernst Mayr (1990o, 1991h, 1998c) regarding the basic structure of the living world. There are two major groups of organisms, the prokaryotes (bacteria) and eukaryotes (organisms with nucleated cells). The bacteria comprise the eubacteria and the archaebacteria from which latter group the eukaryotes derived an important part of their genome. In reliance on the joint possession of derived (synapomorphous) characters and ignoring autapomorphous characters the microbiologist C. Woese established three groups ("empires"): (1) the eubacteria, (2) archaebacteria and (3) eukaryotes. Mayr pointed out that the acquisition of nucleus, chromosomes and other characteristics by the eukaryotes was perhaps the most important evolutionary event in the whole history of life, separating the prokaryotes (all bacteria) from the eukaryotes. Moreover, genetic studies showed that eubacteria and archaebacteria are far more similar (i.e., more closely related) to each other than the archaebacteria are to the fully evolved eukaryotes. The principle of balance clearly favors combining eubacteria and archaebacteria in the empire Prokaryota. There is a huge gap between these and any protist, plant or animal belonging to the Eukaryota. Their world is entirely different from the world of the two kinds of bacteria, the Prokaryota (a "paraphyletic" group under cladistic principles). According to Mayr (1998c) only a two-empire classification correctly reflects the structure of the living world (see also Mayr 2001f: 45–46).

## Zoological Nomenclature

It is largely unknown that for many years Mayr was one of the leaders in this neglected and little respected field—the system of scientific names applied to taxonomic units of extant or extinct animals. Most of those zoologists who showed an interest were occupied only with the paragraphs of the rules, and not with the basic meaning of the Code of Nomenclature and its rules, i.e., conveyance of information. Eventually Mayr was deeply engaged and developed what one might call a philosophy of nomenclature, publishing actively in this field during the 1950s and 1960s (Fig. 4.2). His main concern has always been the stability of names under which animals are known. The names are more important than the nomenclatural rules. According to his recommendations the International Commission of Zoological Nomenclature approved, at the XIVth International Congress of Zoology at Copenhagen (1953), a Preamble of the Code of Nomenclature stating that its primary aim is "to promote stability and universality in the scientific names of animals, and to ensure that each name is unique and distinct." The whole purpose of giving names to objects is to facilitate communication. Every time a name is changed, all the information that previously had been gathered under this name will be lost or at least become very difficult to retrieve. Hence a change of name is a violation of easy communication.

Stability of names can be achieved either through a strict priority approach (that is the oldest name ever given to a taxon should be the only one valid) or by means of conservation of well-known names (even if older, but overlooked ones, senior synonyms, are found in the literature). During the early 20th century, the defenders of the strict "Law" of Priority assured that there were undoubtedly only a limited number of such forgotten names, that all such names would be found soon and given their rightful position to achieve a stable nomenclature. Alas, the period of name changing never ended. In particular the names of popular groups of animals, like birds, mammals, butterflies and others, were used in literally hundreds of publications each year and every change, when a "name digger" had found a senior synonym of a taxon, caused chaos in the literature (see Mayr et al. 1953(a): 212–235 and Bock 1994b on the Principle of Priority).

Finally, zoologists of various countries protested in print against certain rules of nomenclature, especially the "Law of Priority" and demanded relief. In March 1950, K. P. Schmidt of the Field Museum in Chicago published a protest in *Science* asking for the production of a sensible code of nomenclature at the forthcoming International Congress of Zoology at Copenhagen (1953) emphasizing, in particular, "the desirability of conserving familiar names and avoiding confusing changes," even if this impinged on the rights of authors whose names were not used. The needs of all zoologists for maximum ease of communication were deemed more important. Schmidt was immediately supported by many American zoologists including the ichthyologist Carl L. Hubbs and Ernst Mayr who, at that time, explored the foundations of nomenclature for his textbook, *Methods and Principles of Systematic Zoology* (Mayr, Linsley and Usinger 1953a). He also wrote a specialized document on "The status of 'nomina nuda' listed in synonymy" and sent it to the International Commission in August 1950. It was published in German translation in 1951(h), in English in 1953. When Mayr visited Europe during the summer of 1951, he discussed the problems of nomenclature with Dr. Henning Lemche of the Zoological Museum at Copenhagen in preparation of the 1953 Zoological Congress. On February 13, 1952, Mayr sent him a detailed memorandum on various points that ought to be discussed at the Congress (copy in the Stresemann Papers, Nomenklatur 2, Signatur IV, Bestand Zool. Museum, Museum für Naturkunde Berlin):

"1. It seems to me that the international rules should have a preamble in which it is stated clearly that it is the function of the rules to safeguard a stable nomenclature. This preamble should also state that it should be the function of the Commission to use its plenary powers whenever the application of the rules seems to be in conflict with this basic principle." "2. There seems no reason whatsoever to change the family name every time the name of the type genus is changed; 3. The present Article 25 is rather chaotic. [...] I would suggest that this article be divided into two of which the first be entitled 'The Rule of Availability' and the second 'The Rule of Priority;'" 4. For no reason whatsoever the Zoological Congress in Paris

(1948) introduced "absolute line and page priority. This retroactive change has catastrophic effects on the stability of nomenclature and ought to be rejected categorically; 5. An equally detrimental change was adopted [in Paris] with respect to the definition of secondary homonyms. [...] 7. I feel very strongly that the Commission should adopt a ruling or a definite article in the rules which would state clearly that every name that has been in general usage for more than 30 years becomes automatically a *nomen conservandum*. Such a name cannot be changed without approval of the International Commission," etc.

These suggestions influenced the applications of several German Zoological and Paleontological Societies to the International Commission of Zoological Nomenclature as published in *Senckenbergiana* (vol. 33: 193–196, 1952) as well as the program of a Colloquium on Zoological Nomenclature held in Copenhagen (29 July–4 August 1953) just prior to the Congress (Fig. 10.2). At this meeting, Mayr quickly took over the leadership of the American group and made most of the crucial proposals. He was particularly proud of the introduction of the Preamble (quoted above). Subsequently, including future congresses, this Preamble which stated the principle of stability, was often decisive in the articulation of new paragraphs in the Rules. Mayr was not a member of the International Commission at

**Fig. 10.2.** Ernst Mayr as a member of the Colloquium on Zoological Nomenclature, International Zoological Congress, Copenhagen (August 1953). *Front row* (from *right* to *left*) H. Stammer, E. Hering, E. Mayr, R. Meinertzhagen, J. R. Dymond, R. L. Usinger, H. Boschma. *Second row* (from *right* to *left*) C. L. Hubbs, W. J. Follett, G. W. Sinclair, Mrs. K. Palmer, C. W. Sabrosky, L. B. Holthuis, L. D. Brongersma. (Archive, Museum of Natural History Berlin, Orn. 219)

Copenhagen but since his basic understanding of the rules of nomenclature had impressed everybody, he was soon elected as a Commissioner and served with distinction from 1954 until 1976. The *Copenhagen Decisions on Zoological Nomenclature* (Hemming 1953) document that Mayr was the driving force at this meeting. His proposal for "The Principle of Conservation" on page 119 of this document reads:

"Any specific name which has been published for not less than 60 years and which has been adopted for a given taxonomic unit (a) in at least ten publications, and (b) by two or more authors shall be deemed to be worthy of conservation in the interest of stability. [...] Such a name is not to be replaced by a senior subjective synonym where the latter has not been applied to the taxonomic unit concerned during the preceding 50 years."

He also participated in the formulation of an alternative draft submitted by W.I. Follet, E. Mayr, R.V. Melville and R.L. Usinger and printed on page 122 of the *Copenhagen Decisions*. Stresemann who had participated in the Copenhagen Congress and the Colloquium on Zoological Nomenclature wrote to R. Moreau (England): "I am just back from a three weeks journey which brought me first to Copenhagen; Meinertzhagen may have told you already of the splendid results of our efforts to stabilize nomenclature by overthrowing the tyranny of absolute priority. We had to struggle very hard for reaching our goal and without the cleverness of Ernst Mayr would probably never have attained it" (25 August 1953).

The *Copenhagen Decisions* served as the working basis for the "Section of Nomenclature" at the XVth International Congress of Zoology held in London (16–23 July 1958), where Mayr again played a leading role, rhetorically skilful and effective in discussions. Due to his efforts the Congress adopted a *Statute of Limitation* (Article 23b of the Code) giving automatic protection to names that had been in unchallenged use for 50 or more years ("50-year rule"). In November 1969, the International Commission adopted a revised wording of Article 23b again making clear that the well-used junior name is automatically protected. By its provisions ("at least five different authors and at least ten publications") it excluded *de facto* all areas of animal taxonomy that are inactive (Mayr et al. 1971e). The compromise wording adopted in 1972 emphasized that a prior but unused name should not upset a synonym in established usage, and the 1985 Code describes (Article 79c) criteria constituting *prima facie* grounds for the Commission's conserving a name by the suppression of a disused senior synonym (Melville 1995). Bock (1994b) and Hemming (1995) summarized the general history of the International Commission of Nomenclature and the various meetings during the International Congresses of Zoology.

The only textbook in the field of nomenclature during the 1950s was *Methods and Principles of Systematic Zoology* by Mayr et al. (1953a) with its well-reasoned promotion of stability. It was read by every young taxonomist and eventually more and more zoologists realized the absurdity of the "Law" of Priority. In his *Principles of Systematic Zoology* Mayr (1969b) included on over 80 pages not only the text of the Code of Nomenclature but also a detailed interpretation of all its Rules. This is the only such detailed commentary and far more informative than the explanations given in the Code itself. The second edition (1991a) of the *Principles* does not contain the text of the Code and only an abbreviated commentary with a detailed discussion of *nomina oblita* and *nomina nuda* (p. 392 ff). The authors explain clearly the difference between a valid name and an available one as well as many other nomenclatural issues.

Mayr has the great merit to have attained a regulation against the revival of long-forgotten names ("name digging"), although his original wording of this article has been adapted and revised later. Even more important was his success in castigating strict adherence to priority. Therefore the steps to be taken are carefully considered today when a senior synonym is encountered. Mayr always believed clear communication to be more important than the revival of forgotten names.

The Nomenclature Committee of the AMNH of which Mayr was a member and at times the chairman published a series of comments and proposals in the Bulletin of Zoological Nomenclature during the early 1950s regarding the stability of nomenclature, the validity of certain taxonomic names, nomina conservanda, the designation of neotypes, rules for naming families and suprageneric categories of lower rank, the validation and emendation of certain names, substitute names, the revision or clarification of certain articles of the Code, the specific name and nominate subspecies. As an active member of the International Commission on zoological nomenclature (until 1979), Mayr commented on the applicability of the gender rules (1958j), on the equal and exchangeable status of patronymics with the terminations *–i* and *–ii* (1958m), and on the composite nature of Linnaeus' *Helix vivipara* (1960b). In 1972, he proposed several amendments to the International Code regarding the formation of family-group names, the law of homonymy, the choice of a type genus and others (1972i). Over many years, he also published in the Bulletin of Zoological Nomenclature, like other commissioners, brief and more extensive nomenclatorial comments on the proposals submitted by other zoologists.

In later years, Mayr published on individual provisions of the Code such as the problem of secondary homonymy (1987b) raised already in his memorandum of 1952 (see above). Secondary or subjective junior homonyms result when a specialist combines two or more genera containing species with the same specific name. Other specialists may not agree with this action (and much less with any renaming of the supposedly "homonymous" species). At the Congress in London (1958) Mayr had protested strenuously against Article 59b of the Code addressing this problem and in 1987 suggested to expand Article 79c by applying it not only to unused senior synonyms but also to unused senior homonyms. In some cases of obvious homonymy Mayr supplied new names for certain taxa of birds (e.g., *Pachycephala lorentzi* Mayr 1931 and *Dicaeum hypoleucum pontifex* Mayr 1946). Mayr (1983c, 1989h) further discussed the philosophical question whether the type method of zoological nomenclature for attaching labels (names) to definite objects would be suitable in philosophy, i.e., to attach names to ideologies, to research traditions, and evolving scientific concepts. His conclusion was that the type method is rather unsuitable for this purpose. Instead of coining a new term every time the meaning of an evolving concept changes (e.g., the concept of "Darwinism"), the definition should be modified and the continuity of the term preserved. Certain concepts of nomenclature changed over time, resulting in conflicts between the rules of naming by the early post-Linnaeans and the taxonomists in the late 19th century who formulated the codes of nomenclature.

Mayr was also active in cases of synonymy, not as a priority buff, but because he was seeking the right solution to a complex question. In a series of brief articles published between 1931 and 1986, he clarified the nomenclatural status of certain names of birds based on his superb knowledge of the ornithological literature and his familiarity with the Code of Nomenclature. Examples are *Calao* Bonnaterre (1931g), *Egretta brevipes* (1933e), *Rhamphozosterops* (1933o), several taxa of *Aplonis* starlings (1935a), *Limosa lapponica menzbieri* (1936e), the family name Meliphagidae (1944i), the Fijian Mountain Lorikeet *Vini amabilis* (1945a), the African *Ploceus graueri* (1945g), the Tibetan shrike *Lanius tephronotus* (1947g), *Turdus musicus* (1952h), *Accipiter ferox* (1956o), *Heteralocha* (1956p), the pigeons *Treron griseicauda* and *T. pulverulenta* (1956n), *Ducula (Serresius) galeata* (1957h), the snipes *Gallinago* (1963d) and the name of the fossil rail *Rallus hodgeni* of New Zealand (1986p).

And, of course, Mayr participated actively in application to the International Commission of Zoological Nomenclature for the protection of names threatened by "name diggers." The work connected with almost all of these applications took as much time as writing a scientific paper. Hence it is entirely justified to mention them here and to list their titles in the bibliography. In addition, other zoologists asked Mayr to support their applications as a co-author. His applications in the interest of nomenclatural stability include the following: To suppress J. R. Forster's (1794) names for several Australian birds which had been overlooked universally (1952b), to suppress the names *Drosophila brouni* (1954a), *Turdus musicus* in favor of *T. philomelos* (1957c), the genus name *Microura* in favor of *Pnoepyga* (1961e, 1963e), and to suppress eight dubious names based on young or damaged birds or on egg shells (1962c); to end confusion and establish the current usage of the generic names *Tanagra* and *Euphonia* (1963i), to accept *Cardinalis* Bonaparte, 1838 for the congeneric Northern and Middle American species of *Richmondena and Pyrrhuloxia* (1964h), and to validate the names *Cacatua* Brisson (1964p) and *Lorius* Vigors (1968d). The last proposals refer to two generic names introduced into the literature by Brisson (1760) which were Mayr's early concern in a letter to Erwin Stresemann dated 2 July 1931. At that time the nomenclatural stability of the names of hundreds of genera of birds was threatened if Brisson's names would be declared invalid unless accompanied by valid species names; Brisson's generic names were finally accepted in 1955 (Bock 1994b; Melville 1995).

# 11 History and Philosophy of Biology—Mayr's Third Synthesis

During the last quarter of his life, Ernst Mayr contributed importantly to the fields of history and philosophy of biology based on his previous work in systematic and evolutionary biology. The titles of his major books documenting these contributions are *The Growth of Biological Thought* (1982d), *Toward a New Philosophy of Biology. Observations of an Evolutionist* (1988e), *This is Biology. The Science of the Living World* (1997b), and *What Makes Biology Unique? Considerations on the Autonomy of a Scientific Discipline* (2004a). With these accomplishments he became a foremost historian of biology and one of the leaders of a modern philosophy of biology. According to Mayr in biological science, especially evolutionary biology, most major progress is made by the introduction of new concepts like the concepts of natural selection, geographical variation and isolating mechanisms rather than new facts or discoveries. This illustrated for him that biology differs in several important ways from the physical sciences. In biology probabilistic generalizations, "rules" and concepts exist rather than laws as in the physical sciences which are dominated by essentialistic philosophy. In biology population thinking is all-important emphasizing the uniqueness of individuals. Other characteristics of living organisms are complexity and organization, the possession of a genetic program, and their historical nature. Since the late 19th century biology, for Mayr, struggled to become autonomous and independent from the physicalist approaches.

History of science explores the history of scientific problems and their solutions as well as the controversies in current biology. According to Mayr it functions as a tool for concept analysis and clarification of the structure of biology. He saw the development of science as an increasing emancipation of scientific knowledge from religious, philosophical and other ideological beliefs. The close connections between the past and the present are one of the main characteristics of Mayr's historical approach (Junker 1995, 1996). As an evolutionary biologist he applied Darwinian principles to the study of developmental historiography, the study of those aspects of the past that help our understanding of the science of the present. Mayr's approach thus differed in many ways from lexicographic, chronological, biographical or sociological histories of science; he (Mayr 1990c) did not see his analysis of biology as Whig history as others had claimed. When he read, in 1958, Arthur Lovejoy's *Chain of Being* (1936), he "suddenly saw a theme for how to deal with history that goes way beyond being descriptive, namely that in the history of science there are certain themes, concepts, problems, ideas, and they all have an evolution just exactly like organisms have an evolution" (1994m: 373).

Mayr rejected the view of the present being the inevitable outcome of a triumphant historical process. The scientific process obeys the principles of Darwinian variational evolution without teleological components. In his interpretation, the process of "natural selection" applies to the evolution of scientific concepts, because there is an interplay between variation and selection in the history of ideas as it is in organic nature.

While histories of several individual biological disciplines were available, no developmental historiography existed for biology as a whole. To fill this gap in the literature was the object of Mayr's historical handbook, *The Growth of Biological Thought* (1982d), which covers evolutionary biology in the first volume, but also functional biology in the form of genetics. A second volume was to cover physiology and other functional areas of biology, but did not appear (see below). Mayr emphasized that the treatment of the development of concepts dominating modern biology does not necessitate exploring every temporary development or blind alley that left no impact on the subsequent history of biology. He was particularly interested in the history of scientific problems and their solutions, in tracing an idea, concept or controversy back to its sources devoting sufficient attention to the general context of different time levels to cast additional light on the varying surrounding situations. Mayr began his work on the history of biology from the perspective of someone who had been very active in shaping that history. Some historians have faulted this approach to history, even labeling it as "whiggish," a criticism that was rejected vehemently by Mayr (1990c) from his personal point of view. He defended and summarized his historical methods as follows:

- Developmental historiography requires an understanding of the present, but each time period is also studied in its own merits;
- Errors of earlier authors should be pointed out as in any scientific discussion and evaluation of their work is permitted;
- Historians of ideas must be selective in order to follow the history of particular ideas or concepts. They must also break complex systems into their components;
- Developmental historiography must be historical or "vertical" rather than "horizontal," the latter describing the events of only a single moment of time or very narrow time period.

Since the professionalization of the history of science during the mid-20th century, there are "historical" historians (mostly trained in the humanities or social sciences) and "scientific" historians (trained in the natural sciences, specifically in the life sciences in our case). The former state that the only legitimate way of doing historiography is that of describing each period in great detail, trying to present the total picture including the social and economic situation, whereas the "scientific" historiographer is interested in the development of ideas, from their origin through all their permutations up to the present day. As Mayr (1990c) stated, if the term "whiggishness" is used to describe the nature of historical work, it should be applied only to genuine cases of whiggishness sensu Butterfield (1931) and not to developmental historiography which certainly counts as genuine history.

Hull (1994) wrote that Mayr's historical publications are not "wiggish" in several basic meanings. However, "whiggism" is also taken to denote writing history of science as if it homed in on our present day beliefs. These histories make greater concessions to the present-day reader (Hull, l.c.). Major faults of historiography which have been known for a long time and which Mayr tried to avoid include (1) bias against some theory or author, (2) chauvinism by nationality or field, (3) priorities that are neglected or falsified, and (4) finalistic interpretations. It should be noted that Mayr started publishing more actively in the fields of history and philosophy of biology during the Darwin centennial of 1959, that is relatively late in his career and, at the same time, continued as an active worker in the fields of systematics, species, and evolution (Fig. 4.2).

Mayr's friends and students in history of science presented him with a festschrift on the occasion of his 75th birthday which was published in the series "Studies in History of Biology" in 1979. When his major work on *The Growth of Biological Thought* (1982d) had appeared, the Society of the History of Science awarded him their highest award, the Sarton Medal, in 1986. Three universities bestowed upon him honorary degrees in philosophy of biology, *vice* Guelph University (Canada) in 1982, University of Konstanz (Germany) in 1994, and Rollins College, Florida (USA) in 1996. Two festschriften with contributions by historians and philosophers of science appeared when Ernst Mayr turned 90 and 100 years old in 1994 and 2004, respectively (Greene and Ruse 1994; Ayala 2004).

## History of Biology

Mayr's earliest historical work was conducted not for the sake of history but to analyze certain ornithological problems. In his doctoral dissertation (1926e) on the range expansion of the Serin finch (*Serinus serinus*), he included a chapter discussing historical records on the early occurrence and colonization of particular regions by this species. An entry in his notebook of 1926/1927 indicates that among other topics, he planned to write "a historical paper on color change of feathers without molt". This problem had intrigued ornithologists during the mid-19th century, until proven wrong several decades later. Another entry of 1927 was "a historical survey of the ornithological exploration of the world. [...] Note Buffon's ideas on the polar origin of life, wavelike dispersion, etc. [...] Refer to the writings of Alexander Humboldt [...] Lyell, Schmarda, Sclater, Moritz Wagner, Darwin." The publications of these naturalists figure prominently in his writings many years later. As with Mayr's interests in the fields of systematics, genetics, zoogeography, and evolution, the above data indicate that the roots of his interests in history of biology also go back to the mid-1920s, when he was a student under Erwin Stresemann whose approach to science was very broad including the historical development of ornithology (p. 41).

Mayr's expedition notebooks from New Guinea (1929) reveal the fact that one of his early research plans for the years after his return to Germany had been to test the territory theory with respect to birds in the region around Berlin. This plan, however, did not materialize. He took the topic of territory up again in New York when, during the early 1930s, he cooperated with young field ornithologists, members of the Linnaean Society (see p. 109). In conjunction with this work he wrote his first strictly historical article on "Bernard Altum and the territory theory" (1935c). In 1868, Altum had clearly recognized that many male birds set up a territory which they defend against conspecific males (this, incidentally, had been established independently already by Aristotle, by Adam von Pernau in Germany during the early 18th century and was again discovered by British ornithologists during the early 20th century).

**Early historical studies.** When in 1946 the Academy of Natural Sciences of Philadelphia awarded Mayr the Leidy medal for his book *Systematics and the Origin of Species* (1942e) he compared, in his acceptance speech, the activities of the naturalists of the 19th and 20th centuries. Hundred years ago natural history was largely descriptive and analytical, he said, building a vast store of facts. Today, increasingly often "How?" and "Why?" questions are being asked and naturalists attempt to interpret the gathered facts and to correlate them with those from neighboring sciences like zoogeography, ecology, genetics, and ethology. Mayr here spoke for the first time of the "non-dimensional species" of the naturalist, non-interbreeding entities in a local area (in sympatry) and at a single time level, and called attention to the conceptual shift from the typological (morphological) to the biological species concept. Species (taxa) are two-dimensional and often polytypic (species and subspecies). Allopatric forms are difficult to handle taxonomically but have helped to solve the problem of geographical speciation.

Mayr's (1954i) review of W. Zimmermann's book, *History of Evolution* (1953) is intriguing because he outlined here how a history of the modern evolutionary synthesis ought to be written and gave, so to speak, a blueprint of his later historical handbook (1982d). At the same time he stated why he thought that the evolutionary synthesis was truly a synthesis:

"If I were to write a history of this field, I would try to show how growing maturity in the contributing fields eventually permitted this synthesis. After an introductory chapter devoted to the period before Darwin, I would try to demonstrate how the publication of the *Origin of Species* stimulated an unprecedented amount of fact searching and theory building in biology. [...] Nearly every theory was partly right, and partly wrong, and, by discarding the 'wrong' parts, it became possible to synthesize the theory that is now almost universally held among evolutionists."

Many historical aspects of the study of individual and geographical variation in animals are discussed in conjunction with his book projects (1942e, 1953a, 1963b). A penetrating analysis followed of the theoretical views on biological species and speciation of the entomologist Karl Jordan (1861–1959) who, around the turn of the 20th century, discussed the gist of the new or population systematics and, with Edward Poulton (1856–1943), promoted the biological species concept and geographical speciation (Mayr 1955e, 1990g). This article on the work of K. Jordan (and E. Poulton) was one of Mayr's earliest attempts to point out how great a contribution systematists had made (see also Mayr 1973i) to the evolutionary synthesis and, more broadly speaking, to the current concepts of evolutionary biology. Unfortunately, Jordan published most of his theoretical discussions as parts of entomological monographs rather than in separate articles.

Mayr was "filled with admiration bordering on awe when one compares Jordan's discussions around 1900 on the subjects of the mode of speciation, the existence of natural selection, and the meaning of mimicry and polymorphism with those of most of his contemporaries. Jordan's work is most important for the historian of biology because he developed many concepts in the 1890s and the first decade of the 20th century which modern authors think to have originated during the evolutionary synthesis or even later. Sometimes some of these were ascribed to me and I was particularly anxious to establish Jordan's priority."

We should note here that there is a difference between having the priority for first talking about certain ideas and really being influential. K. Jordan may have had priority in this case but very few people actually followed him.

**The Darwin centennial.** In 1959, the centennial year of the publication of Darwin's *Origin*, various societies, academies, and universities organized symposia celebrating this event. As a result Mayr undertook several historical studies and published discussions of the role of "Isolation as an evolutionary factor" (1959a), on "Darwin and the evolutionary theory in biology" (1959b), "Agassiz, Darwin and evolution" (1959c), and "The emergence of evolutionary novelties" (1959d). Writing to J. B. S. Haldane:

"No one seems to give us poor evolutionists a chance to be lazy and inert in this year 1959. Traveling from one evolution conference to the next, I feel like the old-time Vaudeville performer who traveled from convention to convention and from county fair to county fair. At least he had the advantage of showing the same tricks to ever-new audiences, while I am supposed to say something new each time because every word I utter is going to be published" (April 13, 1959).

The Darwin year of 1959 was decisive in initiating Mayr's third career, that of a historian-philosopher of biology (Burkhardt 1994; Junker 1995, 1996), although the transition was gradual beginning with his move to Harvard in 1953. The first contribution just mentioned, was presented at a Centennial symposium organized during the annual meeting of the American Philosophical Society in April 1959. It is a historical analysis on the views of the pioneers of evolution on the importance of geographical isolation, with an emphasis where they were "right" and where they "missed the boat" and why as seen from the current perspective. It is a study of the life-history of an idea. Whereas Darwin, Wallace and Weismann minimized the role of isolation, Moritz Wagner thought that isolation is a *conditio sina qua non* for evolution to take place. When Mayr wrote this article in 1959, Darwin's *Notebooks* were still unknown which later showed that he had supported geographical speciation between 1837 and the 1840s (Sulloway 1979). Mayr's comments on Darwin's principle of divergence were tentative and are superceded by his later article of 1992(n). The first author who made a clear distinction between phyletic evolution and multiplication of species (through geographical isolation of populations) was the ornithologist Henry Seebohm (1887) who discussed these problems in a thoroughly modern manner. The concept of species as an interbreeding community became general among naturalists and systematists around 1900 (Karl Jordan, Edward Poulton, David Starr Jordan). Shortly afterwards the saltationist views of the geneticists Bateson and de Vries led to a deep schism between the experimentalists on one hand and the naturalists on the other, which was only bridged by the evolutionary synthesis during the 1930s and 1940s. Geographical isolation is now seen as an important boundary condition for speciation to take place in conjunction with natural selection and mutation rather than in competition with these factors as some early pioneers had believed is the case.

The editor of the Harvard Library Bulletin, G. William Cottrell[1], suggested to Mayr in 1958 that he prepare for this periodical a paper on Louis Agassiz (1807–1873) who had been the founder and director of Harvard's Museum of Comparative Zoology and a lifelong opponent of Darwin (Mayr 1959c). Agassiz was educated in Switzerland and Germany during a period that was dominated by romantic ideas and by a largely metaphysical approach to nature, especially Plato's essentialism. Four main concepts determined his thinking: (1) A rational plan of the universe, (2) typological thinking, (3) discontinuism, and (4) an ontogenetic concept of evolution. The gradual change of one type (species) into another over time was so inconceivable to Agassiz that it never entered his mind at all. It was not religious scruples that prevented Agassiz from becoming an evolutionist but rather a framework of ideas that could not be combined with evolutionism. Mayr dedicated this paper to Erwin Stresemann on the occasion of his 70th birthday.

In 1970, after he had stepped down as Director of the Museum of Comparative Zoology, Mayr began working on his history of ideas in biology (1982d), as mentioned in his letters to Stresemann ("I am busy with my history of ideas in biology," 23 November 1970, and "I am working on a history of ideas in biology and hope to have completed the manuscript until my retirement on 30 June 1975," 26 December 1971). During the years of writing this large manuscript he published several historical articles on Lamarck (1972f; discussed by Burkhardt 1994), an essay review of books on the history of genetics (1973i), a historical survey of American ornithology (1975c), and a brief review of the history of evolution (1977f). In addition, Mayr contributed several articles to the growing "Darwin industry" (see below) and organized, in 1974, two conferences to discuss the history of the evolutionary synthesis.

In his review of the recent historiography of genetics (1973i), Mayr emphasized the contributions of naturalist-systematists during the history of Mendelism and population genetics, a subject which is dealt with again and in more detail in Mayr (1982d).

As I mentioned above (p. 41), Erwin Stresemann in Berlin had been interested in the history of ornithology since the 1910s and had published on and off in this field. When, after 1945, other research was difficult because of political and logistic postwar conditions, he turned his attention primarily to historical topics and

[1] A major assistant in editing the English translation of E. Stresemann's *Ornithology from Aristotle to the Present* (1975) and, during the 1980s, of the last volumes of Peters' *Check-list of Birds* of the World.

considered writing a modern book on the development of ornithology (letter to Mayr dated 12 February 1946). Mayr invited him several months later to include this book in the series "The Bird Student's Library" which Oxford University Press was planning at that time, with Mayr as editor. However, this series did not materialize and Stresemann's volume was published in Germany in 1951. Ever since that time Mayr had tried to launch an English edition and eventually was successful in 1975. There was relatively little American ornithology in Stresemann's volume and Mayr agreed to prepare an epilogue entitled "Materials for a History of American Ornithology" (1975c). In this wide-ranging survey Mayr discussed the history of museums and of other centers of research in North America, field studies, the role of ornithological societies and of amateurs and women in ornithology, the history of American ornithological journals and the significance of technical advances like bird banding and photography.

The Academy of Natural Sciences of Philadelphia held a special symposium in 1976 to celebrate the 200th anniversary of the founding of the United States. Mayr (1977f) was invited to present a historical review of the study of evolution: No theory of evolution was proposed before 1800 when it was advanced by Lamarck. The main reasons for this long delay, Mayr said, were the adherence to the account of Creation in Genesis and the philosophy of typological essentialism, according to which the realities of this world consist of fixed, discontinuous essences. The belief in these dogmas was gradually undermined by the observations of naturalists during the 17th and 18th centuries and by the liberating thoughts of the philosophy of enlightenment. Evolution is "change in the adaptation *and* in the diversity of populations of organisms." Adaptation and speciation are the main components of evolution and characterize its dual nature. The widespread definition of evolution as "change in gene frequencies in populations" (which Mayr unthinkingly had adopted for many years, see p. 269) is unacceptable as it refers to a result of evolution, an indirect by-product of the superior reproductive success of certain individuals. The second aspect of evolution, the origin of diversity, is widely neglected by evolutionists, particularly geneticists. Around 1900 genetics, in the form of Mendelism, was opposed to Darwinism and natural selection. Naturalists continued to believe in natural selection, but also in soft inheritance emphasizing the gradualness of evolution and geographical speciation. The breach between geneticists and naturalists was healed during the evolutionary synthesis.

Aspects of the history of behavior studies were reviewed (1977c, 1982q, 1983i) and the three major theories during the 18th century listed: (1) The animal as a machine without a soul (Descartes, Buffon, La Mettrie and the reflexologists of the 20th century); (2) The thinking animal, starting with its mind a tabula rasa, learning by experience throughout life and making use of these experiences in a rational manner (all those who anthropomorphized animals, like A. E. Brehm during the 1860s–1870s, were close to this concept as well as the American behaviorists of the 20th century), and (3) The instinct school, from H. S. Reimarus (1762) to B. Altum (1868) to ethology (Whitman, Heinroth, Lorenz, Tinbergen). After feuding for 200 years these three schools have synthesized their valid components, beginning in the 1950s. Mayr (1982q) wrote a foreword for a reprint of the book of the physico-theologist H. S. Reimarus (1762), the first critical and summarizing study of animal behavior. The two groups of authors of the 16th–19th centuries who published on animal behavior were (1) Philosophers like Descartes and Condillac who were especially interested in the comparison man–animal. They solved their problems strictly deductive-rationalistically; and (2) True naturalists like J. H. Zorn, Reimarus, Gilbert White, Huber, Kirby, Spence who were physico-theologists.

***The Growth of Biological Thought. Diversity, Evolution and Inheritance* (1982d).** When this large book appeared during the 100th anniversary of Darwin's death, it was widely acclaimed and reviewed in numerous journals, newspapers, and magazines. Reviewers described it as written by a master of detail, interpretation and synthesis; magisterial, breathtaking, a monumental tome, a work of immense erudition and scholarship. Even two reviews of the book reviews appeared by M. Ruse ("Admayration," Quarterly Rev. Biol. 60: 183–192, 1985) and W. F. Bynum (Nature 317: 585–586, 1985). D. Futuyma wrote in his review:

"One cannot help standing in awe of the Germanic capacity for vast, all-embracing synthesis: consider the lifelong devotion of Goethe to *Faust*, or Wagner's integration of the arts into a *Gesamtkunstwerk* in which all of human history and experience is wrought into epic myth. It is perhaps in this tradition that Ernst Mayr's *The Growth of Biological Thought* stands: a history of all of biology, a *Ring des Nibelungen* complete with leitmotivs such as the failures of reductionism, the struggle of biology for independence from physics, and the liberation of population thinking from the bonds of essentialism. Mayr's goal is to draw from the successful ideas of the past an integrated history of all of biology and its implications for the philosophy of science" (*Science* 216: 842, 1982).

Several reviewers noted the "autobiographical" character of the book, because Mayr had himself "done a considerable amount of research in most areas covered by this volume" (p. VII). He was thus able to place his own relevant publications in the context of the overall history of evolutionism and to trace the historical roots of his work and opinions. Appropriately, he dedicated this large work to his wife Gretel. The volume is a history of the problems, concepts and ideas which evolutionary biology has dealt with since the time of Aristotle. The problems are the natural units of the evolution of biological thought which are here followed through the centuries in a clear exposition: "It is the principle objective of this volume to discover for each branch of biology and for each period what the open problems were and what proposals were made to solve them. [...] At its best this approach would portray the complete life history of each problem of biology" (p. 19).

The book is organized by topics—(1) diversity of life, (2) evolution, (3) variation and its inheritance—but within each of these sections, the subject matter is treated chronologically. Part I of this volume deals with natural history, the study of organic diversity, a history of systematics as the science of diversity. Darwin's *Origin*, biogeography, ethology, and ecology developed out of natural history. The core of Part II on evolution is the three chapters on Charles Darwin and his five theses which are followed by a discussion of the events leading to the evolutionary synthesis of the 1940s and of post-synthesis developments. In Part III the work of Mendel's forerunners, of several species hybridizers and of plant breeders is explained. These workers did not think in terms of variable populations describing their results only qualitatively without calculating ratios. The species question and his teacher F. Unger's interest in variants inspired Gregor Mendel's work. He adopted the method of population analysis and studied tens of thousands of seeds and plants. Although cited several times, his publication of 1866 was overlooked until it was rediscovered in 1900. Soft and hard inheritance and the work of A. Weismann and H. de Vries during the late 19th century are discussed in detail by Mayr as well as the rise of genetics and the chemical basis of inheritance including the discovery of the structure of DNA in 1953. The author exposed in this volume the historical and philosophical relations between systematics, evolution and genetics. In addition, general historical issues are discussed in three introductory chapters and in the epilogue.

Mayr had made particularly good progress with the manuscript during a half year's stay in Germany (Würzburg and Tübingen) in 1977. Originally he had planned to follow up the first volume covering mostly the history of evolutionary biology (ultimate causations) with a second volume dealing with the history of functional biology (proximate causations).

"I stayed in Tübingen for about half a year in 1980 working very hard on the history of physiology and embryology but after about seven months of work I decided to abandon the project. When there were controversies in 1800 or 1850 in the field of physiology I simply didn't know enough of physiology to get the proper feeling for the nature of the controversy nor did I usually know how it finally came out. I could have produced something strictly descriptive but not the kind of interpretive story as I had done with the volume devoted to systematics, evolution, and genetics."

Despite the fact that he had abandoned the project of a second volume already in 1980, in the Preface of the published volume of 1982, Mayr still expressed "his hope to deal with the biology of 'proximate' (functional) causations in a later volume that will cover physiology in all of its aspects, developmental biology, and neurobiology" (pp. VII–VIII). Evidently the Preface had been written several years prior to 1982 and was published unchanged.

## From the Greeks to Darwin

The Greek philosopher Plato (ca. 427–347 B.C.) was a disaster for biology, Mayr (1982d) concluded. His essentialistic concepts influenced this branch of the natural sciences adversely for centuries. The rise of modern biological thought is, in part, the emancipation from Platonic thinking.[2] On the other hand, Aristotle (384–322 B.C.) has made greater contributions to biology than any other thinker before Darwin. As an empiricist, Aristotle was the founder of the comparative method and

[2] The history of essentialism is rather complex, because several early thinkers in Europe had already objected to Platonic idealism a long time before Darwin (Winsor 2006: 3).

of life history studies. He was interested in the phenomenon of organic diversity and searched for causes, asking why-questions. However, his belief in a more or less perfect world precluded any thought in evolution.

Specialization by biologists on particular groups of animals like birds (Turner, Belon) or fishes (Rondelet) occurred during the 16th century, when the encyclopedists Gessner and Aldrovandi published series of large volumes. During the time of Linnaeus (1707–1778) the emphasis on classifying and name-giving led to a near obliteration of all other aspects of natural history until G. Buffon's *Histoire naturelle* (1749–1783) reversed this trend. He painted vivid word-pictures of the living animals with no concern for classification or identification. But Linnaeus and Buffon approached each other in later years: Linnaeus liberated his views on the fixity of species and Buffon defined species as reproductive communities. The latter's views on species and classification (all characters should be taken into consideration) was adopted by many scientists around 1800. The history of taxonomy is a history of concepts like species, relationship, delimitation of higher taxa, reliable characters, ranking of taxa and others. As Mayr emphasized, this field has rarely been given credit for its role in initiating new approaches in biology.

The 18th century concept of the *scala naturae*, the scale of perfection, received its final blow, when Cuvier in 1812 showed that there are four distinct phyla of animals without any connection. In search of a higher order in nature some scientists of the early 19th century experimented with "quinarian" classifications (based on the number 5), while others preferred the number 4. This approach to classification was already obsolete, when Charles Darwin introduced classifications based on common descent with his book, *On the Origin of Species* (1859).

## Darwin and Darwinism

During his historical studies Mayr developed an almost unbounded admiration for Charles Darwin (1809–1882) and discussed his work and its influence on modern thought in numerous articles and books. He emphasized Darwin's statement "I was a born naturalist." As a boy and student Darwin was interested in many activities like collecting, fishing, and hunting. He observed the habits of birds, collected insects and learned a great deal from Professor Henslow. When Darwin joined the "Beagle" in 1831, he was already a very capable naturalist. The Galapagos experience and, upon his return to England, his discussions with the ornithologist John Gould in March 1837 converted him to become an evolutionist, especially through his surprising discovery that most animals on Galapagos, although of American character, are endemic to this group of islands, in that they are not found in either North or South America. These endemic animals (whether species or subspecies) must have originated ("evolved") on the Galapagos Islands from American immigrants (Sulloway 1982: 349). For the next 20 years, Darwin collected evidence for the theory of evolution, but did not publish anything on this subject because of the need to accumulate massive support for his materialistic (nomological) theories and, Mayr (1972d) felt, because of Darwin's awe of Charles Lyell's opposition to the idea of the nonfixity of species. Darwin began writing the manuscript of his big species book in May 1856 after Lyell had urged him to publish his ideas lest he be scooped by someone else. This indeed almost happened, when A. R. Wallace sent Darwin a manuscript in 1858 which, in 1859, led to the publication of Darwin's *Origin*, a story recounted many times.

The "First Darwinian Revolution" of the late 19th century did not involve only a new theory of evolution, but an entirely new conceptual world, consisting of numerous separate concepts and beliefs. Among these are (1) an evolving rather than a constant and created world which is, in addition, not of recent but of great age, (2) the refutation of both catastrophism and a steady-state world, (3) refutation of the concept of an automatic upward evolution (cosmic teleology), (4) rejection of creationism, (5) replacement of the philosophies of essentialism and nominalism by population thinking, and (6) the abolition of anthropocentrism; Man was relegated to his place in the organic world. Because this intellectual revolution entailed the rejection of at least six widely held basic beliefs, it had a tremendous relevance outside of science causing a broad opposition to Darwin's theses. Opponents were especially orthodox Christians, natural theologians, lay-persons, philosophers, physical scientists, and non-Darwinian biologists (Mayr 1972d). Darwin's conceptualizations affected the thinking of average people and their worldview more than the work of Copernicus, Newton, Marx or Einstein.

**Darwin's evolutionary paradigm.** Darwin's theoretical views of evolution comprised a set of at least five separate theses rather than one unified theory (Mayr 1982d,e, 1985i, 1991g, 1997b, 2001f who spoke of five separate "theories"):

(1) *Evolution as such.* This is the thesis that the world is not constant nor recently created nor perpetually cycling but rather is steadily changing and that organisms are transformed in time.
(2) *Common descent.* This is the thesis that every group of organisms descended from a common ancestor and that all groups of organisms, including animals, plants, and microorganisms, ultimately go back to a single origin of life on earth.
(3) *Multiplication of species.* This thesis explains the origin of the enormous organic diversity. It postulates that species multiply, either by "splitting" into daughter species or by "budding," that is, by the establishment of geographically isolated founder populations that evolve into new species.
(4) *Gradualism.* According to this thesis, evolutionary change takes place through the gradual change of populations and not by the sudden (saltational) production of new individuals that represent a new type.
(5) *Natural selection.* According to this thesis, evolutionary change comes about through the abundant production of genetic variation in every generation. The relatively few individuals who survive, owing to a particularly well-adapted combination of inheritable characters, give rise to the next generation.

Analysis of these five theses constitute Mayr's most significant contribution to the history of evolutionary biology. These different theses still describe very well the five major areas of evolutionary biology and should be kept separate in theoretical discussions of the Darwinian paradigm to avoid misunderstandings. Additional theses which Darwin proposed were *sexual selection* (differential success in reproduction), *pangenesis* (the hypothesis that all parts of the body contribute genetic material to the reproductive organs, and particularly to the gametes), and *character divergence* (differences developing in two or more related "varieties" or species in their area of coexistence, owing to the selective effect of competition; Mayr 1992n, 1994g).

Even though Darwin's five main theses of evolution are closely related to each other, they do not form a unity, as shown by their differing fate after 1859 (Table 11.1). Most authors who had accepted the first thesis (*Evolution as such*) rejected one or several of Darwin's other four theses. This shows that the five theses are not one indivisible whole; the core thesis of natural selection was not generally accepted until the 1940s during the evolutionary synthesis. Mayr (1977a, 1988e: 95–115) analyzed the impact of Malthus (1798) on Darwin's thinking, the developments in Darwin's theoretical ideas between the return from the *Beagle* voyage (October, 1836) and the forming of natural selection theory (September–November, 1838), the question of the target of selection (the individual or the species), and the nature of the struggle for existence (fierce or benign). As Mayr stated in his autobiographical notes, he was then "particularly interested in the question to what extent Darwin's discovery of the principle of natural selection grew out of his natural history studies (internal causation) and to what extent it was derived from the *Zeitgeist* (external causation)." The widespread contemporary opposition to Darwin's theories indicated to Mayr that no external influence of the *Zeitgeist* on Darwin's theorizing was involved.

**Table 11.1.** The composition of the evolutionary theories of various evolutionists. All these authors accepted a fifth component, that of evolution as opposed to a constant, unchanging world (from Mayr 1982d: 506; 1991g: 37)

| | Common descent | Multiplication of species | Grad-ualism | Natural selection |
|---|---|---|---|---|
| Lamarck | No | No | Yes | No |
| Darwin | Yes | Yes | Yes | Yes |
| Haeckel | Yes | ? | Yes | In part |
| Neo-Lamarckians | Yes | Yes | Yes | No |
| T.H. Huxley | Yes | No | No | (No)[a] |
| De Vries | Yes | No | No | No |
| T.H. Morgan | Yes | No | (No)[a] | Unimportant |

a. Parentheses indicate ambivalence or contradiction

**Fig. 11.1.** Darwin's explanatory model of evolution through natural selection (from Mayr 1997b: 190)

Mayr had conceptualized the difference between Darwin's population thinking and typological essentialism during the late 1940s (p. 361) and had explained it in a short paper (1959b) on "Darwin and the evolutionary theory in biology." For many years this was the most widely quoted reference by other authors to these so drastically different ways of viewing the world. Population thinking takes into consideration the uniqueness of each individual in a population of organisms and unlimited variation which, in conjunction with natural selection, may lead to further evolutionary change. On the other hand, typologists assume that the unchanging "essence" of each species determines the extent of its variation. For typologists, the presumably fixed limits of variation preclude evolution from occurring (except through saltations).

Darwin's inference from the first three facts in Fig. 11.1 was the "struggle for existence" and resulted from his reading Malthus (l.c.) on September 28, 1838. Mayr felt strongly that the origin of Darwin's "population thinking" was from his reading the literature of animal and plant breeders during the six months before the Malthus event. Darwin had learned that every individual in the herd was different from every other one and extreme care had to be used in selecting the sires and dams from which to breed the next generation. All that Darwin needed to know was that individual variation occurred (step 1 of natural selection). A correct theory of the origin of variation (non-existing at the time) was not a prerequisite for the establishment of the thesis of natural selection. Selection of individuals with particular heritable qualities (step 2), continued over many generations, automatically leads to evolution (inference 3). This is natural selection (selective demands) as a cause or mechanism. Natural selection as an outcome is an a posteriori phenomenon of probabilistic nature. It is the survival of those individuals that have attributes giving them superiority in the particular context. Natural selection is the nonrandom retention (survival) of some of the new genetic variants. Darwin said: "The preservation of favorable variations and the rejection of injurious variations, I call Natural Selection" (1859: 81), natural selection as an outcome. Although natural selection is an opportunistic optimization process, it does not necessarily lead to perfection. There are numerous constraints. Also, selection as a process obviously was unable to prevent the extinction of the majority of species which originated during the history of life.[3]

Immediate support for Darwin's concept of natural selection came only from several naturalists, especially Wallace, Bates, Fritz Müller, Weismann, Alfred Newton, and initially Tristram (who interpreted the substrate adaptations of North African larks as due to natural selection in 1859 before the publication of the *Origin*, but regretted this interpretation after the publication of the *Origin of Species*).

[3] See Bock (1993) for a discussion of natural selection as (1) a cause or mechanism, (2) a process, or (3) an outcome.

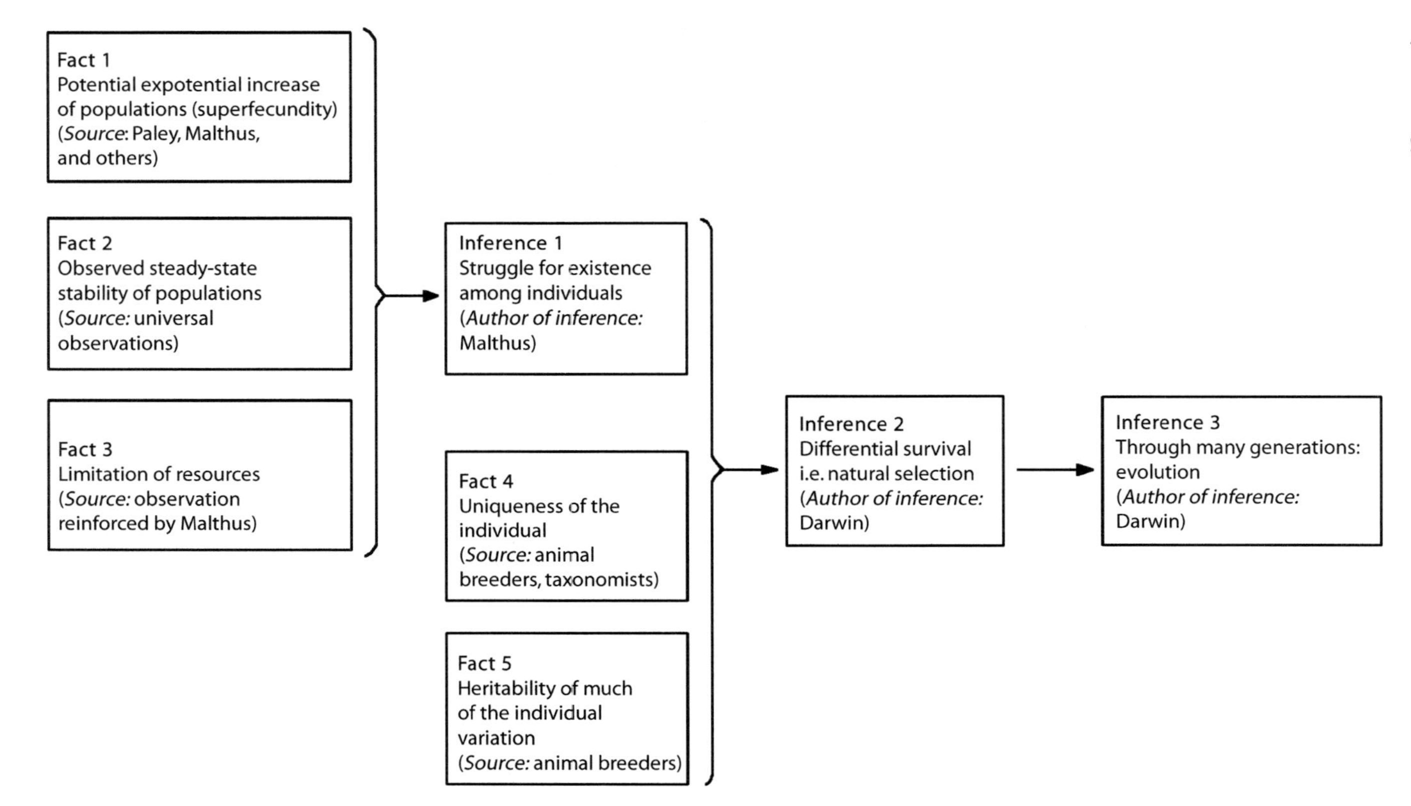
Fact 1
Potential expotential increase of populations (superfecundity)
(*Source*: Paley, Malthus, and others)
Fact 2
Observed steady-state stability of populations
(*Source:* universal observations)
Fact 3
Limitation of resources
(*Source:* observation reinforced by Malthus)
Inference 1
Struggle for existence among individuals
(*Author of inference:* Malthus)
Fact 4
Uniqueness of the individual
(*Source:* animal breeders, taxonomists)
Fact 5
Heritability of much of the individual variation
(*Source:* animal breeders)
Inference 2
Differential survival i.e. natural selection
(*Author of inference:* Darwin)
Inference 3
Through many generations: evolution
(*Author of inference:* Darwin)

Most other zoologists strongly resisted this concept until the 1930s–1940s, although some "minor" selection was "permitted," such as, e.g., the mimicry of the cuckoo's eggs to those of the host species. Darwin called an individual's varying success in reproduction "sexual selection" and devoted almost two thirds of the text of his *Descent of Man* (1871) to its discussion. This important process was largely ignored during the ensuing 100 years, even though Mayr (1963b: 199–201) called attention to the importance of "selection for reproductive success."

**Facsimile edition of the *Origin* (1964).** When, during the early 1960s, Mayr began studying Darwin's thoughts in detail, he had difficulties referring to the first edition of the *Origin*, which had become rare. Virtually every reprint was either one of the sixth edition or, if it was the first edition, the paging had been changed. He succeeded in persuading Harvard University Press to produce, in 1964, an inexpensive facsimile of the first edition which became a surprising success having "sold 4,000 copies in the hardback edition in the 7 years since it was published" (Mayr 1971h: 273). The 13th printing of the paperback edition came out in 1994 and continued to sell often a thousand copies per year. Mayr (1964l) wrote a detailed "Introduction" (20 pages) to this facsimile edition emphasizing that Darwin's book indeed "shook the world." Every modern discussion of man's future, the population explosion, the purpose of man and the universe, and of man's place in nature rests on Darwin. He was wrong in his views on inheritance and the origin of variation, confused about "varieties" and species, and unable to elucidate speciation, but he discovered the basic mechanism of evolutionary change–natural selection. This "force" not only eliminates the unfit (or less fit) but is also positive and constructive accumulating the beneficial and, at the same time, is probabilistic. Darwin had accepted an inheritance of acquired characters from the first edition on and not as the result of the later attack by H. C. F. Jenkin (1833–1885). Opposing claims notwithstanding, Darwin was a major philosopher, as Mayr (l.c.) emphasized. He applied routinely the hypothetico-deductive and comparative methods as well as "population thinking" rejecting Plato's essentialism (or typology) which had dominated science for over 2,000 years.

The criticism of Darwin's style of writing by some authors is "not only irrelevant but largely undeserved. Indeed some of his writing has considerable literary quality, for instance p. 130, 310–311, 489–490" (1964l, p. XXI). Some authors criticized that Darwin did not give sufficient credit to his precursors. Up to a point this is true, because the *Origin* has no bibliography or other references to the literature (which, however, was not customary in those times). For example, Darwin knew Leopold von Buch's theory of geographic speciation (1819, 1825), as documented by one of his *Notebooks*, yet there is no reference to L. von Buch on page 301 of the *Origin* in connection with an argument apparently adapted directly from this forerunner (1964l, p. XXI).

**Darwin studies.** Since the Darwin jubilee in 1959 a veritable "Darwin industry" has developed which no single person can fully deal with any longer. In his book of 1982 (d) Mayr presented his own interpretation of the major questions of the Darwin research and discussed other aspects of Darwin's thoughts in various articles.

The concept of organic evolution includes two independent processes, (1) transformation (anagenesis, phyletic evolution, the "vertical," adaptive component) and (2) diversification (cladogenesis, speciation, the "horizontal" component of change in differentiating populations and incipient species). At first Darwin was aware of this difference but later did not sufficiently stress the independence of these processes. He developed the theory of geographic speciation on islands and believed that his "principle of character divergence" (ecological differentiation supposedly explaining sympatric speciation) would overcome the difficulty that on continents barriers are apparently lacking. Darwin more or less retained his theoretical views from the 1840s on, even though he changed his mind regarding the importance of certain factors, e.g., geographical isolation and soft inheritance. He reversed himself with respect to geographical isolation during the 1850s, when he accepted sympatric speciation through ecological, seasonal or behavioral specialization treating speciation increasingly as a process of adaptation and no longer considered species as reproductively isolated populations.

Darwin's theses on evolution are summarized by Mayr in a small book, *One Long Argument. Charles Darwin and the Genesis of Modern Evolutionary Thought* (1991g) and related to the development of the modern theory of evolution. Darwin's *Origin* was "one long argument" against special creation, not one in favor of natural selection. His scientific method was the time-honored method of the best naturalists going continually back and forth between making observations, posing questions, establishing hypotheses or models, testing them by making further observations, and so forth (the hypothetico-deductive method). Perhaps Darwin was the first naturalist who used this method so consistently and with as much success. With his talents and interests he became a bridge-builder between various scientific fields.

Soon after 1859 most scientists accepted Darwin's theses of evolution as such, common descent and multiplication of species which Mayr (1991g) termed the "First Darwinian Revolution" (the "Second Darwinian Revolution" refers to the Evolutionary Synthesis of the 1940s, when the geneticists and naturalists reached a consensus, p. 183)[4]. The *Origin* was a superb treatment of the thesis of common descent and a great plea for the efficacy of natural selection, but, according to Mayr, it was vague and contradictory both on the nature of species and on the mode of speciation. Darwin's theses challenged the traditional religious and philosophical views of his time, although not every one of his theses was in conflict with all of them. The concept of natural selection was evidently not a reflection of the industrial revolution and the socioeconomic situation because this concept was almost unanimously rejected by Darwin's contemporaries. Ideological factors, like the philosophy of essentialism, also had a powerful effect on the rejection of several of his theses. Darwin accepted the strict working of "natural laws" at the physiological level, but was aware of chance (stochastic) processes

---

[4] Occasionally, Mayr (1977a, 1982d: 116–117) also referred to Darwin's theories of common descent and natural selection as the "first" and "second" Darwinian revolution, respectively.

at the organismic level. Darwin's views on classification, the "Natural System," are being discussed controversially. Whereas Mayr (1969b, Mayr and Bock 2002h) contended that Darwin had proposed to take into consideration both common descent (genealogy) and morphological divergence of sister taxa, Ghiselin (2004) quoted from Darwin's publications suggesting that classifications should reflect exclusively genealogy.

The term "Darwinism" has several different meanings: Darwinism as selectionism is true today and since the Evolutionary Synthesis (Synthetic Darwinism, Junker 2004). During the late 19th century, this was also true for Weismann, Wallace and several other early naturalists. August Weismann, the most important evolutionist immediately after Darwin, showed (for animals) that the germ line and the somatic line are strictly separated and, therefore, he rejected the inheritance of acquired characters (Mayr 1985h). Darwinism as variational evolution refers to the non-essentialistic tenets of this theory. Darwinism as the creed of the Darwinians refers to their rejection of special creation; he who believed in the origin of the diversity of life through natural causes, was a Darwinian. Darwinism signified the destruction of the previously ruling worldview. Neo-Darwinism as practiced by August Weismann since 1883 is Darwinism without 'soft' inheritance, the inheritance of acquired characters.

**Book reviews.** Mayr's reviews of newly published books on Darwin and Darwinism are more than mere descriptions of their contents, because he used these opportunities to point out to his readers the continuing relevance of Darwin's ideas which far transcends evolutionary biology. In 1971(j) he referred to the publication of Darwin's correspondence and portions of the Darwin papers in the library of Cambridge University (including the *Notebooks*). At first Mayr had thought that nominalism ("only individuals exist") had something to do with Darwin's introduction of population thinking, but he did not find any support for this suggestion in the writings of any author influenced by nominalism. Instead, population thinking probably originated among the naturalists [who began to collect "series" (population samples) of specimens during the mid-19th century] and among the animal and plant breeders. Mayr here quoted Darwin's letter to Asa Gray (20 July 1856): "My notions about *how* species change are derived from long-continued study of the works of [...] agriculturists and horticulturists."

In his review of David Hull's book, *Darwin and his Critics* Mayr (1974k) emphasized that Darwin, in 1859, forced his readers to choose one of three alternatives: (1) Continuing creation of species and their adaptations, (2) teleological evolution, or (3) evolution by random variation and natural selection with no recourse to any supernatural intervention, even at the beginning. In each of 12 detailed contemporary reviews of the *Origin* assembled in Hull's book creationism is prominent. Mayr admitted that Darwin provided little evidence for the change of one species into another in succeeding geological strata or the production of new structures and taxonomic types by natural selection (historical evolution). Also, Darwin left unanswered the origin of life and of new genetic variation, and he was confused about certain problems. The four favorable reviewers of the *Origin* still did not feel comfortable with the concept of natural selection and were not truly thinking in terms of variable populations.

Darwin's theses on evolution were received much more favorably in North America and in Germany than in France and Britain. *Naturphilosophie*, early evolutionary interpretations, and a wave of extreme materialism had prepared for Darwin's enthusiastic reception in Germany.

In 1978(f) and 1980(b) Mayr spoke for the first time of Darwin's several evolutionary theses listing separately (1) his thesis that life is constantly changing, evolving, (2) the thesis of life's common descent including Man and speciation, and (3) the concept of natural selection. In another invited essay "Darwinistic misunderstandings" of several continental European workers are clarified (1982m) showing that their criticism deals with the beliefs of atomistic genetics and does not target genuine Darwinian evolutionism. Occasionally, the "death of Darwin and his evolutionary theses" or "Darwinism: this century's mistake" have been pronounced. Such publications immediately excited Mayr's (1984h, 1984i) response who patiently pointed out again the basic facts of Darwinism and the numerous mistakes and misunderstandings of the respective authors.

In his review of P. Bowler's book, *The non-Darwinian Revolution*, Mayr (1990a) refuted the author's thesis that because natural selection was generally rejected during the 1860s–1870s (and until about 1900), therefore the evolutionary revolution of those times was "non-Darwinian" (!?). This author and many other historians ignored the complexity of Darwin's theoretical framework and in particular his five main evolutionary theses. The total victory of evolutionism (the "First Darwinian Revolution") during the late 19th century was transformational evolution. Non-Darwinian elements of the developing proposals in those times were the teleological, saltational and Lamarckian components of the anti-Darwinian reaction. Mayr concluded that Bowler's notion of a "non-Darwinian revolution" is a myth.

In Mayr's reviews of two Darwin biographies, by G. Himmelfarb and J. Browne, respectively, the accounts on Darwin's life and personality, on his intellectual development, working habits, his relations to his family, his interaction with friends and opponents and other aspects of his life as well as the emergence of evolutionary thought and the reception of the *Origin* (1859) are praised (1959g, 2002f). On the other hand, Mayr (1959g) criticized severely the second part of Himmelfarb's biography dealing with Darwinian evolutionary theory which the author—as a historian—has misunderstood in large part. Janet Browne is also a historian, not an evolutionist, and therefore evidently felt that it was not her task to analyze Darwin's evolutionary paradigm, which is not discussed, even though her biography comprises two large volumes. Short treatments of Darwin's theoretical ideas are included in Mayr's books (1982d, 1991g) and articles, but a comprehensive modern analysis is lacking.

**The Evolutionary Synthesis.** The "Fisherian synthesis" of mathematical population genetics during the 1920s (R.A. Fisher 1930, S. Wright 1931, and J.B.S. Haldane 1932) had solved one of the two great problems of evolution, the interaction between genetical changes and selective demands leading to phyletic evolution—that is evolution as Natural Selection or evolution as such. The other main problem of evolution, the origin of diversity through the multiplication of species, was solved during the "Evolutionary Synthesis" between 1937 and 1950 (p. 183). Mayr summarized the events leading up to the evolutionary synthesis as follows (1982d: 535–570; see Table 8.1).

During the late 19th century, a widening gap had developed between the experimental biologists including the geneticists and the naturalists (most zoologists, botanists, and paleontologists) who worked with whole organisms. Both groups were interested in evolution but had difficulties communicating with each other. The naturalists dealing with populations studied mainly diversity, its origin and meaning; the systematists were intrigued by the species problem and the paleontologists by evolutionary trends and the origin of the higher taxa. By contrast, the geneticists dealing with genes focused on evolutionary changes within populations, transformational evolution, disregarding diversity, the origin of taxa. After 1900 and the rediscovery of Mendel's rules Bateson and de Vries proposed typological mutationism (saltationism) as an explanation of the origin of species, whereas the naturalists were impressed by and emphasized gradual variation and speciation (either sympatric or allopatric). The naturalists continued to believe in soft inheritance, but acknowledged natural selection as a major evolutionary force.

Advances made in both groups during the first decades of the 20th century prepared an eventual reconciliation of the two opposing camps. According to Mayr population geneticists (Haldane, R.A. Fisher, Wright, Chetverikov, Timoféeff-Ressovsky) showed that (1) there is no soft inheritance, (2) recombination and small mutations are the most important source of genetic variation in populations, (3) continuous phenotypic variation is not in conflict with particulate inheritance, and (4) natural selection is an effective evolutionary cause ("Fisherian synthesis"). Population systematics of the naturalists had been in existence since the early 19th century and new studies of the early 20th century on fishes, moths, and mice could easily be translated into population genetics. The naturalists studied "series" as population samples, geographical gradients of populations, analyzed adaptive variation statistically and studied geographical speciation (Mayr 1963b).

Geneticists and naturalists reconciled their differences during the short period of the Evolutionary Synthesis (1937–1950), when a small group of evolutionists in North America and Europe were able to build bridges among different fields and to remove misunderstandings. These "architects" bridged the gap between the gene-frequency approach of the reductionist population geneticists and the population thinking of the naturalists, especially the study of species and their variation. The main architects were Dobzhansky, Mayr, Simpson, Huxley, Stebbins in North America and several workers in Germany and Russia (Mayr 1999a). At an international conference in Princeton, New Jersey, in January 1947 there was general agreement among the participating geneticists and naturalists on the gradualness (continuity) of evolution (but with varying rates), the importance of natural selection, and the populational aspect of the origin of diversity. A true synthesis had occurred between the two very different research traditions (Jepsen, Simpson, and Mayr 1949). Some differences that remained at that time included the problem of the target of selective demands which, for the population geneticists continued, until the 1980s, to be the gene, whereas the naturalists insisted that it was the individual as a whole.

In May and October 1974, Mayr invited, with the support of the American Academy of Arts and Sciences, a number of leading evolutionists to two conferences to discuss the genesis and historical course of the evolutionary synthesis.[5] The published proceedings (Mayr and Provine 1980e) are important but represent only a beginning in the study of this period. Numerous questions remain open (1982d: 568, 1993a, Smocovitis 1997), e.g., what was the particular role played by each of the "architects" in North America and Europe? Which insights or which particular phenomena permitted an agreement to be reached? The beginning of the synthesis was certainly marked by T. Dobzhansky's *Genetics and the Origin of Species* (1937) where he discussed population genetics, variation in natural populations, selection, isolating mechanisms, and species as natural units. This book provided the bridge between genetics and systematics. Dobzhansky was trained in Russia as an entomological taxonomist and combined his knowledge in this field with his experience as an experimental geneticist in Morgan's laboratory. Recent historical research has suggested that the Evolutionary Synthesis was not restricted to the Anglo-American region but was an international process in which not only B. Rensch but also several other European geneticists and systematists in Germany participated, especially E. Baur, N. Timoféeff-Ressovsky, and W. Zimmermann (Mayr 1999a, Junker and Engels 1999, Reif et al. 2000, Junker 2004). Timoféeff-Ressovsky had brought Russian population genetics to Germany in 1925, just as Dobzhansky had "imported" it to North America in 1927. The Evolutionary Synthesis was a major step in the maturation and implementation of Darwin's theory of evolution and the most decisive event in the history of evolutionary biology since the publication of the *Origin of Species* in 1859 (1982d: 569, 1993a, 1997g).

**Post-synthesis developments.** The controversies among evolutionary biologists since the evolutionary synthesis during the second half of the 20th century have taken place and are taking place largely, but not completely within the framework of modern synthetic Darwinism, e.g., discussions on the occurrence of sympatric speciation, the existence or not of cohesive domains within the genotype, the relative frequency of complete stasis in the evolutionary history of species, the rate of speciation, and others. The development of the technique of starch gel electrophoresis permitted the study of molecular evolution and the construction of a "molecular clock." The work of M. Lerner provided massive evidence for gene interaction, reinforced by Dobzhansky's research on "synthetic lethals." This spelled the end of the faith in constant fitness values of genes (Mayr 1963b, chapter 10). The target of selective demands is the whole organism rather than genes. Therefore a separation into "internal" and "external" selective demands is not feasible and the effects of recombination and gene regulation are at least as

[5] Sewall Wright was not among those invited because his contributions were part of the earlier "Fisherian synthesis" of the 1920s and early 1930s, not of the Evolutionary Synthesis (1937–1950), as defined by most authors.

important for selection as are mutations. Selective demands encounter numerous constraints and cannot produce perfection. Every genotype is a compromise between various selective demands. The "balance" school (Dobzhansky, Lerner, Mayr) considers the genotype a harmoniously balanced system of many genes with the heterozygotes often superior to the homozygotes. Fitness values of genes depend on their "genetic milieu" and hence on the resulting individual phenotype (Chetverikov). A high level of allelic polymorphism was confirmed through enzyme electrophoresis since 1966 and proved the enormous genetic variability of natural populations.

The genetic studies of speciation in Hawaiian *Drosophila* by H. Carson in 1975 convincingly confirmed Mayr's theory of peripatric speciation from founder populations. Sympatric speciation is considered to be possible in certain families of freshwater fishes and host-specific insects. The greatest unsolved problem in speciation research remains that of the genetic basis of speciation. Perhaps only a small number of genes are involved. Paleontologists traditionally concentrated on the "vertical" component of evolution until Eldredge and Gould (1972) proposed their model of "punctuated equilibria" which was based on Mayr's (1942e, 1954c) model of geographic speciation in small isolated populations. The claim of Eldredge and Gould (l.c.) that the new species originate by saltations as "hopeful monsters" (*sensu* R. Goldschmidt), not as a populational process, was later retracted.

## Biographical studies

After he had written a few biographical contributions from the 1930s to the 1960s, Mayr published numerous historical accounts on the lives of biologists during the 1970s (ten), 1980s (ten), 1990s (twenty-seven) and 2000s (eighteen). Among these are (1) brief obituary notices or personal memories of friends and colleagues, (2) biographies or memories of biologists which include summaries of their scientific work, and (3) biographies of biologists with fairly extensive discussions of their scientific contributions (see listing below). Among these are many brief memories or obituaries of ornithologist friends in Germany (E. Stresemann, B. Rensch, G. Schiermann, R. Zimmermann) and in the United States (D. Amadon, H. Birckhead, W. Drury, E. T. Gilliard, L. Sanford) as well as formal contributions to the *Dictionary of Scientific Biography* (1975, 1976, 1990) for F. Chapman, D. Davis, K. Jordan, G. Noble, R. Ridgway, M. Sars, C. Semper, K. P. Schmidt, E. Stresemann, and C. O. Whitman. Although Mayr was limited in each of these latter cases to only one thousand words, even this short treatment required a lot of research. Among additional obituaries of ornithologists are those of E. Hartert, J. Delacour, D. Lack, and A. Wetmore most of whom Mayr had known well over many years. I included in the text of this biography E. Mayr's memories of friends from the ten summers he spent at Cold Spring Harbor (A. Buzzati-Traverso, E. Caspari, M. Delbrück, Th. Dobzhansky, C. Stern, and B. Wallace) and of several others (R. Goldschmidt, J. Moore, F. Weidenreich).

*(1) Brief memories, obituaries or biographies of friends, colleagues or other biologists:*

Amadon 1998b, Birckhead 1945b, Buzzati-Traverso, here p. 246, Caspari, here p. 247, Dathe 1992h, Drury 1998a, Delbrück, here p. 247, Fisher, R. A., here p. 238, Gilliard 1969f, Goldschmidt, here p. 195, Griscom 1995k, here p. 109, Haldane, here p. 237, Hamerstrom, F. and F. 1992l, Hartert 1934c, Huxley, J., here p. 237, Lorenz 1997l, here p. 127, Moore, J. A. here p. 297, Rensch 1992d, here p. 43, Sanford 1997l, here p. 101, Schiermann 1997l, here p. 46, Schüz 1993e, Schuster 1955d, Simpson 1980o, Stebbins 2002d, Stern, 1980p, here p. 249, Stresemann, 1949j, 1969j, 1975a, b, 1990l, here p. 41, Timeféeff-Ressovsky 1993j, Wallace, R. A., 1994n, Wallace, B. here p. 248, Weidenreich, here p. 330, Wickler 1996f, Zimmermann 2003, here p. 20.

*(2) Biographies or memories of biologists including brief summaries of their scientific work:*

Chapman 1990e, Davis 1990f, Delacour 1986f, de Vries 1982d, Dobzhansky, here p. 133, Fleming 1989f, here p. 263, Jordan, K. 1990g, Lack 1973j, Mendel 1982d, Murphy 1974a, Noble 1990h, Ridgway 1975j, Romer 1990i, Sars 1975d, Semper 1975e, Schmidt, K. P. 1990j, Stresemann 1973e, 1990k, 1997l, Wagner, M. 1996g, Wetmore 1979c, White, M. J. D. 1985g, Whitman, C. O. 1976j.

*(3) Biographies of biologists and fairly extensive discussions of their scientific work:*

Darwin 1991g, Haldane 1992f, 1993h, 1995d, Jordan, K. 1955e, Miller, A. H. 1973d, Weismann 1982d, 1985h.

*(4) Autobiographical accounts:*

1930f, 1932e, 1938f, 1943h, 1980n, 1981b, 1992i, 1997d, 1999j, 2004b

Mayr repeatedly discussed the work of Charles Darwin as summarized above and mentioned the conceptual contributions of numerous biologists in his historical handbook (1982d). As part of the celebrations of the 100th anniversary of the birthday of the geneticist J. B. S. Haldane, Mayr (1992f) reviewed his classic book, *The Causes of Evolution* (1932) from the point of view of the historian, based on a reprint edition (1990), and compared Haldane's evolutionary views with the current consensus (see also Mayr 1993h). The main purpose of Haldane's volume was to synthesize Mendelian genetics and Darwinian evolution. Like Fisher (1930), Haldane also refuted soft inheritance so effectively that Dobzhansky (1937) devoted little space to this topic. Haldane argued against the belief of the Mendelians in "mutation pressure" as a factor that could override natural selection. Because the importance of this evolutionary factor had not yet been accepted generally, Haldane included an entire long chapter to its discussion and stated that in addition to the origin of genetically-based phenotypic variation, "natural selection is the main cause of evolutionary change in species as a whole." He was aware of "neutral characters" whose incorporation into the phenotype he explained by pleiotropy of many genes, and he pioneered the beginning of discussions of the evolution of altruism.

Fisher, Haldane, and Wright were mostly dealing with evolutionary change as such, that is with the causes and maintenance of adaptedness of populations. However, they did not have the background to deal with the evolutionary biology of biodiversity and its origin, geographical variation and speciation ("It was the major contribution of the naturalists to have brought the evolution of biodiversity into the evolutionary synthesis;" Mayr 1992f, p. 181). Another evolutionary cause relatively neglected by Haldane and most other evolutionists in those decades was sexual selection. However, on the whole, Haldane's interpretation of evolution was very sound. In his arguments against creationism, he pointed out the frequency of extinction, and the frequency and awfulness of parasitism.

In his essay on *Daedalus* (1923) the young Haldane presented a view into the future and discussed some of his deepest beliefs and hopes (Mayr 1995d). He was optimistic about a better future for mankind and was convinced that science, particularly the science of biology, would lead us to this desired goal. For Haldane science was "an endless frontier." He realized that conventional sources of energy, like oil and gas, are limited and predicted that other sources, like wind and sun light, will be utilized increasingly in the future. He also prophesied the shift from agricultural to an almost completely industrialized society; mankind will be completely urbanized, he thought. He was convinced that "the centre of scientific interest lies in biology" and discussed eugenics. Mayr (l.c.) believed that in our days, Haldane would have been less optimistic in view of overpopulation, pollution, destruction of the environment, and other reasons. He would have been pleased to see that biology had indeed become the queen of sciences, as he had predicted, especially in view of the spectacular achievements of molecular biology. He would have thought a good deal about developments relating to human health, particularly the study of human nutrition, one of the most neglected fields of preventive medicine.

## Philosophy of Biology

Mayr's interest in philosophy also goes back to his youth. In his father's library, philosophical titles filled several shelves and he always referred to philosophy with great respect. Discussions with fellow students in Greifswald in 1924 often concerned philosophical topics. One of Mayr's notebooks of those years contains the following passage which refers to himself (transl.):

"The more he thought about the problem of the position of man in nature the more he saw that there is only one solution: Man and animals form one unit! This is the absolutely necessary consequence of all biological laws and knowledge. Thus he finally had a basis for his worldview" (Weltanschauung).

At the University of Berlin in 1925/1926, Mayr took courses in the history of philosophy and a seminar in Immanuel Kant's *Critique of Pure Reason* but the teachers made no attempt to indicate any connections between these philosophical topics and biological problems. In his PhD examination in positivism he passed with an A because he had been well prepared. The only books Mayr carried with him on his expeditions to the forests of New Guinea in 1928–1929 were Driesch's *Philosophie des Organischen* (1899) and Bergson's *Schöpferische Entwicklung* (1911), as he mentioned in a letter to E. Stresemann dated 24 March 1934 (Haffer 1997b). By the time he returned to Germany, he had concluded that neither Driesch nor Bergson was the answer to his search (2004a: 2). He said that during the following 15 years or so he more or less ignored philosophy. However, his discussions on the nature of biological species, classification, and evolution during the early 1940s have a most important bearing on the philosophy of biology. It is essential to realize the central bearing of Mayr's empirical work on the systematics and biogeography of birds on his future thinking on the philosophy of science.

As editor of the newly founded journal *"Evolution"* (1947–1949) Mayr corresponded with numerous zoologists on their work and, at the same time, began to think in more detail about underlying historical and philosophical aspects. In his letters to Professor Stresemann he commented on maturation of the species concept and criticized that "the school of symbolic logic, started by the mathematicians Bertrand Russell and Woodger, is now trying to invade biology with a strictly static and formalistic philosophy. As one would expect from such a philosophy they deny the existence of species. [...] A small minority of biologists want to take biology 'back to nature' " (24 October 1949) and "I am afraid philosophy has been rather a handicap in biological research as the writings of Schindewolf, Beurlen, Dacqué and others [in Germany] prove" (20 November 1947).

Mayr never wrote a textbook on the philosophy of biology but his books, *This is Biology* (1997b) and *What Makes Biology Unique?* (2004a) as well as many papers on evolutionary theory summarize his thinking in this field and provide a good overview. However, in view of his advanced age he was unable to develop a coherent philosophy of science that included all aspects of biology (Bock 2006). Mayr expressed his regret on not having published a book on the philosophy of biology in his video interview in November 2003 (Bock and Lein 2005), and said that he simply ran out of time.

## Population Thinking

In several letters to Erwin Stresemann during the late 1940s, Mayr asked why the Anglo-Saxons (Darwin, Wallace) rather than the continental European scientists had found the right solution for the great problem of evolution and proposed that they had always been interested in the study of populations and of variation unencumbered by the idealistic philosophy of continental, especially German biologists. As typologists, they had considered all biological phenomena as type phenomena (letters dated 14 March 1949 and 14 December 1950). Mayr repeatedly referred to the importance of "population thinking" (the uniqueness of individuals) in his letters, e.g.:

"It is interesting to see in which way the methods of research have changed during our lifetime! Local populations are now studied increasingly. It is no longer stated *the* Yellowhammer does this or that, but 25% of the males in the Song Sparrow population of central Ohio have been nonmigratory and 75% migratory, etc. Such variations within populations in particular have been underestimated or entirely neglected by earlier biologists. [...] I consider the replacement of the 'type' as the object of study by the population as the most basic revolution in biology" (9 August 1948). "As soon as you think of species or subspecies or any other type of biological phenomenon as populations it is easy to see how selection can affect it. It is also easy to see how the bridge can be made from one of these populations to the next one. [...] In Germany the question was always asked as to what does *the* thrush do or *the* chaffinch or *the* nightingale. In other words, the type was investigated rather than the population of which it is composed" (14 March 1949). "In connection with my book on the principles and methods of taxonomy, which I am now working on, I had to do a little thinking about the history of the field. [...] it became abundantly clear to me that the world of biology owes to taxonomy one of its greatest concepts, namely, the *population concept* [emphasis added]. The anatomists (and many paleontologists) are still confirmed typologists and so were the geneticists until they were awakened by three students who had been trained by taxonomists: Goldschmidt, Sumner, and Dobzhansky. The taxonomists had started to think in these terms easily fifty to 75 years earlier by collecting 'series' and by collecting in adjacent localities" (28 April 1950). "Actually, no two populations of a species are ever quite the same, they differ in their winter quarters, in the beginning of the breeding season, in the average number of eggs they lay, young they raise, and preferred nesting sites, etc. In some cases these differences are gliding [clinal], in others they are quite abrupt. To attach names to these differences is usually not helpful" (4 June 1968).

Typologists and populationists may be distinguished in this way: For a *typologist* (*essentialist*) the underlying essence of a phenomenon (well-defined, fixed, unchangeable, and separated from others by decided gaps) has reality and variation is irrelevant and no more than deviation from the type ("noise"). In *population thinking* variation is considered the reality (all individuals are unique), while the statistics and in particular the mean values of samples and populations are the abstractions. Natural selection and gradual evolution are meaningless for the typologist to whom the concept of race is also an all-or-none phenomenon. As early as 1924 Mayr had opposed the typological dogma of the creationist origin of animal species in a letter to Erwin Stresemann (p. 28).

Although the principles of "new systematics" as applied by Stresemann, Rensch, and Mayr since the 1920s are based on the study of populations rather than types and Mayr's species definition of 1942 refers to populations, none of them conceptualized "population thinking" based on the uniqueness of individuals until Mayr did so during the late 1940s. He established the fact that Darwin was the first to apply population thinking when he proposed the concept of natural selection (p. 350).

"The replacement of typological thinking by population thinking is perhaps the greatest conceptual revolution that has taken place in biology" (1963b: 5). Mayr's conceptualization of population thinking is a, perhaps *the*, major contribution to the philosophy of biology. He mentioned population thinking for the first time in print in his textbook, *Methods and Principles of Systematic Zoology* (1953a; see above p. 321, also p. 312) and formally introduced this concept in his article on "Darwin and the evolutionary theory in biology" (1959b).

During the 1950s, Mayr became acquainted with the literature of philosophy of science and was disappointed. This was a philosophy of logic, mathematics, and the physical sciences and had nothing to do with the concerns of biologists (1995j, 1997b: XI). When Snow (1959) spoke of two cultures, science and humanities, physics was for him the exemplar for science. For some physicists "all biology is a dirty science" because nice clean laws without exceptions do not exist in large parts of biology. This reflects the nature of biological phenomena, but some workers thought that biology could be reduced completely to physics.

The rise of a philosophy of biology began with the evolutionary synthesis as a contributing factor, when teleological and typological (essentialistic) concepts were eliminated from biology and continued especially during the 1960s with Mayr being a driving force. His books and articles in this field were mostly written from the point of view of evolutionary biology and provide a conceptual framework of biology and the basis for a new philosophy of science that incorporates the approaches of all sciences, including physics and biology. His publications in philosophy of biology were written in a lucid graceful style accessible to all interested biologists and knowledgeable persons. His main objectives were to help his readers to "gain a better understanding of our place in the living-world, and of our responsibility to the rest of nature" (1997b: XV).

## "The New versus the Classical in Biology"

The above account documents the early origin of Mayr's genuine interest in philosophical aspects of biology, particularly of systematics and evolution during the modern synthesis of the 1940s ("internal causation" of his interest). This interest grew stronger during the 1950s after he had joined the staff of the MCZ at Harvard University, when he dealt increasingly with general problems of evolutionary biology at a time when molecular biology, the new glamor field, threatened to dry up nearly all research funds available and classical organismic biology began to lose ground. Mayr became spokesman of systematics and evolutionary biology in the United States, especially after he had assumed the directorship of the MCZ in 1961. He criticized the generally unbalanced situation in American science such as, for example, the replacement of seven of eight retiring professors of biology at Harvard University by molecular biologists. Anyone in a more traditional area of biology had considerable difficulties in attracting promising students, and whole fields, especially of invertebrate zoology, became depleted of specialists. In view of this situation he wrote an editorial on *"The New versus the Classical in Biology"*

(1963k) which he ended stating "the new should supplement the classical and not totally displace it." Much to his surprise, it was published in the journal *Science* and even reprinted in an Italian newspaper.

Mayr analyzed various philosophical topics including the dual nature of biology, meaning its division into two areas, functional-physiological biology and evolutionary biology (1961c). Advances in both of these fields characterize the progress of a balanced biology. His emphasis of evolutionary biology and organismic biology vis-à-vis chemistry/physics-oriented functional biology during the 1960s may be seen, at least in part, as his strategy to counteract the pressure exerted by molecular biology on organismic biology (partially "external causation" of Mayr's interest in this aspect of the philosophy of biology[6]).

## Autonomy of Biology

During the 1950s Mayr saw clearly that any approach to a philosophy of science, essentially based on logic, mathematics, and physics rather than including also the specifically unique concepts of biology, would be unsatisfactory. One cannot talk about a philosophy of science when one only deals with the phenomena of the physical sciences. Among the differences between biological and nonbiological systems Mayr mentioned the enormous complexity of biological systems, the historical nature of organisms and of their genetic programs, that causality in organismic and evolutionary biology is predictive only in a statistical sense, teleonomic processes and behaviors, and population thinking versus essentialism. He began to write a series of essays over the next decades which he expanded in part and assembled in four volumes which make his contributions widely available (1988e, 1991g, 1997b, 2004a).

In his famous article on "Cause and effect in biology" Mayr (1961c) pointed out that no biological phenomenon is explained until both its functional-environmental (proximate) causations and its historical-evolutionary (ultimate) causes are determined. A northern warbler starts its fall migration on a particular night because of certain physiological-environmental (proximate) causes and because of the general genetic disposition of the bird (ultimate or historical-evolutionary causes). This is the basis of Mayr's other central contribution to philosophy of biology—that of "dual causation"—and is the major foundation for his correct position of biology being an autonomous science, i.e., a science that is fundamentally *not* reducible to the physical sciences. Other topics discussed were teleology and the autonomy of biology. As to the latter Mayr emphasized the differences between the inanimate, chemical-physiological world and the living world, where time and history are legitimate components for a philosophy of biology. Other biological aspects discussed were the frequent occurrence in biology of indeterminacy in general and emergence in particular. "Emergence" is the emergence of new properties at higher levels of integration in complex biological systems (the whole is more than the sum of its parts). Although many of the topics of this article

[6] As to Mayr's use of the expressions "internal" and "external causation" see p. 349

were not new—for example, ornithologists had previously distinguished between proximate and ultimate causations–these aspects were articulated more forcefully than in the earlier literature. Simpson (1963) used Mayr's article and also made a vigorous plea for the recognition of the fact that the principles of the physical sciences are simply inadequate for the explanation of the phenomena of life.

In his "Introduction" to a facsimile of the first edition of Darwin's *Origin* (1959) Mayr (1964l) pointed out that Darwin had been not only an extraordinary scientist but also a great philosopher. His refutation of creationism, teleology, essentialism, and deterministic physicalism were fundamental contributions to a philosophy of biology, as mentioned above (p. 348). During the AAAS meeting in December 1965 Mayr strongly urged the development of a philosophy of biology in a deliberately provocative manner (not published until 1969g).

Mayr's continuing preoccupation with the methodology and conceptual framework of biology is reflected in two further articles (1969c,d) where he discussed the genetic program which controls the development of individual organisms, but does not, as such, participate in the development. This machinery of translating the genetic program may be considered a general theory of development. Development is nothing but the decoding or translation of the genetic program of the zygote interacting with the environment in the making and subsequent life of the phenotype, the individual. Functional phenomena can be dissected into their physical-chemical components, but their integration to achieve novel insights at a higher level of organization is mostly unsuccessful. Reductionism as a philosophical approach, either to lower levels of organisms or to the physical sciences, is totally unimportant for a study of nearly all evolutionary problems. When Mayr objected to reductionism to the lower levels of organization, he did so in the general philosophical approach and not as the useful method of analysis which he termed simply "analysis." Unless one makes this distinction between philosophical reductionism and analysis (usually called reduction in philosophy), Mayr's position can be misunderstood.

In a series of publications (1982d, 1985c), Mayr continued to discuss the various ways in which biology differs from the physical sciences arguing for an extension of the philosophy of science to embrace biology as an independent branch of the sciences at least equal to physics and chemistry. The experiment is not the only method of science—observation, comparison, and the construction of historical narratives are also legitimate as are the important methods in the physical sciences (astronomy, geology, oceanography, meteorology) and in the biological sciences (comparative anatomy, systematics, evolutionary biology, biogeography, ecology, ethology). Experiment and observation have both their place in the sciences. Of course, all processes in living organisms are consistent with the laws of physics and chemistry. Special aspects of the world of life are uniqueness and variability, reproduction, metabolism, the possession of a genetic program, historical nature, and natural selection. In retrospect every biological process or behavior pattern can be explained causally, but in reality looking forward there are always several possibilities. For Mayr causality is an a-posteriori process (1985c: 49). He was explicit in his criticism of logical positivism, essentialism, physicalism, and vitalism.

The principles he listed (1982d: 75–76) as a good basis for a philosophy of biology include:

(1) Theories of physics and chemistry alone are insufficient fully to understand organisms.
(2) The historical nature of organisms must be considered, in particular their possession of a genetic program.
(3) Individuals at most levels are unique and form populations.
(4) There are two biologies, functional and evolutionary biology.
(5) The history of biology has been dominated by the establishment of concepts (their maturation, modification and occasionally their rejection).
(6) The complexity of living systems is hierarchically organized and the higher levels are characterized by the emergence of novelties.
(7) Observation and comparison are methods in biological research that are as scientific and heuristic as the experiment.
(8) The insistence on the autonomy of biology does not mean an endorsement of vitalism or any other theory that is in conflict with the laws of chemistry and physics.

Four areas with many open questions characterize the current frontier of biology: (1) the structure and functioning of the genome, (2) the ontogeny of organisms from the zygote to the adult stage, (3) the functioning of the brain with its billions of nerve cells, and (4) the interactions of different organisms in a complex ecosystem.

A philosophy of biology must include a consideration of all concepts of functional and evolutionary biology, systematics, behavioral biology and ecology. At the same time it should stay equally far away from vitalism and from a physicalist reductionism that is unable to do justice to specifically biological phenomena and systems, Mayr stated. He did not establish another (third) philosophy of biology between vitalism and physicalism, but had in mind a "biological reductionism" capable of explaining also specifically biological phenomena (Junker 2007).

## Teleology

In view of the composite nature of the concept of teleology the literature on this subject had been very confused until Mayr clarified this situation in several articles and chapters in his books (1974e, 1982d, 1984b, 1988e, 1992b, 1998e, 2004a). He distinguished four or five categories of teleology as follow:

(1) Teleomatic processes: Automatically achieved, like a rock falling to the ground. Such processes have an endpoint but they never have a goal.
(2) Teleonomic processes: A process or behavior that owes its goal-directedness to the influence of an evolved program. The controlling genetic program is constantly adjusted by the selective value of the achieved endpoint. The goal is already coded in the program that directs these activities.

(3) Purposive behavior in thinking organisms; widespread in mammals and birds (when, e.g., jays bury acorns and return later to these caches).
(4) [Adapted features: Mayr thinks that the word teleological would not seem appropriate for phenomena that do not involve movements, like adapted features.]
(5) Cosmic teleology: Change in the world was assumed to be due to an inner force or tendency toward progress and to ever-greater perfection (finalism). Modern science refuted cosmic teleology during the first half of the 20th century. The seemingly upward trend in organic evolution is explained by natural selection that favors the rise of ever better-adapted species (Darwin). Whether this upward trend is progress is still controversial.

## Reductionism and Emergence

The analysis of biological phenomena in terms of chemical and physical laws is called reduction. Mayr (1982d, 2004a) distinguished three categories of reductionism:

(1) *Constitutive reductionism* asserts that the material composition of organisms is exactly the same as that found in the inorganic world. This, of course, is accepted by modern biologists, none of whom is a vitalist.
(2) *Explanatory reductionism* claims that one cannot understand a whole until one has dissected it into its smallest components at the lowest level of integration. Biologists, however, know that dealing with the separated components of a complex biological system yields no information about their interactions (the whole is more than the sum of its parts). Emergent properties often appear in the upper levels of complex biological systems ("emergence").
(3) *Theory reductionism* postulates that the theories and generalizations formulated in biology are only special cases of theories and laws formulated in the physical sciences, in other words biological theories can be "reduced" to physical theories. It is only in functional biology of proximate causations that theory reduction is occasionally feasible, but no principle of evolutionary biology of ultimate causation can ever be reduced to the laws of physics or chemistry.

Mayr concluded his discussions stating that reductionism can be ignored in the construction of any philosophy of biology, at least in his rather restricted meaning of this concept. Note here Mayr's distinction between Explanatory Reductionism and Analysis mentioned above (p. 365).

In a letter to the journal *Nature*, Mayr (1988b) objected to the claims of Steven Weinberg that the SSC, the Superconducting Supercollider, would provide a much greater understanding of the universe. Reduction into smaller and smaller parts is a form of analysis, but it very rarely sheds much light on the nature of the higher level systems (with which Weinberg agreed). Yet he still believed that a further understanding of the nature of the atomic forces would be contributing to an understanding of the deepest riddles of the universe. Mayr rejected this claim and said that whatever would be found on the components of the atomic nucleus would not shed any light on the so-called middle world (mesocosmos), the world between the atom and the solar system. At the same time this article recorded Mayr's objection to the expenditures which the White House was willing to make for the study of physicists. The SSC would have probably cost ten billion dollars and the small result that was expected, in Mayr's opinion, did not justify this large sum. His objection may have contributed to the later disapproval of this project by the U.S. Congress. The same campaign against excessive government spending led Mayr later to object to the project of the Search for Extraterrestrial Intelligence (SETI), as mentioned below (p. 374).

## The Roots of Dialectical Materialism

Mayr's essay on the *Roots of dialectical materialism* (1997j) was written as a contribution to a memorial festschrift of the Russian Marxist theoretician K. M. Zavadsky who, during the 1960s, had asked the American historian of biology Mark Adams whether Ernst Mayr was a Marxist (which he was not), because "his writings are pure dialectical materialism." This remark had puzzled Mayr ever since and he began searching for the reasons of Zavadsky's comment. Mayr's essay (1997j) was the result of that search which showed certain similarities between the thinking of naturalists and that of Friedrich Engels and Karl Marx explaining Zavadsky's remark.

Dialectical materialism, a general philosophy of nature, was founded by Engels and Marx in the late 19th century, in part as a result of their own ideas and in part based on the analogous thinking of contemporary naturalists, especially Darwin regarding evolution. Among the principles of this philosophy are many with which Mayr had been familiar, since his youth, as principles of natural history. He listed six of them:

(1) "The universe is in state of perpetual evolution. This, of course, had been an axiom for every naturalist at least as far back as Darwin but as a general thought going back to the age of Buffon.
(2) Inevitably all phenomena in the inanimate as well as the living world have a historical component.
(3) Typological thinking (essentialism) fails to appreciate the variability of all natural phenomena including the frequency of pluralism and the widespread occurrence of heterogeneity.
(4) All processes and phenomena including the components of natural systems are interconnected and act in many situations as wholes. Such holism or organicism has been supported by naturalists since the middle of the 19th century.
(5) Reductionism, therefore, is a misleading approach because it fails to represent the ordered cohesion of interacting phenomena, particularly of parts of larger systems. Feeling this way about reductionism I have for many years called attention to the frequency of epistatic interactions among genes and to the

general cohesion of the genotype–Dialectical materialism emphasizes that there is a hierarchy of levels of organization, at each of which a different set of dialectical processes may be at work. This is the reason why reduction is often so unsuccessful.

(6) The importance of quality. The qualitative approach, for instance, is the only meaningful way to deal with uniqueness."

This kind of thinking by 19th century naturalists continued throughout the 20th century. But Engels would not have supported all the modern views held by Marxists, as shown by the case of Lysenko between about 1935 and 1965. His Lamarckian pseudo-science had nothing to do with dialectical materialism. Some modern Marxists are apparently opposed to the Darwinian principle of the uniqueness of the individual, presumably because it is in conflict with the principle of equality. However, as Mayr (l.c.) pointed out, genetic uniqueness and civic equality are two entirely different things. In view of the diverse abilities of human individuals it was necessary to provide diverse opportunities, as Haldane (1949) had already recognized. To insist, as many Marxists do, that all individuals are identical would be a falling back to classical essentialism. Human heterogeneity was not in any way in conflict with dialectical materialism.

## Natural Selection

In one of his latest contributions, Mayr (2004a) discussed natural selection as survival selection. Selection as a mechanism refers to true selection of the best or to "culling" (eliminating) the inferior individuals leaving all the rest for breeding. This means that in a real population in harsh years only the best individuals survive, in a mild year only the worst individuals are culled and most individuals survive, not only the best are "selected." The process of mild elimination leaves a large reservoir of variation available. This view of selection by eliminating the inferior members of a population provides a better explanation for this second cause of evolution than a "selection for the best," Mayr stated and continued very correctly that the expression "Survival of the fittest" is misleading to the extreme.

As to the controversial subject of kin selection Mayr (1990r, 2004a) distinguished between casual and social groups. The former, like temporary fish schools or flocks of birds, are never targets of selection, whereas social groups may have a high fitness value and may be the target of kin selection. The members of such groups have a potential for kin selection, if the groups are clearly delimited and compete with other such social groups, for example in early humans. According to Mayr, species as entities do not answer to selection. When one species replaces another, the individuals of these species interact and compete, not the species. Such events should be designated as species turnover or species replacement, not as species selection.

## The Ontological Status of Species

Since the 1970s a controversy is raging in the literature whether a species taxon as a whole is an ontological individual or a class. At first Mayr accepted the notion that a species taxon is a causal individual (1976h) but later changed his position somewhat and argued that a species is a variable *biopopulation* leaning toward its interpretation as a group and not an individual (1984g, 1987d, 1996e, 2004a). He stated that a biopopulation has the spatio-temporal properties, internal cohesion, and potential for change of a historical individual, but he preferred the designation 'population' because this term conveys the impression of the multiplicity and composite nature of a species. What Mayr evidently preferred was an ontological category for species somewhat intermediate between an ontological individual and a typological class. This is more or less the alternative proposition of those authors who oppose the interpretation of species as ontological individuals. They point out that species are neither typological classes nor ontological individuals but non-typological classes possessing family resemblance (Bock 1986, 2000):

I. Essentialism (refers to classes or groups)

1. Typological essentialism (Plato) assumes the reality of immutable essences underlying species and other objects.
2. Family group essentialism (Aristotle) refers to sets (e.g., species) of varying characteristics. The individual members of such a class (species taxon) share many, but not always all, or even most, characteristics. Under this view species taxa are classes possessing family group resemblances. The intrinsic isolating mechanisms of a species taxon constitute the family group essence of that class. Family resemblance essences vary geographically and over time. A species taxon does not have the type and degree of organization of an individual. It can be destroyed only when a large part of it is removed or almost all of its members.

II. Individualism (refers to ontological individuals)

Individuals are so organized that removal of certain very small portions will result in their destruction (e.g., the heart or the pituitary gland in the case of a human individual).

These views probably come close to Mayr's thinking, although in no paper did he ever clearly accept them. The question of whether species are classes or ontological individuals has relevance for the discussion of whether micro- and macroevolution are "decoupled," that is whether evolutionary theory is hierarchically structured or not. Causes can act only on ontological individuals, not on groups. Hence a hierarchically structured causal evolutionary theory requires that species and higher taxa are real ontological individuals (which, however, does not appear to be the case; Bock 2000).

## Evolution and Ethics

Evolutionism does not provide precisely defined ethical norms, nevertheless the acceptance of evolution leads to definite constraints. Three of the five Darwinian theses (p. 348) have ethical implications (Mayr 1984c):

(1) Evolution as such: The constant world of Christianity created 6,000 years ago (which will end on the day of judgment not far away) was replaced by Darwin with an evolving world of great age. Humans in a world controlled by God had no obligation for the future, neither with respect to the world's environment nor for its fauna and flora. Darwinism deprived man's morality of its very foundations and a new morality, a considerably different set of ethical norms from the Christian ones was required. The new world view of *evolutionary humanism* is conscious of the evolutionary past of humans and is aware of their responsibility for the community as a whole and for posterity. Man's challenge is to develop an ethic that can cope with the new situation.
(2) Darwin's thesis of common descent deprived man of his unique position in the world. He is little more than one of the species that originated through the evolutionary process and for this reason has no right to exterminate even the least product of evolution on earth, that is any other species of animals or plants. Man's ethic itself is a heritage of his primate ancestry.
(3) The thesis of natural selection, the struggle among the individuals of the same species resulting in differential reproduction. Altruistic behavior is quite often of high selective advantage when it enhances the reproductive success of a given genotype. Well-delimited social groups of early humans composed of closely related and cooperating individuals may be subject to kin selection (see above).

Most of man's behavior is guided by norms that were acquired after birth by education or self-learning. However, there are behavioral tendencies that involve a genetic component. Man's ethical norms are part of his cultural traditions. As to a religion without God (J. Huxley) *evolutionary humanism* is based on a feeling of solidarity with and loyalty toward mankind. Our most basic ethical principle should be to do everything toward the maintenance and future of mankind, the current caretaker of all of nature on our fragile globe. "An understanding of evolution gives us a world view that can serve as a sound basis for the development of an ethical system that is both appropriate for the maintenance of a healthy human society and that also provides for the future of mankind in a world preserved by the guardianship of man" (1984c: 46). This attitude requires a change in man's value systems which should include a concern of the community, the groups of which we are members, the largest being the human species, and all of nature. Many ethical values are 'density-dependent' (Mayr 1973a,b) like the freedom of unlimited reproduction and freedom of movement. Mayr pointed out that all primitive cultures have developed population control, pre-natal or post-natal control measures. Some form of birth-control is required. Overpopulation is the core problem which faces mankind and a change in attitude toward family size is needed. Voluntary birth-control appears to be not enough and Mayr recommended a set of incentives to be built into the tax system, pension system, and welfare system in the hope that this will lead to zero population growth in the United States (and elsewhere).

## The Biological Future of Mankind

Ever since his days as a student did Mayr favor positive eugenics, although he realized that it is difficult to achieve. In view of the growing overpopulation of the world "the time will come, and perhaps sooner than we think, when parents will have to take out a license to produce a child" and "positive eugenics is of great importance for the future of mankind and all roadblocks must be removed that stand in the way of intensifying research in this area" (in a letter dated April 14, 1971).

In his comments on J. B. S. Haldane's essay on "Daedalus or Science and the Future" of 1925 Mayr (1995d) pointed out that eugenics was universally popular during the early 20th century, from far-right to far-left writers. Enough was known about inheritance that in theory a genetic improvement of mankind seemed possible and Haldane speculated about the production of "ectogenic" test-tube babies (!) causing a scandal. His friend Aldous Huxley was inspired by Haldane's scheme of eugenics to elaborate this scenario in his book, *Brave New World* (1932).

In a lecture which Mayr gave at the Jungius Society of Hamburg on the biological future of man (1974f), he made clear that any truly biological improvement of man was only possible through eugenics, but left it open whether this was feasible or not. He emphasized that ethical values are nothing absolute but conditioned by circumstances.

"There were a lot of students in the audience, and in particular a strong delegation of the Spartakus Bund, a communist organization. They tried to refute me by invoking the communist paradise in Russia, but in answering them, pointing out that I had been in Russia and none of them had, and also refuting in detail every single other one of their arguments. With much applause from the other audience, I silenced them so completely that they decided it was better strategy not to answer me again. It simply would have given me more opportunity to describe the futility of Soviet communism."

(The "Biology Section" of the Spartakus Bund in Hamburg commented on Mayr's lecture in their pamphlet "The Red Handlense" (Die rote Lupe), Nr. 6, October 1973).

## Realism and Liberal Education

Most people adopt commonsense realism because it works (Mayr 1997i). They accept that there is an outside world and that it is more or less as our sense organs tell us: This is the middle world (mesocosmos) which extends from the atom to the solar system. The microcosmos is the world of the atom and elementary particles and the macrocosmos is the world outside the solar system. The micro-

and macrocosmos are irrelevant to the lives of humans, Mayr stated. However, many of the electronic devices we use all the time depend on the microcosm. Useful as these devices may be, they are not essential to the lives of humans.

Until about 50 years ago, physics was the dominant science as reflected in the philosophy of science. Now it is frequently stated that this is the age of biology which is due not only to the victorious march of molecular biology but also to the existence of a unified theory of evolution and advances in other branches of biology, e.g., neurobiology. A new philosophy of biology is developing, largely based on evolutionary thinking. Theories in biology are based on concepts and this is also the case for our worldview which is based on the concepts of democracy, freedom, altruism, competition, progress, and responsibility. Mayr suggested in this essay more room in the liberal education curriculum for the study of concepts that make up our worldview and a more detailed analysis of the concepts that are the basis of our belief in democracy.

The gap between the sciences and liberal arts (humanities) is more or less filled by biology, because there is an unbroken chain from physics across biology to the "softest" branches of the humanities. A line of demarcation between science and the humanities could be drawn right through the middle of biology, placing evolutionary biology with the humanities and functional biology with the sciences. Evolutionary biology shares with the humanities, particularly history, a number of attributes: Uniqueness of the treated entities, inability to predict, frequency of tentative (subjective) inferences, and relevance of religion and morality. These aspects are of great importance for liberal education (1997i: 294).

## The Advance of Science and Scientific Revolutions

Science makes steady advances leading to an ever improving understanding of the world. What is controversial is how these advances occur, through scientific revolutions or paradigm changes (Kuhn 1962) separated by periods of normal science or through Darwinian evolutionary epistemology. Kuhn's thesis reflects essentialistic-saltationist thinking for which gradual evolution is unacceptable. The Darwinian epistemologists introduced an entirely different conceptualization for theory change in biology, usually referred to as evolutionary epistemology. According to Mayr (1994i, 2004a) several scientific paradigms may coexist simultaneously for long periods of time and biological revolutions are not necessarily separated by long periods of normal science. Rather, there are always minor revolutions and theory changes of various magnitudes going on in any particular area of the sciences. Theory change in biology fits Darwinian evolutionary epistemology, consisting of variation (the continuous proposal of new theories) and selection (the survival of the successful theories). There is no clear cut difference between revolutions and periods of normal science, and one finds a complete gradation between minor and major changes (revolutions).

Changes caused by the discoveries of new facts usually have only little impact on a paradigm. The situation is different with the development of new concepts, for example when Darwin's theorizing forced the inclusion of humans in the tree of common descent, it caused indeed an ideological revolution.

Mayr (1994i) summarized his discussion stating:

(1) There are indeed major and minor revolutions in the history of biology.
(2) Yet even the major revolutions do not necessarily represent sudden, drastic paradigm shifts. An earlier or the subsequent paradigm may co-exist for long periods. They are not necessarily incommensurable.
(3) Active branches of biology seem to experience no periods of "normal science." There is always a series of minor revolutions between the major revolutions. Periods without such revolutions are found only in inactive branches of biology, but it would seem inappropriate to call such quiet periods "normal science."
(4) The descriptions of Darwinian evolutionary epistemology seem to fit the theory of change in biology better than Kuhn's description of scientific revolutions. Active areas of biology experience a steady proposal of new conjectures (Darwinian variation) and some of them are more successful than others. One can say these are "selected," until replaced by still better ones.
(5) A prevailing paradigm is likely to be more strongly affected by a new concept than by a new discovery.

## Extraterrestrial Intelligence?

The project of the "Search for Extraterrestrial Intelligence" (SETI) is being pursued mostly by astronomers and physicists. Some biologists ask more modestly "Is there other life somewhere in the universe?" Venus and Mars have at some stage of their development most likely been suitable for the origin of life, presumably of a bacteria-like kind of life. Living molecular assemblages might have originated on other planets. "So what?" Mayr asked. If life has originated somewhere, it is highly improbable that intelligence followed (Mayr 1985f, 1988e, 1992g, 1993d, 1995f, g, h, l, 1996a, 2001f, 2004a). Only one of approximately 50 billion species that have lived on earth was able to generate civilizations. Among the ca 20 civilizations, only one developed electronic technology. This indicates that the acquisition of high intelligence on another planet is utterly improbable. If one multiplies the various improbabilities with each other, one finds an improbability of astronomical dimensions. Mayr deplored that NASA awarded $100 million or more for some astronomers and physicists to listen into space with the question in their minds "Is anyone out there?" He felt that this tax money was wasted (1993d).

# 12 Summary: Appreciation of Ernst Mayr's Science

As a young naturalist Ernst Mayr became familiar with the birds of his native Germany in the years after World War I and spent every free minute of his time in the woods and fields watching falcons, thrushes and warblers and along the lakes and rivers observing ducks and plovers. Following a family tradition he entered medical school in 1923, but soon came under the influence of Dr. Erwin Stresemann (1889–1972) in Berlin, the country's leading ornithologist, who saw in this young student "a rising star, of fabulous systematic instinct." He enticed him to switch to zoology by offering to place him on an expedition to the tropics after finishing his PhD dissertation. This had been Mayr's dream since he was a boy and he could not resist. He moved to the University of Berlin and accepted a dissertation topic from Erwin Stresemann passing his PhD examination *summa cum laude* in June 1926.

The foundation of all of Mayr's later theoretical interests in species and speciation, evolution, and the history and philosophy of biology was laid during his student days. Already in 1924, when he was 19 years old, he discussed with Stresemann the species question, the analysis of phylogenetic relations among closely related bird species, rates of differentiation, and convergence (from a Lamarckian point of view). The subject of human inheritance and genetics fascinated him since he was a medical student. Already in 1927 Mayr deplored in print "how little geneticists and systematists cooperate even today." Therefore he was enthusiastic when, after moving to New York in 1931, he read an article by T. Dobzhansky, a naturalist and geneticist from Russia, on the geographical variation in lady-beetles and exclaimed: "Here is finally a geneticist who understands us taxonomists!" They cooperated closely during the following years, especially after Dobzhansky had moved to Columbia University in New York City in late 1939, when they became close friends. From then on they discussed problems of the new systematics, genetical aspects of speciation and other topics of mutual interest.

The *Fisherian synthesis* of the 1920s and early 1930s united the Darwinian theory of natural selection with modern genetics solving one of the two main problems of evolution, the problem of phyletic evolution of individual populations, through the work of the mathematical population geneticists R. A. Fisher (1930), S. Wright (1931), and J. B. S. Haldane (1932). The other main problem of evolution which the geneticists had left open, the origin of biodiversity or multiplication of species, was solved during the *evolutionary synthesis* (1937–1950), when Dobzhansky (1937), Mayr (1942e), Huxley (1942), Simpson (1944), Rensch (1947), Stebbins (1950) and several other workers in the Old World established the synthetic theory of evolution. With the publication of his book, *Systematics and the Origin of Species*

(1942e), which synthesized new systematics, evolution and population genetics, Mayr assumed a central position in the evolutionary synthesis and evolutionary studies in the United States. He emphasized repeatedly how large a role systematics played in the evolutionary synthesis, because this is so often ignored. The naturalists-systematists solved the problem of speciation, but they also showed that there was a smooth, unbroken connection between evolution at the species level and evolution of higher taxa and of major evolutionary innovations (macroevolution). There were no discontinuities (saltations), because the units of evolution are populations and populations can only change gradually, more slowly or faster depending on the size of the population.

Mayr's work as a curator of ornithology at the American Museum of Natural History in New York (1931–1953) on the patterns of geographical variation and speciation in the birds of New Guinea, Melanesia, Polynesia, Micronesia—the first two areas he visited on expeditions during the late 1920s—formed the empirical basis for his theoretical work on species, speciation and general problems of evolution in animals. Evolution is gradual and continuously progressing, yet the species of a local fauna, the products of evolution, are sharply separated by unbridgeable gaps. Building on the work of several earlier systematists, Mayr solved this apparent contradiction with the theory of allopatric speciation, the differentiation of geographically separated (allopatric) populations and their later contact and overlap of their ranges without hybridization after genetic-reproductive isolation between them had been completed. In later years he agreed that in some groups of animals species may also arise without geographic barriers (in sympatry). In the field of zoogeography he discussed the composition, origins, history and boundaries of faunas and established the theory of island biogeography already in 1933(j). He was a prime mover of evolutionary studies in the United States since the mid-1940s, one of the founders of the Society for the Study of Evolution and the founding editor of its journal *Evolution* (1947–1949). During these years he also discussed many problems with a bearing on the history and philosophy of biology like historical aspects of the study of geographical variation, the theoretical species concept, and the concept of "population thinking," on the basis of which Charles Darwin had conceptualized his thesis of natural selection.

Mayr's career is an example for important theoretical work in the biological sciences being based on empirical research in ornithology. He published numerous papers and three books as an ornithologist during the 1920s to the 1950s. Among the almost 300 titles on birds just over half are descriptive faunistic works and taxonomic revisions; about 30% treat natural history topics (life history, migration, and molt), and the rest refer to biogeography and speciation (Gill 1994). Early in his ornithological career Mayr also developed an interest in zoological nomenclature, particularly in the stability of scientific names of birds and other animals. During the 1950s and 1960s, he published actively in this field. Beginning with his work on *Systematics and the Origin of Species* (1942e) ornithological papers dropped to about 25% and less over the next several decades.

In 1953 Mayr accepted the position of an Alexander Agassiz Professor of Zoology at the Museum of Comparative Zoology, Harvard University, Cambridge,

Massachusetts. During the following years he discussed the role of species in nature, sibling species, isolating mechanisms, the process of speciation, the interaction of genes and their varying selective values, the unity of the genotype, the two-step nature of natural selection (accident and design) thus establishing evolutionary biology as a separate field of enquiry in the United States. As spokesman of systematics and evolutionary biology Mayr emphasized the division of biology into functional biology analyzing proximate causations and organismic-evolutionary biology studying ultimate causations. A balanced biology advances both fronts, that of molecular biology and physiology as well as that of organismic-systematic-evolutionary biology. In his books on *Animal Species and Evolution* (1963b) and *Populations, Species and Evolution* (1970e) Mayr united systematics, population genetics and evolutionary biology, but he never discussed in any detail or conducted functional-physiological studies himself.

From the late 1950s, and especially after his "retirement" as Professor of Zoology in 1975, Mayr turned increasingly to studies of the history and philosophy of biology publishing numerous works, particularly on Charles Darwin and his five theses of evolution (instead of one), and a large book on the conceptual history of biology (1982d). Already in 1949 he had written to Erwin Stresemann: "It is important to emphasize the evolutionary angle as a counterbalance against the assertions of the physicists and chemists who see nothing in this branch of research" (25 January 1949). Mayr discussed the many peculiarities of life such as historical contingency, genetic programs, diversity, individuality, classification, and others which necessitate a modern philosophy of biology excluding essentialism, physicalist reductionism, finalism, and determinism. He emphasized the distinctness (autonomy) of biology as a science from physics and forged a third synthesis connecting systematic biology, evolutionary biology, and the history and philosophy of biology in his recent books *Toward a New Philosophy of Biology* (1988e), *One Long Argument. Charles Darwin and the Genesis of Modern Evolutionary Thought* (1991g), *This is Biology* (1997b), and *What Makes Biology Unique?* (2004a).

We look back: In ornithology and the new systematics, Mayr was a student of Erwin Stresemann whose insights (and those of B. Rensch) he extended over the entire field of zoology and appreciably enlarged through the inclusion of population genetics. As a lifelong naturalist, Ernst Mayr was always interested in the diversity of the living world, its origin and biological meaning as well as in the clarification of evolutionary concepts and processes. Although he was not the originator of the biological species concept, he formulated its familiar definition which was widely adopted and he placed the biospecies concept at the center of the synthetic theory of evolution. As a torch bearer of Darwinism he decisively influenced the field of evolutionary biology as well as those of the history and philosophy of biology.

He always read very widely and was interested not only in facts but also in the underlying theories and principles. He was a very hard worker, as shown when he finished his PhD dissertation and the necessary courses in 16 months and when, as director of the MCZ, he often got up at 4:30 a.m. to work on his manuscripts. His excellent memory was a great help in recalling information whenever needed. The breadth of his biological knowledge and interests enabled him to forge several major critical syntheses in his published work: As an effective bridge-builder he united systematics with population genetics and evolution (1942e) and analyzed the status of the synthetic theory of evolution (1963b, 1970e); 20 years later he forged a synthesis of these fields with the history of biology (1982d), and finally he united all of these disciplines with the philosophical foundations of biology (1988e, 1997b, 2004a). He was convinced that science based on observation and reason leads to progress, he was convinced of the uniqueness of individual organisms (population thinking in contrast to essentialism) and he emphasized the historical viewpoint in biology.

Mayr (1995m) himself listed four factors which he thought contributed most to his later success as a biologist:

(1) His training under Erwin Stresemann who had adopted the principles of the new systematics as developed by K. Jordan, E. Poulton, and various continental European taxonomists since about 1900. He taught Mayr how to apply these principles.
(2) His early familiarity with the patterns of geographical variation and speciation of animals inhabiting island regions (New Guinea and Solomon Islands), comparable to Darwin and Wallace.
(3) His work with the magnificent collections of the American Museum of Natural History (New York) and the freedom to study anything he wanted to study.
(4) The opportunity to benefit from two great intellectual traditions, that of the English speaking world and that of continental Europe (comparable to the history of T. Dobzhansky).

Essential is the fact that Ernst Mayr has been a naturalist and observer all his life who endeavored to comprehend the living world in all its relations (diversity, populations, inferences from observations, and evolutionary aspects). He reached distinction in six different fields: ornithology, systematics, zoogeography, evolutionary biology, history of biology, and philosophy of biology. The breadth of Mayr's work, his competence, erudition and scholarship in so many fields of knowledge are truly amazing. He was able to combine successfully different views as they developed in Europe with those in North America. Rooted in two cultures, he found a broad basis for solutions to problems and inspiration for new ideas. However, this is only one of the reasons for his successful career in the United States, the other one being his rare ability to synthesize critically the knowledge gained in distant fields of research. In the sense of emergence his synthetic works are more than the sum of their parts.

Mayr has been called the "Sage of twentieth-century biology," "Grand Master" and "Darwin of the 20th century," "the world's greatest evolutionary biologist," "the foremost evolutionary biologist and a celebrated historian of science" and one of the founders of a modern philosophy of biology. He has won the three highest awards in biology—The Triple Crown—Balzan Prize, Japan Prize, and Crafoord Prize. Honorary PhD degrees, medals and special awards, and numerous honorary society memberships came to him from universities and societies around the world totaling 17 honorary PhD degrees, 35 medals, and 52 honorary society memberships. Yet he was most proud of having the library at the Museum of Comparative Zoology named after him, as he considered the collected knowledge stored in libraries the most important part of human culture.

Ernst Mayr himself classified a scientist's achievements which may lie in several different areas:

"As an *innovator* (new discoveries, new theories, new concepts), as a *synthesizer* (bringing together scattered information, sharing relationships and interactions, particularly between different disciplines, like genetics and taxonomy), as a *disseminator* (presenting specialized information and theory in such a way that it becomes accessible to non-specialists [popularizer is a misleading term]), as a *compiler* or *cataloguer*, as an *analyst* (dissecting complex issues, clarifying matters by suggesting new terminologies, etc.), and in other ways" (see Provine 2005).

In the sense of this classification, Mayr was a *compiler* or *cataloguer* in much of his ornithological work at the American Museum which formed the empirical basis for his later theoretical studies. He published numerous articles and books on the birds of New Guinea, the Philippines and the islands of the SW Pacific based on the biological species concept. He never considered this taxonomic work as an end in itself but always as a means to go beyond it. Since his student days he pursued a synthesis of systematics, evolution and genetics, as documented by his letter to E. Stresemann of May 1924, when he was 19 years old (p. 27), by his discussions in 1927 (p. 45) and his letter to Th. Dobzhansky in 1935 (p. 185). His synthesis of systematics, evolutionary biology, natural history and population genetics in his book, *Systematics and the Origin of Species* (1942e) was a major achievement followed, in 1963 and 1970, by his magisterial syntheses of evolutionary biology (*Animal Species and Evolution*; *Populations, Species, and Evolution*). Mayr made "the species problem" a central concern of evolutionary biology. Other major syntheses were his textbook, the *Principles of Systematic Zoology* (1969b) as well as his comprehensive works on the history of biology (1982d) and on the philosophy of biology (1997b, 2004a). Mayr was clearly a major *synthesizer* of biology and also a master *analyst*. He dissected such complex concepts as population thinking, Darwin's five theses of evolution, chance and necessity in evolution, functional (proximate) and historical (ultimate) causations, biological classification and cladification, teleology and many others; and he proposed new terminologies clarifying many complex issues (1978c). For numerous existing evolutionary concepts Mayr was an effective *disseminator* like biological species concept, gene pool, isolating mechanisms of species, geographical (allopatric) speciation and others. New theories which he proposed as an *innovator* include his theory of island biogeography (1933j, 1940i) and his founder principle or theory of genetic revolutions (1954c). The latter is quite controversial but has stimulated a large amount of research. Also his syntheses include various new insights like his concept of population thinking and his discussions of the autonomy of biology. Mayr's roles as *visionary*, *organizer* and *community architect* are obvious from his functions as president of various scientific societies and director of the Museum of Comparative Zoology (Harvard University) as well as founding editor of the journal *Evolution*. He was also a museum curator and administrator, teacher, writer, and editor. Innumerable biologists have been enormously influenced by Mayr's work.

It is probably true as Bock (2004c) suggested that Ernst Mayr would not have been as successful in Germany as he was in the United States. If he had stayed in Europe as a curator, the wide gap between museum workers and academia in Germany at that time would have made a transition to a university institute quite difficult, but most importantly he would have been drafted into the army with little chance to survive World War II. Advantages in the United States included his freedom of research, the close communication between museum curators and the academia, and his friendly cooperation with the geneticist-naturalist Th. Dobzhansky who had immigrated to the United States from Russia in 1927.

One of the foundations of Ernst Mayr's success in science was his unshakeable self-confidence from his youth. This includes his undertaking a one-man exped-

**Fig. 12.1.** Ernst Mayr at the age of 97 years in the dinosaur hall of the Museum of Natural History Berlin, Germany celebrating the $75^{th}$ anniversary of his PhD examination (June 1926). Persons to the right of E. Mayr (*first* and *second rows*) are Prof. Otto Kraus (Hamburg, half-hidden), Mayr's daughter Mrs. Christa Menzel (USA), Prof. Walter Sudhaus (Berlin), Prof. A.-B. Ischinger (Berlin), Dr. Ellen Strong (USA), and Dr. Matthias Glaubrecht (Berlin). Photograph taken by M. Lüdicke (Berlin) on 25 June 2001

ition to New Guinea in 1928, his assuming a leadership role in evolutionary studies in the United States during the mid-1940s, his spokesmanship for systematics and evolutionary biology during the 1960s and his later leadership roles in the fields of history and philosophy of biology. Yet he also knew his limits, for example when he restricted his book on species and evolution (1963b) to animal species rather than also including plants, as several friends had suggested. He was a vigorous campaigner and an indefatigable street fighter for his ideas which he presented, rephrased and discussed again from a different angle and in a different context, and which he defended always with clarity, authority, and wisdom in a lucid, elegant, entertaining and engaging style as well as in an often brilliant manner of exposition. His contributions will continue to spawn discussions among biologists, historians and philosophers of biology. Mayr was willing to discuss matters with anybody, whether an experienced colleague or a hesitant graduate student. And he answered enquiries without delay and promptly reviewed manuscripts in detail for friends, students and journal editors and supported students and young colleagues in many other ways. He and his wife Gretel selflessly invested much time and money to support European ornithologists during the immediate postwar years of the mid- and late-1940s. He was also generous with his money received for academic prizes which he gave to purposes of education and conservation. Everybody who corresponded with him knows how helpful and stimulating his detailed letters were. His social skills enabled him to establish excellent relationships with the native tribes in the areas of New Guinea and the Solomon Islands that he visited in the late 1920s and with many colleagues during his professional years as well as to forge friendships later in his life with people that were much younger than himself. He influenced several generations of biologists and philosophers of biology worldwide with his ideas and his publications. He lived a full life and kept his mind busy all the time with projects and new plans until the month before he died on February 3rd, 2005 at the age of 100 years and seven months. With his passing, the light of the "rising star" of 1924 was not extinguished, because it kindled a fireworks of ideas which will continue to radiate within biological science for a long time to come.

# Acknowledgments

I am deeply grateful to Ernst Mayr with whom I corresponded since the early 1960s. I was always astonished that despite his busy schedule as Director of the Museum of Comparative Zoology, he would take time to answer my enquiries in detail. I met him at Harvard University in 1968 and discussed aspects of speciation and evolution with him when we met several times in the United States and in Germany in later years. I also thank him for his contributions to this volume and for his permission to review material in the Mayr Papers at the Ernst Mayr Library (Museum of Comparative Zoology, Harvard University) as well as portions of his correspondence at the Harvard University Archives. During visits to Cambridge and Bedford, Massachusetts in September 2003 and October 2004, we discussed an early version of this manuscript. I was also permitted to use a 12-page manuscript by Margarete (Gretel) Mayr, Ernst's wife, on "Our life in America during the Second World War." Their daughters Mrs. Christa Menzel and Mrs. Susanne Mayr Harrison also supplied additional information. Mrs. Harrison mentioned to me details of the family's frequent visits to Cold Spring Harbor (New York) during the 1940s and Mrs. Menzel and her husband Gerhart Menzel kindly showed me The Farm near Wilton, New Hampshire. Mrs. Harrison also provided prints and electronic files of family photographs and reviewed various parts of the manuscript. I thank the nephews Dr. Otto Mayr and Dr. Jörg Mayr who maintain a family archive in Lindau, Germany, for historical photographs and data regarding the family. Ernst Mayr's niece Roswitha Kytzia (née Mayr) and her husband Peter Kytzia (Hammersbach near Hanau, Germany), who served as E. Mayr's main family contact in Germany during the last 20 years, also furthered this project and helped with various information.

At the American Museum of Natural History (New York), Ms. Mary LeCroy of the Department of Ornithology kindly placed at my disposal relevant documents such as the unpublished reports of members of the Whitney South Sea Expedition (1920–1940), the annual reports of the department during the 1930s and its historical correspondence files. She also helped in locating historical photographs, answering questions regarding several collecting sites in New Guinea, and sent information on certain birds that Mayr had obtained on one of the Solomon Islands in 1929. She also sent me taxonomic notes on several subspecies of birds described by Mayr and later synonymized by other authors. I am very grateful for her help and substantial contributions. I take this opportunity to also thank the curators of the Department of Ornithology (AMNH) for their support during many visits in the course of the last 40 years.

Walter Bock (Columbia University, New York) answered many questions based on his intimate knowledge of Ernst Mayr's work. He spent much time editing my manuscript and commenting on numerous aspects discussed. Mrs. Amélie Koehler (Freiburg), M. Glaubrecht (Museum of Natural History, Berlin), and Ms. M. LeCroy (New York) also read most of this manuscript and suggested many improvements of the text. R. Bruckert (Département Mamifères et Oiseaux, Museum National d'Histoire Naturelle, Paris) provided from his data file a printout of taxonomic bird names introduced by E. Mayr and R. Creath (Arizona State University, Tempe, Arizona) permitted me to quote from his interview with Marjorie Sturm, Mayr's administrative assistant while he was Director of the Museum of Comparative Zoology, Harvard University. S. Eck (deceased; formerly at the Staatliches Museum für Tierkunde Dresden) reported on Mayr's high school in Dresden and procured copies of several articles published in local journals; F. Steinheimer (Berlin) informed me about certain correspondence between E. Hartert and E. Mayr held by the Natural History Museum, London; J. Neumann (Neubrandenburg) provided information on several Saxon ornithologists of the 1920s; O. Kraus (Zoological Institute, University of Hamburg) informed me on the early history of the International Commission of Zoological Nomenclature; W. Meise (deceased; formerly in Hamburg) permitted to quote from his correspondence; Ms. Alison Pirie (Department of Ornithology; Museum of Comparative Zoology, Harvard University) was very helpful during my visit to Cambridge and Bedford. Ms. M. Macari (Cold Spring Harbor, New York) sent information on Cold Spring Harbor Biological Laboratory and the Ernst Mayr Dining Room. G. Stresow (Diessen, Germany) told me of his memories of E. Mayr in New York during the early 1930s. I am also grateful to T. Junker (Frankfurt a.M.) for discussions and copies of certain useful documents and to U. Kutschera (Department of Biology, University of Kassel) who helped in various ways during the final stages of the preparation of this book. My wife Maria Haffer accepted many inconveniences while I was working on this project.

The following persons in charge of the respective archives permitted access to correspondence and my quoting selected paragraphs: P. J. Becker (Manuscript Division, Staatsbibliothek Preussischer Kulturbesitz Berlin), H. Landsberg and S. Hackethal (Historische Bild- und Schriftgutsammlungen, Museum für Naturkunde Berlin), Ms. M. Gachette (Harvard Archives, Pusey Library), Mrs. C. A. Rinaldo (Ernst Mayr Library, Museum of Comparative Zoology, Harvard University), and Mrs. S. Snell (Archive, Natural History Museum, London). The Library Department (American Museum of Natural History, New York) authorized the publication of various photographs and Karen Klitz (Archives; Museum of Vertebrate Zoology, University of California, Berkeley) permitted me to quote from a letter by E. Mayr to J. Grinnell.

# List of Abbreviations

| | |
|---|---|
| AAAS | American Association of the Advancement of Science |
| AMNH | American Museum of Natural History, New York |
| AOU | American Ornithologists' Union |
| BOC | British Ornithologists' Club |
| BOU | British Ornithologists' Union |
| DOG | Deutsche Ornithologische Gesellschaft (1875–1945) |
| DO-G | Deutsche Ornithologen-Gesellschaft (1850–1875; 1949–present) |
| IOC | International Ornithological Congress |
| JfO, J. Orn. | Journal für Ornithologie, Journal of Ornithology (as of 2004) |
| MCZ | Museum of Comparative Zoology, Harvard University |
| MPG | Max Planck Gesellschaft |
| Nov. Zool. | Novitates Zoologicae (Museum W. Rothschild) |
| transl. | Indicates the author's translation of a quote in German (Mayr's own translations of German texts are unmarked) |

# References


Ali S (1985) The fall of a sparrow. Oxford University Press, New Delhi

Altum B (1868) Der Vogel und sein Leben (1st edn., 168 pp), Münster (6th edn., 300 pp, 1898; 11th edn, 1937)

Amadon D (1942a) Birds collected during the Whitney South Sea Expedition, XLIX. Notes on some non-passerine genera, 1. American Museum Novitates, no. 1175

Amadon D (1942b) Birds collected during the Whitney South Sea Expedition. L. Notes on some non-passerine genera, 2. American Museum Novitates, no. 1176

Amadon D (1943) Birds collected during the Whitney South Sea Expedition, 52. Notes on some non-passerine genera, 3. American Museum Novitates, no. 1237

Amadon D (1957) Remarks on the classification of the perching birds [order Passeriformes]. Proc. Zool Soc Calcutta, Mookerjee Memorial Volume, pp 259–268

Amadon D (1966) The superspecies concept. Syst Zool 15:245–249

Amadon D, Short LL (1992) Taxonomy of lower categories—suggested guidelines. Bull Brit Orn Club, Centenary Suppl. 112A:11–38

Appel TA (2000) Shaping biology. The National Science Foundation and American biological research, 1945–1975. Johns Hopkins University Press, Baltimore

Ashlock PD (1972) Monophyly again. Syst Zool 21:430–438

Audley-Charles MG (1987) Dispersal of Gondwanaland: relevance to evolution of the angiosperms. In: Whitmore TC (ed) Biogeographical evolution of the Malay Archipelago. Oxford University Press, Oxford, pp 5–25

Ayala FJ (2004) (ed) Ernst Mayr 1904. Ludus Vitalis XII:1–245

Backer CA (1936) Verklarend Woordenboek der wetenschappelijke namen van de in Nederland en Nederlandsch-Indie in het wild groeiende en in tuinen en parken gekweekte varens en hoogere planten. Batavia P, Noordhoff N. V. Reprint, 2000, L. J. Veen, Antwerpen

Baker RM (1951) The avifauna of Micronesia, its origin, evolution, and distribution. University of Kansas Publications, Mus Nat Hist 3(1):1–359

Barluenga M, Stölting KN, Salzburger W, Muschnik M, Meyer A (2006) Sympatric speciation in Nicaraguan crater lake cichlid fish. Nature 439:719–723

Barrow MV Jr (1998) A passion for birds. American ornithology after Audubon. Princeton University Press, Princeton

Bates HW (1862) Contributions to an insect fauna of the Amazon Valley. Lepidoptera: Heliconidae. Trans Linn Soc Lond 23:495–566

Bates HW (1864) On the variation of species [extracted from *The Naturalist on the River Amazons*, vol 1, pp 255–265]. The Entomologists' Annual for 1864:87–94

Bates HW (1865) Contributions to an insect fauna of the Amazons Valley: Coleoptera, Longicornes. Ann Mag Nat Hist III. 15:382–394

Baur E, Fischer E, Lenz F (1923) Menschliche Erblichkeitslehre, 2nd edn. J. F. Lehmanns Verlag, Munich

Beatty J (1994) The proximate/ultimate distinction in the multiple careers of Ernst Mayr. Biol Philos 9:333–356

Beehler BM, Pratt TK, Zimmerman DA (1986) Birds of New Guinea. Princeton University Press, Princeton
Behrmann W (1919) Detzners Forschungen in Neuguinea. Zeitschrift der Gesellschaft für Erdkunde zu Berlin 1919, pp 371–376
Bergson H (1912) Schöpferische Entwicklung. Jena, Fischer
Beurton PJ (1995) Ernst Mayr und der Reduktionismus. Biol Zentralblatt 114:115–122
Beurton PJ (2002) Ernst Mayr through time on the biological species concept: a conceptual analysis. Theory Biosci 121:81–98
Biskup P (1968) Hermann Detzner: New Guinea's first coast watcher. J Papua New Guinea Soc 2:4–21
Bock WJ (1956) A generic review of the family Ardeidae (Aves). Am Mus Novitates 1779, 49 pp
Bock WJ (1959) Preadaptation and multiple evolutionary pathways. Evolution 13:194–211
Bock WJ (1979) The synthetic explanation of macroevolutionary change: a reductionist approach. Bull Carnegie Mus 13:20–69
Bock WJ (1986) Species concepts, speciation, and macroevolution. In: Iwatsuki K, Raven PH, Bock WJ (eds) Modern aspects of species. University of Tokyo, Tokyo, pp 31–57
Bock WJ (1990) Special review: Peters JL (ed) Check-list of birds of the world and a history of avian check-lists. Auk 107:629–639
Bock WJ (1992a) The species concept in theory and practice. Jpn J Zool 9:697–712
Bock WJ (1992b) Methodology in avian macrosystematics. Bull Brit Ornithological Club, Centenary Supplement, 112A:53–72
Bock WJ (1993) Selection and fitness: definitions and uses: 1859 and now. Proc Zool Soc Calcutta, Haldane Commemorative Volume, pp 7–26
Bock WJ (1994a) Ernst Mayr, naturalist: his contributions to systematics and evolution. Biol Philos 9:267–327
Bock WJ (1994b) History and nomenclature of avian family-group names. Bull Am Mus Nat Hist 222:1–281
Bock WJ (1995a) Ernst Mayr–1994 laureate of the International Prize for Biology. In: Arai R, Kato M, Doi Y (eds) Biodiversity and evolution. The National Science Museum Foundation, Tokyo, pp 13–24
Bock WJ (1995b) The species concept versus the species taxon: their roles in biodiversity analyses and conservation. In: Arai R, Kato M, Doi Y (eds) Biodiversity and evolution. The National Science Museum Foundation, Tokyo, pp 47–72
Bock WJ (2000) Towards a new metaphysics: the need for an enlarged philosophy of science. Biol Philos 15:603–621
Bock WJ (2004a) Presidential address: three centuries of international ornithology. Proceedings of the 23rd International Ornithological Congress. Acta Zool Sinica 50(6):779–855
Bock WJ (2004b) Affinities of *Carpospiza brachydactyla*. J Ornithol 145:223–226
Bock WJ (2004c) Ernst Mayr at 100: a life inside and outside of ornithology. The Auk 121:637–651 (revised and reprinted in Bock and Lein 2005:2–16)
Bock WJ (2004d) Species: the concept, category and taxon. J Zool Syst Evol Res 42:178–190
Bock WJ (2005) Ernst Mayr—teacher, mentor, friend. J Biosci 30:422–426
Bock WJ (2006) Ernst Walter Mayr, 5 July 1904–3 February 2005. Biograph Mem Fellows R Soc Lond 52:167–187
Bock WJ (2007) Explanations in evolutionary theory. J Zool Syst Evol Res 45:89–103
Bock WJ, Farrand J (1980) The number of species and genera of Recent birds. Am Mus Novitates 2703:1–29
Bock WJ, Lein MR (2005) (eds) Ernst Mayr at 100. Ornithologist and naturalist. Ornitholog Monogr 58, 109 pages (with interview remarks by Ernst Mayr on Video CD-ROM)

Bogert C (1937) Birds collected during the Whitney South Sea Expedition, XXXIV. The distribution and the migration of the Long-tailed Cuckoo (*Urodynamis taitensis* Sparrman). Am Mus Novitates 933, 12 pp
Brisson MJ (1760) Ornithologia sive synopsis methodica. 6 volumes. Paris
Buch L von (1819) Allgemeine Uebersicht der Flora auf den Canarischen Inseln. Abhandlungen der Königlichen Akademie der Wissenschaften in Berlin (1816–1817) 337–384
Buch L von (1825) Uebersicht der Flora auf den canarischen Inseln. In: Physicalische Beschreibung der Canarischen Inseln. Berlin (French translation Description physique des Iles Canaries, Paris 1836), pp 105–200
Buffon GL (1749) Histoire naturelle, générale et particulière, vol 1, Paris
Burkhardt RW Jr (1992) Huxley and the rise of ethology. In: Waters KC, Van Helden A (eds) Julian Huxley. Biologist and statesman of science. Rice University Press, Houston, Texas, pp 127–149
Burkhardt RW Jr (1994) Ernst Mayr: biologist-historian. Biol Philos 9:359–371
Burret M (1933) Neue Palmen aus Neu-Guinea. Notizblatt des Botanischen Gartens und Museums zu Berlin-Dahlem 11:704–713
Butterfield H (1931) The Whig interpretation of history. London, Bell
Cain J (1993) Common problems and cooperative solutions: organizational activities in evolutionary studies, 1937–1946. Isis 84:1–25
Cain J (1994) Ernst Mayr as community architect: launching the Society for the Study of Evolution and the journal *Evolution*. Biol Philos 9:387–427
Cain J (2000a) Towards a greater degree of integration: the Society for the Study of Speciation, 1939–41. Brit J Hist Sci 33:85–108
Cain J (2000b) For the promotion and integration of various fields: first years of *Evolution*, 1947–1949. Arch Nat Hist 27:231–259
Cain J (2001) The Columbia Biological Series, 1894–1974: a bibliographical note. Arch Nat Hist 28:353–366
Cain J (2002a) Co-opting colleagues: appropriating Dobzhansky's 1936 lectures at Columbia. J Hist Biol 35:207–219
Cain J (2002b) Epistemic and community transition in American evolutionary studies: the Committee on Common Problems of Genetics, Paleontology, and Systematics (1942–1949). Studies in History and Philosophy of Biological and Biomedical Sciences 33:283–313
Cain J (2004) Launching the Society of Systematic Zoology in 1947. In: Williams DM, Forey PL (eds) Milestones in systematics. Syst Assoc Spec Vol Ser 67:19–48
Chapin JP (1932) The birds of the Belgian Congo, vol 1. Bull Am Mus Nat Hist 65:1–756
Chapman FM (1924) Criteria for the determination of subspecies in systematic ornithology. Auk 41:17–29
Chapman FM (1935) The Whitney South Sea Expedition. Science 81:95–97
Clode D, O'Brien R (2001) Why Wallace drew the line: a re-analysis of Wallace's bird collections in the Malay Archipelago and the origins of biogeography. In: Metcalf I, Smith JMB, Morwood M, Davidson I (eds) Faunal and floral migrations and evolution in SE Asia-Australasia. A. A. Balkema Publishers, Lisse, pp 113–121
Coates BJ, Bishop KD, Gardner D (1997) A Guide to the birds of Wallacea. Sulawesi, the Moluccas and Lesser Sunda Islands, Indonesia. Dove Publications, Alderley, Queensland, Australia
Cooper WT, Forshaw JM (1977) The birds of paradise and Bower birds. Collins, Sydney
Coyne JA, Barton NH, Turelli M (1997) A critique of Sewall Wright's shifting balance theory of evolution. Evolution 51:643–671
Coyne JA (2005) Ernst Mayr (1904–2005). Science 307:1212–1213
Coyne JA, Orr HA (2004) Speciation. Sinauer Associates, Sunderland, MA

Cracraft J (2002) Gondwana genesis. A combination of molecular data, anatomical evidence, and knowledge of ancient geography is providing new answers to the contentious issue of when–and where–modern birds arose. Nat Hist 102:64–72

Darwin C (1859) On the origin of species by means of natural selection. A facsimile of the first edition with an introduction by Ernst Mayr (1964) Harvard University Press, Cambridge, MA

Delacour J, Vaurie C (1957) A classification of the oscines (Aves). Los Angeles County Museum, Contributions in Science 16:6 p

Diamond JM (1966) Zoological classification system of a primitive people. Science 151:1102–1104

Diamond JM (1977) Continental and insular speciation in Pacific land birds. Syst Zool 26:263–268

Dieckmann U, Doebeli M, Metz JAJ, Tautz D (2004) (eds) Adaptive speciation. Cambridge University Press, Cambridge

Dobzhansky Th (1933) Geographical variation of lady-beetles. Am Naturalist 67:97–126

Dobzhansky Th (1935) A critique of the species concept in biology. Philos Sci 2:344–355

Dobzhansky Th (1937) Genetics and the origin of species. Columbia University Press, New York

Dobzhansky Th (1940) Speciation as a stage in evolutionary divergence. Am Nat 74:312–321

Doughty C, Day N, Plant A (1999) Birds of the Solomons, Vanuatu and New Caledonia. A. C. Black, London

Driesch H (1899) Philosophie des Organischen. Quelle and Meyer, Leipzig

Dunn ER (1922) A suggestion to zoogeographers. Science 56:336–338

Dunn ER (1931) The herpetological fauna of the Americas. Copeia 1931:106–119

DuPont JE (1976) South Pacific birds. Delaware Museum of Natural History, Monograph Series no. 3

Edwards SV, Kingan SB, Calkins JD, Balakrishnan CN, Jennings WB, Swanson WJ, Sorenson MD (2005) Speciation in birds: genes, geography, and sexual selection. Proc Nat Acad Sci USA 102, suppl. 1:6550–6557

Ekman P (1998) (ed) In: Reprint edition of C. Darwin: The expression of the emotions in man and animals (see Afterword). Oxford University Press, Oxford, pp 363–387

Eldredge N, Gould SJ (1972) Punctuated equilibria: an alternative to phyletic gradualism. In: Schopf TJM (ed) Models in paleobiology. Freeman, Cooper, and Co., San Francisco, pp 82–115

Farrand J Jr (1991) The Bronx County Bird Club. Memories of ten boys and an era that shaped American birding. Am Birds 45:372–381

Fisher RA (1930) The genetical theory of natural selection. Clarendon Press, Oxford

Flannery T (1995) Mammals of New Guinea (revised and updated edition). Cornell University Press, Ithaca, New York

Fleischer RC, McIntosh CE (2001) Molecular systematics and biogeography of the Hawaiian avifauna. Stud Avian Biol 22:51–60

Frith CB, Beehler BM (1998) The birds of paradise *Paradisaeidae*. Bird families of the world. Oxford University Press, Oxford

Futuyma DJ (1998) Wherefore and whither the naturalist? Am Nat 151:1–6

Futuyma DJ (2006) Ernst Mayr, genetics and speciation. Trends Ecol Evol 21:7–8

Gavrilets S (2005) Adaptive speciation—it is not that easy: a reply to Doebeli et al. Evolution 59:696–699

Gay P (1968) Weimar culture: the outsider as insider. Harper and Row, New York

Ghiselin MT (1969) The triumph of the Darwinian method. University of California Press, Berkeley

Ghiselin MT (2004) Mayr and Bock versus Darwin on genealogical classification. J Zool Syst Evol Res 42:165–169
Gill FB (1994) Ernst Mayr, the ornithologist. Evolution 48:12–18
Gilliard ET (1969) Birds of paradise and Bower birds. Weidenfeld and Nicolson, London
Glaubrecht M (2002) The experience of nature: from Salomon Müller to Ernst Mayr, or the insights of travelling naturalists toward a zoological geography and evolutionary biology. Verhandlungen zur Geschichte und Theorie der Biologie 9:245–282
Glaubrecht M (2004) Leopold von Buch's legacy: treating species as dynamic natural entities, or why geography matters. Am Malacolog Bull 19:111–134
Görnitz K (1923) Ueber die Wirkung klimatischer Faktoren auf die Pigmentfarben der Vogelfedern. J Ornithol 71:456–511
Goldschmidt R (1933) Some aspects of evolution. Science 78:539–547
Goldschmidt R (1935) Geographische Variation und Artbildung. Naturwissenschaften 23:169–176
Goldschmidt R (1940) The material basis of evolution. Yale University Press, New Haven, CT
Gould SJ (1984) Balzan Prize to Ernst Mayr. Science 223:255–257
Gould SJ (2002) The structure of evolutionary theory. Harvard University Press, Cambridge, MA
Grant PR (2002a) Geographical speciation. Review of The birds of northern Melanesia, by E. Mayr and J. Diamond. Evolution 56:1880–1882
Grant PR (2002b) Founder effects and silvereyes. Proc Nat Acad Sci USA 99:7818–7820
Grant V (1994) Evolution of the species concept. Biologisches Zentralblatt 113:401–415
Greene JC (1999) Debating Darwin. Adventures of a scholar. Regina Books, Claremont, CA
Greene JC, Ruse M (1994) (eds) Ernst Mayr at ninety. Biol Philos 9:263–427
Greenway JC Jr (1973) Type specimens of birds in the American Museum of Natural History, part 1. Bull Am Mus Nat Hist 150:207–346
Gulick JT (1872) The variation of species as related to their geographical distribution, illustrated by the Achatinellidae. Nature 6:222–224
Gulick JT (1873) On diversity of evolution under one set of external conditions. J Linn Soc Lond 11:496–505
Haeckel E (1899) Die Welträthsel. Gemeinverständliche Studien über monistische Philosophie. Emil Strauß, Bonn
Haffer J (1967) Speciation in Colombian forest birds west of the Andes. Am Mus Novitates 2294:1–57
Haffer J (1969) Speciation in Amazonian forest birds. Science 165:131–137
Haffer J (1974) Avian speciation in tropical South America. Publications of the Nuttall Ornithological Club No. 14, 390 pp
Haffer J (1975) Avifauna of northwestern Colombia, South America. Bonner Zoologische Monographien, Nr. 7, 182 pp
Haffer J (1977) Secondary contact zones of birds in northern Iran. Bonner Zoologische Monographien, Nr. 10, 64 pp
Haffer J (1992) The history of species concepts and species limits in ornithology. Bull Brit Ornithological Club, Centenary Supplement 112A:107–158
Haffer J (1994a) Die Seebohm-Hartert Schule der europäischen Ornithologie. J Ornitholog 135:37–54
Haffer J (1994b) Es wäre Zeit, einen allgemeinen Hartert zu schreiben: Die historischen Wurzeln von Ernst Mayrs Beiträgen zur Evolutionssynthese. Bonner Zoologische Beiträge 45:113–123
Haffer J (1995) Ernst Mayr als Ornithologe, Systematiker und Zoogeograph. Biologisches Zentralblatt 114:133–142

Haffer J (1997a) Essentialistisches und evolutionäres Denken in der systematischen Ornithologie des 19. und 20. Jahrhunderts. J Ornithol 138:61–72

Haffer J (1997b) The correspondence between E. Stresemann and E. Mayr during the period 1923–1972. In: Haffer J (ed) Ornithologen-Briefe des 20. Jahrhunderts. We must lead the way on new paths. The work and correspondence of Hartert, Stresemann, Ernst Mayr–international ornithologists. Ökologie der Vögel 19, 980 p. (With contributions by Ernst Mayr). Ludwigsburg, pp 369–771

Haffer J (2000) Erwin Stresemann (1889–1972)—Life and work of a pioneer of scientific ornithology, pp 399–427, In: Haffer J, Rutschke E, Wunderlich K (2000) Erwin Stresemann (1889–1972)—Leben und Werk eines Pioniers der wissenschaftlichen Ornithologie. Acta Historica Leopoldina 34, 465 pp (2nd edn. 2004)

Haffer J (2001a) Ornithological research traditions in central Europe during the 19th and 20th centuries. J Ornithol 142, Suppl. 1:27–93

Haffer J (2001b) Die Stresemannsche Revolution in der Ornithologie des frühen 20. Jahrhunderts. J Ornithol 142:381–389

Haffer J (2001c) Ernst Mayr–Ornithologe, Evolutionsbiologe, Historiker und Wissenschaftsphilosoph. J Ornithol 142:497–502 [Engl. translation in Verhandlungen zur Geschichte und Theorie der Biologie 9:125–132, 2002]

Haffer J (2005a) Zur Biographie des Ornithologen und Harvard-Professors Ernst Mayr (1904–2005). Blätter aus dem Naumann-Museum 24:1–33

Haffer J (2005b) Ernst Mayr—bibliography. Ornitholog Monogr 58:73–108

Haffer J (2006) The history of the biological species concept. Proceedings of the 23rd International Ornithological Congress, Acta Zoologica Sinica 52, Suppl.:415–420

Haffer J (2007a) The development of ornithology in central Europe. J Ornithol 148, Suppl (in press)

Haffer J (2007b) Ergänzungen zum Briefwechsel zwischen Erwin Stresemann (1889–1972) und Ernst Mayr (1904–2005)–Internationale Ornithologen des 20. Jahrhunderts. Ökologie der Vögel 25 (2005) (in press)

Haffer J, Rutschke E, Wunderlich K (2000) Erwin Stresemann (1889–1972)–Leben und Werk eines Pioniers der wissenschaftlichen Ornithologie. Acta Historica Leopoldina 34, 465 pp (2nd edn. 2004)

Haldane JBS (1932) The causes of evolution. Longmans, Green and Co., London

Haldane JBS (1949) Human evolution: past and future. In: Jepsen GL, Mayr E, Simpson GG (eds) Genetics, paleontology and evolution. Princeton University Press, Princeton, pp 405–418

Hartert E (1903–1922) Die Vögel der paläarktischen Fauna. 3 volumes. Friedländer, Berlin

Hartert E (1930) On a collection of birds made by Dr. Ernst Mayr in northern Dutch New Guinea; List of the birds collected by Ernst Mayr. Novitates Zool 36:18–19; 27–128

Harwood J (1993) Styles of scientific thought. The German genetics community 1900–1933. University of Chicago Press, Chicago

Harwood J (1994) Metaphysical foundations of the evolutionary synthesis: a historiographical note. J Hist Biol 27:1–20

Heberer G (1943) (ed) Die Evolution der Organismen. Gustav Fischer Verlag, Stuttgart

Hemming F (1953) (ed) Copenhagen decisions on zoological nomenclature. Approved and adopted by the Fourteenth Congress of Zoology, Copenhagen, August 1953. Int Trust Zool Nomencl London, XXIX + 135 pp

Hemming F (1995) Towards stability in the names of animals. A history of the International Commission on Zoological Nomenclature 1895–1995. Int Trust Zool Nomencl London (c/o The Natural History Museum)

Hennig W (1950) Grundzüge einer Theorie der phylogenetischen Systematik. Deutscher Zentralverlag, Berlin

Hennig W (1966) Phylogenetic systematics. University of Illinois Press, Chicago
Hesse R (1924) Tiergeographie auf ökologischer Grundlage. Gustav Fischer Verlag, Stuttgart
Hey J (2006) On the failure of modern species concepts. Trends Ecol Evol 21:447–450
Hoffmann B (1919) Führer durch unsere Vogelwelt zum Beobachten und Bestimmen der häufigsten Arten durch Auge und Ohr. 2nd edn., 1921. Teubner, Leipzig
Hofsten N von (1916) Zur älteren Geschichte des Diskontinuitätsproblems in der Biogeographie. Zoologische Annalen 7:197–353
Hope G (1996) Quaternary change and the historical biogeography of Pacific islands. In: Keast JA, Miller SE (eds) The origin and evolution of Pacific Island biotas, New Guinea to Eastern Polynesia: patterns and processes. CRC Press, Boca Raton, pp 165–190
Hull DL (1988) Science as a process. An evolutionary account of the social and conceptual development of science. University of Chicago Press, Chicago
Hull DL (1994) Ernst Mayr's influence on the history and philosophy of biology: a personal memoir. Biol Philos 9:375–386
Huxley J (1940) (ed) The new systematics. Clarendon Press, Oxford
Huxley J (1942) Evolution. The modern synthesis. Allen and Unwin, London
Immelmann K, Kalberlah H-H, Rausch P, Stahnke A (1978) Sexuelle Prägung als möglicher Faktor innerartlicher Isolation beim Zebrafinken. J Ornitholog 119:197–212
Irwin DE, Bensch S, Irwin JH, Price TD (2005) Speciation by distance in a ring species. Science (Washington) 307:414–416
Jahn I (1998) (ed) Geschichte der Biologie, 3rd edn. Gustav Fischer Verlag, Stuttgart
Jepsen GL (1949) Foreword. In: Jepsen GL, Mayr E, Simpson GG (eds) Genetics, paleontology and evolution. Princeton University Press, Princeton, pp V–X
Jordan DS (1905) The origin of species through isolation. Science 22:545–562
Jordan K (1896) On mechanical selection and other problems. Novitates Zool 3:426–525
Jordan K (1905) Der Gegensatz zwischen geographischer und nichtgeographischer Variation. Zeitschrift für wissenschaftliche Zoologie 83:151–210
Junker T (1995) Vergangenheit und Gegenwart: Bemerkungen zur Funktion von Geschichte in den Schriften Ernst Mayrs. Biologisches Zentralblatt 114:143–149
Junker T (1996) Factors shaping Ernst Mayr's concepts in the history of biology. J Hist Biol 29:29–77
Junker T (2003) Ornithology and the genesis of the synthetic theory of evolution. Avian Sci 3:65–73
Junker T (2004) Die zweite Darwinsche Revolution. Geschichte des Synthetischen Darwinismus in Deutschland 1924 bis 1950. Basilisken-Presse. Marburg, 633 pp
Junker T (2006) Der Darwinismus als internationales Netzwerk: Die 1930er und 1940er Jahre. Verhandlungen zur Geschichte und Theorie der Biologie 12:19–33
Junker T (2007) Ernst Mayr (1904–2005) and the new philosophy of biology. J Gen Philos Sci 38 (in press)
Junker T, Engels E-M (1999) (eds) Die Entstehung der Synthetischen Theorie. Beiträge zur Geschichte der Evolutionsbiologie in Deutschland 1930–1950. Verhandlungen zur Geschichte und Theorie der Biologie 2. Berlin
Kay LE (1985) Conceptual models and analytical tools: the biology of physicist Max Delbrück. J Hist Biol 18:207–246
Keast JA (1961) Bird speciation on the Australian continent. Bull Mus Comp Zool 123:305–495
Keast JA (1983) In the steps of Alfred Russel Wallace: biogeography of the Asian-Australian Interchange Zone. In: Sims RW, Price JH, Whalley PES (eds) Systematics Association Special Volume No. 23. Evolution, time and space: the emergence of the biosphere. Academic Press, New York, pp 367–407

Keast JA (1996) Avian geography: New Guinea to the eastern Pacific. In: Keast JA, Miller SE (eds) The origin and evolution of Pacific island biotas, New Guinea to Eastern Polynesia: patterns and processes. SPB Academic Publishing, Amsterdam, pp 373–398
Keast JA, Miller SE (1996) (eds) The origin and evolution of Pacific island biotas, New Guinea to Eastern Polynesia: patterns and processes. SPB Academic Publishing, Amsterdam
Kleinman K (1993) His own synthesis: Corn, Edgar Anderson, and evolutionary theory in the 1940s. J Hist Biol 32:293–320
Knapp S, Mallet J (2003) Refuting refugia? Science 300:71–72
Kohler RE (2002) Landscapes and labscapes. Exploring the lab-field border in biology. University of Chicago Press, Chicago
Kottler MJ (1978) Charles Darwin's biological species concept and theory of geographic speciation: the transmutation notebooks. Ann Sci 35:275–297
Kramer G (1948) Review of E. Mayr's Systematics and the Origin of Species. Ornithologische Berichte 1:149–164
Krementsov NL (1994) Dobzhansky and Russian entomology: the origin of his ideas on species and speciation. In: Adams MB (ed) The evolution of Theodosius Dobzhansky. Essays on his life and thought in Russia and America. Princeton University Press, Princeton, pp 31–48
Kuhn Th (1962) The structure of scientific revolutions. University of Chicago Press, Chicago
Kutschera U (2004) Species concepts: leeches versus bacteria. Lauterbornis 52:171–175
Kutschera U, Niklas KJ (2004) The modern theory of biological evolution: an expanded synthesis. Naturwissenschaften 91:255–276
Lack D (1944) Ecological aspects of species-formation in passerine birds. Ibis 86:260–286
Lack D (1947) Darwin's finches. Cambridge University Press, Cambridge
Lack D (1949) The significance of ecological isolation. In: Jepsen GL, Simpson GG, Mayr E (eds) Genetics, paleontology and evolution. Princeton University Press, Princeton, pp 299–308
Lack D (1971) Ecological isolation in birds. Blackwell Scientific Publications, Oxford
Lack D (1976) Island biology illustrated by the landbirds of Jamaica. University of California Press, Berkeley
Landsberg H (1995) Ernst Mayer [sic] in Berlin–vom vielversprechenden jungen Mann zum anerkannten Systematiker im American Museum of Natural History. Biologisches Zentralblatt 114:123–132
Lanyon WE (1995) Ornithology at the American Museum of Natural History. In: Davis WE Jr, Jackson JA (eds) Contributions to the history of North American ornithology. Mem Nuttall Ornithological Club 12:113–144
LeCroy M (1989) A wing for the birds. Nat Hist 89:90–91
LeCroy M (2005) Ernst Mayr at the American Museum of Natural History. Ornitholog Monogr 58:30–49
Lein MR (2005) Ernst Mayr as a life-long naturalist. Ornitholog Monogr 58:17–29
Löhrl H (1960, 1961) Vergleichende Studien über Brutbiologie und Verhalten der Kleiber *Sitta whiteheadi* Sharpe und *Sitta canadensis* L. J Ornithol 101:245–264; 102:111–132
Lönnberg E (1926) Einige Bemerkungen über den Einfluss der Klimaschwankungen auf die afrikanische Vogelwelt. J Ornithol 74:259–273
Lönnberg E (1927) Some speculations on the origin of the North American ornithic fauna. K. Sven. Vetenskapsakad. Handl., ser. 3, 4(6):1–24
Lönnberg E (1929) The development and distribution of the African fauna in connection with and depending upon climatic changes. Arkiv för Zool 21A:1–33
Lorenz K (1941) Vergleichende Bewegungsstudien an Anatinen. J Ornitholog 89, Ergänzungsband III (Festschrift Oskar Heinroth) pp 194–293
Lovejoy AO (1936) The great chain of being. Harvard University Press, Cambridge, MA

Lovette IJ (2005) Glacial cycles and the tempo of avian speciation. Trend Ecol Evol 20:57–59
MacArthur RH, Wilson EO (1963) An equilibrium theory of insular biogeography. Evolution 17:373–387
MacArthur RH, Wilson EO (1967) The theory of island biogeography. Princeton University Press, Princeton
MacKinnon J, Phillipps K (1993) A field guide to the birds of Borneo, Sumatra, Java and Bali. Oxford University Press, Oxford
MacKinnon J, Phillipps K (2000) A field guide to the birds of China. Oxford University Press, Oxford
Malthus TR (1798) An essay on the principle of population, as it affects the future improvement of society. Johnson, London
Marchant S (1972) A critical history of "Emu." Emu 72:51–69
Markl H (1984) Vielfalt und Anpassung. Der deutsch-amerikanische Zoologe Ernst Mayr, einer der größten Evolutionsbiologen seit Darwin, erhält den Balzan-Preis. Die Zeit, Nr. 8 (17. Februar), Seite 64
Mathews GM (1927, 1930) Systema Avium Australasianarum. London
McKinney HL (1972) Wallace and natural selection. Yale University Press, New Haven
Mearns B, Mearns R (1998) The bird collectors. Academic Press, London
Melville RV (1995) Towards stability in the names of animals. A history of the International Commission on Zoological Nomenclature 1895–1995. The Natural History Museum, London
Mertens R (1928) Über den Rassen- und Artenwandel auf Grund des Migrationsprinzipes, dargestellt an einigen Amphibien und Reptilien. Senckenbergiana 10:81–91
Meyer A (2007) Die Entstehung neuer Arten–Darwins Geheimnis der Geheimnisse. Ernst Mayr Lecture 2006; Berlin-Brandenburgische Akademie der Wissenschaften, Berichte und Abhandlungen (in press)
Meyer AB, Wiglesworth LW (1898) The birds of Celebes and the neighbouring islands. Fiedländer, Berlin
Miller AH (1941) Speciation in the avian genus *Junco*. University of California Publications in Zoology 44:173–434
Moore JA (1994) Some personal recollections of Ernst Mayr. Evolution 48:9–11
Morley RJ (2000) Origin and evolution of tropical rain forests. Wiley, New York
Morse RA (2000) Richard Archbold and the Archbold Biological Station. University Press of Florida, Gainesville
Mousson A (1849) Die Land- und Süsswasser-Mollusken von Java. Schulthess, Zürich
Murphy RC (1922) The Whitney South Sea Expedition of the American Museum of Natural History. Science 56:701–704
Murphy RC (1924) The Whitney South Sea Expedition. A sketch of the bird life of Polynesia. Nat Hist 24:539–553
Murphy RC (1932) Moving a museum. Natural History 32:497–511
Murphy RC (1938) The need for insular exploration as illustrated by birds. Science 88:533–539
Murphy RC (1951) Obituary of Leonard Cutler Sanford. Auk 68:409–410
Nice MM (1937, 1943) Studies in the life history of the song sparrow, volumes 1–2. Transactions of the Linnaean Society of New York 4:1–247, 6:1–328
Nice MM (1979) Research is a passion with me. Toronto, Amethyst Communication Inc
Nichols JT (1931) A theoretical discussion of the history of bird migration, by Mayr and Meise. Auk 48:302–304
Nicolai J (1977) Intraspezifische Selektion und die Wechselbeziehungen zwischen natürlicher Auslese und geschlechtlicher Zuchtwahl. Die Vogelwarte 29, Sonderheft (Ökophysiologische Probleme in der Ornithologie) 120–127

Olson SL (1975) The South Pacific gallinules of the genus *Pareudiastes*. Wilson Bull 87:1–5

Orr HA (2005) The genetic basis of reproductive isolation: insights from *Drosophila*. Proc Natl Acad Sci USA 102, suppl. 1:6522–6526

Palumbi SR, Lessios HA (2005) Evolutionary animation: how do molecular phylogenies compare to Mayr's reconstruction of speciation patterns in the sea? Proc Natl Acad Sci USA 102, suppl. 1:6566–6572

Paterson HEH (1985) The recognition concept of species. In: Vrba ES (ed) Species and speciation. Transvaal Museum Monograph No. 4:21–29. Transvaal Museum, Pretoria

Pennisi E (2006) Speciation standing in place. Science 311:1372–1374

Peters JL (1931–1986) Check-list of birds of the world, 16 volumes. Museum of Comparative Zoology, Cambridge, MA

Pitelka FA (1986) Rollo Beck—old-school collector, member of an endangered species. Am Birds 40:385–387

Poulton EB (1903) What is a species? Reprinted in E. Poulton (1908) Essays on evolution. Proc Entomological Soc Oxford, London, pp LXXVI–CXVI

Pratt HD, Bruner PI, Berrett DG (1987) A field guide to the birds of Hawaii and the tropical Pacific. Princeton University Press, Princeton

Pratt TK (1982) Biogeography of birds in New Guinea. In: Gressit JL (ed) Biogeography and ecology of New Guinea. Monogr Biol 42:815–836

Provine WB (1986) Sewall Wright and evolutionary biology. University of Chicago Press, Chicago

Provine WB (1989) Founder effects and genetic revolutions in microevolution and speciation: a historical perspective. In: Giddings LV, Kaneshiro KY, Anderson WW (eds) Genetics, speciation, and the founder principle. Oxford University Press, Oxford, pp 43–75

Provine WB (1994) The origin of Dobzhansky's Genetics and the origin of species. In: Adams MB (ed) The evolution of Theodosius Dobzhansky. Essays on his life and thought in Russia and America. Princeton University Press, Princeton, pp 99–114

Provine WB (2004) Ernst Mayr: genetics and speciation. Genetics 167:1041–1046

Provine WB (2005) Ernst Mayr, a retrospective. Trends Ecol Evol 20:411–413

Reif W-E, Junker T, Hossfeld U (2000) The synthetic theory of evolution: general problems and the German contribution to the synthesis. Theory Biosci 119:41–91

Reimarus HS (1762) Allgemeine Betrachtungen über die Triebe der Thiere. Hamburg. Reprint Hamburg 1982

Rensch B (1928) Grenzfälle von Rasse und Art. J Ornithol 76:222–231

Rensch B (1929) Das Prinzip geographischer Rassenkreise und das Problem der Artbildung. Borntraeger, Berlin

Rensch B (1933) Zoologische Systematik und Artbildungsprobleme. Verhandlungen der Deutschen Zoologischen Gesellschaft 1933:19–83

Rensch B (1934) Kurze Anweisung für Zoologisch-Systematische Studien. Akademische Verlagsgesellschaft, Leipzig

Rensch B (1936) Die Geschichte des Sundabogens. Eine tiergeographische Untersuchung. Borntraeger, Berlin

Rensch B (1939) Typen der Artbildung. Biol Rev 14:180–222

Rensch B (1947) Neuere Probleme der Abstammungslehre. Die Transspezifische Evolution. Enke. Stuttgart, (Engl. translation, Evolution above the Species Level. New York, Columbia, 1959)

Rensch B (1979) Lebensweg eines Biologen in einem turbulenten Jahrhundert. Fischer, Stuttgart

Ripley SD, Birckhead H (1942) Birds collected during the Whitney South Sea Expedition. 51. On the fruit pigeons of the *Ptilinopus purpuratus* group. Am Mus Novitates 1192:1–14

Robson GC, Richards OW (1936) The variations of animals in nature. Longmans and Green, London
Rothschild M (1983) Dear Lord Rothschild. Birds, butterflies, and history. Hutchinson, London
Rothschild W (1899) Ein neuer interessanter Vogel aus Neuguinea. Ornithologische Monatsberichte 7:137
Rothschild W, Dollman G (1933) A new tree-kangaroo from the Wondiwoi Mountains, Dutch New Guinea. Proc Zool Soc Lond 359:540–541
Rothschild W, Jordan K (1903) A revision of the lepidopterous family Sphingidae. Novitates Zoologicae 9, Supplement
Rümmler H (1932) Ueber die Schwimmratten (Hydromyinae), zugleich Beschreibung einer neuen *Leptomys* Thos., *L. ernstmayri,* aus Neuguinea. Aquarium (Berlin) 1932:131–135
Ruse M (1999) Mystery of mysteries. Is evolution a social construction? Harvard University Press, Cambridge
Salomonsen F (1972) New pigeons from the Bismarck Archipelago (Aves, Columbidae). Steenstrupia (Zoological Museum, University of Copenhagen) 2:183–189
Salvadori T (1880–1882) Ornitologia della Papuasia e delle Molucche, 3 volumes. Torino
Scharnke H (1931) Ornithologische Beobachtungen in der Umgebung von Greifswald. Mit Benutzung der Aufzeichnungen von Herbert Kramer und Ernst Mayr. Dohrniana (Abhandlungen und Berichte der Pommerischen Naturforschenden Gesellschaft) 11:40–86
Schilthuizen M (2001) Frogs, flies and dandelions. Speciation—the evolution of new species. Oxford University Press, Oxford
Schodde R (2005) Ernst Mayr and southwest Pacific birds: inspiration for ideas on speciation. Ornitholog Monogr 58:50–57
Schodde R, Mason IJ (1999) The directory of Australian birds. Passerines. Canberra
Seebohm H (1881) Turdidae. In: Catalogue of Birds in the British Museum, vol. 5. Trustees of the British Museum (Natural History), London
Seebohm H (1887) The geographical distribution of the Family Charadriidae. Sotheran, Manchester
Serventy DL (1950) Taxonomic trends in Australian ornithology—with special reference to the work of Gregory Mathews. Emu 49:257–267
Sheldon FH, Winkler DW (1993) Intergeneric phylogenetic relationships of swallows estimated by DNA-DNA hybridization. Auk 110:798–824
Shermer M, Sulloway FJ (2000) The grand old man of evolution. An interview with evolutionary biologist Ernst Mayr. Skeptic 8(1):76–82
Short LL (1969) Taxonomic aspects of avian hybridization. Auk 86:84–105
Short LL, Schodde R, Horne JFM (1983) Five-way hybridization of Varied Sitellas *Daphoenositta chrysoptera* (Aves: Neosittidae) in central Queensland. Austr J Zool 31:499–516
Simpson GG (1940) Mammals and land bridges. J Wash Acad Sci 30:137–163
Simpson GG (1943) Turtles and the origin of the fauna of Latin America. Am J Sci 241:413–429
Simpson GG (1944) Tempo and mode in evolution. Columbia University Press, New York
Simpson GG (1947) Holarctic mammalian faunas and continental relationships during the Cenozoic. Bull Geol Soc Am 58:613–688
Simpson GG (1961) Principles of animal taxonomy. Columbia University Press, New York
Simpson GG (1963) Biology and the nature of science. Science 139:81–88
Smith JJ (1934) Neue Orchideen Papuasiens aus den Sammlungen von Dr. E. Mayr und G. Stein. Botanische Jahrbücher für Systematik, Pflanzengeschichte und Pflanzengeographie 66:161–215

Smith JJ (1936) Ericaceae. Nova Guinea. Résultats des expéditions scientifiques à la Nouvelle Guinée 18:89–121
Smocovitis VB (1994a) Disciplining evolutionary biology: Ernst Mayr and the founding of the Society for the Study of Evolution and *Evolution* (1939–1950). Evolution 48:1–8
Smocovitis VB (1994b) Organizing evolution: founding the Society for the Study of Evolution (1939–1950). J Hist Biol 27:241–309
Smocovitis VB (1997) Unifying biology. The evolutionary synthesis and evolutionary biology. Princeton University Press, Princeton
Snow CP (1959) The two cultures and the scientific revolution. Cambridge University Press, New York
Snow DW (1973) Robert Cushman Murphy and his Journal of the Tring trip. Ibis 115:607–611
Steadman DW (2006) Extinction and biogeography of tropical Pacific birds. University of Chicago Press, Chicago, XIV + 594 pp
Stebbins GL (1950) Variation and evolution in plants. Columbia University Press, New York
Stegmann B (1938) Principes généraux des subdivisions ornithogéographiques de la région paléarctique. Faune de l'URSS. Acad Sci URSS (1)2:156 pp
Stein G (1932) Einige neue Beuteltiere aus Neuguinea. Zeitschrift für Säugetierkunde 7:254–257
Steinheimer F (2004) Ernst Mayr und die Nymphenrallen–eine ornithologische Anekdote aus Neuguinea. Ornithologischer Anzeiger 43:93–102
Steyermarck JA (1979) Flora of the Guyana highland: endemicity of the generic flora of the Venezuelan tepuis. Taxon 28:45–54
Stickney EH (1943) Birds collected during the Whitney South Sea Expedition. 53. Northern shore birds in the Pacific. Am Mus Novetates 1248:9
Stresemann E (1919a) Zur Frage der Entstehung neuer Arten durch Kreuzung. Club van Nederlandsche Vogelkundigen, Jaarbericht 9:24–32
Stresemann E (1919b) Ueber die europäischen Baumläufer. Verhandlungen der Ornithologischen Gesellschaft in Bayern 14:39–74
Stresemann E (1920) Die taxonomische Bedeutung qualitativer Merkmale. Ornithologischer Beobachter 17:149–152
Stresemann E (1921) Die Spechte der Insel Sumatra. Archiv für Naturgeschichte 87, Abteilung A, pp 64–120
Stresemann E (1923) Dr. Bürgers' ornithologische Ausbeute im Stromgebiet des Sepik. Ein Beitrag zur Kenntnis der Vogelwelt Neuguineas. Archiv für Naturgeschichte. Ser. A, vol 89, Heft 7, pp 1–96, Heft 8, pp 1–92
Stresemann E (1926) Stand und Aufgaben der Ornithologie 1850 und 1925. J Ornithol 74:225–232
Stresemann E (1927) Die Wanderungen der Rotschwanzwürger (Formenkreis *Lanius cristatus*). J Ornithol 75:68–85
Stresemann E (1927–1934) Sauropsida: Aves. In: Kükenthal W, Krumbach Th (eds) Handbuch der Zoologie, vol 7, part 2. W de Gruyter and Co., Berlin, 890 pp
Stresemann E (1930) Welche Paradiesvogelarten der Literatur sind hybriden Ursprungs? Novitates Zool 36:6–15
Stresemann E (1939) Die Vögel von Celebes, Teil 1 und 2 (Zoogeographie). J Ornithol 87:299–425
Stresemann E (1943) Ökologische Sippen-, Rassen- und Artunterschiede bei Vögeln. J Ornithol 91:305–324
Stresemann E (1951) Die Entwicklung der Ornithologie von Aristoteles bis zur Gegenwart. Berlin, Peters (English translation Ornithology from Aristotle to the Present. Harvard University Press, Cambridge, 1975

Stresemann E (1959) The status of avian systematics and its unsolved problems. Auk 76:269–280
Sulloway FJ (1979) Geographic isolation in Darwin's thinking: the vicissitudes of a crucial idea. Stud Hist Biol 3:23–65
Sulloway FJ (1982) Darwin's conversion: The *Beagle* voyage and its aftermath. J Hist Biol 15:325–396
Sulloway FJ (1996) Born to rebel. Birth order, family dynamics, and creative lives. Pantheon Books, New York
Szalay FS, Bock WJ (1991) Evolutionary theory and systematics: relationships between process and patterns. Zeitschrift für Zoologische Systematik und Evolutionsforschung 29:1–39
Tautz D (2003) Splitting in space. Nature 421:225–226
Van Royen P (1965) Sertulum Papuanum 14: an outline of the flora and vegetation of the Cyclop Mountains. Nova Guinea, Botany 21:451–469
Vaurie C (1949) A revision of the Dicruridae. Bull Am Mus Nat Hist 93:199–342
Vaurie C (1959, 1965) The birds of the Palaearctic Fauna, 2 volumes. Witherby, London
Vink W (1965) Botanical exploration of the Arfak Mountains. Nova Guinea, Botany 22:471–494
Voigt A (1921) Excursionsbuch zum Studium der Vogelstimmen, 8th edn. Quelle und Meyer, Leipzig
Vuilleumier F (2001–2002) The American Museum of Natural History. Calypso Log 2001, pp 30–35, 2002, pp 12–15
Vuilleumier F (2005a) Dean of American ornithologists: the multiple legacies of Frank M. Chapman of the American Museum of Natural History. Auk 122:389–402
Vuilleumier F (2005b) Ernst Mayr's biogeography: a lifetime of study. Ornitholog Monogr 58:58–72
Vuilleumier F, LeCroy M, Mayr E (1992) New species of birds described from 1981 to 1990. Bull Brit Ornithologists' Club, Centenary Supplement, 112A:267–309
Wagner H, Reiner H (1986) The Lutheran Church in Papua New Guinea. Lutheran Publishing House, Adelaide
Wagner M (1868) Die Darwin'sche Theorie und das Migrationsgesetz der Organismen. Leipzig, Dunker and Humblot (English translation The Darwinian Theory and the Law of the Migration of Organisms. London, 1873)
Wallace AR (1864) The Malayan Papilionidae or swallow-tailed butterflies, as illustrative of the theory of natural selection. In: Wallace AR (ed) Contributions to the theory of natural selection. Macmillan, London, pp 130–200
Wallace AR (1869) The Malay Archipelago. Tuttle Publishing, North Clarendon, VT
Walsh B (1863) Observations on certain North American Neuroptera. Proc Entomolog Soc Philadelphia 2:167–272
Walsh B (1864) On phytophagic varieties and phytophagic species. Proc Entomolog Soc Philadelphia 3:403–430
Watling D (2001) A guide to the birds of Fiji and western Polynesia. Environmental Consultants Ltd., Suva, Fiji
Watson EL (1991) Houses for science. A pictorial history of Cold Spring Harbor Laboratory. Cold Spring Harbor Laboratory Press, New York
Wetmore A (1951) A revised classification for the birds of the world. Smithsonian Miscellaneous Collections 117(4):1–22
White CMN, Bruce MD (1986) The birds of Wallacea (Sulawesi, The Moluccas and Lesser Sunda Islands, Indonesia). BOU Check-list No. 7. British Ornithologists' Union, London, 524 pp
Whittemore AT (1993) Species concepts: a reply to Ernst Mayr. Taxon 42:573–583

Wiglesworth LW (1898) On formulae for indicating the variation of a species within itself. Abhandlungen und Berichte des Zoologisch-Anthropologischen Museums Dresden VII (2):32–33
Wilkins AS (2002) Interview with Ernst Mayr. BioEssays 24(10):960–973
Wilkins AS (2007) Ernst Mayr and the evolution of Neodarwinism. (in press)
Williams GC (1966) Adaptation and natural selection. Princeton University Press, Princeton
Wilson EO (1994) Naturalist. Island Press, Washington, DC
Wilson EO (1998) Consilience: The Unity of Knowledge. New York, Alfred A. Knopf
Winsor MP (2006) Linnaeus's biology was not essentialist. Ann Missouri Bot Gard 93:2–7
Wright S (1931) Evolution in Mendelian populations. Genetics 16:97–159

**Non-printed sources held by:**

(1) Staatsbibliothek zu Berlin (Preussischer Kulturbesitz), Handschriften-Abteilung:

(a) Ernst Mayr Papers (letters of E. Stresemann to E. Mayr 1924–1972),

(b) Erwin Stresemann Papers (letters of E. Mayr to E. Stresemann 1923–1972).

(2) Museum für Naturkunde Berlin (Humboldt University), Historische Bild- und Schriftgut-Sammlung. Bestand: Zoologisches Museum (ZMB).

Signatur S III, E. Mayr: Schriftwechsel und Sammellisten seiner Expeditionen nach Neu-Guinea.

Signatur S III, E. Mayr: Personalakte.

(3) Harvard University Archives, Cambridge, Massachusetts; Pusey Library, Ernst Mayr Papers (Correspondence and autobiographical notes).

(4) Museum of Comparative Zoology, Harvard University, Cambridge University (Cambridge, Massachusetts); Archives, Ernst Mayr Papers.

# Part IV
# Appendices

# Appendix 1
## Curriculum Vitae
(including honorary degrees, honorary memberships, medals, and other special awards)

**Last title:** Alexander Agassiz Professor of Zoology, Emeritus, Museum of Comparative Zoology, Harvard University
**Special fields:** Ornithology, systematics, zoogeography, evolution, history and philosophy of biology
**Born:** Kempten, Germany on 5 July 1904, naturalized U.S. citizen
**Married:** Margarete Simon (29 February 1912–23 August 1990)
**Children:** Christa, Susanne; 5 grandchildren, 10 great-grandchildren
**Died:** Bedford, Massachusetts, on 3 February 2005

## Education

**1925** Cand. med., University of Greifswald
**1926** PhD, University of Berlin (Zoology; 24 June)

## Expeditions

**1928** Rothschild Expedition to Dutch New Guinea
**1928–29** Expedition to Mandated Territory of New Guinea (University of Berlin)
**1929–30** Whitney South Sea Expedition to Solomon Islands (American Museum of Natural History, AMNH, New York)

## Positions

**1926–32** Assistant Curator, Zoological Museum, University of Berlin, Germany
**1931–32** Visiting Research Associate, Department of Ornithology, AMNH, New York
**1932–44** Associate Curator, Whitney-Rothschild Collection, Department of Ornithology, AMNH, New York
**1944–53** Curator, Whitney-Rothschild Collection, Department of Ornithology, AMNH, New York (to 30 June)
**1953–75** Alexander Agassiz Professor of Zoology, Harvard University (from 1 July)
**1961–70** Director, Museum of Comparative Zoology, Harvard University
**1975–2005** Alexander Agassiz Professor of Zoology, Emeritus, Harvard University

## Honorary Degrees

**1957** PhD, Uppsala University (Sweden), systematics
**1959** D.Sc., Yale University, systematics

**1959** D.Sc., University of Melbourne (Australia), evolution
**1966** D.Sc., Oxford University (England), ornithology
**1968** D.Phil., University of Munich (Germany), evolution
**1974** D.Phil., University of Paris VI (Sorbonne), evolution
**1979** D.Sc., Harvard University, evolution
**1982** D.Sc., Cambridge University (England), evolution
**1982** D.Sc., Guelph University (Canada), philosophy of biology
**1984** D.Sc., University of Vermont, evolution
**1993** PhD, University of Massachusetts, Amherst, systematics
**1994** D.Sc., University of Vienna (Austria), ornithology
**1994** D.Phil., University of Konstanz (Germany), philosophy of biology
**1995** D.Sc., University of Bologna (Italy), evolution
**1996** D.Sc., Rollins College, Florida, philosophy of biology
**1997** Honoris Causa Degree, Museum National d'Histoire Naturelle, Paris, systematics
**2000** D.Phil., Humboldt University of Berlin (Germany), systematics

Countries: Sweden, USA (5), Australia, Great Britain (2), Germany (3), Canada, Austria, Italy, France (2)

## Lectureships and Visiting Professorships

**1941** Jesup Lecturer, Columbia University, New York
**1947** Lecturer, Philadelphia Academy of Sciences
**1949; 1974** Visiting Professor, University of Minnesota
**1950–53** Lecturer, Columbia University
**1951** Visiting Professor, University of Pavia, Italy
**1952** Visiting Professor, University of Washington
**1967** Life Sciences Lecturer, University of California, Davis
**1971** Maytag Visiting Professor, Arizona State University, Tempe
**1972** Visiting Professor, University of California, Riverside
**1977** Alexander von Humboldt Awardee, Würzburg University, Germany
**1978** Visiting Professor, University of California, San Diego
**1985** Messenger Lecturer, Cornell University
**1987** Hitchcock Professor, University of California

## Awards and Recognitions (35)

**1946** Leidy Medal, Academy of Natural Sciences, Philadelphia
**1958** Wallace-Darwin Medal, Linnean Society of London
**1965** Brewster Medal, American Ornithologists' Union
**1966** Verrill Medal, Peabody Museum, Yale University
**1967** Daniel Giraud Eliot Medal, National Academy of Sciences
**1969** Centennial Medal, American Museum of Natural History, New York
**1969** National Medal of Science
**1971** Walker Prize, Museum of Science, Boston, MA
**1972** Molina Prize, Accademia delle Scienze, Bologna, Italy
**1977** Linnean Medal (Zoology), Linnean Society, London
**1977** Coues Prize, American Ornithologists' Union

**1978** Prêmio Jabuti, Brazil, CBL[1]
**1978** Medal, Collège de France
**1980** Mendel Medal, Leopoldina Academy, Halle (Germany
**1983** E. Eisenmann Medal, Linnaean Society of New York
**1983** Balzan Prize (Switzerland/Italy)
**1984** Darwin Medal, Royal Society (England)
**1986** Award for Service to the Systematics Community by the Association of Systematics Collection
**1986** Sarton Medal (History of Science)
**1989** Alexander von Humboldt Stiftung, Medal
**1991** Phi Beta Kappa Book Prize
**1992** Naming of Ernst-Mayr-Dining-Room at Cold Spring Harbor Laboratory
**1994** Salvin-Godman Medal, British Ornithologists' Union
**1994** International Prize for Biology, Japan Prize
**1994** Dedication of the Ernst Mayr Library at the Museum of Comparative Zoology, Harvard University
**1995** Walk of Fame, Rollins College
**1995** Benjamin Franklin Medal (American Philosophical Society)
**1996** George Gaylord Simpson Award (Society for the Study of Evolution)
**1997** Establishment of the Ernst Mayr Lectureship at the Berlin-Brandenburgische Akademie
**1998** Lewis Thomas Prize
**1999** Crafoord Prize, Stockholm (Sweden)
**2000** Golden Plate Award (American Academy of Achievement)
**2000** Biologist of the Year 2000 (Washington)
**2004** Treviranus Medal (Verein deutscher Biologen)
**2004** Gold Medal from the Accademia dei Lincei, Rome, Italy

## Honorary Society Memberships (52)

**1939** Royal Australian Ornithological Union, Corresponding Member
**1941** Deutsche Ornithologen-Gesellschaft, Honorary Fellow
**1943** American Society of Naturalists, Honorary Fellow
**1944** New York Zoological Society, Fellow; 1987, Scientific Fellow
**1945** Netherlands Ornithological Society, Corr. Member; 1953, Honorary Fellow
**1948** Société Ornithologique de France, Honorary Foreign Member
**1948** Zoological Society of London, Corresponding Member
**1949** Academy of Natural Sciences of Philadelphia, Correspondent
**1950** Royal Society of New Zealand, Honorary Fellow
**1951** Botanical Gardens of Indonesia, Honorary Fellow
**1951** Ornithologische Gesellschaft in Bayern, Corresponding Member; 1976,
**1952** Linnean Society of London, Foreign Member
**1952** South African Ornithological Society, Corresponding Member
**1954** American Academy of Arts and Sciences, Fellow

[1] "Jabuti" is an Indian name for a turtle. The prize is given by an organization of publishers (CBL) to the author of the best book translated into Portuguese published in Brazil during the preceding year.

**1954** National Academy of Sciences, Member
**1955** K. Vetenskaps Societeten, Uppsala, Honorary Member
**1955** Zoological Society of India, Corr. Fellow; 1961, Honorary Member
**1956** British Ornithologists' Union, Honorary Member
**1956** Dansk Ornithologisk Forening, Honorary Member
**1958** American Association for the Advancement of Science, Fellow
**1962** Zoological Society of India (Calcutta)
**1962** Sociedad Venezolana de Ciencias Naturales, Corresponding Member
**1963** Asociación Ornitológica del Plata (Buenos Aires), Honorary Member
**1963** Societas Scientiarum Fennica (Helsingfors), Honorary Member
**1965** American Philosophical Society, Member
**1968** Sociedad Colombiana de Naturalistas, Honorary Member
**1970** Zoologische Gesellschaft, Germany, Honorary Member
**1971** Academia de Ciencias Físicas, Matemáticas y Naturales (Caracas, Venezuela), Foreign Corresponding Member
**1972** Department of Ornithology, AMNH, Curator Emeritus
**1972** Deutsche Akademie der Naturforscher Leopoldina, Halle (Germany)
**1972** Société Zoologique de France, Honorary Member
**1975** Sociedad Española de Ornitología (Madrid), Honorary Member
**1975** Nuttall Ornithological Club, Honorary Member
**1976** Society of Systematic Zoology, Honorary Member
**1976** Linnaean Society of New York, Honorary Member
**1977** Senckenbergische Gesellschaft, Frankfurt, Corresponding Member
**1977** Bayerische Akademie, Munich, Corresponding Member
**1978** Académie des Sciences, etc., Toulouse, Corresponding Member
**1980** Accademia Nazionale dei Lincei, Foreign Member
**1981** Italian Zoological Society, Foreign Member
**1984** Zoological Society of London, Honorary Member
**1986** American Society of Zoologists, Honorary Member
**1988** Royal Society, Foreign Member
**1989** Académie des Sciences, Paris, Associate
**1993** Center for the Philosophy of Science, Pittsburgh, Honorary Fellow
**1994** Russian Academy of Science, Moscow; Honorary Member
**1994** Berlin-Brandenburgische Akademie; Honorary Member
**1998** Gesellschaft Naturforschender Freunde zu Berlin, Corresponding Member
**2002** Verband Deutscher Biologen und biowissenschaftlicher Fachgesellschaften, Honorary Member
**2003** Gesellschaft für Biologische Systematik, Honorary Member
**2003** Sociedade Fritz Müller de Ciências Naturais, Brazil, Honorary Member
**2003** Darwin-Gesellschaft, Honorary Member

Appendix 1



# Society Offices

Linnaean Society of New York
Editor of the Proceedings and Transactions 1934–1941

American Ornithologists' Union
Vice President, 1953–1956; President, 1956–1959

Nuttall Ornithological Club
Councilor, 1954–1955; vice-president, 1955–1957; president, 1957–1959

International Commission on Zoological Nomenclature
Commissioner, 1954–1976

American Society of Naturalists
President, 1962–1963

Society for the Study of Evolution
Secretary, 1946; Editor, 1947–1949; President, 1950

Society of Systematic Zoology
President, 1966

11th International Zoological Congress
Vice President, 1958

13th International Ornithological Congress
President, 1962

International Society for the History, Philosophy, and Social Studies of Biology
Honorary President, 1990

# Appendix 2
## Bibliography of Ernst Mayr

This bibliography is based on a list compiled by Ernst Mayr and his secretaries over many years. The closing date for this list is January 2007, and I believe that it includes all of Mayr's last publications. I added some of his book reviews and several of his articles, particularly in the field of zoological nomenclature, and combined translations of his books and reprints of certain papers with their respective originals. I also listed reviews of Mayr's books in alphabetical order of their author's names after the respective titles. In most cases of coauthored publications, Mayr was the senior author, but I have not checked this, nor is this indicated in the list. Within a given year, the publications are listed in no particular sequence. Consecutive numbering of the list is indicated by numbers in the margins for every 50th title.

### 1923

1923a Die Kolbenente (*Nyroca rufina*) auf dem Durchzuge in Sachsen. Ornithologische Monatsberichte 31:135–136

1923b Der Zwergfliegenschnäpper bei Greifswald. Ornithologische Monatsberichte 31:136

### 1924

1924a Zwergmöven im Greifswalder Bodden. Ornithologische Monatsberichte 32:107 (with W. Klein)

1924b Maßnahmen zum Schutz der Trappe (*Otis tarda*). Mitteilungen des Landesvereins Sächsischer Heimatschutz 13:298–302

1924c Zum Vorkommen des Bienenfressers im Bielatale [with H. Förster and R. Zimmermann]. Mitteilungen des Vereins sächsischer Ornithologen 1:119–120

### 1925

1925a Erkennungsmerkmale des Sandregenpfeifers. Ornithologische Monatsberichte 33:130

1925b Zur Verbreitung des Girlitz in Norddeutschland. Ornithologische Monatsberichte 33:131

1925c Ein Vergleich der Vogelwelt Vorpommerns und Sachsens. Mitteilungen des Vereins sächsischer Ornithologen 1 (Sonderheft):64–70

### 1926

1926a Beteiligt sich das Buchfinkenmännchen am Nestbau? Ornithologische Monatsberichte 34:56

1926b Zu P. Wemers Angaben über den Nestbau des Buchfinken. Ornithologische Monatsberichte 34:88

1926c Das Freibrüten der Mehlschwalbe, *Delichon urbica* L., an den Kreidefelsen von Stubbenkammer auf Rügen. Ornithologische Monatsberichte 34:114

1926d Zum Vorkommen der Weidenmeise in der Mark. Ornithologische Monatsberichte 34:151

1926e Die Ausbreitung des Girlitz (*Serinus canaria serinus* L.). Ein Beitrag zur Tiergeographie. Journal für Ornithologie 74:571–671

1926f Unveröffentlichte Beobachtungen Kaluzas. Bericht des Vereins Schlesischer Ornithologen 12:1–4

1926g Zum Vorkommen des Seidenschwanzes, *Bombycilla garrulus* (L.) in Sachsen. Mitteilungen des Vereins Sächsischer Ornithologen 1 (8):212

1926h Nochmals das Vorkommen des Bienenfressers im Bielatale [with H. Förster and R. Zimmermann]. Mitteilungen des Vereins sächsischer Ornithologen 1:176

1926i Zur Ausbreitung des Girlitz. Ornithologische Monatsberichte 34:184–185

**1927**

1927a Zur Ausbreitung des Girlitz. Ornithologische Monatsberichte 35:42

1927b Beiträge zur Systematik der Afrikanischen *Serinus*-Arten. Ornithologische Monatsberichte 35:47–48

1927c Frühes Rufen der Rohrdommel. Ornithologische Monatsberichte 35:88

1927d Schlangenhäute als Nestmaterial. Ornithologische Monatsberichte 35:146–148

1927e Neue Formen von *Prunella rubeculoides*. Ornithologische Monatsberichte 35:148–149

1927f Die Schneefinken (Gattungen *Montifringilla* und *Leucosticte*). Journal für Ornithologie 75:596–619

1927g Zur Verbreitung von *Serinus donaldsoni buchanani* Hart. Ornithologische Monatsberichte 35:181

1927h Review of *Excursion ornithologique dans la région de Camaret (Finistère)*, by J. Rapine. Ornithologische Monatsberichte 35:30

1927i Review of *A study of the neotropical finches of the genus* Spinus, by W. E. C. Todd. Ornithologische Monatsberichte 35:31

1927j Review of *The races of domestic fowl*, by A. Jull. Ornithologische Monatsberichte 35:123–124

1927k Review of *Some climatic limits in the extension of birds in eastern Europe*, by V. V. Stantchinsky. Ornithologische Monatsberichte 35:125

1927l Review of *The birds of Latium, Italy*, by H. G. Alexander. Ornithologische Monatsberichte 35:151–152

1927m Review of *Attentiveness and inattentiveness in the nesting behaviour of the house wren*, by S. P. Baldwin and S. C. Kendeigh. Ornithologische Monatsberichte 35:152

1927n Review of *Die Frage der Grenzbestimmung zwischen Kreide und Tertiär in zoogeographischer Betrachtung*, by G. Pfeffer. Ornithologische Monatsberichte 35:184–185

**1928**

1928 Weidenmeisen-Beobachtungen (*Parus atricapillus salicarius* Brehm). Journal für Ornithologie 76:462–470

## 1929

1929a Zeitschriftenverzeichnis des Museums für Naturkunde [with W. Meise]. Mitteilungen aus dem Zoologischen Museum in Berlin 14 (Sonderheft):1–187

1929b [Letter from New Guinea]. Ornithologische Monatsberichte 37:62–63

## 1930

1930a Beobachtungen über die Brutbiologie der Grossfußhühner von Neuginea (*Megapodius*, *Talegallus*, und *Aepypodius*). Ornithologische Monatsberichte 38:101–106

1930b *Loboparadisea sericea aurora* subsp. nova. Ornithologische Monatsberichte 38:147–148

1930c Theoretisches zur Geschichte des Vogelzuges. Der Vogelzug 1:149–172 (with W. Meise)

1930d Was sind *Gerygone arfakiana* Salvad. und *Gerygone rufescens* Salvad.? Ornithologische Monatsberichte 38:176–178

1930e Die Unterarten des Kragenparadiesvogels (*Lophorina superba*). Ornithologische Monatsberichte 38:178–180

1930f My Dutch New Guinea Expedition, 1928. Novitates Zoologicae 36:20–26

## 1931

1931a *Ptiloprora plumbea granti*, subsp. nov. and *Pachycephalopsis hattamensis axillaris*, subsp. nov. Bulletin of the British Ornithologists' Club 51:59

1931b Birds collected during the Whitney South Sea Expedition. XII. Notes on *Halcyon chloris* and some of its subspecies. American Museum Novitates, no. 469:1–10

1931c Anatomie und systematische Stellung der Salvadori-Ente (*Salvadorina waigiuensis* Rothsch. und Hartert). Ornithologische Monatsberichte 39:69–70

1931d Birds collected during the Whitney South Sea Expedition. XIII. A systematic list of the birds of Rennell Island with descriptions of new species and subspecies. American Museum Novitates, no. 486:1–29

1931e Birds collected during the Whitney South Sea Expedition. XIV. The relationships and origin of the birds of Rennell Island. With notes on the geography of Rennell Island and the ecology of its bird life [by H. Hamlin]. American Museum Novitates, no. 488:1–11

1931f Birds collected during the Whitney South Sea Expedition. XV. The parrot finches (genus *Erythrura*). American Museum Novitates, no. 489:1–10

1931g The generic name *Calao*. The Auk 48:600–601

1931h *Rhamphozosterops sanfordi* genus et spec. nov. Ornithologische Monatsberichte 39:182

1931i Birds collected during the Whitney South Sea Expedition. XVI. Notes on fantails of the genus *Rhipidura*. American Museum Novitates, no. 502:1–21

**[50]** 1931j Birds collected during the Whitney South Sea Expedition. XVII. The birds of Malaita Island (British Solomon Islands). American Museum Novitates, no. 504:1–26

1931k Deutung der De Vis'schen Vogelnamen. Mitteilungen aus dem Zoologischen Museum in Berlin 16:913–917

1931l Die Vögel des Saruwaged- und Herzoggebirges (NO-Neuguinea). Mitteilungen aus dem Zoologischen Museum in Berlin 17:639–723

1931m Die Syrinx einiger Singvögel aus Neu-Guinea. Journal für Ornithologie 79:333–337

1931n Wörter der Nissan-Sprache. Zeitschrift für Eingeborenensprachen 21:252–256

## 1932

1932a Birds collected during the Whitney South Sea Expedition. XVIII. Notes on Meliphagidae from Polynesia and the Solomon Islands. American Museum Novitates, no. 516:1–30

1932b Birds collected during the Whitney South Sea Expedition. XIX. Notes on the Bronze Cuckoo *Chalcites lucidus* and its subspecies. American Museum Novitates, no. 520:1–9

1932c Birds collected during the Whitney South Sea Expedition. XX. Notes on thickheads (*Pachycephala*) from the Solomon Islands. American Museum Novitates, no. 522:1–22

1932d Birds collected during the Whitney South Sea Expedition. XXI. Notes on thickheads (*Pachycephala*) from Polynesia. American Museum Novitates, no. 531:1–23

1932e A tenderfoot explorer in New Guinea. Reminiscences of an Expedition for Birds in the Primeval Forests of the Arfak Mountains. *Natural History Magazine* 32:83–97. [An abbreviated German translation appeared under the title "Paradiesvogeljagd auf Neuguinea" in "Blau-Gold. Nachrichten- und Erinnerungsblätter des Staatsgymnasiums zu Dresden-Neustadt," Jahrgang 6 (Heft 11):22–26, 1933]

1932f Review of *Reisen in den Britischen Salomonen*, by Eugen Paravicini. Zeitschrift der Gesellschaft für Erdkunde zu Berlin Nr. 9/10:383–384

1932g Berichtigungen und Verbesserungen zu meiner Arbeit: Die Vögel des Saruwaged- und Herzoggebirges. Mitteilungen aus dem Zoologischen Museum in Berlin 18:169

## 1933

1933a Birds collected during the Whitney South Sea Expedition. XXII. Three new genera from Polynesia and Melanesia. American Museum Novitates, no. 590:1–6

1933b Der Formenkreis *Zosterops minor*. Ornithologische Monatsberichte 41:53–54

1933c Zur Besiedlungsgeschichte von Biak. Ornithologische Monatsberichte 41:54–55

1933d Birds collected during the Whitney South Sea Expedition. XXIII. Two new birds from Micronesia. American Museum Novitates, no. 609:1–4

1933e The type of *Egretta brevipes*. The Auk 50:206–207

1933f Birds collected during the Whitney South Sea Expedition. XXIV. Notes on Polynesian flycatchers and a revision of the genus *Clytorhynchus* Elliot. American Museum Novitates, no. 628:1–21

1933g Zur systematischen Stellung von *Paramythia* De Vis. Ornithologische Monatsberichte 41:112–113

1933h On a collection of birds, supposedly from the Solomon Islands. The Ibis, series 13, vol. 3:549–552

1933i Birds collected during the Whitney South Sea Expedition. XXV. Notes on the genera *Myiagra* and *Mayrornis*. American Museum Novitates, no. 651:1–20

1933j Die Vogelwelt Polynesiens. Mitteilungen aus dem Zoologischen Museum in Berlin 19:306–323

1933k Birds collected during the Whitney South Sea Expedition. XXVI. Notes on *Neolalage banksiana* (Gray). American Museum Novitates, no. 665:1–5

1933l Birds collected during the Whitney South Sea Expedition. XXVII. Notes on the variation of immature and adult plumages in birds and a physiological explanation of abnormal plumages. American Museum Novitates, no. 666:1–10

1933m Note critique sur les sous-espèces de *Goura victoria* [with J. Berlioz]. *L'Oiseau* No. 4:751–754

1933n [Index to] Birds collected during the Whitney South Sea Expedition. I-XXV. Articles from the American Museum Novitates 1924–1933: v–xx (with Murphy, Mathews, Hartert, and Hamlin)

1933o *Rhamphozosterops* versus *Cinnyrorhyncha*. The Ibis, series 13, vol. 3:389–390

## 1934

1934a Birds collected during the Whitney South Sea Expedition. XXVIII. Notes on some birds from New Britain, Bismarck Archipelago. American Museum Novitates, no. 709:1–15

1934b Birds collected during the Whitney South Sea Expedition. XXIX. Notes on the genus *Petroica*. American Museum Novitates, no. 714:1–19

1934c Ernst Johann Otto Hartert [Obituary]. The Auk 51:283–285

1934d Blackcap in Atlantic in June. British Birds 28:119–120

1934e Morse's American bird lists of 1789 and 1793. By L. Nelson Nichols with interpolated comment by E Mayr. Abstracts of the Proceedings of the Linnaean Society of New York 43, 44:27–33

## 1935

1935a Zur Nomenklatur einiger *Aplonis*-Arten. Mitteilungen aus dem Zoologischen Museum in Berlin 20:334–336

1935b Avocet recorded for North Carolina. The Auk 52:185

1935c Bernard Altum and the territory theory. Proceedings of the Linnaean Society of New York 45, 46:24–38

1935d How many birds are known? Proceedings of the Linnaean Society of New York 45, 46:19–23

1935e The ornithological year 1933 in the New York City region. Compiled by Ernst Mayr. Supplemented and edited by John F. and Richard G. Kuerzi. Proceedings of the Linnaean Society of New York 45, 46:74–100

1935f Notes from Beaverkill, Sullivan Co., N. Y. Proceedings of the Linnaean Society of New York 45, 46:102

1935g Results of the Archbold Expeditions. No. 6. Twenty-four apparently undescribed birds from New Guinea and the D'Entrecasteaux Archipelago [with A. L. Rand]. American Museum Novitates 814:1–17

1935h Birds collected during the Whitney South Sea Expedition. XXX. Descriptions of twenty-five new species and subspecies. American Museum Novitates, no. 820:1–6

## 1936

1936a Vorläufiges über die Ergebnisse der Archbold-Rand Neu-Guinea-Expedition von 1933 [with A. L. Rand]. Ornithologische Monatsberichte 44:41–44

1936b Birds collected during the Whitney South Sea Expedition. XXXI. Descriptions of twenty-five species and subspecies. American Museum Novitates, no. 828:1–19

1936c Results of the Archbold Expeditions. No. 10. Two new subspecies of birds from New Guinea [with A. L. Rand]. American Museum Novitates, no. 868:1–3

1936d New subspecies of birds from the New Guinea region. American Museum Novitates, no. 869:1–7

1936e Portenko on *Limosa lapponica*. The Auk 53:367–368

1936f Neue Unterarten von Vögeln aus Neu-Guinea [with A. L. Rand]. Mitteilungen aus dem Zoologischen Museum in Berlin 21:241–248

1936g *Melidectes belfordi kinneari*, subsp. nov. Bulletin of the British Ornithologists' Club 57 (399):42–43

**1937**

1937a Notes on the genus *Sericornis* Gould. American Museum Novitates, no. 904:1–25

1937b Review of *Check-list of Birds of the World*, vol. III by J. L. Peters. The Auk 54:550–551

1937c Results of the Archbold Expeditions. 14. Birds of the 1933–1934 Papuan expedition [with A. L. Rand]. Bulletin of the American Museum of Natural History 73:1–248

1937d The homing of birds. Bird-Lore 39:5–13 ***[100]***

1937e Birds collected during the Whitney South Sea Expedition. XXXII. On a collection from Tanna, New Hebrides. American Museum Novitates, no. 912:1–4

1937f Birds collected during the Whitney South Sea Expedition. XXXIII. Notes on New Guinea birds. I. American Museum Novitates, no. 915:1–19

1937g Birds collected during the Whitney South Sea Expedition. XXXV. Notes on New Guinea birds. II. American Museum Novitates, no. 939:1–14

1937h Birds collected during the Whitney South Sea Expedition. XXXVI. Notes on New Guinea birds. III. American Museum Novitates, no. 947:1–11

1937i Preface to *Hawks*, by Ellsworth D. Lumley. Conservation Series, Publ. No. 66:1–11

1937j Additional notes from Litchfield Co., Conn. [with J. F. and R. G. Kuerzi]. Proceedings of the Linnaean Society of New York 48:98

1937k [Description of new subspecies]. In: H. Birckhead, The Birds of the Sage West China Expedition. American Museum Novitates, no. 966:1–17

**1938**

1938a Forms of *Mesia argentauris* [with J. C. Greenway, Jr.]. Proceedings of the New England Zoölogical Club 17:1–7

1938b The birds of the Vernay-Hopwood Chindwin Expedition. The Ibis: 277–320

1938c Birds collected during the Whitney South Sea Expedition. XXXVIII. On a collection from Erromanga, New Hebrides. American Museum Novitates, no. 986:1–3

1938d Birds of the Crane Pacific Expedition [with S. Camras]. Zoological Series of Field Museum of Natural History 20:453–473

1938e The proportion of sexes in hawks. The Auk 55:522–523

1938f Leben in New York. "Blau-Gold. Nachrichten- und Erinnerungsblätter des Staatsgymnasiums zu Dresden-Neustadt." Jahrgang 11 (Heft 20):16–18

1938g Notes on a collection of birds from south Borneo. Bulletin of the Raffles Museum, Singapore 14:5–46

1938h Birds on an Atlantic crossing. Proceedings of the Linnaean Society of New York 49:54–58

1938i A review of the genus *Acanthiza* Vigors and Horsfield [with D. L. Serventy]. The Emu 38:245–292

1938j Birds collected during the Whitney South Sea Expedition. XXXIX. Notes on New Guinea birds. IV. American Museum Novitates, no. 1006:1–16

1938k Birds collected during the Whitney South Sea Expedition. XL. Notes on New Guinea birds. V. American Museum Novitates, no. 1007:1–16

1938l *Aethostoma celebense connectens*, subsp. nov. Ornithologische Monatsberichte 46:157

1938m Das Zahlenverhältnis der Geschlechter bei den Vögeln. Der Vogelzug 9:184–185

1938n The sex ratio in birds (summary), page 231. IXe Congrès Ornithologique International (J. Delacour, ed.). Rouen, France

## 1939

1939a The sex ratio in wild birds. The American Naturalist 73:156–179

1939b Die Grössenvariation von *Lanius schach hainanus*. Ornithologische Monatsberichte 47:63–64

1939c Zoological results of the Denison-Crockett Expedition to the South Pacific for the Academy of Natural Sciences of Philadelphia, 1937–1938. Part I.- The birds of the island of Biak [with R. Meyer de Achauensee]. *Proceedings of the Academy of Natural Sciences of Philadelphia* 91:1–37

1939d Ein neuer Wespenbussard von den Philippinen. Ornithologische Monatsberichte 47:74–76

1939e The winter quarters of *Chalcites malayanus minutillus* (Gould). The Emu 39:128–129

1939f Zoological results of the Denison-Crockett South Pacific Expedition for the Academy of Natural Sciences of Philadelphia, 1937–1938. Part IV.- Birds from northwest New Guinea [with R. Meyer de Schauensee]. Proceedings of the Academy of Natural Sciences of Philadelphia 91:97–144

1939g Zoological results of the Denison-Crockett South Pacific Expedition for the Academy of Natural Sciences of Philadelphia, 1937–1938. Part V.- Birds from the western Papuan Islands [with R. Meyer de Schauensee]. Proceedings of the Academy of Natural Sciences of Philadelphia 91:145–163

## 1940

1940a Birds collected during the Whitney South Sea Expedition. XLI. Notes on New Guinea birds. VI. American Museum Novitates, no. 1056:1–12

1940b Birds collected during the Whitney South Sea Expedition. XLII. On the birds of the Loyalty Islands. American Museum Novitates, no. 1057:1–3

1940c Speciation phenomena in birds. American Naturalist 74:249–278 [Reprinted in part (pages 250–256) under the title "Toward a modern species definition" *in* E. Mayr, Evolution and the Diversity of Life, pages 481–484, 1976]

1940d Notes on Australian birds. I. The genus *Lalage*. The Emu 40:111–117

1940e *Pericrocotus brevirostris* and its double. The Ibis: 712–722

1940f The Vernay-Cutting Expedition to Northern Burma.- Part I [with J. K. Stanford]. The Ibis: 679–711

1940g Birds collected during the Whitney South Sea Expedition. XLIII. Notes on New Guinea birds. VII. American Museum Novitates, no. 1091:1–3

1940h Borders and subdivision of the Polynesian region as based on our knowledge of the distribution of birds. Pages 191–195 *in* Proceedings of the Sixth Pacific Science Congress, vol. 4

1940i The origin and the history of the bird fauna of Polynesia. Pages 197–216 *in* Proceedings of the Sixth Pacific Science Congress, vol. 4 [Reprinted in E. Mayr, Evolution and the Diversity of Life, pages 601–617, 1976]

## 1941

1941a The Vernay-Cutting Expedition to Northern Burma.- Part II, III, IV, and V [with J. K. Stanford]. The Ibis series 14, vol. 5:56–105, 213–245, 353–378, and 479–518

1941b Taxonomic notes on the birds of Lord Howe Island. The Emu 40:321–322

1941c Wanderung oder Ausbreitung? Zoogeographica 4:18–20

1941d A new race of *Coracina caledonica* [with S. D. Ripley]. The Auk 58:250

1941e Birds collected during the Whitney South Sea Expedition. XLIV. Notes on the genus *Lalage* Boie. American Museum Novitates, no. 1116:1–18

1941f List of New Guinea birds. A systematic and faunal list of the birds of New Guinea and adjacent islands. American Museum Natural History, XI + 260 pp.
Reviews: Peters, J. L. (1942) in The Auk 52:452–453; Stresemann, E. (1942) in Ornithologische Monatsberichte 50:23–24

1941g What is an Artenkreis? Copeia (2):115–116

1941h Birds collected during the Whitney South Sea Expedition. XLV. Notes on New Guinea birds. VIII. American Museum Novitates, no. 1133:1–8

1941i The origin of gaps between species. The Collecting Net 16:137–143

1941j Birds collected during the Whitney South Sea Expedition. XLVI. Geographical variation in *Demigretta sacra* (Gmelin) [with D. Amadon]. American Museum Novitates, no. 1144:1–11

1941k Birds collected during the Whitney South Sea Expedition. XLVII. Notes on the genera *Halcyon, Turdus* and *Eurostopodus*. American Museum Novitates, no. 1152:1–7

1941l Red-wing observations of 1940. Proceedings of the Linnaean Society of New York 52–53:75–83

1941m Ueber einige Raubvögel der Kleinen Sunda-Inseln. Ornithologische Monatsberichte 49:42–47 ***[150]***

1941n Die geographische Variation der Färbungstypen von *Microscelis leucocephalus*. Journal für Ornithologie 89:377–392

**1942**

1942a *Stachyris leucotis obscurata*, new name. The Auk 59:117–118

1942b Birds collected during the Whitney South Sea Expedition. XLVIII. Notes on the Polynesian species of *Aplonis*. American Museum Novitates, no. 1166:1–6

1942c Review of *Genetics and the origin of species*, by Theodosius Dobzhansky. 2d. ed. (New York:Columbia University Press, 1941). Journal of the New York Botanical Garden 43:158–159

1942d Speciation in the Junco. Review of *Speciation in the avian genus Junco*, by Alden H. Miller [University of California Publications Zoology 44 (1941):173–434]. Ecology 23:378–379

1942e Systematics and the Origin of Species. New York: Columbia University Press, XII + 334 p. [Reprinted in 1944, 1947 and 1949 [Paperback reprint with a new introduction by the author (pages IX–XI), Dover Publications, New York, 1964; paperback edition with an introduction by Niles Eldredge (pages XV–XXXVII), Columbia University Press, New York, 1982; paperback reprint with a new introduction by the author (pages XIII–XXXV), Harvard University Press, Cambridge, Mass., 1999]
Reviews: Anon. (1943) in Quarterly Review of Biology 18:270; Beebe, W. (1943) in New York Times Book Review (January 17); Blair, A. P. (1943) in Natural History Magazine (18 March); Bonet, F. (1943) in Ciencia (Revista hispano-americana de Ciencias puras y aplicadas) 4:261–262; Emerson, A. E. (1943) in Ecology 24:412–413; Epling, C. (1945) in Chronica Botanica 9:222–223; Griscom, L. (1943) in The Wilson Bulletin 55:136–137; Haldane, J. B. S. (1944) in Britsh Medical Journal (15 April); Hubbs, C. L. (1943) in American Naturalist 77:173–178; Huxley, J. (1943) in Nature 151:347–348; Jepsen, G. L. (1943) in American Journal of Sciences 241:521–528; K., C. H. (1943) in Annals of the Entomological Society of America, p. 138–139;

Kramer, G. (1948) in Ornithologische Berichte 1:149–164; D. L[ack] (1943) in The Ibis 85:360–361; McClung, C. E. (1943) in Science 97:424–425; Miller, A. H. (1943) in The Auk 60:289–291; Ownbey, M. (1943) in Madrono 7:94–96; Pough, R. H. (1943) in Audubon Magazine (May–June); Rhoades, M. M. (1943) in Journal of the New York Botanical Garden (December); Ripley, D. in Bird Banding 14:89–90; Schmidt, K. P. (1943) in Copeia 1943:198–199; S[clater], W. L. (1943) in The Geographical Journal 102:87–88; Sturtevant, A. H. (1943) in The Journal of Heredity (October), p. 307–308; Sutton, G. M. (1943) in New York Herald Tribune (Weekly Book Review, May 9); Wright, H. F. (1944) in School of Science and Mathematics; Zirkle, C. (1944) in Isis 35, Part 1 (No. 99):44–45

1942f Index and summary of birds collected during the Whitney South Sea Expedition XXVI–L [with D. Amadon, C. Bogert, R. C. Murphy, S. D. Ripley]. Articles from the American Museum Novitates 1933–1942:v–xix.

1942g Review of *Zoogeographical studies of the Tsugitata Mountains of Formosa* by T. Kano. Geographical Review 32:693–694

1942h Review of *Begründung einer statistischen Methode in der Regionalen Tiergeographie* by S. Ekman. Geographical Review 32:694–695

1942i [Letter to the Editor.] Vernay-Cutting Expedition to Northern Burma. The Ibis series 14, vol. 6:525–526

## 1943

1943a The zoogeographic position of the Hawaiian Islands. The Condor 45:45–48 [Reprinted and revised under the title "The ornithogeography of the Hawaiian Islands" in E. Mayr, Evolution and the Diversity of Life, pages 653–658, 1976]

1943b New species of birds described from 1938–1941 [with J. T. Zimmer]. *The Auk* 60:249–262

1943c A new swallow-shrike. The Auk 60:268

1943d What genera belong to the family Prionopidae? The Ibis 85:216–218

1943e Criteria of subspecies, species and genera in ornithology. The Annals of the New York Academy of Sciences 44:133–139

1943f Notes on the generic classification of the swallows, Hirundinidae [with J. Bond]. The Ibis 85:334–343

1943g Notes on Australian birds. II. The Emu 43:3–17

1943h A journey to the Solomons. Natural History 52:30–37, 48

## 1944

1944a Bird watching, a hobby and a science. Review of *A Guide to Bird Watching*, by J. J. Hickey. The Auk 61:151–152

1944b Review of *The Birds of Burma* by B. E. Smythies. The Auk 60:292

1944c Review of *The Book of Indian Birds* by S. Ali. The Auk 60:287

1944d Wallace's Line in the light of recent zoogeographic studies. *The Quarterly Review of Biology* 19:1–14 [Reprinted in E. Mayr, Evolution and the Diversity of Life, pages 626–645, 1976]

1944e The birds of Timor and Sumba. Bulletin of the American Museum of Natural History 83:123–194 [Reprinted in part (p. 181–186) under the title "Land Bridges and Dispersal Faculties" in E. Mayr, Evolution and the Diversity of Life, pages 618–625, 1976]

1944f Chromosomes and phylogeny. Review of *Contributions to the Genetics, Taxonomy and Ecology of* Drosophila pseudoobscura *and Its Relatives* by Th. Dobzhansky and C. Epling. *Science* 100:11–12

1944g Remarks on Hobart Smith's analysis of the Western King Snakes. The American Midland Naturalist 31:88–95

1944h The number of Australian bird species [with D. L. Serventy]. The Emu 44:33–40

1944i The family name of the Australian honeyeaters. The Auk 61:465

1944j Experiments on sexual isolation in *Drosophila*. I. Geographic strains of *Drosophila willistoni* [with T. Dobzhansky]. Proceedings of the National Academy of Science USA 30:238–244

1944k Timor and the colonization of Australia by birds. The Emu 44:113–130

1944l Review of *The sensory basis of bird navigation* by D. R. Griffin. Bird Banding 15:161–162

1944m Birds collected during the Whitney South Sea Expedition. 54. Notes on some genera from the Southwest Pacific. American Museum Novitates No. 1269:1–8

1944n Review of *Ecological aspects of species-formation in Passerine birds* by D. Lack. The Wilson Bulletin 56:223–224

1944o The Pacific World [with many coauthors]. Edited by Fairfield Osborn. W. W. Norton, New York

1944p Committee on Common Problems of Genetics, Paleontology and Systematics. Bulletins 1–4 (editor), National Research Council, Washington, D. C.

1944q The downy plumage of the Australian dabchick. The Emu 44:231–233

1944r Review of *Studies in the Life History of the Song Sparrow*, Vol. II: *The Behavior of the Song Sparrow and Other Passerines* by M. M. Nice. Audubon Magazine 46:60

**1945**

1945a The correct name of the Fijian Mountain Lorikeet. The Auk 62:141–142

1945b Hugh Birckhead [Obituary note]. The Auk 62:346

1945c Review of *The Avifaunal Survey of Ceylon* by H. Whistler. The Auk 62:321–322

1945d Experiments on sexual isolation in *Drosophila*. IV. Modification of the degree of isolation between *Drosophila pseudoobscura* and *Drosophila persimilis* and of sexual preferences in *Drosophila prosaltans* [with T. Dobzhansky]. Proceedings of the National Academy of Science USA 31:75–82

1945e The family Anatidae [with J. Delacour]. The Wilson Bulletin 57:3–54 [Spanish translation in El Hornero 9:24–79, 1949]

1945f Bird geography in the Southwest Pacific. Audubon Magazine 47:159–165

1945g Note on the status of *Ploceus graueri*. Bulletin of the British Ornithologists' Club 65:41–42

1945h Birds of paradise. Natural History Magazine 54:264–276

1945i "Introduction" (page 69) and "Some evidence in favor of a recent date" (pages 70–83). Symposium on age of the distribution pattern of the gene arrangements in *Drosophila pseudoobscura*. Lloydia 8:69–83

1945j Birds collected during the Whitney South Sea Expedition. 55. Notes on the birds of Northern Melanesia. 1. American Museum Novitates, no. 1294:1–12

1945k Bird habitats of the Southwest Pacific. Audubon Magazine 47:207–211

1945l Bird conservation problems in the Southwest Pacific. Audubon Magazine 47:279–282

1945m Notes on the taxonomy of the birds of the Philippines [with J. Delacour]. Zoologica 30:105–117

[200] 1945n Birds of the Southwest Pacific. A Field Guide to the Birds of the Area between Samoa, New Caledonia, and Micronesia. New York: The Macmillan Co., XIX + 316 pp. [2nd edition, 1949; reprint paperback edition with b/w plates, Wheldon and Wesley, Ltd., Codicote, Hertfordshire, United Kingdom, 1968; reprint paperback edition with colored plates, Charles E. Tuttle Co., Rutland, Vermont, 1978]
*Reviews:* Chapman, F. M. (1945) in Natural History 54:100; Griscom, L. (1945) in The Auk 62:319–321; San José Mercury Herald (Febr. 4, 1945); New York Herald Tribune (January 28, 1945); Stresemann, E. (1947) in Ornithologische Berichte 1:31–32

1945o Review of *Tempo and Mode in Evolution* by G. G. Simpson (Columbia University Press, New York, 1944). Natural History Magazine 54:149–150

## 1946

1946a The number of species of birds. The Auk 63:64–69

1946b What needs to be learned about Australian ducks. The Emu 45:229–232

1946c Birds collected during the Whitney South Sea Expedition. 56. Evolution in the *Rhipidura rufifrons* group [with M. Moynihan]. American Museum Novitates, no. 1321:1–21

1946d Experiments on sexual isolation in *Drosophila*. VI. Isolation between *Drosophila pseudoobscura* and *Drosophila persimilis* and their hybrids. Proceedings of the National Academy of Science USA 32:57–59

1946e Review of *The Birds of Northern Thailand* by H. G. Deignan. The Auk 63:106

1946f A new name for a Philippine flowerpecker. Zoologica (New York) 31:8

1946g Experiments on sexual isolation in *Drosophila*. VII. The nature of the isolating mechanisms between *Drosophila pseudoobscura* and *Drosophila persimilis*. Proceedings of the National Academy of Science USA 32:128–137

1946h History of the North American bird fauna. The Wilson Bulletin 58:3–41 [Reprinted and abridged in E. Mayr, Evolution and the Diversity of Life, pages 565–588, 1976]

1946i A revision of the Striped-crowned Pardalotes [with K. A. Hindwood]. The Emu 46:49–67

1946j Supplementary notes on the family Anatidae [with J. Delacour]. The Wilson Bulletin 58:104–110

1946k Birds of the Philippines [with J. Delacour]. New York: Macmillan, New York. XV + 309 pp. Review: Stresemann, E (1948) in Ornithologische Berichte 1:258

1946l The naturalist in Leidy's time and today. Proceedings of the Academy of Natural Sciences of Philadelphia 98:271–276

## 1947

1947a Foreword. Evolution 1:i–ii

1947b Notes on tailorbirds (*Orthotomus*) from the Philippine Islands. Journal of the Washington Academy of Sciences 37:140–141

1947c A parrot new for Australia. The Emu 47:54–55

1947d A review of the Dicaeidae [with D. Amadon]. American Museum Novitates, no. 1360:1–32

1947e Ecological factors in speciation. Evolution 1:263–288 [Reprinted and revised under the title "Sympatric speciation" in E. Mayr, Evolution and the Diversity of Life, pages 144–175, 1976]

1947f A catastrophic decrease in a starling population. The Wilson Bulletin 59:37

1947g On the correct name of the Tibetan shrike usually called *Lanius tephronotus*. Journal of the Bombay Natural History Society 47:125–127

1947h [To the editor; reply to C. H. B. Grant]. The Auk 64:178

**1948**

1948a Geographic variation in the Reed-warbler. The Emu 47:205–210

1948b The new Sanford Hall. Natural History 57:248–254

1948c The bearing of the new systematics on genetical problems: The nature of species. Advances in Genetics 2:205–237 [Reprinted in part (pages 227–231) under the title "Sibling or Cryptic Species among Animals" in E. Mayr, Evolution and the Diversity of Life, pages 509–514, 1976]

1948d Evolution in the family Dicruridae (Birds) [with C. Vaurie]. Evolution 2:238–265

1948e Climatic races in plants and animals. Evolution 2:375–376

1948f Repeated anting by a song sparrow. The Auk 65:600

1948g Gulls feeding on flying ants. The Auk 65:600

**1949**

1949a Geographical variation in *Accipiter trivirgatus*. American Museum Novitates, no. 1415:1–12

1949b Review of *A Handbook of the Birds of Western Australia*, by D. L. Serventy and H. M. Whittell. Natural History Magazine 58:150

1949c Birds collected during the Whitney South Sea Expedition. 57. Notes on the birds of Northern Melanesia. 2. American Museum Novitates, no. 1417:1–38

1949d A second Australian record of the Corn Crake (*Crex crex*). The Emu 48:243

1949e Enigmatic sparrows. The Ibis 91:304–306

1949f Speciation and systematics. Pages 281–298 *in* Genetics, Paleontology, and Evolution. (G. L. Jepsen, E. Mayr and G. G. Simpson, Eds.) Princeton University Press, Princeton, New Jersey

1949g Genetics, Paleontology, and Evolution. Edited by G. L. Jepsen, G. G. Simpson, and E. Mayr: Princeton University Press, Princeton, New Jersey, XIV + 474 pp.
Review: Washburn, S. L. (1950), Amer. J. Phys. Anthropol. 8:245

1949h Ornithologie als Biologische Wissenschaft. Festschrift zum 60. Geburtstag von Erwin Stresemann. Edited by E. Mayr and E. Schüz. Carl Winter, Heidelberg, XII + 291 pp

1949i Artbildung und Variation in der *Halcyon-chloris* Gruppe. Pages 55–60 *in* Ornithologie als biologische Wissenschaft. Festschrift zum 60. Geburtstag von Erwin Stresemann. (E. Mayr and E. Schüz, Eds.). Carl Winter, Heidelberg

1949j Einführung [with E. Schüz], p. VII–IX in Ornithologie als biologische Wissenschaft. Festschrift zum 60. Geburtstag von Erwin Stresemann (E. Mayr and E. Schüz, Eds.). Carl Winter, Heidelberg

1949k Speciation and selection. Proceedings of the American Philosophical Society 93:514–519

1949l Comments on evolutionary literature. Evolution 3:381–386

1949m The species concept: semantics versus semantics. Evolution 3:371–372

1949n Importance des caractères biologiques dans la systématique [with J. Delacour] Pages 374–376 *in* XIIIe Congrès International de Zoologie (Paris 1948)

### 1950

1950a A new Cuckoo-shrike from the Solomon Islands. The Auk 67:104
1950b Taxonomic notes on the genus *Neositta*. The Emu 49:282–291
1950c The role of the antennae in the mating behavior of female *Drosophila*. Evolution 4:149–154
1950d A new bower bird (*Archboldia*) from Mount Hagen, New Guinea [with E. T. Gilliard]. American Museum Novitates, no. 1473:1–3
1950e Isolation, dispersal and evolution. Review of *Symposium on Satpura Hypothesis* edited by Hora et al. Evolution 4:363
1950f Polymorphism in the Chat genus *Oenanthe* (Aves) [with E. Stresemann]. Evolution 4:291–300

### 1951

1951a A classification of Recent birds [with D. Amadon]. American Museum Novitates, no. 1496:1–42
***[250]*** 1951b Comments on evolutionary literature. 2. Evolution 5:85–89
1951c Bearing of some biological data on geology. Bulletin of the Geological Society of America 62:537–546
1951d Review of *Birds of Paradise and Bower Birds* by T. Iredale. The Emu 50:214–216
1951e New species and subspecies of birds from the highlands of New Guinea [with E. T. Gilliard]. American Museum Novitates, no. 1524:1–15
1951f Comments on evolutionary literature. 3 [with H. Lewis]. *Evolution* 5:185–190
1951g Taxonomic categories in fossil hominids. Cold Spring Harbor Symposia on Quantitative Biology 15:109–118 [Reprinted and revised in E. Mayr, Evolution and the Diversity of Life, pages 530–545, 1976]
1951h Über den Rechtsstand in Synonymielisten [sic] enthaltener Nomina Nuda [with E. Stresemann]. Senckenbergiana 32:211–218. [English version: The status of "nomina nuda" listed in synonymy. Bulletin of Zoological Nomenclature 10:315–322, 1953]
1951i What is *Sylvia ticehursti* Meinertzhagen? [with R. Meinertzhagen]. Bulletin of the British Ornithologists' Club 71:47–48
1951j Notes on some pigeons and parrots from Western Australia. The Emu 51:137–145
1951k Comments on evolutionary literature. 4. Evolution 5:418–421
1951l Speciation in birds. Pages 91–131 *in* Proceedings of the Xth International Ornithological Congress (S. Höestadius, Ed.). Almquist and Wiksell, Uppsala, Sweden

### 1952

1952a The Ribbon-tailed Bird of Paradise (*Astrapia mayeri*) and its allies [with E. T. Gilliard]. American Museum Novitates, no. 1551:1–13
1952b Proposed use of the plenary powers to suppress for nomenclatorial purposes a paper by Forster (J. R.) containing new names for certain Australian birds published in 1794 in volume 5 of the "Magazin von merkwürdigen neuen Reise Beschreibungen" [with D. Amadon, J. Delacour, L. Glavert, R. C. Murphy, D. L. Serventy and H. M. Whittell]. Bulletin of Zoological Nomenclature 9:45–46
1952c Notes on evolutionary literature. 5. Evolution 6:138–144
1952d Introduction and Conclusion. Pages 85, 255–258 *in* The problem of land connections across the South Atlantic with special reference to the Mesozoic. Bulletin of the American Museum of Natural History 99:79–258

1952e Six new subspecies of birds from the highlands of New Guinea [with E. T. Gilliard]. American Museum Novitates, no. 1577:1–8

1952f Comments on evolutionary literature. 6. Evolution 6:248–252

1952g Review of *The Avifauna of Micronesia, Its Origin, Evolution, and Distribution* by Rollin H. Baker. The Auk 69:331–332

1952h *Turdus musicus* Linnaeus. The Ibis 94:532–534

1952i Geographic variation and plumages in Australian bowerbirds (Ptilonorhynchidae) [with K. Jennings]. American Museum Novitates, no. 1602:1–18

1952j German experiments on orientation of migrating birds. Biological Review 27:394–400

1952k Altitudinal hybridization in New Guinea honeyeaters [with E. T. Gilliard]. The Condor 54:325–337

1952l Comments on evolutionary literature. 7. Evolution 6:449–454

1952m Foreword. In *A Checklist of Genera of Indian Birds* by Biswamoy Biswas. Records of the Indian Museum 50:1–2

1952n Support to stabilise the nomenclature of the Mallophaga. Bulletin of Zoological Nomenclature 6:210

1952o On the proposal for the adoption of a "declaration." Bulletin of Zoological Nomenclature 6:244

1952p On the question whether the presence or absence of a diacritic mark should be held to make two names consisting of otherwise identical words distinct names for the purposes of the law of homonymy. Bulletin of Zoological Nomenclature 6:255

1952q Comments on the identity of *Muscicapa novahollandia* Latham, 1790. Bulletin of Zoological Nomenclature 9:48

1952r Validation of "adippe." Bulletin of Zoological Nomenclature 9:137

**1953**

1953a Methods and Principles of Systematic Zoology [with E. G. Linsley and R. L. Usinger]. McGraw-Hill, New York, IX + 336 pp. [Russian and Arabic translations of this book have appeared.]
Reviews: Storer, R. W. (1957) in The Auk 74:110; Weber, H. (1954) in Berichte über die wissenschaftliche Biologie 88:145–146

1953b Concepts of classification and nomenclature in higher organisms and microorganisms. Annals of the New York Academy of Sciences 56:391–397

1953c Thoughts on State bird books. Linnaean News-Letter 7:2

1953d Geographic and individual variation in the Shrike-tit (*Falcunculus frontatus*). The Emu 53:249–252

1953e Taxonomic notes on *Oreoica gutturalis*. The Emu 53:252–253

1953f Comments on evolutionary literature. 8. Evolution 7:273–281

1953g Additional notes on the birds of Bimini, Bahamas. The Auk 70:499–501

1953h The south-west Australian races of the spotted scrub-wren, *Sericornis maculatus* [with R. Wolk]. Western Australian Naturalist 4:66–70

1953i Review of *Classification of the Anatidae based on the cyto-genetics* by Y. Yamashina [Papers from the Coordinating Committee Res. Genetics 3 (1952):1–34]. Bird Banding 24:126–127

1953j Review of *Vom Vogelzug. Grundriss der Vogelzugskunde* by E. Schüz. The Auk 70:381–382

1953k La Specie. La Ricerca Scientifica 23 (Suppl.):9–59

1953l The status of "nomina nuda" listed in synonymy. Bulletin of Zoological Nomenclature 10:315–322 (see 1951h)

1953m Means for securing stability in nomenclature. Bulletin of Zoological Nomenclature 8:71–72

1953n Preamble of the Rules of Nomenclature [with J. T. Zimmer, G. H. H. Tate and C. H. Curran]. Bulletin of Zoological Nomenclature 8:76

1953o Plenary Powers of the Commission [with G. H. H. Tate, J. T. Zimmer and C. H. Curran). Bulletin of Zoological Nomenclature 8:99

1953p Designation of neotypes (with J. T. Nichols, G. H. H. Tate and J. T. Zimmer]. Bulletin of Zoological Nomenclature 8:132–133

1953q Rules of naming families and suprageneric categories of lower rank (with E. H. Colbert, J. T. Nichols, G. H. H. Tate and J. T. Zimmer). Bulletin of Zoological Nomenclature 8:241

1953r Nomenclature of orders and higher groups [with E. H. Colbert, J. T. Nichols, G. H. H. Tate and J. T. Zimmer]. Bulletin of Zoological Nomenclature 10:23

1953s Emendation of zoological names [with E. H. Colbert, J. T. Nichols, G. H. H. Tate and J. T. Zimmer]. Bulletin of Zoological Nomenclature 10:127–128

1953t Substitute names [with E. H. Colbert, J. T. Nichols, G. H. H. Tate and J. T. Zimmer]. Bulletin of Zoological Nomenclature 10:187

1953u [Differential diagnosis] [with E. H. Colbert, W. J. Gertsch, G. H. H. Tate and J. T. Zimmer]. Bulletin of Zoological Nomenclature 10:287

***[300]*** 1953v Supplementary statement. Bulletin of Zoological Nomenclature 10:288

1953w Comment. Bulletin of Zoological Nomenclature 10:334

1953x Names published in the synonymy of other names [with E. H. Colbert, J. T. Nichols, G. H. H. Tate and J. T. Zimmer]. Bulletin of Zoological Nomenclature 10:343

1953y [On specific names] [with J. T. Zimmer, G. H. H. Tate and C. H. Curran]. Bulletin of Zoological Nomenclature 10:452–453

1953z [On the term "nominate"] [with G. H. H. Tate, J. T. Zimmer and C. H. Curran]. Bulletin of Zoological Nomenclature 10:454

## 1954

1954a Proposed use of the plenary powers to suppress the specific name, "*brouni*" Hutton, 1901, as published in the combination "*Drosophila brouni*," for the purpose of preserving the specific name "*immigrans*" Sturtevant, 1921, as published in the combination "*Drosophila immigrans*" (Class Insecta, order Diptera) [with J. T. Patterson, Marshall P. Wheeler, and Warren P. Spencer]. Bulletin of Zoological Nomenclature 9:161–162

1954b Notes on Australian whistlers (Aves, *Pachycephala*). American Museum Novitates, no. 1653:1–22

1954c Change of genetic environment and evolution. Pages 157–180 *in* Evolution as a Process (Julian Huxley, A. C. Hardy, and E. B. Ford, Eds.). Allen and Unwin, London [Reprinted under the title "Change of environment and speciation" in E. Mayr, Evolution and the Diversity of Life, pages 188–210, 1976]

1954d Binominal nomenclature. Florida Naturalist 27:7–8

1954e Geographic speciation in tropical echinoids. Evolution 8:1–18

1954f Birds of Central New Guinea. Results of the American Museum of Natural History Expeditions to New Guinea in 1950 and 1952 [with E. T. Gilliard]. Bulletin of the American Museum of Natural History 103:311–374

1954g The tail molt of small owls [with M. Mayr]. The Auk 71:172–178
1954h Notes on nomenclature and classification. Systematic Zoology 3:86–89
1954i Review of *Evolution: Die Geschichte ihrer Probleme und Erkenntnisse* by W. Zimmermann (Karl Alber, Freiburg and München, 1953). The Scientific Monthly 79:57–58
1954j Report of the Standing Committee on distribution of terrestrial faunas in the inner Pacific. Pages 5–11 *in* Proceedings of the 7th Pacific Science Congress, vol. 4
1954k Fragments of a Papuan ornithogeography. Pages 11–19 *in* Proceedings of the Seventh Pacific Science Congress, vol. 4 [Reprinted in E. Mayr, Evolution and the Diversity of Life, pages 646–652. 1976]
1954l On the origin of bird migration in the Pacific. Pages 387–394 *in* Proceedings of the 7th Pacific Science Congress, vol. 4

**1955**

1955a Birds collected during the Whitney South Sea Expedition. 63. Notes on the birds of Northern Melanesia. 3. Passeres. American Museum Novitates, no. 1707:1–46
1955b Comments on some recent studies of song bird phylogeny. The Wilson Bulletin 67:33–44
1955c Review of *The Literature of Australian Birds: a History and Bibliography of Australian Ornithology* by H.M. Whittell. The Auk 72:99
1955d Dr. Ludwig Schuster [Obituary notice]. The Auk 72:323
1955e Karl Jordan's contribution to current concepts in systematics and evolution. The Transactions of the Royal Entomological Society of London 107:45–66. [Reprinted (pages 51–57) as "Karl Jordan and the biological species concept," (pages 58–63) as "Karl Jordan on speciation," and (pages 45–51; 64–65) as "Karl Jordan on the theory of systematics and evolution" in E. Mayr, Evolution and the Diversity of Life, pages 485–492, 135–143, and pages 297–306, 1976]
1955f Origin of the bird fauna of Pantepui [with W.H. Phelps, Jr.]. Pages 399–400 *in* Acta XI Congressus Internationalis Ornithologici (A. Portmann and E. Sutter, Eds.). Birkhaeser Verlag, Basel, Switzerland
1955g Biological Materials. Part I. Preserved materials and museum collections [with R. Goodwin]. National Academy of Sciences National Research Council, Publication 399
1955h Integration of genotypes: Synthesis. Cold Spring Harbor Symposia on Quantitative Biology 20:327–333

**1956**

1956a Is the Great White Heron a good species? The Auk 73:71–77
1956b Geographical character gradients and climatic adaptation. Evolution 10:105–108 [Reprinted and revised in E. Mayr, Evolution and the Diversity of Life, pages 211–217, 1976]
1956c The interpretation of variation among the Yellow Wagtails. British Birds 49:115–119
1956d Sequence of passerine families (Aves) [with J.C. Greenway]. Breviora, no. 58:1–11
1956e Die systematische Stellung der Gattung *Fringilla*. Journal für Ornithologie 97:258–263
1956f Results of the Archbold Expeditions. No. 74. The birds of Goodenough Island, Papua [with H.M. Van Deusen]. American Museum Novitates, no. 1792:1–8
1956g The species as a systematic and as biological problem. Pages 1–12 *in* Biological Systematics. Proceedings of the 16th Annual Biology Colloquium, Oregon State College, Corvallis, Oregon

1956h Summation: Systematics and modes of speciation. Pages 45–51 *in* Biological Systematics. Proceedings of the 16th Annual Biology Colloquium, Oregon State College, Corvallis, Oregon

1956i Suspected correlation between blood-group frequency and pituitary adenomas [with L. K. Diamond, R. P. Levine, and M. Mayr). Science 124:932–934

1956j Geographic variation and hybridization in populations of Bahama snails (*Cerion*) [with C. B. Rosen]. American Museum Novitates, no. 1806:1–48

1956k Review of *Die Vögel Hessens* by Ludwig Gebhardt. The Auk 73:461

1956l Gesang und Systematik. Beiträge zur Vogelkunde 5:112–117

1956m Geographic speciation in marine invertebrates. Pages 145–146 *in* Proceedings of the XIV International Congress of Zoology, Danish Science Press, Copenhagen

1956n The names *Treron griseicauda* and *Treron pulverulenta*. Bulletin of the British Ornithological Club 76:62

1956o *Accipiter ferox* Gmelin. Bulletin of Zoological Nomenclature 12:122

1956p Validation of "*Heteralocha*" Cabanis. Bulletin of Zoological Nomenclature 12:141

## 1957

1957a New species of birds described from 1941 to 1955. Journal für Ornithologie 98:22–35

1957b Birds collected during the Whitney South Sea Expedition. 64. Notes on the birds of Northern Melanesia. 4. The genus *Accipiter*. American Museum Novitates, no. 1823:1–14

1957c Proposed use of the plenary powers to suppress the specific name "*musicus*" Linnaeus, 1758, as published in the combination "*Turdus musicus*" and to approve a neotype for "*Turdus iliacus*" Linnaeus, 1758, the Eurasian Redwing (Class Aves) [with C. Vaurie]. Bulletin of Zoological Nomenclature 13:177–182

1957d Die denkmöglichen Formen der Artentstehung. Revue Suisse de Zoologie 64:219–235

1957e The Species Problem. Edited by Ernst Mayr. American Association for the Advancement of Science, no. 50, Washington, DC, IX + 395 pp

1957f Species concepts and definitions. Pages 1–22 *In* The Species Problem (E. Mayr, Ed.) American Association for the Advancement of Science, no. 50, Washington, DC [Reprinted in E. Mayr, Evolution and the Diversity of Life, pages 493–508, 1976 and in Grene, M. and Mendelsohn, E. eds., Boston Sudies in the Philosophy of Science 27:353–371, 1976]

1957g Difficulties and importance of the biological species concept. Pages 371–388 *in* The Species Problem (E. Mayr, Ed.). American Association for the Advancement of Science, no. 50, Washington, DC

1957h The name of the giant pigeon of the Marquesas Islands. The Ibis 99:521

1957i Report of the Committee on Systematic Biology of the American Institute of Biological Sciences. 22 pp

**[350]** 1957j Comments on Copenhagen Decision 54. Bulletin of Zoological Nomenclature 15A:143–144

## 1958

1958a Research in conservation. The Bulletin of the Massachusetts Audubon Society 42:67–70

1958b Evolutionary aspects of host specificity among parasites of Vertebrates. Pages 7–14 *in* First Symposium on Host Specificity among Parasites of Vertebrates, Imprimerie Paul Attinger, Neuchatel

1958c Concluding remarks. Pages 311–315 *in* First Symposium on Host Specificity among Parasites of Vertebrates, Imprimerie Paul Attinger, Neuchatel

1958d The correct gender of generic names ending in *-rhynchus*, *-rhamphus*, *-gnathus*. The Auk 75:225

1958e The sequence of the songbird families. The Condor 60:194–195

1958f The evolutionary significance of the systematic categories. Uppsala Universitets Årsskrift (1958) 6:13–20

1958g Behavior and systematics. Pages 341–362 *in* Behavior and Evolution (A. Roe and G. G. Simpson, Eds.). Yale University Press, New Haven, Connecticut [Reprinted and revised in E. Mayr, Evolution and the Diversity of Life, pages 677–693, 1976]

1958h What is a species? Kagaku 28:170–173

1958i Review of *A company of birds* by Loke Wan Tho. The Auk 75:361–362

1958j [Applicability of gender rules]. Bulletin of Zoological Nomenclature 15A:270–271

1958k The Principle of Conservation; Draft proposal. Bulletin of Zoological Nomenclature 15B:626–627

1958l The Principle of Conservation: Alternative draft [with W. I. Follett, R. V. Melville and R. L. Usinger]. Bulletin of Zoological Nomenclature 15B:628–629

1958m [Status of specific names based on modern patronymics]. Bulletin of Zoological Nomenclature 15B:684

1958n Discussions. *In* Concepts of Biology: Condensed transcript of the conference (R. W. Gerard, Ed.). Behavioral Science 3:103–195. [Also published in Concepts of Biology, Publication 560, National Academy of Sciences, National Research Council, Washington, DC]

**1959**

1959a Isolation as an evolutionary factor. Proceedings of the American Philosophical Society 103:221–230 [Reprinted under the titles "Darwin and isolation," and "Darwin, Wallace, and the origin of isolating mechanisms," in E. Mayr, Evolution and the Diversity of Life, pages 120–128 and 129–134, 1976]

1959b Darwin and the evolutionary theory in biology. Pages 3–12 *in* Evolution and Anthropology: A Centennial Appraisal. (B. J. Meggers, Ed.). The Anthropological Society of Washington, D.C. [Reprinted under the title "Typological versus population thinking" in E. Mayr, Evolution and the Diversity of Life, pages 26–29, 1976; also reprinted in Sober, Elliot, ed., *Conceptual Issues in Evolutionary Biology*, MIT Press, Cambridge, MA, 1986 (1st edition), pp. 14–18, 1995 (2nd edition), pp. 157–160, 2006 (3rd edition), pp. 325–328]

1959c Agassiz, Darwin, and Evolution. Harvard Library Bulletin 13:165–194 [Reprinted and revised in E. Mayr, Evolution and the Diversity of Life, pages 251–276, 1976]

1959d The emergence of evolutionary novelties. Pages 349–380 *in* The Evolution of Life: Evolution after Darwin, vol. 1 (S. Tax, Ed.). University of Chicago Press, Chicago [Reprinted and revised in E. Mayr, Evolution and the Diversity of Life, pages 88–113, 1976]

1959e Trends in avian systematics. The Ibis 101:293–302

1959f Where are we? Cold Spring Harbor Symposia on Quantitative Biology 24:1–14 [Reprinted in E. Mayr, Evolution and the Diversity of Life, pages 307–328, 1976]

1959g Concerning a new biography of Charles Darwin, and Its scientific shortcomings. Review of *Darwin and the Darwinian Revolution*, by G. Himmelfarb [Doubleday and Co., New York 1959). Scientific American 201 (November):209–216

1959h Comment [on blood-group correlations]. Journal of Medical Education 34:421

1959i Review of *Fundamentals of Ornithology* by J. Van Tyne and A. J. Berger [John Wiley and Sons, New York, 1959). The Wilson Bulletin 71:391–394

1959j *Turdus musicus*. Opinion 551. Opinions and Declarations Rendered by the International Commission on Zoological Nomenclature. London: International Trust for Zoological Nomenclature. Bull. Zool. Nomencl. 17:201–206

1959k Comments on the proposal relating to *Beraea* Stephens, 1833. Bulletin of Zoological Nomenclature 17:68

## 1960

1960a Check-list of birds of the world, vol. 9. Edited by Ernst Mayr and James C. Greenway, Jr. Museum of Comparative Zoology, Cambridge, Massachusetts. [Family Motacillidae, (with C.H. Vaurie, C.M.N. White, and J.C. Greenway, Jr.), pages 129–167; Family Campephagidae, (with J.L. Peters and H. Friedmann), pages 167–221]

1960b Is the Linnean species composite? [*Helix vivipara*] Document 16. Bulletin of Zoological Nomenclature 17:128–130

1960c Queensland collectors of the Godeffroy Museum. The Emu 60:52

1960d The distribution and variation of *Mirafra javanica* in Australia [with A. McEvey]. *The Emu* 60:155–192

1960e The meaning of adaptation. Chairman's introduction to the symposium on adaptive evolution. Pages 495–498 in Proceedings XII International Ornithological Congress (G. Bergmann, K.O. Donner, and L. van Haartman, Eds.). Tilgmannin Kirjapaino, Helsinki

1960f Gustav Kramer [Obituary]. The Auk 77:117

## 1961

1961a Review of *Atlas der Verbreitung Palaearktischer Vögel* edited by E. Stresemann and L.A. Portenko (1960). The Auk 78:103

1961b Review of *Darwin's Biological Work: Some Aspects Reconsidered* edited by P.R. Bell (Cambridge University Press, Cambridge, Massachusetts, 1959). Isis 52:433–435

1961c Cause and effect in biology: Kinds of causes, predictability, and teleology are viewed by a practicing biologist. *Science* 134:1501–1506 [Reprinted in E. Mayr, Evolution and the Diversity of Life, pages 359–371, 1976 and in E. Mayr, Toward a New Philosophy of Biology, pages 24–37, 1988]

1961d The classification of the Red-tipped Pardalotes. The Emu 61:201–202

1961e *Pnoepyga* (Hodgson, 1844) Proposed validation under the Plenary Powers (Class Aves). Z. N. (S.) 1457. Bulletin of Zoological Nomenclature 18:209–210

1961f Comments on the proposal relating to *Gari* Schumacher, 1817. Bulletin of Zoological Nomenclature 18:298

1961g Review of *Charles Darwin: The Founder of the Theory of Evolution and Natural Selection*, by Gerhard Wichler (Pergamon Press, New York, 1961). Science 134:607

## 1962

1962a On cause and effect in biology. [Answer to letters.] Science 135:972–981

1962b Accident or design: The paradox of evolution. Pages 1–14 *in* The Evolution of Living Organisms. Proceedings of the Darwin Centenary Symposium of the Royal Society of Victoria, Melbourne 1959. Melbourne University Press, Melbourne. [Reprinted in E. Mayr, Evolution and the Diversity of Life, pages 30–43, 1976; German translation: Zufall oder Plan, das Paradox der Evolution. Pages 21–35 *in* Evolution und Hominisation (G. Kurth, Ed.). Gustav Fischer, Stuttgart, 1962]

1962c Eight dubious species of birds: Proposed use of the Plenary Powers to place these names on the Official Index. Z. N. (S.) 1033. Bulletin of Zoological Nomenclature 19:23–26

1962d Check-list of Birds of the World, vol. 15. Edited by E. Mayr and J. C. Greenway, Jr., Museum of Comparative Zoology, Cambridge, Massachusetts. [Grallinidae (pages 159–160), Artamidae (pages 160–165), Ptilonorhynchidae (pages 172–181), Paradisaeidae (pages 181–204)]

1962e Origin of the human races. A new thesis of mankind's ancient unity and of a corresponding antiquity of racial diversity of man. Review of *The Origin of Races*, by C. Coon [Knopf, New York, 1962)]. Science 138:420–422

1962f Review of *A Synopsis of the Birds of India and Pakistan*, by S. D. Ripley. The Auk 79:720

1962g Review of *Geschichte der Zoologie und der Zoologischen Anstalten in Jena* 1779–1919 by G. Uschmann. Isis 53:241

1962h Report of the by-laws committee [with other authors]. Bulletin of Zoological Nomenclature 19:358–374

**1963**

1963a Annual Report 1961–1962. Museum of Comparative Zoology. Harvard University, Cambridge, Massachusetts, 32 pp

1963b Animal Species and Evolution. Belknap Press of Harvard *und* University Press, Cambridge, Massachusetts, XIV + 797 pp. [German edition: Artbegriff und Evolution. Translation by G. Heberer and G. Stein. Parey Verlag, Hamburg and Berlin, 617 pp., 1967; Yugoslav edition: Zivotinjske Vreste i Evolucija. VUK Karadzic Animal, Beograd, 1970]

*Reviews:* R. D. Alexander (1963) in Systematic Zoology 12:202–204; A. J. Cain (1965) in The Auk 85:654–657; T. Dobzhansky (1963) in Amer. J. Phys. Anthropol. 21:387–389; L. C. Dunn (1964) in Isis 55:225–227; G. Heberer (1964) in Berichte über die wissenschaftliche Biologie, Abt. A, vol. 205:169–170; R. Mertens (1964) in Natur und Museum 94:171; G. Osche (1971) in Zeitschrift für Tierpsychologie 28:546–554; K. C. Parkes (1964) in The Wilson Bulletin 76:193–203; F. Pirchner (1970) in Wiener Tierärztliche Monatsschrift Nr. 1; G. L. Stebbins (1964) in Evolution 18:134–137; L. H. Throckmorton and J. L. Hubby (1963) in Science 140:628–631

1963c Comments on the taxonomic position of some Australian genera of songbirds. The Emu 63:1–7

1963d *Gallinago* versus *Capella*. The Ibis 105:402–403 **[400]**

1963e Preservation of *Pnoepyga*. Bulletin of Zoological Nomenclature 20:15–17

1963f Names given to hybrids. Bulletin of Zoological Nomenclature 20:50–51

1963g Comments on the proposed suppression of *Cypraea piperita* Gray, 1825. Bulletin of Zoological Nomenclature 20:104

1963h Review of *The Francolins, a study in speciation*, by B. P. Hall [Bulletin of the British Museum (Natural History) Zoological Series 10 (1963):105–204]. The Ibis 105:421–422

1963i *Tanagra* Linnaeus, 1764, and *Tanagra* Linnaeus, 1766 (Aves) Proposed use of the Plenary Powers to end confusion. Z. N. (S). 1182. Bulletin of Zoological Nomenclature 20:296–302

1963j Introduction. Pages VII–VIII *in* Phylogeny and Evolution in Crustacea (H. B. Whittington and W. D. I. Rolfe, Eds.). Museum of Comparative Zoology, Cambridge, Massachusetts

1963k The new versus the classical in science. Science 141:765. [Reprinted in *La Stampa*, Anno 97, Numero 252, October 24, 1963:9]

1963l The fauna of North America, its origin and unique composition. Pages 3–11 *in* Proceedings of the XVIth International Congress of Zoology, Washington D. C., vol. 4

1963m The statute of limitation and chilopod nomenclature. Annals and Magazine of Natural History ser. 13, 6 (1963):509–510

1963n The biology, genetics, and evolution of Man. Review of *Mankind Evolving*, by Th. Dobzhansky [Yale University Press, New Haven, 1962]. The Quarterly Review of Biology 38:243–245

1963o Gotemba meeting on cooperative biological research by Japanese and American scientists. American Institute of Biological Sciences Bulletion 13:90

1963p Review of *Fauna und Flora der Adria* by Rupert Riedl. Science 142:659

1963q The taxonomic evaluation of fossil hominids. Pages 332–346 *in* Classification and Human Evolution (S. L. Washburn, Ed.). Aldine Press, Chicago

1963r The role of ornithological research in biology.Pages 27–38 *in* Proceedings XIII International Ornithological Congress (C. G. Sibley, Ed.). American Ornithologists' Union, Washington, DC

1963s Reply to criticism by R. D. Alexander. Systematic Zoology 12:204–206

## 1964

1964a Annual Report 1962–1963. Museum of Comparative Zoology. Harvard University, Cambridge, Massachusetts, 32 pp

1964b Preface (pages IX–XI) to *Systematics and the Origin of Species*. Paperback edition, Dover Publications, New York

1964c Inferences concerning the Tertiary American bird faunas. Proceedings of the National Academy of Science USA 51:280–288 [Reprinted in E. Mayr, Evolution and the Diversity of Life, pages 589–600, 1976]

1964d Evolutionary theory today. Pages 86–95. In McGraw-Hill Yearbook of Science and Technology. McGraw-Hill, New York

1964e Check-list of birds of the world, vol. 10. Edited by E. Mayr and R. A. Paynter, Jr. Museum of Comparative Zoology, Cambridge, Massachusetts

1964f Introduction [with R. A. Paynter, Jr.]. Pages V–VI *in* Check-list of Birds of the World, vol. 10 (E. Mayr and R. A. Paynter, Jr., Eds.). Museum of Comparative Zoology, Cambridge, Massachusetts

1964g Evolution of Neotropical cricetine rodents. Review of *Evolution of Neotropical Cricetine Rodents* by Philip Hershkovitz [Fieldiana: Zoology, vol. 46, 1963]. Evolution 18:132

1964h *Cardinalis* Bonaparte, 1838 (Aves) Proposed validation under the Plenary Powers. Z. N. (S.) 1608. Bulletin of Zoological Nomenclature 21:133–136

1964i Comments on *Thamnophis sirtalis*. Bulletin of Zoological Nomenclature 21:190

1964j Comments on the proposed stabilization of *Macropus* Shaw, 1790. (Z.N. (S.) 1584). Bulletin of Zoological Nomenclature 21:250–251

1964k The evolution of living systems. Proceedings of the National Academy of Science 51:934–941 [Reprinted in Science Teacher 31:13–17 and in E. Mayr, Evolution and the Diversity of Life, pages 17–25, 1976]

1964l Introduction [pages vii–xxvii], Bibliography [pages 491–495], and Subject Index [pages 497–513]. In *On the Origin of Species by Means of Natural Selection, or the Preservation of Favoured Races in the Struggle for Life* by Charles Darwin [John Murray, London, 1859]. A Facsimile of the First Edition. Harvard University Press, Cambridge, Massachusetts

1964m Review of *Die Einbürgerung von Säugetieren und Vögeln in Europa*, by Günther Niethammer. The Auk 81:98

1964n The new systematics. Pages 13–32 *in* Taxonomic Biochemistry and Serology. (C. A. Leone, Ed.) Ronald Press, New York

1964o Schizophrenia as a genetic morphism [with Julian Huxley, Humphry Osmond, and Abram Hoffer]. Nature 204:220–221

1964p The name *Cacatua* Brisson, 1760 (Aves) Proposed validation under the Plenary Powers. Z. N. (S.) 1647 [with A. Keast and D. L. Serventy]. Bulletin of Zoological Nomenclature 21:372–374

1964q On the taxonomy of *Cuculus pallidus* (Latham). The Emu 64:41

1964r From molecules to organic diversity. Federation Proceedings 23:1231–1235 [Reprinted and revised in E. Mayr, Evolution and the Diversity of Life, p. 64–72, 1976]

1964s Nearctic Region (pages 514–516) and Neotropical Region (pages 516–518). *In* A New Dictionary of Birds (A. L. Thomson, Ed.). Thomas Nelson, London [Reprinted *in* A Dictionary of Birds, edited by B. Campbell and E. Lack, pages 379–381; 381–382, T. and A. D. Poyser, Calton, United Kingdom, 1985]

1964t Review of *Hundert Jahre Evolutionsforschung. Das wissenschaftliche Vermächtnis Charles Darwins* edited by Gerhard Heberer und Franz Schwanitz (Gustav Fischer, Stuttgart, 1960). The Quarterly Review of Biology 39:384

## 1965

1965a Annual Report 1963–1964. Museum of Comparative Zoology, Harvard University, Cambridge, Massachusetts, 40 pp

1965b Classification and phylogeny. American Zoologist 5:165–174

1965c *Dromaius* Vieillot, 1816 (Aves) Proposed addition to the official list Z. N. (S.) 1668 [with D. L. Serventy and H. Condon]. Bulletin of Zoological Nomenclature 22:63–65

1965d Review of *Insects of Campbell Island*, by J. L. Gressit et al. (Bernice P. Bishop Museum, Honolulu, 1964). Science 148:823–824

1965e Discussion of Part II, Evolution of Proteins 1. Pages 197–198 *in* Evolving Genes and Proteins (V. Bryson and H. J. Vogel, Eds.). Academic Press, New York

1965f Discussion of Part IV, Evolution of Proteins II. Pages 293–294 *in* Evolving Genes and Proteins. (V. Bryson and H. J. Vogel, Eds.). Academic Press, New York

1965g Selektion und die gerichtete Evolution. Die Naturwissenschaften 52:173–180. [Translated and adapted in E. Mayr, Evolution and the Diversity of Life, pages 44–52, 1976]

1965h Foreword. Pages VII–IX. *in* Charles Darwin: A Scientific Biography by Sir Gavin de Beer. Anchor Books,Garden City, New York

1965i The Origin of Adaptations. Review of *The Origin of Adaptations* by V. Grant [Columbia University Press, New York, 1963). Evolution 19:134–136

1965j Review of *Die Vögel Deutschlands* by G. Niethammer, H. Kramer, and H. E. Wolters (Akademische Verlagsgesellschaft, Frankfurt, 1964). The Auk 82:299

1965k Numerical phenetics and taxonomic theory. Systematic Zoology 14:73–97

1965l Relationship among Indo-Australian Zosteropidae (Aves). Breviora, no. 228:1–6

1965m Evolution at the species level. Pages 316–324 *in* Ideas in Modern Biology (J. A. Moore, Ed.). Natural History Press, Garden City, New York

1965n The nature of colonizations in birds. Pages 30–47 *in* The Genetics of Colonizing Species (H. G. Baker and G. L. Stebbins, Eds.) Academic Press, New York [Reprinted in E. Mayr, Evolution and the Diversity of Life, pages 659–672, 1976]

***[450]*** 1965o Summary. Pages 553–562 *in* The Genetics of Colonizing Species (H. G. Baker and G. L. Stebbins, Eds.). Academic Press, New York

1965p Avifauna: Turnover on islands. Science 150:1587–1588

1965q What is a fauna? Zoologisches Jahrbuch für Systematik 92:473–486 [Reprinted and revised in E. Mayr, Evolution and the Diversity of Life, p. 552–564, 1976]

1965r Classification, identification and sequence of genera and species. L'Oiseau et la Revue Française d'Ornithologie 35:90–95

1965s Comments [on "Explanation in nineteenth century biology"]. Pages 151–155 *in* Proceedings of the Boston Colloquium for the Philosophy of Science, 1962–1964, vol. 2 (R. S. Cohen and M. W. Wartofsky, Eds.). Humanities Press, New York. [Comments on E. Mendelsohn, "Physical Models and Physiological Concepts: Explanation in Nineteenth Century Biology," l. c., pages 127–150] [Reprinted in E. Mayr, Evolution and the Diversity of Life, p. 372–376, 1976]

1965t Review of *Die Ornithologen Mitteleuropas* by L. Gebhardt. Isis 56:377

1965u Polymorphism, polytypism, and monotypism. International Social Science Journal 17:123

1965v Races in animal evolution. International Social Science Journal 17:121–122

**1966**

1966a Annual Report 1964–1965. Museum of Comparative Zoology, Harvard University, Cambridge, Massachusetts, 41 pp

1966b The proper spelling of taxonomy. Systematic Zoology 15:88

1966c Hummingbird caught by sparrow hawk. The Auk 83:664

1966d Review of *Birds around the World* by D. Amadon. Natural History Magazine 75:64–66

1966e Review of *Europe: A Natural History. The Continents We Live On* by Kai Curry-Lindahl. The Quarterly Review of Biology 41:202

1966f Birds and Science. Pages 24–31 *in* Birds and our Lives (A. Stefferud and B. Hines, Eds.). U.S. Department of the Interior, Washington, DC

1966g Comment on proposed emendation of the Code to cover designation of types from doubtfully syntypical material. Z.N. (S.) 1571. Bulletin of Zoological Nomenclature 23:195

**1967**

1967a Annual Report 1965–1966. Museum of Comparative Zoology, Harvard University, Cambridge, Massachusetts, 49 pp

1967b Evolutionary challenges to the mathematical interpretation of evolution. Pages 47–58 *in* Mathematical Challenges to the Neo-Darwinian Interpretation of Evolution. (P. S. Moorhead and M. M. Kaplan, Eds.) Wistar Institute Symposium Monograph, no. 5. Philadelphia [Reprinted in E. Mayr, Evolution and the Diversity of Life, pages 53–63, 1976]

1967c The origin of the bird fauna of the South Venezuelan highlands [with W. H. Phelps, Jr.]. Bulletin of the American Museum of Natural History 136:269–327 [Spanish translation: Origen de la avifauna de las altiplanicies del sur de Venezuela. Boletín de la Sociedad Venezolana de Ciencias Naturales 29:314–401, 1971]

1967d Subfamily Pachycephalinae, whistlers or thickheads. Pages 3–52 *in* Check-list of Birds of the World, vol. 12 (R. A. Paynter, Jr., Ed.). Museum of Comparative Zoology, Cambridge, Massachusetts

1967e Family Zosteropidae, White-eyes, Indo-Australian Taxa. Pages 289–337 *in* Check-list of birds of the world, vol. 12 (R. A. Paynter, Jr., Ed.). Museum of Comparative Zoology, Cambridge, Massachusetts

1967f The challenge of island faunas. Australian Natural History 15:369–374

1967g Comment on the proposed preservation of *Pan* from Oken, 1816. Z.N.(S.) 482. Bulletin of zoological Nomenclature 24:66

1967h Comments on the proposed suppression of *Cornufer unicolor* Tschudi 1838 (Amphibia). Z.N.(S.) 1749 [with P. J. Darlington, R. F. Inger, and E. E. Williams]. Bulletin of Zoological Nomenclature 24:192

1967i Comment on the proposed designation of a type-species for *Patanga* Uvarov, 1923. Z.N.(S.) 1761. Bulletin of Zoological Nomenclature 24:275

1967j Review of *Perspectives de la Zoologie Européenne* by J. Leclerq and P. Dagnelie. Isis 63:216

1967k Biological man and the year 2000. Daedalus 96:832–836

**1968**

1968a Annual Report 1966–1967. Museum of Comparative Zoology, Agassiz Museum, Harvard University, Cambridge, Massachusetts, 57 pp

1968b The role of systematics in biology: The study of all aspects of the diversity of life is one of the most important concerns in biology. Science 159:595–599 [Reprinted with slight changes as "Introduction: The Role of Systematics in Biology" (pages 4–15) *in* Systematic Biology. Proceedings of an International Conference, National Academy of Science, Washington, D.C., 1969; and reprinted in E. Mayr, Evolution and the Diversity of Life, p. 416–424, 1976; Dutch translation: De Rol van de Systematiek in de Biologie. Vakblad voor Biologen 12:283–290]

1968c The sequence of genera in the Estrildidae (Aves). Breviora, no. 287:1–14

1968d *Larius* Boddaert, 1783 (Aves) Proposed suppression under the plenary powers. Z. N. (S.) 1833 [with H. T. Condon]. Bulletin of Zoological Nomenclature 25:52–54

1968e Bryozoa *versus* Ectoprocta. Systematic Zoology 17:213–216

1968f Discussion [of "Biological Aspects of Race in Man"]. Pages 103–105 *in* Science and the Concept of Race [M. Mead, T. Dobzhansky, E. Tobach, and R. E. Light, Eds.). Columbia University Press, New York

1968g Verhalten als auslesende Kraft. Umschau in Wissenschaft und Technik 13:415–416

1968h Review of *Perspectives de la Zoologie Européenne* by Jean Leclercq and Pierre Dagnelie. Isis 58:424–425

1968i Illiger and the biological species concept. Journal of the History of Biology 1:163–178

1968j Theory of biological classifications. Nature 220:545–548 [Reprinted and revised in E. Mayr, Evolution and the Diversity of Life, pages 425–432, 1976]

1968k Family Fringillidae, Finches. Pages 202–206 *in* Check-list of Birds of the World, vol. 14 (R. A. Paynter, Jr., Ed.). Museum of Comparative Zoology, Cambridge, Massachusetts

1968l Family Estrildidae [with R. A. Paynter, Jr., and M. A. Traylor]. Pages 306–390 *in* Check-list of Birds of the World, vol. 14 (R. A. Paynter, Jr., Ed.). Museum of Comparative Zoology, Cambridge, Massachusetts

**1969**

1969a Annual Report 1967–1968. Museum of Comparative Zoology, Agassiz Museum. Harvard University, Cambridge, Massachusetts, 66 pp

1969b Principles of Systematic Zoology. McGraw-Hill, New York, XI + 428 pp. [German edition: Grundlagen der Zoologischen Systematik. Translation by Otto Kraus, Parey-Verlag, Hamburg and Berlin, 1975; also Polish and French translations, 1975]
Reviews: Bezzel, E. (1976) in Journal für Ornithologie 117:389; Gill, F. B. (1971) in The Auk 88:190–192; Michener, C. D. (1969) in Systematic Zoology 18:232–234; Moss, W. W. (1971) in Systematic Zoology 20:108–109; Regenfuss, H. (1978) in Zeitschrift für Tierpsychologie 48:222–224; Schwoerbel, J . (1976) in Berichte Biochemie und Biologie 430:201

1969c Comments on "Theories and hypotheses in biology." Pages 450–456 *in* Proceedings of the Boston Colloquium for the Philosophy of Science 1966/1968 (R. S. Cohen and M. W. Wartofsky, Eds.). Boston Studies in the Philosophy of Science, vol. 5. Reidel, Dordrecht. [Reprinted under the title "Theory formation in developmental biology" in E. Mayr, Evolution and the Diversity of Life, pages 377–382, 1976]

1969d Scientific explanation and conceptual framework. Journal of the History of Biology 2:123–128

1969e Bird speciation in the tropics. Biological Journal of the Linnean Society 1:1–17 [Reprinted in E. Mayr, Evolution and the Diversity of Life, pages 176–187, 1976]

1969f Foreword. Pages XI–XVI *in* Birds of Paradise and Bower Birds by E. T. Gilliard. Weidenfeld and Nicholson, London, 485 pp

1969g Discussion: Footnotes on the philosophy of biology. Philosophy of Science 36:197–202

1969h The biological meaning of species. Biological Journal of the Linnean Society 1:311–320 [Reprinted in E. Mayr, Evolution and the Diversity of Life, pages 515–525, 1976]

1969i Grundgedanken der Evolutionsbiologie. Naturwissenschaften 56:392–397 [Translated and abridged as "Basic concepts of evolutionary biology" in E. Mayr, Evolution and the Diversity of Life, pages 9–16, 1976]

1969j Erwin Stresemann zum 80. Geburtstag. Journal für Ornithologie 110:377–378

1969k The Museum of Comparative Zoology and its role in the Harvard community. Cambridge, Massachusetts

**1970**

1970a Annual Report 1968–1969. Museum of Comparative Zoology, Agassiz Museum, Harvard University, Cambridge, Massachusetts, 70 pp

***[500]*** 1970b Species, speciation and chromosomes. Pages 1–7 *in* Comparative Mammalian Cytogenetics. (K. Benirschke, Ed.). Springer-Verlag, New York

1970c Essais sur l'Évolution. Review of *Essais sur l'Évolution* by Th. Dobzhansky and E. Boesinger [Masson and Cie, Paris, 1968)]. Evolution 24:480–481

1970d Ecological upset: Crown-of-thorns. Massachusetts Audubon 55:38–39

1970e Populations, Species, and Evolution. An abridgement of *Animal Species and Evolution.* Belknap Press of Harvard University Press, Cambridge, Massachusetts, XV + 453 pp. [French edition, 1975].
Reviews: F. J. Ayala (1970) in Amer. Scientist 58:698; J. Hamilton (1970) in Amer. Biol. Teacher (November); A. Löwe (1971) in The Bryologist 74:226–227; in Times Literary Supplement (14 May 1971); W. Henke (1972) in Homo 23; L. Macomber (1971) in Annals of the Association of American Geographers 65:492–494; P. Stys (1971)

in Acta Soc. Zool. Bohemoslov. 35:79; E. Edelson (1971) in Book World, 10 January 1971, page 4. [Chapter 2 of this book entitled "Species concepts and their application" was reprinted in Sober, Elliot, ed., 1986, *Conceptual Issues in Evolutionary Biology*, MIT Press, 1st edition, pp. 531–540]

1970f Species Taxa of North American Birds: A Contribution to Comparative Systematics [with L. L. Short]. Publications of the Nuttall Ornithological Club, No. 9. Review by B. L. Monroe, Jr. (1972) in The Auk 89:203–207

1970g Evolution und Verhalten. Verhandlungen der Deutschen Zoologischen Gesellschaft 64:322–336

1970h The diversity of life [with R. D. Alexander, W. F. Blair, P. Illg, B. Schaeffer and W. C. Steer]. Pages 20–49 *in* Biology and the Furure of Man (P. Handler, Ed.). Oxford University Press, Oxford. [Reprinted in The Environmental Challenge (W. H. Johnson and W. C. Steere, Eds.), pages 20–49, Holt, Rinehart and Winston, New York, 1974]

1970i Foreword. Page V *in* An atlas of speciation in African passerine birds. B.P. Hall and R.E. Moreau. London, British Museum (Natural History)

## 1971

1971a Evolution vs. special creation. The American Biology Teacher 33:49–50

1971b Was Virchow right about Neandertal? [with B. Campbell] Nature 229:253–254

1971c Evolution und Verhalten. Umschau in Wissenschaft und Technik 20:731–737

1971d New species of birds described from 1956 to 1965. Journal für Ornithologie 112:302–316

1971e Stability in zoological nomenclature [with G. G. Simpson and E. Eisenmann]. Science 174:1041–1042

1971f Letter to the editor [on the subject of race]. Perspectives in Biology and Medicine 14:505–506

1971g Methods and strategies in taxonomic research. Systematic Zoology 20:426–433

1971h Essay review: Open problems of Darwin research. Studies in History and Philosophy of Science 2:273–280

1971i Comment on the proposed designation of a type-species for *Rybaxis* Saulcy, 1876 Z.N.(S.) 1882. Bulletin of Zoological Nomenclature 28:15

1971j Review of *The Life and Letters of Charles Darwin* edited by F. Darwin. Studies in History and Philosophy of Science 2:273

## 1972

1972a Das Leben – ein Zufall? Zum Bestseller eines philosophierenden Biologen [=J. Monod]. Die Brücke zur Welt, Sonntagsbeilage zur Stuttgarter Zeitung, Samstag, 15. Januar 1972, no. 11, page 49

1972b Consequences of recent research findings. Pages 530–538 *in* Genetic Factors in "Schizophrenia". (A. R. Kaplan, Ed.). Charles C. Thomas, Springfield, Illinois

1972c Continental drift and the history of the Australian bird fauna. The Emu 72:26–28

1972d The nature of the Darwinian revolution: Acceptance of Evolution by Natural Selection Required the Rejection of many Previously Held Concepts. Science 176:981–989. [Reprinted and revised in E. Mayr, Evolution and the Diversity of Life, p. 277–296, 1976. Romanian edition: Natura revolutiei Darwiniste In Istoria stiintei si reconstructia ei conceptuala. Edited by Elie Parvu. Bucuresti: Editura stiintifica si enciclopedica, pages 192–214]

1972e Review of *Oskar Heinroth. Vater der Verhaltensforschung, 1871–1945*, by Katharina Heinroth, with a preface by Konrad Lorenz [Große Naturforscher, Bd. 35 (Wissenschaftliche Verlagsgesellschaft, Stuttgart, 1971)]. Isis 63:127–128

1972f Lamarck revisited. Journal of the History of Biology 5:55–94 [Reprinted and abridged in E. Mayr, Evolution and the Diversity of Life, p. 222–250, 1976]

1972g Sexual selection and natural selection. Pages 87–104 *in* Sexual Selection and the Descent of Man 1871–1971 (B. Campbell, Ed.). Aldine Publishing Co., Chicago [Reprinted and abridged in E. Mayr, Evolution and the Diversity of Life, pages 73–87, 1976]

1972h Geography and ecology as faunal determinants. Pages 549–561 *in* Proceedings of the XV International Ornithological Congress (K. H. Voous, Ed.). E. J. Brill, Leiden, The Netherlands

1972i Six proposed amendments to the International Code. Z.N.(S.) 2005. Bulletin of Zoological Nomenclature 29:99–101

## 1973

1973a Comment [on "Population and behavior"]. Pages 115–118 *in* Ethical Issues in Biology and Medicine: A Symposium on the Identity and Dignity of Man. (P. N. Williams, Ed.). Schenkman, Boston: Schenkman

1973b Comment [on "Problems of Population Control"]. Pages 156–157 *in* Ethical Issues in Biology and Medicine: A Symposium on the Identity and Dignity of Man. (P. N. Williams, Ed.). Schenkman, Boston: Schenkman

1973c The last International Congress of Zoology? Science 180:882–883

1973d Alden Holmes Miller (1906–1965). Biographical Memoirs of the National Academy of Sciences USA 43:177–214

1973e Erwin Stresemann. The Ibis 115:282–283

1973f The importance of systematics in modern biology [in Russian]. Journal of General Biology 34:323–330

1973g Sistematica biologica. Pages 818–823 *in* Enciclopedia del Novecento, vol. 6. Istituto dell'Enciclopedia Italiana, Rome

1973h Descent of man and sexual selection. Pages 33–48 *in* Atti del Colloquio Internazionale Sul Tema: "Le Origine dell'Uomo" (28–30 October 1971). Accademia Nazionale dei Lincei, Rome

1973i Essay review: The recent historiography of genetics Review of *Origins of Mendelism* by R. C. Olby (Constable, London 1966) and *The Origins of Theoretical Population Genetics* by W. B. Provine (University of Chicago Press, Chicago, 1971)]. Journal of the History of Biology 6:125–154 [Reprinted and revised in E. Mayr, Evolution and the Diversity of Life, pages 329–353, 1976]

1973j In appreciation–David Lack. The Ibis 115:432–434

1973k Museums and biological laboratories. Breviora, no 416:1–7 [Reprinted in E. Mayr, Toward a New Philosophy of Biology, pages 289–294, 1988]

1973l Ethics and man's evolution. The Technology and Culture Seminar at MIT (Address delivered 6 December 1972). The Technology and Culture Seminar, Cambridge, Massachusetts

1973m The evolution of man [in Russian]. Priroda 12:36–44, Part 1 (1973); Priroda 13:36–43, Part 2 (1974)

## 1974

1974a Robert Cushman Murphy (1887–1973). Pages 131–135 *in* Year Book of the American Philosophical Society 1973

1974b The diversity of life [with R.D. Alexander, W.F. Blair, P. Illg, B. Schaeffer, and W.C. Steers]. Pages 20–49 *in* The Environmental Challenge. Holt, Rinehart and Winston, New York [Reprint of 1970h]

1974c The challenge of diversity. Taxon 23:3–9 [Reprinted in E. Mayr, Evolution and the Diversity of Life, pages 408–415, 1976]

1974d Comment [on Freeman, D. 1974. The Evolutionary Theories of Charles Darwin and Herbert Spencer, Current Anthropology 15:211-221]. Current Anthropology 15:227–228

1974e Teleological and teleonomic: a new analysis. Pages 91–117 *in* Methodological and Historical Essays in the Natural and Social Sciences. (R.S. Cohen and M.W. Wartofsky, Eds.). Boston Studies in the Philosophy of Science, vol. 14. Reidel, Dordrecht [Reprinted in E. Mayr, Evolution and the Diversity of Life, pages 383–404, 1976; revised version printed in E. Mayr Toward a New Philosophy of Biology, pages 38–66, 1988]

1974f Die biologische Zukunft des Menschen. Pages 58–70 *in* Grenzen der Menschheit. Jungius Gesellschaft vol. XX, Hamburg

1974g Cladistic analysis or cladistic classification? Zeitschrift für zoologische Systematik und Evolutionsforschung 12:94–128 [Reprinted in E. Mayr, Evolution and the Diversity of Life, p. 433–476, 1976]

1974h Comment on the proposed addition to the official list of generic names in zoology of *Lachryma* Sowerby (Mollusca) Z.N.(S.) 2018. Bulletin of Zoological Nomenclature 30:140

1974i The definition of the term disruptive selection. Heredity 32:404–406

1974j Behavior programs and evolutionary strategies. Natural selection sometimes favors a genetically "closed" behavior program, sometimes an "open" one. American Scientist 62:650–659 [Reprinted in E. Mayr, Evolution and the Diversity of Life, pages 694–711, 1976]

1974k Evolution and God. Review of *Darwin and His Critics: The Reception of Darwin's Theory of Evolution by the Scientific Community* by David L. Hull [Harvard University Press, Cambridge, Massachusetts, 1973]. Nature 240:285–286 ***[550]***

1974l Review of *Le Polymorphisme dans le règne animal.* Edited by M. Lamotte. Mémoire de la Société Zoologique de France, no. 37 (1974). Systematic Zoology 23:559–560

## 1975

1975a Erwin Stresemann. Verhandlungen der Deutschen Zoologischen Gesellschaft 67: 411–412

1975b Foreword. Pages IX–X *in* Ornithology from Aristotle to the Present by Erwin Stresemann (G.W. Cottrell, Ed.). Translated by H. J. and C. Epstein. Harvard University Press, Cambridge, Massachusetts

1975c Epilogue: Materials for a history of American ornithology. Pages 365–396 and 414–419 *in* Ornithology from Aristotle to the Present by Erwin Stresemann (G.W. Cottrell, Ed.). Translated by H. J. and C. Epstein. Harvard University Press, Cambridge, Massachusetts

1975d Sars, Michael. Pages 106–107 *in* Dictionary of Scientific Biography, vol. 12 (C.C. Gillispie, Ed.). Charles Scribner's Sons, New York

1975e Semper, Carl Gottfried. Page 299 *in* Dictionary of Scientific Biography, vol. 12 (C. C. Gillispie, Ed.) Charles Scribner's Sons, New York

1975f Review of *The Modern Concept of Nature. Essays on Theoretical Biology and Evolution* by H. J. Muller, edited by E. A. Carlson [State University of New York Press, Albany, 1973]. The Quarterly Review of Biology 50:71

1975g Review of *Check-list of Japanese birds*. 5th ed. [Ornithological Society of Japan. Tokyo, 1974.] *Bird Banding* 45:386

1975h Wie weit sind die Grundprobleme der Evolution gelöst? Nova Acta Leopoldina 42:171–179

1975i The unity of the genotype. Biologisches Zentralblatt 94:377–388 [Reprinted in E. Mayr, Toward a New Philosophy of Biology, pages 423–438, 1988]

1975j Ridgway, Robert. Pages 443–444 *in* Dictionary of Scientific Biography, vol. 11 (C. C. Gillispie, Ed.). Charles Scribner's Sons, New York

1975k Comments on the establishment of a neotype for *Pseudogeloius decorsei* (I. Bolivar, 1905) (Insecta, Orthoptera, Pyrgomorphidae). Z.N.(S.) 2046. Bulletin of zoological Nomenclature 32:21–22

1975l Comment on *Cornufer* Tschudi, 1838. Z.N.(S.) 1749. Bulletin of Zoological Nomenclature 32:78–79

## 1976

1976a Species-area relation for birds of the Solomon Archipelago [with J. M. Diamond]. Proceedings of the National Academy of Science USA 73:262–266

1976b Calcium and birds. Linnaean Newsletter 30:2

1976c Verhalten als Schrittmacher der Evolution. Berichte der Physikalisch-Medizinischen Gesellschaft zu Würzburg N. F. 83:131–142

1976d Species-distance relation for birds of the Solomon Archipelago, and the paradox of the great speciators [with Jared M. Diamond and M. E. Gilpin]. Proceedings of the National Academy of Science USA 73:2160–2164

1976e The present state of evolutionary theory. Pages 4–20 *in* Human Diversity: Its Causes and Social Significance (B. D. Davis and P. Flaherty, Eds.) Ballinger Publishing Company, Cambridge, Massachusetts

1976f Birds on islands in the sky: Origin of the montane avifauna of Northern Melanesia [with J. M. Diamond]. Proceedings of the National Academy of Science USA 73:1765–1769

1976g Review of *Kurzer Grundriss der Evolutionstheorie* by N. V. Timoféeff-Ressovsky, N. N. Vorontsov, and A. V. Jablokov. [Genetik–Grundlagen, Ergebnisse und Probleme in Einzeldarstellungen, 7 (Gustav Fischer, Suttgart, 1975)]. Biologisches Zentralblatt 95:237–238

1976h Is the species a class or an individual? Systematic Zoology 25:192

1976i Chairman's introduction. Symposium on value of taxonomic characters in avian classification. Pages 173–175 *in* Proceedings of the 16th International Ornithological Congress (H. J. Frith and J. H. Calaby, Eds.). Australian Academy of Science, Canberra

1976j Whitman, Charles Otis. Pages 313–315 *in* Dictionary of Scientific Biography, vol. 14 (C. C. Gillispie). Charles Scribner's Sons, New York

1976k Comments on *Echis coloratus* Dumeril, Bibron and Dumeril, 1854 (Reptilia). Z.N.(S.) 2064. Bulletin of Zoological Nomenclature 32:199

1976l Comments on the proposal to validate *Cardium californiense* Deshayes, 1839. Z.N.(S.) 2073. Bulletin of Zoological Nomenclature 32:204

1976m Evolution and the Diversity of Life: Selected Essays. The Belknap Press of Harvard University Press, Cambridge, Massachusetts, IX + 721 pp. [The essays included in this volume are reprinted from earlier articles, but note that new "Introductions" precede the book (pages 1–3) and each of its nine sections: Evolution, pages 7–8; speciation, pages 117–119; history, page 221; philosophy, pages 357–358; systematics, page 407; species, pages 479–480; man, page 529; biogeography, pages 549–551; behavior, pages 675–676. Abridged German edition: Evolution und die Vielfalt des Lebens. Springer-Verlag, Berlin 1979]
*Reviews:* Anderson, W. D., Jr. (1978) in Copeia, p. 729–730; Ayala, F. J. (1977) in American Scientist 65:363–364; Bock, W. (1977) in The Auk 94:404; Cain, A. J. (1977) in Nature 268:375–376; Darlington, C. D. "By Descent and Divergence," The Times Literary Supplement, 24 June 1977, p. 765; Hedley, R. (1977) in The New Scientist 74:208; LaBar, M. (1978) "A true believer edits himself," in Ecology 59:1085–1086; Montalenti, G. (1977) in Scientia 112:860–862; Straney, D. O. (1978) in Journal of Mammalogy 59:220–221; Turner, C. (1978) in Annals of Human Biology 5 (3); Yochelsen, E. L. (1977) in Isis 68:634–635.
*Reviews of German edition* (1979) Grebenscikov, I. (1982) in Biologisches Zentralblatt 101:308; Gutmann, W. F. (1981) in Natur und Museum 3:125; Herrman, B. (1982) in Anthropologischer Anzeiger 40:147; J. (1982) in Folia Geobotanica et Phytotaxonomica 17; Thinès, G. (1983) in Behavioural Processes

## 1977

1977a Darwin and natural selection. How Darwin may have discovered his highly unconventional theory. American Scientist 65:321–327 [Reprinted in E. Mayr, Toward a New Philosophy of Biology, pages 215–232, 1988]

1977b Review of *Ontogeny and Phylogeny* by S. J. Gould [The Belknap Press of Harvard University Press, Cambridge, Massachusetts, 1977]. MCZ Newsletter 7:3

1977c Concepts in the study of animal behavior. Pages 1–16 *in* Reproductive Behavior and Evolution (J. S. Rosenblatt and B. R. Komisaruk, Eds.). Plenum Press, New York

1977d On the evolution of photoreceptors and eyes [with L. v. Salvini-Plawen]. Pages 207–263 *in* Evolutionary Biology, vol. 10. (M. K. Hecht, W. Steere, and B. Wallace, Eds.). Plenum Publishing, New York

1977e An almost comprehensive survey of evolution Review of *Evolution.* by T. Dobzhansky, F. Ayala, G. L. Stebbins, and J. W. Valentine [Freeman, San Francisco, 1977]. Evolution 31:913–917

1977f The study of evolution, historically viewed. Pages 39–58 *in* The Changing Scenes in Natural Sciences, 1776–1976 (C. E. Goulden, Ed.). Academy of Natural Sciences, Special Publication 12. Academy of Natural Sciences Special Publication 12, Philadelphia

## 1978

1978a Review of *Reader's Digest Complete Book of Australian Birds* [Reader's Digest, Sydney, 1977]. The Auk 95:211–212

1978b Tenure: A sacred cow? Science 199:1293. [Readers' responses to this note and author's reply in Science 200:601–602, 604, 606]

1978c Origin and history of some terms in systematic and evolutionary biology. Systematic Zoology 27:83–88

1978d Review of *Handbuch der Vögel Mitteleuropas*. Bd. 2: *Charadriiformes* (2. Teil). Edited by U.N. Glutz von Blotzheim, K.M. Bauer und E. Bezzel [Akademische Verlagsgesellschaft, Wiesbaden, 1977]. Journal für Ornithologie 119:241–243

1978e Review of *Handbuch der Vögel Mitteleuropas* (ed. U. Glutz). The Auk 95:615–617

1978f Evolution. Scientific American 239 (March):47–55

1978g Review of *Les Problèmes de l'Espèce dans le Règne Animal*, 2 vols., edited by C. Bocquet, J Génermant, and M. Lamotte. [Mémoires Société Zoologique de France (1976, 1977)]. Systematic Zoology 27:250–252

1978h Review of *Modes of Speciation* by Michael J. D. White. Systematic Zoology 27:478–482

1978i Review of *Handbook of the Birds of Europe, the Middle East and North Africa: The Birds of the Western Palearctic*, vol. 1: *Ostrich to Ducks*. Edited by Stanley Cramp [Oxford University Press, Oxford, 1977]. Quarterly Review of Biology 53:470–471

## 1979

1979a Family Pittidae (pages 310–329), Family Acanthisittidae (pages 331–333), Family Menurae (pages 333–334), Family Atrichornithidae. *In* Check-list of Birds of the World, vol. 8. (M. A. Traylor, Jr., Ed.), Museum of Comparative Zoology, Cambridge, Massachusetts

1979b Review of *Rudolf Leuckart. Weg und Werk* by K. Wunderlich [Biographien bedeutender Biologen, Bd. 2 (Gustav Fischer, Jena, 1978)]. Isis 70:336

1979c Obituary: Alexander Wetmore. The Ibis 121:519–520

## 1980

1980a Review of *The Birds of Paradise and Bowerbirds* by William T. Cooper [David R. Godine, Boston, 1979]. The Auk 97:208

1980b The Evolution of Darwin's Theory. Review of *A Delicate Arrangement: The Strange Case of Charles Darwin and Alfred Russel Wallace* by A. C. Brackman [Times Books, New York, 1980]; *Darwin and the Beagle*, by A. Moorehead (Penguin, London, 1980); *The Voyage of Charles Darwin: His Autobiographical Writings* edited by C. Ralling (Mayflower Books, New York, 1979]; and *The Collected Papers of Charles Darwin* edited by P.H. Barrett, with a Foreword by Th. Dobzhansky [University of Chicago Press, Chicago, 1977]. Washington Post Book World, 22 June 1980, pages 4 and 10

1980c Review of *Grundriss der Populationslehre*, by N. W. Timoféeff-Ressovsky, A. N. Jablokov, N. V. Glotov [Genetik–Grundlagen, Ergebnisse und Probleme in Einzeldarstellungen, Bd. 8 (Gustav Fischer, Jena 1977)]. Biologisches Zentralblatt 99:374–375

1980d Problems of the classification of birds: a progress report. Erwin Stresemann Memorial Lecture. Pages 95–112 in Acta XVII Congressus Internationalis Ornithologici (R. Nöhring, Ed.). Deutsche Ornithologen Gesellschaft, Berlin. [Abridged version printed in E. Mayr, Toward a New Philosophy of Biology, pages 295–311, 1988]

1980e The Evolutionary Synthesis. Perspectives on the Unification of Biology. Edited by Ernst Mayr and William Provine. Harvard University Press, Cambridge, Massachusetts, XI + 487 pp.

*Reviews:* F.J. Ayala (1982) in BioScience 32:351; W.J. Bock (1981) in The Auk 98:644–646; P.J. Bowler (1983) in British Journal for the History of Science 16:283–284; G. Broberg (1981–82) in Lychnos, p. 308–310; A.J. Cain "Intellectual cross-fertilization and evolution," in Nature 290:645–646; M. Cartmill (1981) in The New

England Journal of Medicine (6 August), p. 350–351; D. J. Futuyma (1981) in American Scientist 69:551–552; B. Glass (1981) in Isis 72:642–647; C. Groves (1982) in Mankind 13:430; P. W. Leslie (1982) in Human Biology, P. B. Medawar (1981) in New York Review of Books 28 (2):34–36; M. Ruse (1981) "Origins of the Modern synthesis," in Science 211:810–811; D. B. Wake (1981) in Evolution 35:1256–1257; M. J. D. White (1981) in Paleobiology 7:287–291; G. C. Williams (1981) in The Quarterly Review of Biology 56:448–450

1980f Prologue: Some Thoughts on the History of the Evolutionary Synthesis. Pages 1–48 *in* The Evolutionary Synthesis. Perspectives on the Unification of Biology (E. Mayr and W. B. Provine, Eds.). Harvard University Press, Cambridge, Massachusetts ***[600]***

1980g The Role of Systematics in the Evolutionary Synthesis. Pages 123–136 *in* The Evolutionary Synthesis. Perspectives on the Unification of Biology (E. Mayr and W. B. Provine, Eds.). Harvard University Press, Cambridge, Massachusetts

1980h Botany. Introduction. Pages 137–138 *in* The Evolutionary Synthesis. Perspectives on the Unification of Biology (E. Mayr and W. B. Provine, Eds.). Harvard University Press, Cambridge, Massachusetts

1980i Paleontology. Introduction. Page 153 *in* The Evolutionary Synthesis. Perspectives on the Unification of Biology (E. Mayr and W. B. Provine, Eds.). Harvard University Press, Cambridge, Massachusetts

1980j Morphology. Introduction. Page 173 *in* The Evolutionary Synthesis. Perspectives on the Unification of Biology (E. Mayr and W. B. Provine, Eds.). Harvard University Press, Cambridge, Massachusetts

1980k Germany. Introduction. Pages 279–283 *in* The Evolutionary Synthesis. Perspectives on the Unification *of Biology* (E. Mayr and W. B. Provine, Eds.). Harvard University Press, Cambridge, Massachusetts

1980l France. Introduction. Page 309 *in* The Evolutionary Synthesis. Perspectives on the Unification of Biology (E. Mayr and W. B. Provine, Eds.). Harvard University Press, Cambridge, Massachusetts

1980m The Arrival of Neo-Darwinism in France. Page 321 *in* The Evolutionary Synthesis. Perspectives on the Unification of Biology (E. Mayr and W. B. Provine, Eds.). Harvard University Press, Cambridge, Massachusetts

1980n How I became a Darwinian. Pages 413–423 *in* The Evolutionary Synthesis. Perspectives on the Unification of Biology (E. Mayr and W. B. Provine, Eds.). Harvard University Press, Cambridge, Massachusetts

1980o G. G. Simpson. Pages 452–466 *in* The Evolutionary Synthesis. Perspectives on the Unification of Biology (E. Mayr and W. B. Provine, Eds.). Harvard University Press, Cambridge, Massachusetts

1980p Curt Stern. Pages 424–429 *in* The Evolutionary Synthesis. Perspectives on the Unification of Biology (E. Mayr and W. B. Provine, Eds.). Harvard University Press, Cambridge, Massachusetts

**1981**

1981a Biological classification: Toward a synthesis of opposing methodologies. Science 214:510–516 [Reprinted under title "Toward a synthesis in biological classification" in E. Mayr, Toward a New Philosophy of Biology, pages 268–288, 1988; also reprinted in Sober, Elliot, editor, Conceptual Issues in Evolutionary Biology, 2nd edition, 1995, pp. 277–294]

1981b Evolutionary biology. Pages 147–162 *in* The Joys of Research (W. Shropshire, Jr., Ed.). Smithsonian Institution Press, Washington, DC

1981c La biologie d'évolution. Hermann, Paris 175 pp. [Italian translation Biologia ed Evoluzione. Paolo Boringhieri, Torino, Italia, 1982)

## 1982

1982a Questions concerning speciation. Nature 296:609

1982b Foreword. Pages XI–XII *in* Darwinism Defended: A Guide to the Evolution Controversies by M. Ruse. Addison-Wesley, Reading, Massachusetts

1982c Review of *Extinction: The Causes and Consequences of the Disappearance of Species*, by P. and A. Ehrlich [Random House, New York, 1981]. BioScience 32:349

1982d The Growth of Biological Thought: Diversity, Evolution, and Inheritance. The Belknap Press of Harvard University Press, Cambridge, Massachusetts, IX + 974 pp. [German edition: Die Entwicklung der biologischen Gedankenwelt. Vielfalt, Evolution und Vererbung. Springer-Verlag, Berlin, 1984, 766 pp.; second printing 2002. Translated by Karin de Sousa Ferreira; French edition Histoire de la biologie. Diversité, evolution et heredité. Fayard, Paris, 1989, 894 pp.; Italian edition Storia del Pensiero Biologico. Bollati Boringhieri, Torino 1990]
*Reviews:* R. Albury (1983) in Mankind 13:540; G. Allen (1983) "Essay review: Ernst Mayr and the philosophical problems of biology" in Science and Nature 6:21–29; J. Bernstein (1984) "The evolution of evolution," in The New Yorker, 23 January, p. 98–104; A. Brues (1983) in American Journal of Physical Anthropology 60:545–546; R. Burckhardt, Jr. (1983) "The fruits of evolution," The Times Higher Education Supplement, 23 September, p. 14; W. F. Bynum (1985) "On the written authority of Ernst Mayr," in Nature 317:585–586; S. Cachel (1986) "The growth of biological thought revisited," American Anthropologist 88:452–454; A. J. Cain (1982) "Porcupine biology," in Nature 297:707–709; J. L. Carter and W. V. Mayer (1988) "Beyond the textbook: Books on biology," BioScience 38:490–492; N. Eldredge (1982) "A biological urge to oversimplify," in The Philadelphia Inquirer, November 7, section P, p. 3; D. S. Farner (1983) in The Auk 100:507–509; D. J. Futuyma (1982) "A synthetic history of biology," in Science 216:842–844; M. T. Ghiselin (1983) in Isis 74:410–413; B. Glass (1983) in The Mendel Newsletter 23:5–7; J. L. Gould (1982) in The New York Review of Books, May 23, p. 7 and 32; J. Greene (1992) "From Aristotle to Darwin: Reflections on Ernst Mayr's interpretation in *The Growth of Biological Thought*," in Journal of the History of Biology 25:257–284; M. D. Grmek (1984) in History and Philosophy of the Life Science 6:107–109; C. P. Haskins (1982) "An encyclopedic tour de force" in Natural History (July), p. 70–73; F. L. Holmes (1982) "The evolution of evolution," in Washington Post Book World, June 20, page 4; M. J. Kottler (1983) "A history of biology: Diversity, evolution, inheritance,"in Evolution 37:868–872; F. D. Ledley (1983) in The New England Journal of Medicine 308:1174–1175; Maynard J. Smith (1982) "Storming the fortress," in The New York Review of Books, May 13, pages 41–42; R. Olby (1983) "History laden with history," in New Scientist 97:184–185; W.-E. Reif (1985) in Zentralblatt für Geologie und Paläontologie, Teil II, pages 2–7; J. Roger (1983) in Isis 74:405–410; M. Ruse (1984) in Journal of the History of the Behavioral Sciences 20:220–224; M. Ruse (1985) "Admayration," in The Quarterly Review of Biology 60:183–192; R. J. Russell (1984) in Ethology and Sociobiology 5:63–64; G. G. Simpson (1982) "Autobiology," in The Quarterly Review of Biology 57:437–444; P. R. Sloan (1985) "Essay review: Ernst Mayr on the history

of biology," in Journal of the History of Biology 18:145–153; Tattersall, (1984) The good, the bad, and the synthesis. Amer. Anthropologist 86:86–90; P. Thompson (1984) in Queen's Quarterly 91:198–200; J. W. Valentine (1983) "Mayr again on evolution," BioScience 33:209; R. T. Wright (1983) in Christian Scholar's Review 12:368; L. Wolpert (1983) "The multiplication of the living," in The Times Literary Supplement, 4 March, p. 216

*Reviews of German edition* (1984) B., J. (1985) "Bibel der Biologen," in Bayernkurier Nr. 48, 30. November; B. Barcikowska (1985) in Genetica Polonica 26:419–420; Bäumer, Ä. (1987) in Berichte zur Wissenschaftsgeschichte 10:58–63; G. J. de Klerk (1985) in Theoretical and Applied Genetics 70:573; G. F. (1985) in Universitas 40:833–834; A. Fagot-Largeault (1992) in Archive Phil. (Paris) 55:691–694; W. Herre (1986) in Zeitschrift für zoologische Systematik und Evolutionsforschung 24:84; R. Kittel (1986) in Zoologische Jahrbücher, Abteilung für Anatomie und Ontogenie der Tiere 114:41–42; A. La Vergata (1991) in Nuncius 6 (2):361–364 (Italian translation); S. Lorenzen (1984) "Warum Darwin? Warum Mendel?" in Die Zeit, Nr. 46, 9. November, page 19; H. Markl (1985) in Spektrum der Wissenschaft 1985:135–136; A. C. Matte (1986) in Andrologia 18:200; R. Mocek (1988) in Wissenschaftliche Zeitschrift der Universität Halle 38:140–142; H. Mohr (1986) in Ethology 71:345–346; G. Peters (1986) in Deutsche Literaturzeitung 107:715–718; U. Pieper (1986) in Zeitschrift für Morphologie und Anthropologie 76:242; F. Pirchner (1988) in Zeitschrift für Tierzüchtung 105:484–485; H. Querner (1985) in Die Naturwissenschaften 72:386–388; H. Rath (1985) "Das Buch der Natur neu lesen," Frankfurter Allgemeine Zeitung, 8. Oktober, page L14; H.-J. Rheinberger (1986) "Kritischer Essay. Ernst Mayr: Die Entwicklung der biologischen Gedankenwelt," in Medizinhistorisches Journal 21:172–185; W. Schönborn (1985) in Archiv für Protistenkunde 130:354; G. Zirnstein (1986) in NTM-Schriftenreihe für Geschichte der Naturwissenschaft, Technik und Medizin 23:127

1982e Epilogue. Biological Journal of the Linnean Society 17:115–125 [Abridged and revised version printed under the title "The challenge of Darwinism" in E. Mayr, Toward a New Philosophy of Biology, pages 185–195, 1988]

1982f Adaptation and selection. Biologisches Zentralblatt 101:161–174 [Reprinted in E. Mayr, Toward a New Philosophy of Biology, pages 133–147, 1988]

1982g Review of *Systematics and Biogeography. Cladistics and Vicariance* by G. Nelson and N. Platnick [Columbia University Press, New York, 1981]. The Auk 99:621–622

1982h Review of *Vicariance Biogeography*, edited by G. Nelson and D. E. Rosen (Columbia University Press, New York, 1981). The Auk 99:618–620

1982i Of what use are subspecies? The Auk 99:593–595

1982j Review of *A Concordance to Darwin's Origin of Species*, First Edition. Edited by P. Barrett, D. J. Weinshank, and T. T. Gottleber [Cornell University Press, Ithaca, New York, 1981]. Isis 73:476–477

1982k Der gegenwärtige Stand des Evolutionsproblems. Zweite "Bernhard Rensch-Vorlesung" gehalten am 20. Mai 1981. Pages 1–18 *in* Evolution, Zeit, Geschichte, Philosophie. Universitätsvorträge Westfälische Wilhelms-Universität Münster, Heft 5. Aschendorff, Münster

1982l Processes of speciation in animals. Pages 364–382 *in* Mechanisms of speciation (C. Barigozzi, Ed.). Alan R. Liss, New York [Reprinted in E. Mayr, Toward a New Philosophy of Biology, 1988]

1982m Darwinistische Mißverständnisse. Pages 44–57 *in* Darwin und die Evolutionstheorie (Kurt Bayertz, B. Heidtmann und H.-J. Rheinberger, Eds.). Studien zur Dialektik, Bd. 5. Pahl-Rugenstein, Köln

1982n Review of *Geologen der Goethezeit* (H. Prescher and D. Beeger, Eds.) [Abhandlungen des Staatlichen Museums für Mineralogie und Geologie zu Dresden, 29; VEB Deutscher Verlag für Grundstoffindustrie, Leipzig, 1979]. Isis 73:582

1982o Reflections on human paleontology. Pages 231–237 *in* A History of American Physical Anthropology, 1930–1980 (F. Spencer, Ed.). Academic Press, New York

1982p Speciation and macroevolution. Evolution 36:1119–1132. [Spanish edition: Especiación y macroevolution. Interciencia 8:133–142, 1983; abridged English version printed in E. Mayr, Toward a New Philosophy of Biology, pages 439–456, 1988]

1982q Geleitwort. Pages 9–18 *in* Allgemeine Betrachtungen über die Triebe der Thiere, hauptsächlich über ihre Kunsttriebe by Hermann Samuel Reimarus (facsimile of the second edition, 1762) (J. von Kempski, Ed.). Veröffentlichung der Joachim Jungius-Gesellschaft der Wissenschaften Hamburg, Nr. 46. 2 vols. Vandenhoeck and Ruprecht, Göttingen

## 1983

1983a How to carry out the adaptationist program? The American Naturalist 121:324–334 [Reprinted in E. Mayr, Toward a New Philosophy of Biology, pp. 148–159, 1988]

1983b Review of *The Evolution of Culture in Animals* by John Tyler Bonner (Princeton University Press, Princeton, 1980). *Biologisches Zentralblatt* 102:127

1983c Comments on David Hull's paper on exemplars and type specimens. Pages 504–511 *in* Proceedings of the 1982 Biennial Meeting of the Philosophy of Science Association, vol. 2 (P. D. Asquith and T. Nickles, Eds.). Philosophy of Science Association, East Lansing, Michigan

1983d Darwin, intellectual revolutionary. Pages 23–41 *in* Evolution from Molecules to Man (D. S. Bendall, Ed.). Cambridge University Press, Cambridge, United Kingdom. [Reprinted in E. Mayr, Toward a New Philosophy of Biology, pages 168–184, 1988]

1983e Introduction. Pages 1–21 *in* Perspectives in Ornithology: Essays Presented for the Centennial of the American Ornithologists' Union (A. H. Brush and G. A. Clark, Jr., Eds.). Cambridge University Press, Cambridge, United Kingdom

1983f New species of birds described from 1966 to 1975 [with F. Vuilleumier]. Journal für Ornithologie 124:217–232.

1983g Review of *Philopatry, Inbreeding, and the Function of Sex* by W. M. Shields [State University of New York Press, Albany, 1982] and *General Theory of Evolution* by V. Csanyi [Studica Biologica Hungarica, 18, Akademiai Kiado, Budapest, 1982]. Zeitschrift für Tierpsychologie 63:261.

1983h The joy of birds. Natural History Magazine 92: inside cover

1983i Review of *Allgemeine Betrachtungen über die Triebe der Thiere, hauptsächlich über ihre Kunsttriebe* by Hermann Samuel Reimarus (facsimile of the second edition, 1762) (J. v. Kempski, Ed.) Veröffentlichung der Joachim Jungius-Gesellschaft der Wissenschaften Hamburg, Nr. 46, two volumes [Vandenhoeck and Ruprecht, Göttingen, 1982). Isis 74:596–597

## 1984

1984a The contributions of ornithology to biology. BioScience 34:250–255 [Condensed version published in WingTips 1:70–80]

1984b The concept of finality in Darwin and after Darwin. Scientia 77:97–117 (in English); 119–134 (in Italian) [Reprinted in E. Mayr, Toward a New Philosophy of Biology, pages 233–257, 1988]

1984c Evolution and ethics. Pages 35–46 *in* Darwin, Marx, and Freud: Their Influence on Moral Theory (A. L. Caplan and B. Jennings, Eds.). Plenum Press, New York

1984d The unity of the genotype [reprinted from *Biologisches Zentralblatt*, 1975, with new preface]. Pages 69–84 *in* Genes, Organisms, Populations: Controversies over the Unit of Selection (R. N. Brandon and R. M. Burian, Eds.) MIT Press, Cambridge, Massachusetts.

1984e Evolution of fish species flocks: a commentary. Pages 383–397 *in* Evolution of Fish Species Flocks. (A. A. Echelle and I. Kornfield, Eds.). University of Maine at Orono Press, Orono, Maine [Abridged version printed in E. Mayr, Toward a New Philosophy of Biology, pages 383–397, 1988]

1984f Review of *Dear Lord Rothschild: Birds, Butterflies, and History* by M. Rothschild (Hutchinson, London, 1983). Isis 75:601–603.

1984g Professor Mayr Responds [to a review by M. Ruse of *The Growth of Biological Thought: Diversity, Evolution, and Inheritance*]. Journal of the History of the Behavioral Sciences 20:223–224

1984h The triumph of evolutionary synthesis. [Review of *Beyond Neo-Darwinism: An Introduction to the New Evolutionary Paradigm* edited by M.-W. Ho and P. T. Saunders (Academic Press, London, 1984)]. The Times Literary Supplement, 2 November 1984, pages 1261–1262

1984i Kein Zufall. Evolutionsbiologie: Ein Kommentar zum angeblichen "Irrtum des Jahrhunderts". Die Zeit, 9 November l984, page 84

1984j *Treskiornithidae* Richmond, 1917 (Aves) Application to place on official list of family-group names in zoology and to give precedence over *Plataleinae* Bonaparte, 1838, and other competing family-group names [with E. Eisenmann and K. C. Parkes]. Z.N.(S.) 2136. Bulletin of Zoological Nomenclature 41:240–244

**1985**

1985a Review of *Lehre und Forschung: Autobiographische Erinnerungen*, by Albert Frey-Wyssling [Vorwort von Peter Sitte]. Grosse Naturforscher, Bd. 44. (Wissenschaftliche Verlagsgesellschaft, Stuttgart, 1984). The Quarterly Review of Biology 60:83–84 ***[650]***

1985b Review of *Die Würger der Paläarktis: Gattung Lanius* by E. N. Panow [Die Neue Brehm-Bücherei, Bd. 557 (A. Ziemsen Verlag, Wittenberg Lutherstadt, 1983)]. The Quarterly Review of Biology 60:92

1985c How biology differs from the physical sciences. Pages 43–63 *in* Evolution at a Crossroads: The New Biology and the New Philosophy of Science (D. J. Depew and B. H. Weber, Eds.). MIT Press, Cambridge, Massachusetts

1985d Darwin and the definition of phylogeny. Systematic Zoology 34:97–98

1985e Natürliche Auslese. Naturwissenschaften 72:231–236

1985f The probability of extraterrestrial intelligent life. Pages 23–30 *in* Extraterrestrials: Science and Alien Intelligence (E. Regis, Jr., Ed.). Cambridge University Press, Cambridge, United Kingdom [Reprinted in E. Mayr, Toward a New Philosophy of Biology, pages 67–74, 1988]

1985g Michael James Denham White (1910–1983). Year Book of the American Philosophical Society 1984, 156–159

1985h Weismann and Evolution. Journal of the History of Biology 18:295–329 [Reprinted in E. Mayr, Toward a New Philosophy of Biology, pages 491–524, 1988; German version: August Weismann und die Evolution der Organismen. Freiburger Universitätsblätter 24 (Heft 87/88):61–82, 1985]

1985i Darwin's five theories of evolution. Pages 755–772 *in* The Darwinian Heritage (D. Kohn, Ed.). Princeton University Press, Princeton [Abridged version reprinted in E. Mayr, Toward a New Philosophy of Biology, pp. 196–214, 1988]

**1986**

1986a Biogeographic evidence supporting the Pleistocene Forest Refuge Hypothesis [with R. J. O'Hara]. Evolution 40:55–67

1986b Check-list of Birds of the World, vol. 11 (E. Mayr and G. W. Cottrell, Eds.). Museum of Comparative Zoology, Cambridge, Massachusetts. Introduction (pages v–vii), Sylviidae (pt., pages 3–29), Muscicapidae (pt., pages 295–375), Maluridae (pages 390–409), Acanthizidae (pages 409–464), Monarchidae (pt., pages 464–556), Eopsaltriidae (pages 556–583)

1986c Memorial minute: George G. Simpson [with A. W. Crompton and S. J. Gould]. Harvard Gazette 81, 16 May 1986, page 5

1986d Joseph Gottlieb Kölreuter's contributions to biology. Osiris 2 (2nd series):135–176

1986e Natural Selection: The Philosopher and the Biologist. Review of *The Nature of Selection:* Evolutionary Theory in Philosophical Focus by E: Sober [MIT Press, Cambridge, Massachusetts, 1984]. Paleobiology 12:233–239 [Reprinted in part in E. Mayr, Toward a New Philosophy of Biology, pp. 116–125, 1988]

1986f In Memoriam: Jean (Theodore) Delacour. The Auk 103:603–605

1986g Review of *Evolutionary Theory: The Unfinished Synthesis* by R. G. B. Reid (Cornell University Press, Ithaca, New York, 1985). Isis 77:358–359

1986h Foreword. Pages VII–X *in* Defining Biology: Lectures from the 1890s (J. Maienschein, Ed.). Harvard University Press, Cambridge, Massachusetts

1986i La systématique évolutionniste et les quatre étapes du processus de classification. Pages 143–160 *in* L'Ordre et la Diversité du Vivant. Quel Statut Scientifique pour les Classifications Biologiques? (P. Tassy, Ed.). Fondation Diderot, Fayard, France

1986j Ermahnung und Ermutigung aus berufenem Mund. Review of *Evolution, Genetik und menschliches Verhalten* by Hubert Markl [Piper, München, 1986). Bild der Wissenschaft 9:160–161

1986k Uncertainty in science: Is the giant panda a bear or a raccoon? Nature 323:769–771

1986l What is Darwinism Today? Pages 145–156 *in* Proceedings of the 1984 Biennial Meeting of the Philosophy of Science Association, vol. 2 (P. D. Asquith and P. Kitcher, Eds.). Philosophy of Science Association, East Lansing, Michigan

1986m The death of Darwin? Revue de synthèse 107:229–235 [Reprinted in E. Mayr, Toward a New Philosophy of Biology, pages 258–264, 1988e]

1986n Review of *Handbuch der Vögel der Sowjetunion. Vol. 1: Erforschungsgeschichte, Gaviiformes, Podiciformes, Procellariiformes* (V. D. Il'icev and V. E. Flint, Eds.) [A Ziemsen Verlag, Wittenberg Lutherstadt, 1985). The Quarterly Review of Biology 61:558–559

1986o Review of *Biophilosophical Implications of Inorganic and Organismic Evolution* by Bernhard Rensch (Die Blaue Eule, Essen, 1985). Philosophy of Science 53:612

1986p The correct name of *Rallus hodgeni*. Notornis 33:268

**1987**

1987a The species as category, taxon, and population. Pages 303–320 *in* Histoire du Concept d'Espèce dans les Sciences de la Vie. Fondation Singer-Polignac, Paris, France [See also Brill Catalogue, Leiden 1993; reprinted in E. Mayr, Toward a New Philosophy of Biology, pages 315-334, 1988]

1987b The status of subjective junior homonyms. Systematic Zoology 36:85–86

1987c New species of birds described from 1976 to 1980 [with F. Vuilleumier]. Journal für Ornithologie 128:137–150

1987d The ontological status of species: Scientific progress and philosophical terminology. Biology and Philosophy 2:145–166 [Reprinted in E. Mayr, Toward a New Philosophy of Biology, pages 335-358, 1988]

1987e Answers to these Comments [on 1987d]. Biology and Philosophy 2:212–220

1987f Island Progress. Review of *The Malay Archipelago* by A. R. Wallace, with an introduction by J. Bastin [Oxford University Press, Oxford, 1987]. Nature 328:770–771

1987g Review of *The Fall of a Sparrow* by Sálim Ali (Oxford University Press, Delhi, India, 1985). The Auk 104:364–365

1987h Review of *The Birds of Wallacea (Sulawesi, The Moluccas and Lesser Sunda Islands, Indonesia)* by C. M. N. White and M. D. Bruce (BOU Check-list No. 7, 1986). The Wilson Bulletin 99:514–515

**1988**

1988a Review of *Birds of New Guinea* by B. M. Beehler, T. K. Pratt, and D. A. Zimmermann (Princeton University Press, Princeton, New Jersey, 1986). American Scientist 76:81

1988b The limits of reductionism. Nature 331:475

1988c A response to David Kitts. Biology and Philosophy 3:97–98

1988d Recent historical developments. Pages 31–43 *in* Prospects in Systematics (D. L. Hawksworth, Ed.) The Systematics Association, Special Volume no. 36. Clarendon Press, Oxford

1988e Toward a New Philosophy of Biology: Observations of an Evolutionist. Harvard University Press, Cambridge, Massachusetts, IX + 564 pp. [Most essays in this volume are reprinted from earlier articles, but five essays were not published previously and note that new "Introductions" precede each of the nine sections on philosophy, natural selection, adaptation, Darwin, diversity, species, speciation, macroevolution, and history. German edition: Eine neue Philosophie der Biologie, Vorwort von H. Markl. Translated by Inge Leipold. H. Piper, München and Zürich, 470 pp., 1991 *Reviews:* J. Armstrong (1988) in Perspectives on Science and Christian Faith 19:134; F. J. Ayala (1988) "Concepts of biology," in Science 240:1801–1802; A. Brennan (1988) "Shared concern," Times Higher Education Supplement, 5 August; R. M. Burian (1990) "Essay review: Toward a new philosophy of biology," in Journal of the History of Biology 23:321–328; D. Duman (1988) in The Bloomsbury Review, September/October; H. Eagar (1988) "Evolutionist returns philosophy to the masses," in The Kansas City Star, 28 August; H. Eagar (1988) "Philosophy comes full circle; regular folks can participate," The Maui News, 17 July, D5; N. Eldredge (1988) "Observing evolution," in BioScience 38:566–567; M. Ereshefsky (1990) in Philosophy of Science 57:725–727; M. Feldesman in American Journal of Physical Anthropology 78:625–626; D. Futuyma (1988) "A noted biologist traces his own evolution," The Scientist, 8 August, p. 21; M. Ghiselin (1989) in American Scientist 77:70–71; J. R. Griesemer (1989) in The Quarterly Review of Biology 64:51–55; J. P. Hailman

(1989) in The Auk 106:751–752; D.L. Hull (1990) "Ernst Mayr on the philosophy of biology" in Historical Methods 23:42–45; P. Kitcher (1988) "The importance of being Ernst" in Nature 333:25–26; J. Maienschein (1989) in Isis 80:568–569; J. Marks (1989) in Journal of Human Evolution 18:509–511; M. Ridley (1989) "Correcting a bias" in Times Literary Supplement, 17–23 March, p. 273; D.E. Shaner and R. Hutchinson (1990) in Philosophy East and West 40:264–266; C.U.M. Smith (1989) in Annals of Human Biology 16:382–383; J.M. Smith (1992) "Taking a chance on evolution" in The New York Review of Books, 14 May, p. 34–36; W. Sudhaus (1989) in Ethology 83:350–352; P. Thompsen (1990) in History and Philosophy of the Life Sciences 12:277–278; C. Tudge (1988) "The many kinds of 'Why?'" in New Scientist 119:64; *German edition* (1991) R. Kaspar (1991) "Wer fällt spannender, der Stein oder die Katze? Wie sich die Biologie aus den Fängen der Physik rettet und emanzipiert" in Die Presse; H. Mayer (1992) "Was ist eine Art? Der Evolutionstheoretiker Ernst Mayr als Philosoph der Biologie" in Frankfurter Allgemeine Zeitung, 17. Februar; A. von Schirnding (1991) "Zigeuner des Kosmos. Ernst Mayr entwirft das Weltbild der Evolutionsbiologie" in Süddeutsche Zeitung, 12. November

1988f Indomalayan Region: A substitute name for Wallace's Oriental Region [with 15 coauthors]. The Ibis 130:447–448

1988g Die Darwinsche Revolution und die Widerstände gegen die Selektionstheorie. Pages 221–249 *in* Die Herausforderung der Evolutionsbiologie (H. Meier, Ed.). Piper, München, Zürich. [Also published in Naturwissenschaftliche Rundschau 42:255–265 and in English translation (see 1991f)]

1988h The why and how of species. Biology and Philosophy 3:431–441

## 1989

1989a A strengthened court [letter]. Science 243:590

1989b All about systematics. Current Contents 9 (27 February 1989):14

1989c The contributions of birds to evolutionary theory. Pages 2718–2723 *in* Acta XIX Congressus Internationalis Ornithologici (H. Ouellet, Ed.). Vol. 2. National Museum of Natural Sciences, Ottawa.

1989d Evoluzione. Pages 364–374 *in* Enciclopedia del Novecento, vol. 8. Istituto dell'Enciclopedia Italiana, Rome

1989e Taxonomists and stability. Nature 339:654

1989f Sir Charles Alexander Fleming (9 September 1916–11 September 1987). In Year Book 1988 (American Philosophical Society), Biographical Memoirs: 159–163

1989g Commentary—A new classification of the living birds of the world. The Auk 106:508–512

1989h Attaching names to objects. Pages 235–243 *in* What the Philosophy of Biology Is (M. Ruse, Ed.). Kluwer, Dordrecht

1989i Speciational evolution or punctuated equilibria. Journal of Social and Biological Structures 12:137–158 [Reprinted in E. Mayr, Toward a New Philosophy of Biology, pages 457–488, 1988; also published, with slight revisions, on pages 23–53 *in* The Dynamics of Evolution: The Punctuated Equilibrium Debate in Natural and Social Sciences (A. Somit and S.A. Peterson, Eds.). Cornell University Press, Ithaca, New York]

***[700]*** 1989j Universal and human beginnings. Review of *Origins. The Darwin College Lectures* edited by A.C. Fabian [Cambridge University Press, New York, 1988]. BioScience 39:813–814

1989k Future Challenges of Systematic Ornithology. Journal of the Yamashina Institute for Ornithology 21:154–164

1989l Comment on the proposed confirmation of the spelling of Liparidae Gill, 1861 (Osteichthyes, Scorpaeniformes). Case 2440. Bulletin of Zoological Nomenclature 46:45

## 1990

1990a The Myth of the Non-Darwinian Revolution. Review of *The Non-Darwinian Revolution. Reinterpreting a Historical Myth* by P.J. Bowler [Johns Hopkins University Press, Baltimore, 1988]. Biology and Philosophy 5:85–92

1990b Plattentektonik und die Geschichte der Vogelfaunen. Pages 1–17 *in* Current Topics of Avian Biology (R. van den Elzen, K.-L. Schuchmann, and K. Schmidt-Koenig, Eds.). Proceedings of the International Centennial Meeting of the Deutsche Ornithologen-Gesellschaft, Bonn 1988. Verlag der Deutschen Ornithologen-Gesellschaft, Bonn (with photo of Ernst Mayr on page 18)

1990c When is historiography whiggish? Journal of the History of Ideas 51:301–309

1990d Review of *Handbuch der Vögel der Sowjetunion. Vol. 4: Galliformes, Gruiformes* (R.L. Potapov and V.E. Flint, Eds.). Translated by B. Stephan [A Ziemsen Verlag, Wittenberg Lutherstadt, 1989]. The Quarterly Review of Biology 65:233–234

1990e Chapman, Frank Michler. Pages 152–153 *in* Dictionary of Scientific Biography, vol. 17, supplement II (F.L. Holmes, Ed.). Charles Scribner's Sons, New York

1990f Davis, D. (Delbert) Dwight. Pages 211–213 *in* Dictionary of Scientific Biography, vol. 17, supplement II (F.L. Holmes, Ed.). Charles Scribner's Sons, New York

1990g Jordan, (Heinrich Ernst) Karl. Pages 454–455 *in* Dictionary of Scientific Biography, vol. 17, supplement II (F.L. Holmes, Ed.). Charles Scribner's Sons, New York

1990h Noble, Gladwyn Kingsley. Pages 687–688 *in* Dictionary of Scientific Biography, vol. 18, supplement II (F.L. Holmes, Ed.). Charles Scribner's Sons, New York

1990i Romer, Alfred Sherwood. Pages 752–753 *in* Dictionary of Scientific Biography, vol. 18, supplement II (F.L. Holmes, Ed.). Charles Scribner's Sons, New York

1990j Schmidt, Karl Patterson. Pages 788–789 in Dictionary of Scientific Biography, vol. 18, supplement II (F.L. Holmes, Ed.). Charles Scribner's Sons, New York

1990k Stresemann, Erwin. Pages 888–890 *in* Dictionary of Scientific Biography, vol. 18, supplement II (F.L. Holmes, Ed.). Charles Scribner's Sons, New York

1990l Zum Gedenken an Erwin Stresemann (22. November 1889–20. November 1972) [with G. Mauersberger]. Mitteilungen aus dem Zoologischen Museum in Berlin 65, Supplementheft: Annalen für Ornithologie 13:3–7

1990m Review of *Arguments on Evolution: A Paleontologist's Perspective* by Antoni Hoffman (Oxford University Press, Oxford, 1989). Isis 81:602

1990n Review of *Darwinismus und Botanik: Rezeption, Kritik und theoretische Alternativen im Deutschland des 19. Jahrhunderts* by Thomas Junker (Deutscher Apotheker Verlag, Stuttgart, 1989). Journal of the History of Biology 23:335–336

1990o A natural system of organisms. Nature 348:491

1990p Review of *Die Dohle, Corvus monedula* by R. Dwenger. Die Neue Brehm-Bücherei, Bd. 588 [A. Ziemsen Verlag, Wittenberg Lutherstadt, 1989]. The Quarterly Review of Biology 65:514–515

1990q Review of *Der Graue Kranich, Grus grus*, by H. Prange et al. Neue Brehm-Bücherei, Bd. 229 [ A Ziemsen Verlag, Wittenberg, Lutherstadt, 1989]. The Quarterly Review of Biology 65:515

1990r Myxoma and group selection. Biologisches Zentralblatt 109:453–457

1990s Die drei Schulen der Systematik (The three schools of systematics). Verhandlungsberichte der Deutschen Zoologischen Gesellschaft 83:263–276

## 1991

1991a Principles of Systematic Zoology, 2nd edition [with P.D. Ashlock]. McGraw Hill, New York, XX + 475 pp
*Reviews:* de Queiros, K. (1992) in Systematic Zoology 41:264–266; Groves, C. (1993) in J. Human Evol. 24:251–253; Lundberg, J. G. and McDade, L. A. (1991) in Science 253:458–459; Nelson, G. (1991) in Cladistics 7:305–307

1991b Bureaucratic mischief: Recognizing endangered species and subspecies [with S.J. O'Brien]. Science 251:1187–1188

1991c Response to letters [with S.J. O'Brien]. Science 253:251–252

1991d Die Evolution–eine biologische Tatsache, Natur und Wissenschaft. Frankfurter Allgemeine Zeitung, 27. März 1991, Nr. 73, page N1

1991e Introduction. An overview of current evolutionary biology. Pages 1–14 *in* New Perspectives on Evolution (L. Warren and H. Koprowski, Eds.). Wiley-Liss, New York

1991f The ideological resistance to Darwin's Theory of Natural Selection. Proceedings of the American Philosophical Society 135:123–139 (Translation of 1988g)

1991g One Long Argument: Charles Darwin and the Genesis of Modern Evolutionary Thought. Harvard University Press, Cambridge, Massachusetts, XIV + 195 pp. [German edition ... und Darwin hat doch recht. Charles Darwin, seine Lehre und die moderne Evolutionstheorie. Translated by I. Leipold. Munich and Zurich, Piper. 1994]
*Reviews:* H.J. Birx (1992) in Library Journal 116 (16):132; P.J. Bowler (1991) in Nature 353:713–714; G. Cowley (1992) in New York Times Book Review, 1 March; D.J. Futuyma (1992) in The Quarterly Review of Biology 67:190–191; J.C. Greene (1992) "Warfare of Nature" in Times Higher Education Supplement, 29 May, p. 25; J. Haffer (1994) in Ethology 96:178–179; V.A. Haines (1992) in Contemporary Sociology, November, p. 820–822; S. Herbert (1993) "Essay review" in Isis 84:119–123; R. Lewis (1993) in The American Biology Teacher 55:312–313; J.A. Mills (1994) in Theory and Psychology 4:155–160; J. Poyser (1992) in The Bloomington Voice, 10–17 June; R. Rainger (1993) in Journal of the History of Biology 26:378–380; W.-E. Reif, (1995) in Zentralblatt für Geologie und Paläontologie, II:31–32; M. Ridley (1992) "Population thinker" in Times Literary Supplement, 12 June, p. 5–6; M. Ruse (1993) in American Scientist 81:198–199; M. Ruse (1994) in J. Hist Behav. Sci 30:181–186; J.H. Schwartz (1992) in American Journal of Physical Anthropology 87:499–501; D.R. Weiner (1993) in Isis 84:119–122

1991h More natural classification. Nature 353:122

## 1992

1992a A local flora and the biological species concept. American Journal of Botany 79:222–238

1992b The Idea of teleology. Journal of the History of Ideas 53:117–135

1992c Review of *The Encyclopedia of Evolution: Humanity's Search for Its Origins* by R. Milner (1990). American Scientist 80:192–193

1992d In Memoriam: Bernhard Rensch, 1900–1990. The Auk 109:188

1992e Review of *Handbuch der Vögel der Sowjetunion* by V.D. Il'icev and V.A. Zubakin. Band 6, Teil 1: *Charadriiformes/Lari: Stercorariidae, Laridae (Larinae und Sterninae)* (1990). The Quarterly Review of Biology 67:63–64

1992f Haldane's *Causes of Evolution* after 60 Years. The Quarterly Review of Biology 67:175–186

1992g Lohnt sich die Suche nach extraterrestrischer Intelligenz? Naturwissenschaftliche Rundschau 45:264–266

1992h In Memoriam: Heinrich Dathe, 1910–1991. The Auk 109:648

1992i Controversies in Retrospect. Pages 1–34 *in* Oxford Surveys in Evolutionary Biology, vol. 8 (D. Futuyma and J. Antonovics, Eds.). Oxford University Press, Oxford

1992j Preface. Bulletin of the British Ornithologists' Club, Centenary Supplement 112A:1

1992k Préface, Buffon 88: Actes du Colloque international pour le bicentenaire de la mort de Buffon: 7–10

1992l Preface [Celebration of F. N. and F. Hamerstrom]. The Journal of Raptor Research, 26(3):106–107

1992m New species of birds described from 1981 to 1990 [with F. Vuilleumier and M. LeCroy]. Bulletin of the British Ornithologists' Club,Centenary Supplement 112A:267–309

1992n Darwin's Principle of Divergence. Journal of the History of Biology 25:343–359

1992o Foreword: Charles Darwins Autobiographie. Pages 5–16 *in* Charles Darwin: Mein Leben 1809–1882 (N. Barlow, Ed.). Translated by C. Krüger. Insel Verlag, Frankfurt am Main

1992p [Several quotes from Mayr's letters to J.C. Greene of 1991] in J.C. Greene *From Aristotle to Darwin: Reflections on Ernst Mayr's interpretation in The Growth of Biological Thought.* Journal of the History of Biology 25:257–284

## 1993

1993a What was the evolutionary synthesis? Trends in Ecology and Evolution 8:31–34

1993b Evolution. Pages 63–68 *in* Developing Biological Literacy: A Guide to Developing Secondary and Post-Secondary Biology Curricula. BSCS Innovative Science Education

1993c Proximate and ultimate causations. Biology and Philosophy 8:93–94

1993d Search for intelligence. Science 259:1522–1523

1993e In Memoriam: Ernst Schüz, 1901–1991. The Auk 110:127 ***[750]***

1993f Begrüßungsansprache an die Teilnehmer der ersten Jahresversammlung der Deutschen Gesellschaft für Geschichte und Theorie der Biologie (Marburg, am 26. Juni 1992). Biologisches Zentralblatt 112:98–99

1993g Fifty years of progress in research on species and speciation. Proceedings of the California Academy of Sciences 48:131–140

1993h Haldane and Evolution. Pages 1–6 *in* Proceedings of the Zoological Society, Calcutta, Haldane Commemorative Volume

1993i The resistance to Darwinism and the misconceptions on which it was based. Pages 35–46 *in* Creative Evolution (W. Schopf and J. Campbell, Eds.). Jones and Bartlett, Boston, Massachusetts

1993j Timoffeeff Ressovsky. Pages 178–179 *in* Timoffeeff Ressovsky Reminiscences (N. Vorontsov, Ed.). Moskva, Nayka

1993k Evidence from Professor Ernst Mayr. Select Committee on Science and Technology. Systematic Biology Research. House of Lords Session 1990–91. HMSO, London

**1994**

1994a Driving forces in evolution. An analysis of natural selection. Pages 29–48 *in* Evolutionary Biology of Viruses (S. S. Morse, Ed.). Raven Press, New York

1994b Provisional classifications v. standard avian sequences: Heuristics and communication in ornithology [with W. Bock]. The Ibis 136:12–18

1994c Cladistics and convergence. Trends in Ecology and Evolution 9:149–150

1994d Review of *Styles of Scientific Thought. The German Genetics Community 1900–1933* by J. Harwood [The University of Chicago Press, Chicago, 1993]. Ethology 96:179–181

1994e Dalla storia naturale al DNA. KOS (Rev. Sci. Etica) n.s. 100, v. 10, pages 21–25

1994f Does it pay to acquire high intelligence? Perspectives in Biology and Medicine:150–154

1994g Reasons for the Failure of Theories. Philosophy of Science 61:529–533

1994h *Cotylorhynchus:* Not a mammal. Science 264:1519

1994i The advance of science and scientific revolutions. Journal of the History of the Behavioral Sciences 30:328–334

1994j War Darwin ein Lamarckist? Naturwissenschaftliche Rundschau 47:240–241

1994k Review of *Einführung in die Phylogenetik und Systematik* by W. Sudhaus and K. Rehfeld, (Gustav Fischer, New York, 1992). Systematic Biology 43:463–464

1994l Recapitulation reinterpreted: The somatic program. The Quarterly Review of Biology 69:223–232

1994m Response to Walter Bock, Biology and Philosophy 9:329–331; response to John Beatty, Biology and Philosophy 9:357–358; response to Richard Burkhardt, Biology and Philosophy 9:373–374

1994n Darwin, Charles, pages 125–126; Evolution, pages 237–240; Wallace, Alfred Russel, pages 786–787 *in* The Encyclopedia of the Environment (R. A. Eblen and W. R. Eblen, Eds.). Houghton Mifflin, New York

1994o Review of *Species, Species Concepts and Primate Evolution* edited by W. H. Kimbel and L. B. Martin [Plenum Press, New York]. Quarterly Review of Biology 69:261–262

1994p Evolution—Grundfragen und Missverständnisse (und Replik). Ethik und Sozialwissenschaften, Streitforum für Erwägungskultur EuS 5, Heft 2:203–209 and 270–279

1994q Population thinking and neuronal selection: Metaphors or concepts? International Review of Neurobiology: Selectionism and the Brain 37:27–34

1994r The advance of science and scientific revolutions. Journal of the History and of Behavioral Sciences 30:328–334

1994s Ordering systems [Response to K. Padian]. Science 266:715–716

1994t The number of subspecies of birds [with J. Gerloff]. Bulletin of the British Ornithologists' Club 114:244–248

**1995**

1995a Review of *Charles Darwin: Voyaging* by J. Browne. Newsday, February 12, pages 31–32

1995b Systems of ordering data. Biology and Philosophy 10:419–434

1995c Zur Begegnung zweier Kulturen. Geleitwort. Pages 193–197 *in* Biophilosophie by G. Vollmer. Universal-Bibliothek, Nr. 9386. Stuttgart: Philipp Reclam jun

1995d Haldane's Daedalus. Pages 79–89 *in* Haldane's Daedalus Revisited (K. R. Dronamraju, Ed.). Oxford University Press, Oxford

1995e Editorial. Biodiversity and its molecular foundation. Comptes Rendus de l'Académie des Sciences, Series III, Sciences de la vie 318:727–731

1995f A critique of the Search for Extraterrestrial Intelligence. Can SETI succeed? Not likely. Bioastronomy News 7:2–4

1995g The search for extraterrestrial intelligence: Scientific quest on hopeful folly? The Planetary Report 7:4–7

1995h Observations on SETI. Bioastronomy News 8:1–3

1995i Widerlegt Gutmanns hydraulische Morphologie den Darwinismus? Aufsätze und Reden der Senckenbergischen Naturforschenden Gesellschaft 43:199–211

1995j Foreword, pages VII–XI *in* Concepts, Theories, and Rationality in the Biological Sciences: The Second Pittsburgh-Konstanz Colloquium in the Philosophy of Science (G. Wolters, J. G. Lennox, and P. McLaughlin, Eds.). University of Pittsburgh Press, Pittsburgh, Pennsylvania

1995k Review of *A Biography of Ludlow Griscom* by W. Davis. History and Philosophy of the Life Sciences 17:520

1995l Answer to "The abundance of life-bearing planets." Bioastronomy News 7:3

1995m Species, classification, and evolution. Pages 3–12 *in* Biodiversity and Evolution (R. Arai, M. Kato, and Y. Doi, Eds.). National Science Museum Foundation, Tokyo

1995n Darwin's impact on modern thought. Proceedings of the American Philosophical Society 139:317–325

1995o Howard Barraclough Fell Memorial Minute adopted by the Faculty of Arts and Sciences, Harvard University [with D. L. Pawson and K. J. Boss]. October 17, 1995

**1996**

1996a Search for extraterrestrials criticized. "Chance of success is virtually nil." Harvard University Gazette. 25 January, page 6

1996b The autonomy of biology: The position of biology among the sciences. The Quarterly Review of Biology 71:97–106

1996c The modern evolutionary theory. Journal of Mammalogy 77:1–7

1996d Review of *Darwinism, Evolving Systems Dynamics, and the Genealogy of Natural Selection* by D. J. Depew and B. H. Weber. History and Philosophy of the Life Sciences 18:8–10

1996e What is a Species and what is not? Philosophy of Science 63:262–277

1996f Tribute to "Wolfgang Wickler on his 65th Birthday." Ethology 102:881–882

1996g Moritz Wagner (1813–1887). *In* Dictionnaire du Darwinisme et de l'Evolution (P. Tort, Ed.) Presses Universitaires de France, Paris

1996h Classification *In* Dictionnaire du Darwinisme et de l'Evolution (P. Tort, Ed.), Presses Universitaires de France, Paris

**1997**

1997a Darwinism from France. Review of *Dictionnaire du Darwinisme et de l'Evolution* (P. Tort, ed.) [Presses Universitaires de France]. Science 274:2032 ***[800]***

1997b This is Biology. The Science of the Living World. Cambridge, Massachusetts, Harvard University Press, Cambridge, Massachusetts XV + 327 pp. [German edition Das ist Biologie. Die Wissenschaft des Lebens. Tanslation by J. Wissmann. Spektrum, Heidelberg, 1998; 2nd printing 2000]
*Reviews:* D. Baltimore (1997) in Nature 387:772; H. J. Birx (1997) in Library Journal 122:96; W. Bock (1998) in American Scientist 86:186–187; J. Dupre (1999) in

Philosophy of Science 66:504–506; A. Finkbeiner (1997) in New York Sunday Book Review; M. T. Ghiselin (1998) in Isis 89:150–151; C. D. Grant (2000) in Zygon 35:711–713; J. C. Greene (1999) in Biology and Philosophy 14:103–116; D. G. Homberger (1998) in The Auk 115:1085–1088; B. Marsh (1998) in The Ibis 140:709–710; J. E. Rall (1997) in J. Nervous and Mental Disease 185:707–708; S. Sismondo (2000) in Biology and Philosophy 15:103–106; V. B. Smocovitis (1999) in Journal of the History of Biology 32:385–394; S. Wolf (1998) in Integrative Physiol. and Behav. Sci. 33:76–78; F. M. Wuketits (2004) in Ludus Vitalis (Revista de Filosofía de las Ciencias de la Vida) XII, number 21:149–160

1997c Goldschmidt and the Evolutionary Synthesis: A response. Journal of the History of Biology 30:31–33

1997d Reminiscences from the first curator of the Whitney-Rothschild Collection. BioEssays 19:175–179

1997e Perspective: The objects of selection. Proceedings of the National Academy of Science USA 94:2091–2094

1997f Review of *The Origin and Evolution of Birds* by A. Feduccia. American Zoologist 37:210–211

1997g The establishment of evolutionary biology as a discrete biological discipline. BioEssays 13:263–266

1997h Review of *Buffon. A Life in Natural History* by J. Roger. Los Angeles Times Book Review, August 24, 1997, page 6

1997i Biology, pragmatism, and liberal education. Pages 287–297 *in* Education and Democracy. Re-imagining Liberal Learning in America (R. Orrill, Ed.). The College Board, New York

1997j Roots of dialectical materialism. Pages 12–18 *in* Na Perelome [On the Edge: Soviet Biology in the 1920s–1930s ] (E. I. Kolchinsky, Ed.). St. Petersburg, Russia

1997k Letters from Ernst Mayr to E. Stresemann. Pages 407–771 *in* Ornithologen-Briefe des 20. Jahrhunderts (J. Haffer, Ed.). Ökologie der Vögel, vol. 19, Hölzinger, Ludwigsburg

1997l Foreword (pages 5–6). Reminiscences of K. Lorenz (pages 802–803), L. C. Sanford (pages 822–824), G. Schiermann (pages 824–827), E. Stresemann (pages 848–855) *in* Ornithologen-Briefe des 20. Jahrhunderts (J. Haffer, Ed.), Ökologie der Vögel, vol. 19. Hölzinger, Ludwigsburg

## 1998

1998a Foreword. Pages XI–XIV *in* Chance and Change: Ecology for Conservationists by W. H. Drury, Jr. (J. G. T. Anderson, Ed.). University of California Press, California

1998b Reminiscences of Dean Amadon. Pages 1–2 *in* "A celebration of Dean Amadon, Part one." Linnaean Newsletter , vol. LII, no 2

1998c Perspective: Two empires or three? Proceedings of the National Academy of Science USA 95:9720–9723

1998d Ernst Mayr Lecture: Was ist eigentlich die Philosophie der Biologie? Berlin-Brandenburgische Akademie der Wissenschaften, Berichte und Abhandlungen, Band 5:287–301

1998e The multiple meanings of "teleological." History and Philosophy of the Life Sciences 20:35–40

1998f An Stelle eines Vorworts: Gedanken zum Art-Problem. Zoologische Abhandlungen des Museums für Tierkunde Dresden 50, Supplement (S. Eck, Ed. ["100 Jahre Artkonzepte in der Zoologie"]), pages 6–8

1998g "Como escribir historia de la biología?", "Causa y efecto en biologia," and "Los multiples significadores de 'teleológico." *In* Historia y Explicación en Biología (A. Barahona and S. Martinez, Eds.). Fondación de Cultura Económica, Mexico

## 1999

1999a Thoughts on the Evolutionary Synthesis in Germany. Pages 19–29 *in* Die Entstehung der Synthetischen Theorie (T. Junker and E.M. Engels, Eds.). Verhandlungen der Gesellschaft für Geschichte und Theorie der Biologie 2

1999b [Letters in] *Debating Darwin: Adventures of a Scholar* by J.C. Greene. Regina Books, Claremont, California. [Contains letters from Ernst Mayr to author]

1999c [Quotations in] R.D. Precht: Vorgestellt, Das ist Biologie. Die Wissenschaft des Lebens. Spektrum, Heidelberg-Berlin

1999d Perspectives on Evolution: No competing theories. Kansas Biology Teacher 8:12

1999e Postscript: Understanding Evolution. Trends in Ecology and Evolution 14:372–373

1999f Structure for Theories of Biology. Review of *Sex and Death* by K. Sterelny and P.E. Griffiths. Science 285:1856–1857

1999g Review of *Sudden Origins: Fossils, Genes, and the Emergence of Species* by J.H. Schwartz. BioEssays 21:978–979

1999h An evolutionist's perspective [1999 Crafoord Prize Lecture]. The Quarterly Review of Biology 74:401–403

1999i Taxonomia Evolutiva. Evolución y Filogenia de Arthropoda. Boletín de la Sociedad Entomológica Aragonesa 26:35–39

1999j Rousing the Society's interests in ornithology. Linnaean Newsletter 53:2–3

1999k Introduction, 1999; pages XIII–XXXV. Systematics and the Origin of Species [New paperback edition]. Harvard University Press, Cambridge, Massachusetts

## 2000

2000a In memoriam: Biswamoy Biswas, 1923–1994. The Auk 117:1030

2000b Darwin's influence on modern thought. Scientific American 283:79–83 [German edition: Darwins Einfluss auf das moderne Weltbild. Spektrum der Wissenschaft, September: Portuguese edition: A influencia de Darwin no pensamento moderno, Agalia: Revista Internacional da Associao Galegia Lingua 61:95–103. Reprinted on pages 134–142 *in* The Best American Science Writing 2001 (T. Ferris, Ed.), Harper Collins, New York and *in* R. Robbins and M. Cohen, Darwin, The Bible, and The Debate, Allyn and Bacon, Boston, *in press*]

2000c Review of *Darwin's Spectre. Evolutionary Biology in the Modern World* by M. Ruse. Isis 91:373–374

2000d Biology in the twenty-first century. BioScience 50:895–897

2000e The Biological Species Concept (pages 17–29), A critique from the Biological Species Concept perspective: What is a species, and what is not? (pages 93–100) and A defense of the Biological Species Concept (pages 161–166). *In* Species Concepts and Phylogenetic Theory (Q.D. Wheeler and R. Meier, Eds.), Columbia University Press, New York

2000f Grußwort. *In* Ein Leben für die Biologie(geschichte). Festschrift zum 75. Geburtstag von Ilse Jahn (K.-F. Wessel, J. Schulz, and S. Hackethal, Eds.). Berliner Studien zur Wissenschaftsphilosophie and Humanontogenetik 17:7

**2001**

2001a Is *Spizella taverneri* a species or a subspecies? [with N. K. Johnson]. Condor 103:418–419

2001b Preface, 1998. Pages IX–XIV *in* The Evolutionary Synthesis: Perspectives on the Unification of Biology, 2nd edition (E. Mayr and W. B. Provine, Eds.). Harvard University Press, Cambridge, Massachusetts

2001c A comment on Wu's genic view of speciation. Journal of Evolutionary Biology 14:866–867

2001d Allopatric (page 39), Anagenesis (page 62), Clade (page 384), Cladogenesis (page 384), Holophyly (page 958), Monophyly (page 1238–39), Paraphyly (page 1414), Speciation (pages 1860–1864), Species (pages 1864–1869), Sympatric (page 1904), Taxonomy, evolutionary (pages 1934–1937). *In* Encyclopedia of Genetics (S. Brenner, J. H. Miller, and W. J. Broughton, Eds.). Academic Press, San Diego, California

2001e The philosophical foundations of Darwinism. Proceedings of the American Philosophical Society 145:488–495

2001f What Evolution Is. New York, Basic Books, XVII + 318 pp. [German edition: Das ist Evolution. Translation by S. Vogel. Bertelsmann, Munich, 2003]
*Reviews:* Anon. (2001) in N. Y. Times Book Rev. , Dec. 30, p. 14; L. Betzig (2002) in J. Anthropol. Res. 58:409–410; H. J. Birx (2001) in Library Journal 126:163; H. L. Carson (2002) in BioEssays 25:90–91; T. Flannery (2002) 'A Bird's-Eye View of Evolution,' in The New York Review of Books, July 27, p. 30; W. Kimler (2003) in Configurations 11:272–274; S. Palumbi (2002) in Harvard University Magazine, March–April, p. 26–30; P. Raeburn (2001) 'A friendly textbook' in The New York Times on the Web (December 16); M. Ridgley (2002) in Nature 417:223–224; M. Schilthuizen (2002) in Science 295:50; J. Kurdziel (2003) in Quarterly Review of Biology 78:92–93; M. H. Wolpoff (2003) in Evol. Anthropology 12:53–55
*Reviews of German edition:* J. Voss (2003) in Frankfurter Allgemeine Zeitung, 2. Dezember; A. Meyer (2004) in Die Zeit, 8. Januar 2004, p. 45; M. Glaubrecht (2004) in Bild der Wissenschaft 6/2004, p. 71; D. Korn (2005) in Zentralblatt für Geologie und Paläontologie, Teil II, pages 208–209

2001g The Birds of Northern Melanesia. Speciation, Ecology, and Biogeography [with J. Diamond]. Oxford, Oxford University Press, XXIV + 492 pp
*Reviews:* B. M. Beehler (2001) in Science 294:1007–1008; M. Björklund (2002) in Journal of Evolutionary Biology 15:1095–1096; T. Flannery (2002) 'A Bird's-Eye View of Evolution,' in The New York Review of Books, July 27, p. 26–27, 30; P. R. Grant (2002) in Evolution 56:1880–1882; J. Haffer (2002) in Journal für Ornithologie 143:379–380; A. W. Kratter (2002) in The Auk 119:883–887; T. Price (2002) in Scientific American 90:469–471; S. Pruett-Jones (2002) in Nature 415:959–960

**2002**

2002a Comments by E. Mayr [Response to P. Beurton: Ernst Mayr through time on the Biological Species Concept–a conceptual analysis, Theory in Biosciences 121:81–98.] Theory in Biosciences 121:99–100

2002b Foreword. Pages IX–XIV *in* Acquiring Genomes: A Theory of the Origin of Species by L. Margulis and D. Sagan. Basic Books, New York

2002c Die Autonomie der Biologie. Zweite Walther Arndt Vorlesung. Naturwissenschaftliche Rundschau 55:23–29 [Also published in Sitzungsberichte der Gesellschaft Naturforschender Freunde zu Berlin (N.F.) 40:5–16]

2002d G. Ledyard Stebbins. Biographical Memoirs, Proceedings of the American Philosophical Society 146:129–131

2002e Review of *Adaptive Radiation of Blind Subterranean Mole Rats* by E. Nevo, E. Ivanitzkaya, and A. Beiles. Nature 420:125

2002f The last word on Darwin? Review of *Charles Darwin: The Power of Place* by J. Browne. Nature 419:781–782

2002g Evolution ist eine Tatsache. Laborjournal (May), pages 26–30

2002h Classifications and other ordering systems [with W. J. Bock]. Journal of Zoological Systematics and Evolutionary Research 40:169–194 ***[850]***

2002i The biology of race. Daedalus 131:89–94

## 2003

2003a Erinnerungen an Rudolf Zimmermann. Mitteilungen des Vereins Sächsischer Ornithologen 9:139–140

2003b Foreword. Pages IX–X *in* Tears of the Cheetah by S. J. O'Brien. St. Martin's Press, New York

2003c Dandelions. Carleton-Willard Villager 21(3):8

2003d Grußwort zur Gründung der AG Evolutionsbiologie. Biologen heute 3/467:7

## 2004

2004a What Makes Biology Unique? Considerations on the Autonomy of a Scientific Discipline. Cambridge University Press, New York, XIV + 232 pp. [*German edition:* Konzepte der Biologie. Translation by S. Warmuth and with a Geleitwort, pages XI–XX by M. Glaubrecht. Hirzel, Stuttgart, 2005]
*Reviews:* M. Glaubrecht (2004) "A Centenarian's Summary" in Science 306:614–615; E. A. Smith (2005) "A grandmaster surveys the field," in Trends in Ecology and Evolution 20:163; V. B. Smocovitis (2005) "What made Ernst Unique?" in J. Hist. Biol. 38:609–614; R. G. Bribiescas (2005) in Journal of Mammalian Evolution 12:517–520; C. M. Perrins (2006) in The Ibis 148:579

2004b 80 years of watching the evolutionary scenery. Science 305:46–47

2004c The autonomy of biology. Ludus Vitalis XII, no. 21:15–28

2004d Preface. Pages IX–XII *in* Evolutionary Theory and Processes: Modern Horizons. Papers in Honour of Eviatar Nevo (S. P. Wasser, Ed.). Kluwer Academic Publications, Dordrecht, The Netherlands

2004e La Situación: Un tipo particular de cambio, pages 18–19. *In* Qué es la evolución?. La Vanguardia, 22 August 2004

2004f The impact of new concepts in science. Page 19 *in* Essays by the Recipients of the International Prize for Biology; 1994 Recipient; p. 19. In The Twenty Years of International Prize for Biology (T. Sugimura et al., Eds.). Committee on the International Prize for Biology, Japan Society for the Promotion of Science, Tokyo

2004g Naturschutz und die Waldralle von Neuguinea. Ornithologischer Anzeiger 43:215–216

2004h Darwin tot? Es lebe Darwin! (with A. Meyer). Die Zeit 49:42

## 2005

Foreword, p. XVII. *In* Variation. A central concept in biology. XXII + 568 p. by B. Hallgrimsson and B. K. Hall (eds.), Amsterdam, Elsevier

## List of Books by Ernst Mayr as Author, Co-author, or Editor

Compiled by Ernst Mayr who mentioned to me that this is a more "honest" listing than previous counts omitting most of those co-authored books where he felt he contributed less than half and the *Zeitschriften-Verzeichnis* (1929), but he did include the Princeton volume (1949) and the Stresemann-Festschrift (1949)

1. 1941. *List of New Guinea birds. A systematic and faunal list of the birds of New Guinea and adjacent islands.* American Museum of Natural History, New York. XI + 260 pp
2. 1942. *Systematics and the Origin of Species.* New York, Columbia University Press, XII + 334 pp
3. 1945. *Birds of the Southwest Pacific.* New York, The Macmillan Co., XIX + 316 pp
4. 1946. *Birds of the Philippines.* New York, The Macmillan Co., XV + 309 pp. (with Jean Delacour)
5. 1949. *Genetics, Paleontology, and Evolution.* Edited by Glenn L. Jepsen, George Gaylord Simpson, and Ernst Mayr. Princeton, Princeton University Press, XIV + 474 pp
6. 1949. *Ornithologie als biologische Wissenschaft.* Festschrift zum 60. Geburtstag von Erwin Stresemann. Edited by Ernst Mayr and Ernst Schüz. Heidelberg, Carl Winter, XII + 291 pp
7. 1953. *Methods and Principles of Systematic Zoology.* New York, McGraw-Hill, IX + 336 pp. (with E. G. Linsley and R. L. Usinger)
8. 1957. *The Species Problem.* Edited by Ernst Mayr. Publication No. 50 of the American Association for the Advancement of Science. Washington, D. C., American Association for the Advancement of Science, IX + 395 pp
9. 1963. *Animal Species and Evolution.* Cambridge, Massachusetts, Harvard University Press, XIV + 797 pp
10. 1969. *Principles of Systematic Zoology.* New York, McGraw-Hill, XI + 428 pp
11. 1970. *Populations, Species, and Evolution.* Cambridge, Massachusetts, The Belknap Press of Harvard University Press. An abridgment and revision of *Animal Species and Evolution*, XV + 453 pp
12. 1976. *Evolution and the Diversity of Life. Selected Essays.* Cambridge, Massachusetts and London, The Belknap Press of Harvard University Press, IX + 721 pp
13. 1980. *The Evolutionary Synthesis.* Edited by Ernst Mayr and William Provine. Cambridge, Massachusetts, Harvard University Press, XI + 487 pp
14. 1982. *The Growth of Biological Thought: Diversity, Evolution, and Inheritance.* Cambridge, Massachusetts and London: The Belknap Press of Harvard University Press, IX + 974 pp
15. 1988. *Toward a New Philosophy of Biology: Observations of an Evolutionist.* Cambridge, Massachusetts, Harvard University Press, IX + 564 pp
16. 1991. *Principles of Systematic Zoology.* Revised Edition. New York, McGraw-Hill, XX + 475 pp. (with Peter Ashlock)
17. 1991. *One Long Argument: Charles Darwin and the Genesis of Modern Evolutionary Thought.* Cambridge, Massachusetts, Harvard University Press, XIV + 195 pp
18. 1997. *This is Biology: The Science of the Living World.* Cambridge, Massachusetts, Harvard University Press, XV + 327 pp
19. 2001. *What Evolution Is*, foreword by Jared Diamond. New York, Basic Books, XVII + 318 pp
20. 2001. *The Birds of Northern Melanesia.* Oxford, New York, Oxford University Press, XXIV + 492 pp. (with Jared Diamond)

21. 2004. *What Makes Biology Unique? Considerations on the Autonomy of a Scientific Discipline.* New York, Cambridge University Press, XIV + 232 p

## Coauthorship

In the case of coauthored books and papers it is of interest what percentage was contributed by E. Mayr. He supplied the following information from memory:

### Books

1929. *Zeitschriftenverzeichnis* by E. Mayr and W. Meise: Mayr did somewhat less than half
1946. *Birds of the Philippines* by J. Delacour and E. Mayr: Work was done about half and half
1953. *Methods and Principles of Systematic Zoology* by Mayr, Linsley and Usinger: Mayr wrote about half of the text
1991. *Principles of Systematic Zoology*, 2nd edition by E. Mayr and P. D. Ashlock: Because the latter fell ill, Mayr did most of the work
2001. *Birds of Northern Melanesia* by E. Mayr and J. Diamond: The basic manuscript was Mayr's, but in the end J. D. had done at least half

### Major Papers

1935 ff. Birds of the Archbold Expeditions to New Guinea by Mayr and Rand: The early papers were written mostly by Mayr, the later papers mostly by Rand

1938. Review of *Acanthiza* by E. Mayr and D. L. Serventy: Work was done about half and half by Mayr and Serventy

1944–1946. Experiments with *Drosophila* by Dobzhansky and Mayr or Mayr and Dobzhansky: The methodology was Dobzhansky's, the experiments were done mostly by Mayr

1945. The family Anatidae by J. Delacour and E. Mayr: Outline of the text and the more general discussions by Mayr, most of the factual detail by Delacour

1947. Review of Dicaeidae by E. Mayr and D. Amadon: Work was done about half and half

1948. Evolution in the family Dicruridae by E. Mayr and C. Vaurie: Most of the text by E. Mayr

1949. Polymorphism in *Oenanthe* by E. Mayr and E. Stresemann: Most of the text by E. Mayr

1954. Birds of Central New Guinea by E. Mayr and T. Gilliard: Most of the text by T. Gilliard

1955, 1967. Origin of the birds of Pantepui by E. Mayr and W. H. Phelps, Jr.: Most of the text by Phelps, general evaluation and discussion by E. Mayr

1970. Species taxa of North American birds by E. Mayr and L. L. Short: Most of the text by Short

1976. Papers on Northern Melanesian biogeography with J. Diamond: The latter did more than half of the work

1977. Evolution of eyes by E. Mayr and L. von Salvini-Plawen: The latter did most of the work

# Appendix 3
## Subject Analysis of Ernst Mayr's Publications

*Note: Titles may be indicated under more than one heading*

## (1) Ornithology

### (a) General (Ecology, Behavior, Migration, Expedition Reports, Number of Birds, Sex Ratio, Plumage, and Molt)

1926a–c, 1927c, d, 1929b, 1930a, f, 1932e, 1933l, 1935d, 1937d, 1938e, m, n, 1939a, e, 1941n, 1942i, 1943h, 1944a, 1945h, 1946a, 1948b, f, g, 1952j, m, 1954g, 1963r, 1966c, f, 1976b, 1983e, h

### (b) Faunistics

1923a, b, 1924a, c, 1925a, b, c, 1926d, f, g, 1927a, g, 1928, 1934d, 1935b, e, f, 1937j, 1938h, 1947c, f, 1949d, 1953c, g

### (c) Taxonomy

1927b, e, f, 1930b, d, e, 1931a–d, f, h–m, 1932a–d, g, 1933a, b, d–i, k, m, n, 1934a, b, 1935g, h, 1936a–d, f, g, 1937a, c, e–h, k, 1938a–d, g, i–l, 1939b–d, f, g, 1940a, b, d–g, 1941a, b, d–f, h, k–n, 1942b, f, 1943b, c, g, 1944e, m, q, 1945g, j, m, 1946b, i, 1947b, d, 1948a, 1949a, c, e, 1950a, b, d, 1951e, i, j, 1952a, e, i, 1953d, e, h, 1954b, f, 1955a, 1956f, 1957a, b, 1960d, 1961d, 1963c, 1964q, 1965l, 1967d, e, 1968k, l, 1971d, 1983f, 1987c, 1992m, 1994t

## (2) Zoogeography

1926e, 1930c, 1931e, 1933c, j, 1940h, i, 1941c, 1943a, 1944d, e, k, o, 1945f, i, k, n, 1946h, k, 1951c, 1952d, 1954j, k, l, 1955f, 1963l, 1964c, s, 1965n, p, q, 1967c, f, 1972c, h, 1976a, d, f, 1986a, 1988f, 1990b, 2001g

## (3) Species Problem and Speciation

1927f, 1939c, 1940c, e, 1941g, i, n, 1942d, e, 1943e, 1944g, n, 1945d, 1946c, d, g, l, 1947e, 1948c–e, 1949f, i–k, m, 1950b, f, 1951g, l, 1952a, k, 1953a (p. 1–40), b, k, 1954c, e, 1955e, 1956a, c, g, h, j, m, 1957d–f, g, 1958g, h, 1959a, 1963b, q, 1965l, 1968i, 1969b (p. 23–53), e, h, 1970b, e, f, h, 1976h, m, 1980d, 1982a, d (p. 270–297), i, l, p, 1984e, g, 1987a, d, e, 1988h, 1991a (p. 19–109), b, c, 1992a, i, n, 1993g, 1994o, 1995m, 1996e, 1997b (p. 127–134), 1998f, 2000e, 2001a, c, d, f (p. 161–187), 2001g (p. 119–181), 2002a, e

## (4) Systematics

1942e, 1943d, f, 1944h, 1945e, 1946j, 1949n, 1951g, 1953a, 1955b, e, 1956h, l, 1957i, 1959e, 1964n, o, 1965k, 1968b, 1969b, 1971g, 1973f, g, 1974b, 1976m, 1980g, 1989b, k, 1990s, 1991a, 1995e

## (5) Classification

1951a, 1953b, 1954h, 1956d, e, 1958e, 1960a, b, 1962d, 1964f, 1965b, r, 1967d, e, 1968c, j, 1974g, 1976i, 1979a, 1980d, 1981a, 1986d, i, 1987c, 1989g, 1990o, 1991h, 1994b, c, h, s, 1995b, 1996h, 1998c, 2001d, 2002h

## (6) Zoological Nomenclature

### (a) Synonymy

1931g, 1933e, o, 1935a, 1936e, 1944i, 1945a, g, 1946f, 1947g, 1952q, r, 1954a, 1956n–p, 1957h, 1959k, 1961f, 1963d, e, g, 1964i, j, 1986p

### (b) Theory of Nomenclature

1951h, 1952n–p, 1953a (p. 201–292), l–z, 1954d, h, 1957j, 1958j–m, 1960b, 1962c, 1963f, m, 1966g, 1967h, 1969b (p. 297–380), 1971e, 1972i, 1987b, 1989e, 1991a (p. 383–406)

### (c) Preservation of Names (Stability)

1952b, h, 1953b, 1954a, 1957c, 1959j, 1961e, 1962c, 1963i, 1964h, p, 1965c, 1968d, e, 1974h

### (d) Miscellaneous

1958d, 1966b

## (7) Evolutionary Biology

1942e, 1944p, 1947a, 1949g (editor), k, 1955e, 1956b, i, 1958b, c, f, g, 1959a, b, d, h, 1960e, 1962b, 1963b, j, 1964b, d, k, l, o, r, t, 1965e–g, m, 1967b, 1968f, g, 1969i, 1970e, g, 1971a–c, h, 1972a, g, 1973h, m, 1974c, j, 1975h, 1976c, e, m, 1977c, d, f, 1978c, f, 1980e, g, 1981c, 1982e, f, k, 1983a, 1985e, 1986l, 1989d, i, 1990r, 1991d, e, g, 1993b, 1994a, e, l, p, 1995i, 1996c, 1997e, 1999d, e, h, i, k, 2001d, f, 2002b, g, i, 2004b

## (8) Genetics and *Drosophila* Studies

1944j, 1945d, 1946d, g, 1950c, 1954c, 1955h, 1959f, 1964r, 1965o, 1972b, 1974i, 1975i, 1984d

## (9) History of Biology

1934e, 1935c, 1946l, 1954i, 1955e, 1957f, 1959a, b, f, 1960e, 1964l, 1965s, 1968i, 1969c, 1972d, f, g, 1973i, 1975c, 1976m, 1977a, b, f, 1980e–g, 1982b, d, m, o, 1983a, e, 1984a, 1985d, i, j, 1986l, 1988d, e, g, 1989c, 1990c, 1991f, g, 1992f, i–k, n, o, 1993a, f–i, 1994j, n, r, 1995d, n, 1996g, 1997c, g, k, 1998f, g, 1999a, b, 2000b, 2001b

## (10) Charles Darwin

1959b, c, g, 1961b, g, 1962b, 1963b, 1964d, l, 1965h, 1971h, 1972d, g, 1973h, 1974d, k, 1977a, c, f, 1978f, 1980b, n, 1982b, d, j, m, 1983e, 1984b, h, 1985d, j, 1986e, l, 1988e, g, 1990a, 1991f, g, 1992n, o, 1993i, 1994j, n, 1995a, n, 1996d, 1997a, 1999b, 2000b, c, 2001e, 2002f, 2004a, c

## (11) Philosophy of Biology

1959b, 1961c, 1962a, 1963k, 1965s, 1969d, g, 1973a, b, l, 1974e, f, 1976h, m, 1983d, 1984b, c, g, 1985c, 1987d, e, 1988b, c, e, 1989h, 1992b, 1993c, 1994f, g, i, q, 1995c, j, 1996b, 1997b, i, j, 1998d, e, 2000d, 2001e, 2002c, 2004a, c

## (12) Conservation Biology

1924b, 1937i, 1945l, 1958a, 1970d, 1982c, 1998a, 2004g

## (13) Biographies, Appreciations, Obituaries

1934c, 1945b, 1955d, 1960f, 1969f, j, 1973d, e, j, 1974a, 1975a, b, d, e, j, 1976j, 1979b, c, 1980o, p, 1985h, 1986j, m, 1989f, 1990e–l, 1992d, h, l, 1993e, j, 1994n, 1995o, 1996f, g, 1997l, 1998b, 2000a, f, 2002d, 2003a

## (14) Autobiographical

1930f, 1932e, 1938f, 1943h, 1980n, 1981b, 1992i, 1993g, 1994m, 1997d, k, 1999b, h, j, 2002a, 2003a, 2004b

## (15) Reviews

1927h–n, 1932f, 1937b, 1942c, d, g, h, 1944a–c, f, l, n, r, 1945c, o, 1946e, 1949b, l, 1950e, 1951b, d, f, k, 1952c, f, g, l, 1953f, i, j, 1954i, 1955c, 1956k, 1958i, 1959g, i, 1961a, b, g, 1962e–g, 1963h, n, p, 1964g, m, t, 1965d, i, j, t, 1966d, e, 1967j, 1968h, 1970c, 1971j, 1972e, 1973i, 1974d, k, l, 1975f, g, 1976g, 1977b, e, 1978a, d, e, g, h, i, 1979b, 1980a–c, 1982c, g, h, j, n, 1983c, g, i, 1984f, h, i, 1985a, b, 1986g, n, o, 1987f–h, 1988a, 1989j, 1990a, d, m, n, p, q, 1992c, e, 1994d, k, o, 1995a, k, 1996d, 1997a, f, h, 1999f, g, 2000c, 2002f

## (16) General Topics

(Library catalogue, life on other planets, language, tenure, museum collections, annual reports, etc.) 1929a, 1931n, 1955g, 1963a, k, o, 1964a, 1965a, 1966a, 1967a, k, 1968a, 1969a, k, 1970a, 1973c, k, 1978b, 1985g, 1989a, 1992g, 1993d, k, 1995f–h, l, 1996a

# Appendix 4
## A Chronological List of Published Interviews

1. Kleemann, G. (1981) Darwinist aus Amerika. Im Gespräch mit dem Evolutionsbiologen Ernst Mayr. Stuttgarter Zeitung, 19. Mai 1981
2. Lewin, R. (1982: Biology is not postage stamp collecting. Ernst Mayr, the eminent Harvard evolutionist, explains why he thinks some physical scientists have a problem with evolution. Science 216:718–720
3. Johman, C. A. (1983) Interview with Ernst Mayr. The Harvard evolutionist talks about race, population, and the future of natural selection. Omni Magazine 1, pp. 73–78, 118–119
4. Der Biologe Ernst Mayr provoziert die Philosophen: Alles Leben ist Geschichte. *Bild der Wissenschaft* 22 (1985), Nr. 12:136–142, 146–148 (R. A. Zell)
5. Biology unified; an interview with Ernst Mayr. Scan 1 (July / August 1985):3–4
6. Evolution of Ernst Mayr, film. Dir.: Emmanuel Laurent; Co. La Sept, Les Films du Bouc and I. N. A., 1990
7. A conversation with Ernst Mayr. Harvard University Gazette 87 (No. 11):5–6, 1991 (with Peter Costa)
8. Ernst Mayr, nestor van de moderne evolutie biologie: "Erwordt over evolutie zo ontzettend veel onzin beweerd". *Nieuwe Rotterdamse Courant Handelsblad* (13 August 1992) 3 (with Felix Eijgenraam)
9. A conversation with Ernst Mayr. The American Biology Teacher 54 (7):412–415, 1992
10. Mayr, l'homme qui fait evoluer Darwin. Interview by Corinne Bensimon. In *Eureka* N. S. 3757, 23 June, pp. 27–29, 1993
11. Campbell, N. (1993) An Interview with Ernst Mayr. Biology, 3rd ed., pp 416–419. Benjamin-Cummings Publishing Co., Inc., Redwood City, California
12. Mawatari, S. F. (1995) Interview with Professor Ernst Mayr, the winner of the International Prize for Biology, 1994. Heredity (March) 49:6–11
13. Göldenboog, C. (1995) A meeting with Darwin's current bulldog; Ernst Mayr. Deutschland (June) 3:40–42
14. Yoon, C. K. (1997) Ernst Mayr. Long evolution of "Darwin of 20th Century." Brilliance, longevity and utter confidence are fuel of trailblazer. The New York Times, April 15, 1997
15. Angier, N. (1997) Ernst Mayr at 93. Natural History Magazine (New York) 5/97:4 p
16. Göldenboog, C. (1998) "Um sich ethisch zu verhalten, braucht man keine Religion:" Ein Gespräch mit dem Evolutionsbiologen Ernst Mayr. Psychologie heute, November 1998:40–41
17. Leite, M. (1999) Predominio de Darwin (Interview). Folha de Sao Paulo (29 August 1999), pp. 5, 13
18. Charlesworth, B. (1999) Interview of Ernst Mayr. Human Ethology Bulletin 14, no 3 (September):1–9

19. Shermer, M. and Sulloway, F. J. (2000) The grand old man of evolution. An interview with evolutionary biologist Ernst Mayr. Skeptic 8 (1):76–82
20. Ernst Mayr: What evolution is. Edge 92 (October 31, 2001)
21. Glaubrecht, M. and Wewetzer, H. (2001) "Leben ist mehr als Moleküle." Ernst Mayr, der große alte Mann der Biologie, über das Berlin der 20er, die Gene und die Zukunft der Natur. Der Tagesspiegel, Seite 31, 28. Juni 2001
22. Schwägerl, C. and Müller-Jung, J. (2002) Darwins Apostel. Der fast hundert Jahre alte Biologe Ernst Mayr über den Siegeszug der Evolutionstheorie, über Orpheus und die Zukunft des Menschen im Biotechzeitalter. Abenteurer der Biologie: Ein Besuch bei Ernst Mayr. Frankfurter Allgemeine Zeitung, 12. März 2002, Seite 47
23. Wilkins, A. S. (2002) Interview with Ernst Mayr. BioEssays 24 (10):960–973
24. Dreifus, C. (2002) An insatiably curious observer looks back on a life in evolution. A conversation with Ernst Mayr. The New York Times, April 16, 2002, page F2
25. Friebe, R. (2003) Die Macht des Zufalls, Teil 1 und 2. Interview mit Ernst Mayr. www.netzeitung.de, 17. Juni
26. Bahls, C. (2003) Ernst Mayr, Darwin's disciple. The Scientist 17:17–18 (November 17)
27. Leite, M. (2003) Interview with Dr. Mayr. Memòrias do Presente, Vol. 2. 100 entrevistas do "Mais!:" 1992–2002. Artes do Conhocimento / Adriano Schwartz (org.). Sao Paulo, Brazil: Publifolha, p. 55–63
28. Göldenboog, C. (2003) Gespräch mit dem Evolutionsbiologen und Zoologen Ernst Mayr, pp. 238–260, in C. Göldenboog *Das Loch im Walfisch. Die Philosophie der Biologie*, 2003, Stuttgart, Klett-Cotta
29. DeMarco, P. (2004) Meeting the Minds: Ernst Mayr. Nearly 100, he's still advancing science. The Boston Globe (May 11), p. C2
30. Angelo, C., Leite, and Mirsky, S. (2004) Para Ernst Mayr, biologia nao se reduz às Ciencias fisicas. Folha de Sao Paulo Mais! p. 4–7, 4 July 2004
31. Friebe, R. (2004) Normal ist das alles nicht. Hundert Jahre Geistesarbeit: Der Evolutionsbiologe Ernst Mayr forscht und forscht und forscht. Frankfurter Allgemeine Sonntagszeitung, Nr. 27, Wissenschaft 59. 4 July 2004
32. Ruby, C. (2004) Am Anfang war die Ente–der Biologe Ernst Mayr wird 100 Jahre alt. Transcript of radio interview. WDR 5, Westdeutscher Rundfunk Köln, 5 July 2004
33. Grolle, J. (2004) "Das passt ja wunderbar zusammen," Theorie des Lebendigen. Der Spiegel 28:149–151
34. Mirski, S., Angelo, C. and Leite, M. (2004) The evolution of Ernst: Interview with Ernst Mayr. Posted on Scientific American's website (www.sciam.com), 14 p
35. Amberger, M. (2004) Die Biologie ist keine zweite Physik. Interview mit dem Evolutionsbiologen Professor Ernst Mayr, der am Montag seinen 100. Geburtstag feiert. Die Welt, Samstag, 3. Juli
36. Plüss, M. and Lutz, J. (2004) "Ich weiss nie, ob ich deutsch rede." Für seine Entdeckungen hätte Ernst Mayr den Nobelpreis für Biologie verdient–aber den gibt es ja nicht. Kommende Woche feiert der "Darwin des 20. Jahrhunderts" seinen 100. Geburtstag. Ein Gespräch über Gott und die Zwecklosigkeit der Natur. Weltwoche (Zürich) 72, Nr. 27.04:64–69, 1 July 2004
37. Bock, W., Taylor, S. and Stettenheim, P. (2005) Interview remarks by Ernst Mayr (November 8, 2003). Video CD-ROM included in Ornithological Monograph 58 (the best and most easily accessible Interview source)

# Appendix 5
## A Chronological List of Appreciations and Festschriften

*Historians of Science dedicated a festschrift to E. Mayr on his 75th birthday in 1979. When he was awarded the Balzan Price of Switzerland and Italy in 1983, several newspapers in the United States and in Germany reported in detail about the awardee and his work. Three festschriften appeared on the occasion of his 90th birthday in 1994. From then on appreciations and interviews (see Appendix 7) were published regularly in North American and German newspapers. These festschriften, newspaper articles, and other appreciations as far as known are assembled in the following* chronological *list.*

The authors (1979) To Ernst Mayr on his seventy-fifth birthday, 5 July 1979. In Ernst Mayr Festschrift (W. Coleman and C. Limoges, eds.), Studies in History of Biology 3:VIII–X

Eldredge, N. (1982) *Introduction*, p. XV–XXXVII, in E. Mayr *Systematics and the Origin of Species*, reprint edition. New York (this *Introduction* "contains so many errors and misinterpretations that it is at best a biased commentary on this volume," Bock 1994a:296)

Webster, B, (1983) Ernst Mayr, an ornithologist, gets Linnaean Society Medal. New York Times, March 13th, 1983

c.m. (1983) Balzan-Preis für Ernst Mayr. Frankfurter Allgemeine Zeitung, 17. November 1983

Hapgood, F. (1984) The importance of being Ernst. Science 84:40–46

Markl, H. (1984) Vielfalt und Anpassung. Der deutsch-amerikanische Zoologe Ernst Mayr, einer der größten Evolutionsbiologen seit Darwin, erhält den Balzan-Preis. Die Zeit (Hamburg), Nr. 8 (17. Februar), Seite 64

Moritz, C. (ed., 1984) Mayr, Ernst. Current Biography Yearbook 1984 (New York, H. W. Wilson, p. 258–262

Gould, S. J. (1984) Balzan Prize to Ernst Mayr. Science 223:255–257

Coleman, W. (1987) 1986 Sarton Medal citation (Ernst Mayr). Isis 78:239–241

Laurent, E. (1990) Evolution of Ernst Mayr, film. La Sept, Les Films du Bouc and I. N. A

Hull, D. L. (1990) Ernst Mayr on the philosophy of biology: A review essay. Historical Methods 23:42–45 (Spanish translation in Ludus Vitalis 12 (21):35–41, 2004)

Markl, H. (1991) Vorwort, pp. I–XII; in E. Mayr *Eine neue Philosophie der Biologie*. Munich and Zurich, Piper

Greene, J. and Ruse, M.; eds. (1994) Special issue on Ernst Mayr at Ninety. Biology and Philosophy 9:264–435

Greene, J. (1994) Introduction. Biology and Philosophy 9:265

Bock, W. J. (1994) Ernst Mayr, naturalist: His contributions to systematics and evolution. Biology and Philosophy 9:267–327 [Mayr, E. (1994) Response to Walter Bock. Biology and Philosophy 9:329–331]

Beatty, J. (1994) The proximate/ultimate distinction in the multiple careers of Ernst Mayr. Biology and Philosophy 9:333–356 [Mayr, E. (1994) Response to John Beatty. Biology and Philosophy 9:357–358]

Burkhardt, R. W. (1994) Ernst Mayr: Biologist-historian. Biology and Philosophy 9:359–371 [Mayr, E. (1994) Response to Richard Burkhardt. Biology and Philosophy 9:373–374]

Hull, D. L. (1994) Ernst Mayr's influence on the history and philosophy of biology: A personal memoir. Biology and Philosophy 9:375–386

Cain, J. (1994) Ernst Mayr as community architect: Launching the Society for the Study of Evolution and the journal *Evolution*. Biology and Philosophy 9:387–427

Ruse, M. (1994) Booknotes [on Ernst Mayr and the Evolutionary Synthesis]. Biology and Philosophy 9:429–435

Smocovitis, V. B. (1994a) Disciplining evolutionary biology: Ernst Mayr and the founding of the Society for the Study of Evolution and *Evolution* (1939–1950). Evolution 48:1–8

Smocovitis, V. B. (1994b) Organizing evolution: Founding the Society for the Study of Evolution (1939–1950). Journal of the History of Biology 27:241–309

Moore, J. A. (1994) Some personal recollections of Ernst Mayr. Evolution 48:9–11

Gill, F. B. (1994) Ernst Mayr, the ornithologist. Evolution 48:12–18

Coyne, J. A. (1994) Ernst Mayr and the origin of species. Evolution 48:19–30

Gould, S. J. (1994) Ernst Mayr and the centrality of species. Evolution 48:31–35

Futuyma, D. J. (1994) Ernst Mayr and evolutionary biology. Evolution 48:36–43

Cuevas, J. de (1994) "The heir of Audubon and Agassiz." Harvard Magazine 96 (6):81–82

Rennie, J. (1994) Ernst Mayr, Darwin's current bulldog. Scientific American, August, p. 14–15

Haffer, J. (1994) "Es wäre Zeit, einen 'allgemeinen Hartert' zu schreiben:" Die historischen Wurzeln von Ernst Mayrs Beiträgen zur Evolutionssynthese. Bonner Zoologische Beiträge 45:113 . 123

Haffer, J. (1994) Die "Seebohm-Hartert Schule" der europäischen Ornithologie. Journal für Ornithologie 135:37–54

Gibbons, A. (1994) Ernst Mayr wins the Japan Prize. Science 266:365

Cromie, W. J. (1994) Ernst Mayr wins Japan Prize. Harvard University Gazette, vol. XC:1–2

F.A.Z. (1994) Japan-Preis für Ernst Mayr. Frankfurter Allgemeine Zeitung, 9. November 1994, Seite N1

Bock, W. J. (1995) Ernst Mayr–1994 laureate of the International Prize for Biology, pp. 13–24, in R. Arai, M. Kato and Y. Doi (eds.), *Biodiversity and Evolution*, The National Science Museum Foundation, Tokyo

Beurton, P. J. (1995) Ernst Mayr und der Reduktionismus. Biologisches Zentralblatt 114:115–122

Landsberg, H. (1995) Ernst Mayer [sic] in Berlin–vom "vielversprechenden jungen Mann" zum anerkannten Systematiker im American Museum of Natural History. Biologisches Zentralblatt 114:123–132

Haffer, J. (1995) Ernst Mayr als Ornithologe, Systematiker und Zoogeograph. Biologisches Zentralblatt 114:133–142

Junker, T. (1995) Vergangenheit und Gegenwart: Bemerkungen zur Funktion von Geschichte in den Schriften Ernst Mayrs. Biologisches Zentralblatt 114:143–149

Forneris, L. (1995) Ernst, Mayr, ornitólogo exemplar! Bol. CEO 11:34–39 (Brazil)

*NN* (1997) Ernst Mayr at 93. Natural History No. 5

Padova, T. de (1997) Ein philosophischer Blick auf die Biologie. Ernst Mayr auf der Eröffnungsvorlesung der nach ihm benannten Berliner Lecture. Frankfurter Allgemeine Zeitung 16. Oktober 1997, Seite 31

Haffer, J. (1997) Ornithologen-Briefe des 20. Jahrhunderts. "We must lead the way on new paths." The work and correspondence of Hartert, Stresemann, Ernst Mayr–international ornithologists. Ökologie der Vögel 19:1–980 (Ernst Mayr–ornithologist, systematist and zoogeographer, p. 62–99; the correspondence between E. Stresemann and E. Mayr during the period 1923–1972, p. 369–771; brief biography of Ernst Mayr, p. 804–818)

Wehner, R. (1998) Ernst Mayr–ein Fels in der evolutionsbiologischen Brandung. Berlin-Brandenburgische Akademie der Wissenschaften, Berichte und Abhandlungen 5:283–286

Glaubrecht, M. (1998) Wie das Leben wurde, was es ist. Der Zoologe Ernst Mayr tritt für eine Bio-Philosophie ein–mit der Evolutionstheorie im Zentrum. Der Tagesspiegel 17. September 1998, Seite 35

Daniel, L. (1999) Renowned biologist maps solutions to survival. The Concord Journal, Thursday, February 4th, 1999

*NN* (1999) Ein Ehrenlegionär des Darwinismus. Frankfurter Allgemeine Zeitung, 5. July 1999

Cain, J. (2000) For the 'promotion' and 'integration' of various fields: first years of *Evolution*, 1947–1949

Potier, B. (2001) Ernst Mayr, evolutionary biologist. Harvard University Gazette, September 20th, 2001

Diamond, J. (2001) Foreword in E. Mayr *What evolution is*, p. VII–XII. New York, Basic Books

Haffer, J. (2001) Ernst Mayr–Ornithologe, Evolutionsbiologe, Historiker und Wissenschaftsphilosoph. Journal für Ornithologie 142:497–502 [Engl. translation in Verhandlungen zur Geschichte und Theorie der Biologie 9:125–132, 2002]

G[laubrecht], M. (2001) Architekt für das Haus der Biologie. Vor 75 Jahren promovierte der Evolutionsforscher Ernst Mayr in Berlin. Der Tagesspiegel, 26 . Juni 2001, Seite 29

Cord [Richelmann] (2001) Es war der Girlitz, nicht die Lerche: Ernst Mayr geehrt. Frankfurter Allgemeine Zeitung, Berliner Seiten (BS 2), 26. Juni 2001

Kastilan, S. (2001) "Ein großer Fehler ist, die Natur als Person zu sehen." Ernst Mayr zur Autonomie der Biologie. Die Welt, 27. Juni 2001, Seite 35

Orzessek, A. (2001) Schlaumayrchen in der Saurierhalle. Der Darwin des 20. Jahrhunderts, Ernst Mayr, besuchte Berlin. Stuttgarter Zeitung, 28. Juni 2001

Steinbacher, J. (2001) Ein Kultur-Ereignis in Berlin. Gefiederte Welt 125:355–356

Meyer, A. (2002) Das Rätsel, das Darwin der Nachwelt überließ. Hundert Jahre nach der "Beagle"-Fahrt: Ernst Mayrs bahnbrechende ornithologische Expedition nach Neuguinea. Frankfurter Allgemcine Zeitung, 5. Juli 2002, Seite 44

Meyer, A. (2002) Ernst Mayr ist wahrlich der Darwin unserer Zeit. Der in Kempten geborene Biologe und Evolutionsforscher feiert heute seinen 98. Geburtstag. Ärzte-Zeitung Nr. 124, 5./6. Juli, Seite 13

Beurton, P. J. (2002) Ernst Mayr through time on the biological species concept–a conceptual analysis. Theory in Biosciences 121:81–98

Meyer, A. (2003) Herzlichen Glückwunsch dem Charles Darwin unserer Zeit. Morgen feiert der Kemptener Ernst Mayr in Harvard seinen 99. Geburtstag. Ärzte-Zeitung Nr. 123, 4./5. Juli, S. 13

Göldenboog, C. (2003) *Das Loch im Walfisch. Die Philosophie der Biologie*. Klett-Cotta, Stuttgart. [The book may be considered as a homage of the ninety-nine-year-old Harvard Professor Ernst Mayr]

Pray, L. and C. Bahls (2003) Mechanisms of speciation. The Scientist 17:14–16 (November 17)

Meyer, A. (2004) Der Altmeister erklärt die Evolution. Die Theorien Darwins waren die umwälzendsten aller geistigen Revolutionen in der Geschichte der Menschheit. Der fast 100-jährige Biologe Ernst Mayr hat ihnen sein Lebenswerk gewidmet. Die Zeit, 8. Januar 2004, Seite 45

Jacobsen, H.-J. (2004) Eine Begegnung mit Ernst Mayr. Biologen heute 2004:18

Meyer, A. (2004) Learning from the Altmeister. How reading Ernst Mayr's books (in his bathtub) changed my research. Nature (London) 428:897

Sontag, W. (2004a) Vogelkunde als Jungbrunnen. Erstaunlich viele Ornithologen erreichen ein hohes Lebensalter. Wiener Zeitung, Extra, 27./28. Februar, Seite 5

Sontag, W. (2004b) Die Eigentümlichkeit des Lebens. Der große Biologe Ernst Mayr wird 100 Jahre alt. Wiener Zeitung, Extra, p. 9, 25. Juni 2004
Eck, S. (2004) Professor Dr. Ernst Mayr, Ehrenmitglied des VSO–100 Jahre alt. Mitt Ver. Sächs. Ornithol. 9:277–278
Gedeon, K., Kleinstäuber, G. (2004) Auf den Spuren von Ernst Mayr–Reise des Vereins Sächsischer Ornithologen 2003 nach Westpapua. Mitteilungen des Vereins Sächsischer Ornithologen 9:279–305
Grolle, J. (2004) Theorie des Lebendigen. Der Ornithologe Ernst Mayr fügte Evolutionslehre und Genetik zu einer einheitlichen Wissenschaft der biologische Vielfalt zusammen. Der Spiegel No. 28:148–149
Hölldobler, B. (2004) Ernst Mayr: the doyen of twentieth century evolutionary biology. Naturwissenschaften 91:249–254
Rehfeld, K. (2004) Editorial [über Ernst Mayr]. Naturwissenschaftliche Rundschau 57:353
Glaubrecht, M. (2004a) Ernst Mayr–vom Systematiker zum Begründer einer neuen Biophilosophie. Ein Porträt aus Anlass des 100. Geburtstages des Evolutionsbiologen. Naturwissenschaftliche Rundschau 57:357–368
Glaubrecht, M. (2004b) Der Darwin des 20. Jahrhunderts. GEO-Journal 07 (Juli):78–82, 84
Glaubrecht, M. (2004c) Darwins Erbe. Zum 100. Geburtstag des deutschstämmigen Evolutionsbiologen und Ornithologen Ernst Mayr. Tagesspiegel (Berlin), 5. Juli
Ritter, H. (2004) Lob der Verschiedenheit. Der Aktualisierer der Evolutionstheorie: Zum hundertsten Geburtstag des Biologen Ernst Mayr. Frankfurter Allgemeine Zeitung, 5. Juli 2004 (Nr. 153), p. 31
Günther, M. (2004) Ernst Mayr, Evolutionstheoretiker, wird 100. "Pläne haben sein Leben verlängert." Westdeutsche Allgemeine Zeitung, Wochenende (p. 5), Samstag, 3. Juli 2004 (reprinted in Kieler Nachrichten, Samstag, 17. Juli 2004, 'Journal,' p. 2)
Henderson, M. (2004) Science icon fires broadside at creationists. The Times (London), Saturday, July 03, page 6
Unwin, B. (2004) Bird-watching has dominated my life, says evolution expert. Birdwatching Magazine, July
Mirski, S. (2004) One hundred years of magnitude. Scientific American (August), p. 98
Pennisi, E. (2004) Museums that made a Master. Science 305:37
Kappe, T. (2004) Der Darwin unserer Tage. Ernst Mayr formulierte die heute anerkannte Evolutionstheorie. Am 5. Juli wird der Biologe hundert Jahre alt. Berliner Zeitung, Nr. 153, 3./4. Juli
Bock. W. J. (2004) Ernst Mayr at 100–A life inside and outside of ornithology. The Auk 121:637–651
Bock, W., Fischer, M., Minelli, A., Sperlich, D., Westheide, W. (2004) Herzlichen Glückwunsch, Ernst Mayr! J. Zool. Syst. Evol. Research 42:177
DiConstanzo, J. (2004) Ernst Mayr: Going strong at 100. The Linnaean Newsletter 58 (6), 4 p., September (Linnaean Society of New York)
Haffer, J. and Bairlein, F. (2004) Ernst Mayr–'Darwin of the 20th century.' J. Ornithol. 145:161–162
Haffer, J. (2004) Ernst Mayr–Intellectual leader of ornithology. J. Ornithol. 145:163–176
Irsch, W. (2004) Ernst Mayr–der "Darwin des 20. Jahrhunderts"–100 Jahre alt. Der Falke 51:266, III
Pfeifer, R. (2004) Prof. em. Dr. Dr. h.c. mult. Ernst Mayr. Ornithologischer Anzeiger 43:89–92
Steinheimer, F. (2004) Ernst Mayr und die Nymphenrallen *Rallina forbesi dryas*–eine ornithologische Anekdote aus Neuguinea. Ornithologischer Anzeiger 43:93–102
Grummt, W. (2004) Ernst Mayr 100 Jahre. Gefiederte Welt 128:316
Meier, C. (2004) "Der Jahrhundert-Biologe." In: www.ethlife.ch (Webzeitung der ETH Zürich), 5 July

Amberger, M. (2004) Die Biologie ist keine zweite Physik. Interview mit dem Evolutionsbiologen Professor Ernst Mayr, der am Montag seinen 100. Geburtstag feiert. Die Welt, 3. Juli, Seite 27
Ayala, F. J. (2004) Ernst Mayr and the theory of evolution. Ludus Vitalis (Revista de Filosofía de las Ciencias de la Vida) 12, número 21:3–13
Diamond, J. M. (2004) Ernst Mayr's view of evolution. Ludus Vitalis (Revista de Filosofía de las Ciencias de la Vida) 12, número 21:29–34
Hull, D. L. (2004) Ernst Mayr y la filosofía de la biología. Ludus Vitalis (Revista de la Filosofía de las Ciencias de la Vida) 12, número 21:35–41
Dawkins, R. (2004) An ecology of replicators. Ludus Vitalis (Revista de la Filosofía de las Ciencias de la Vida) 12, número 21:43–52
Ghiselin, M. T. (2004) Mayr on species concepts, categories, and taxa. Ludus Vitalis (Revista de la Filosofía de las Ciencias de la Vida) 12, número 21:109–114
Sober, E. (2004) Evolución, pensamiento poblacional y esencialismo. Ludus Vitalis (Revista de la Filosofía de las Ciencias de la Vida) 12, número 21:115–147
Wuketits, F. M. (2004) *This is Biology*. Ernst Mayr and the autonomy of biology as a science. Ludus Vitalis (Revista de Filosofía de las Ciencias de la Vida) 12, número 21:149–160
Maienschein, J. (2004) Evolution, embryology, and Ernst. Ludus Vitalis (Revista de la Filosofía de las Ciencias de la Vida) 12, número 21:237–245
Provine, W. B. (2004) Ernst Mayr: Genetics and speciation. Genetics 167:1041–1046
Greene, J. C. (2004) Impressions of the Claremont Conference and Ernst Mayr. Reports of the National Center for Science Education 4 (no. 5):34–37
Alayón Garcia, G. (2004) Ernst Mayr: un siglo de vida. Cocuyo (Habana, Cuba) 14:32–33
De Queiros, K. (2005) Ernst Mayr and the modern concept of species. Proc. Natl. Acad. Sci USA 102, suppl. 1:6600–6607
Provine, W. B. (2005) Ernst Mayr, a retrospective. Trends in Ecology and Evolution 20:411–413 [see critical comments by D. J. Futuyma in this journal, vol. 21:7–8, 2006]
Sarkar, S. (2005) In memoriam: Ernst Mayr (1904–2005). J. Biosci. 30 (4)
Mawatari, S. F. (2005) Ernst Mayr in Japan, October 1994. J. Biosci. 30 (4):419–421
Bock, W. J. (2005) Ernst Mayr–teacher, mentor, friend. J. Biosci. 30 (4):422–426
Homberger, D. (2005) Ernst Mayr and the complexity of life. J. Biosci. 30 (4):427–433
Sperlich, D., Fischer, M., Minelli, A. and Westheide, W. (2005) Ernst Mayr died on February 3rd 2005, seven months after his 100th birthday. J. Zool. Syst. Evol. Res. 43:177
Bock, W. J. and Lein, M. R. (eds., 2005) Ernst Mayr at 100. Ornithologist and Naturalist. Ornithological Monograph 58 (with contributions by W. J. Bock, M. LeCroy, M. R. Lein, R. Schodde, F. Vuilleumier, and J. Haffer)
Glaubrecht, M. (2005) Geleitwort, p. XI–XX. In E. Mayr *Konzepte der Biologie*, S. Hirzel Verlag, Stuttgart
Sudhaus, W. (2005) Erinnerungen an unser Ehrenmitglied Ernst Mayr (5. Juli 1904–3. Februar 2005) im Blick auf sein Lebenswerk. Sitzungsberichte der Gesellschaft Narurforschender Freunde zu Berlin (N.F.) 44:133–157
Johnson, K. (2005) Ernst Mayr, Karl Jordan, and the history of systematics. Hist. Sci. 43: 1–35
Kolchinsky E. (2006) Ernst Mayr and modern evolutionary synthesis. KMK Scientific Press Ltd., Moscow, 149 pp. (in Russian)
Bock, W. J. (2006) Ernst Walter Mayr, 5 July 1904–3 February 2005. Biographical Memoirs of Fellows of the Royal Society of London 52: 167–187
Haffer J. (2007a) Ergänzungen zum Briefwechsel zwischen Erwin Stresemann (1889–1972) und Ernst Mayr (*1904)–Internationale Ornithologen des 20. Jahrhunderts. Ökologie der Vögel (in press)

Haffer, J. (2007b) Mayr, Ernst Walter. New Dictionary of Scientific Biography. C. Scribner's Sons, Farmington Hills, Michigan

Junker T. (2007) Ernst Mayr (1904–2005) and the new philosophy of biology. J General Philos Sci 38 (in press)

A Century of Evolution: Ernst Mayr 1904–2005. Linnean Society Day Meeting, 6th October 2006. Speakers were M. Claridge, J. Mallet, S. Knapp, A. Meyer, P. Nosil, B. Emerson, I. Owens, R. Butlin, J. Cain Biodiversity (in press)

Wilkins A. S. (2007) Ernst Mayr and the evolution of 'Neodarwinism.' (in press)

Ernst Mayr Commemoration Volume (2007). Proceedings of National Seminar on Evolutionary Biology and Biotechnology, p. 1–100. Zoological Survey of India. Department of Zoology, University of Calcutta

Bock, W.J. (2007) Ernst Mayr (5 July 1904 – 3 February 2005). Proceedings of the American Philosophical Society 51 (3): 357–370

# Appendix 6
## List of Obituaries


Alayón García G (2005) Ha muerto Ernst Mayr (1904–2005). Cocuyo (Habana, Cuba) 15:32–33

Anon (2005) Ernst Mayr, evolutionary biologist, died on February 3rd, aged 100. The Economist, February 12th, 2005

Bock WJ (2005a) In memoriam: Ernst Mayr, 1904–2005. The Auk 122:1005–1007

Bock WJ (2005b) Ernst Mayr zum Gedenken, 5. Juli 1904–3. Februar 2005. Natur und Museum 135:145–146

Bock WJ (2006) Ernst Walter Mayr, 5 July 1904–3 February 2005. Biographical Memoirs of Fellows of the Royal Society of London 52: 167–187

Bradt S (2005) Ernst Mayr, giant among evolutionary biologists, dies at 100. Harvard University Gazette, 4 February

Coyne JA (2005) Ernst Mayr (1904–2005). Science 307:1212–1213

Curio E (2005) Ernst Mayr–der Darwin des 20. Jahrhunderts. Zoologie 2004/05 (Mitteilungen der Deutschen Zoologischen Gesellschaft) 77–85

Diamond J (2005) Ernst Mayr (1904–2005). Nature 433:700–701

Editor (2005) In Memoriam Dr. Ernst Mayr, editor of *Evolution*, 1947–1949. Evolution 59 (3):702

Haffer J (2005) Prof. Dr. Dr. h.c. mult. Ernst Mayr (1904–2005). Vogelwarte 43:148–150

Junker T (2006) Ernst Mayr (5. Juli 1904–3. Februar 2005). Verhandlungen zur Geschichte und Theorie der Biologie 12:11–14

Junker Th, Hoßfeld U (2005) Ernst Mayr (geb. 5. Juli 1904, gest. 3. Februar 2005). Anzeiger des Vereins Thüringer Ornithologen 5:241–246

Kutschera U (2005a) Vom Ornithologen zum Universalgelehrten. Neue Zürcher Zeitung, 9 February, page 9

Kutschera U (2005b) Ernst Mayr (1904–2005) Systematiker, Evolutionsbiologe und Philosoph. Biologenheute (Mitteilungen des Verbandes deutscher Biologen) Nr. 475 (1-2005):16–19

Lein MR (2006) Ernst Mayr, 1904–2005. Ibis 148:389–391

Markl P (2005) Ein großer Liebhaber der Natur. Zum Tod des Biologen Ernst Mayr, der im Alter von 100 Jahren verstorben ist. Wiener Zeitung (Extra), 18. Februar, Seite 9

Margulis L (2005) Ernst Mayr, biologist extraordinaire. American Scientist 93:200–201 (also in *Métode*, scientific magazine of the University of Valencia, Spain)

Marren P (2005) Obituary–Ernst Mayr. The Independent (London), 10 February

Mawatari SF (2005) Ernst Mayr in Japan, October 1994. J Biosci 30:419–421

Meyer A (2005a) Der Meister des Warum. Ernst Mayr, der Evolutionsbiologe des 20. Jahrhunderts, ist im Alter von 100 Jahren gestorben. Die Zeit (Hamburg), Nr. 7, Seite 38

Meyer A (2005b) "Alles Leben im Lichte der Evolution sehen." Frankurter Allgemeine Zeitung, 9 February, Seite N1

Meyer A (2005c) On the importance of being Ernst Mayr. "Darwin's apostle" died at the age of 100. PLoS Biol. 3 (5), e152, pp 750–752

Meyer A (2005d) Ernst Mayr. Der bedeutendste Evolutionsbiologe des 20. Jahrhunderts und Ehrendoktor der Universität Konstanz starb im Alter von 100 Jahren. Uni'kon 18:35–36
Nevo E (2006) Ernst Mayr (1904–2005). Evolutionary leader, protagonist, and visionary. Theoret Pop Biol 70:105–110
Ruse M (2005) Ernst Mayr 1904–2005. Biol Philos 20:623–631
Sampedro J (2005) Ernst Mayr, 'el Darwin del siglo XX.' El País, 8 de Febrero
Schwägerl C (2005) Der Zusammendenker. Zum Tode des Evolutionsbiologen Ernst Mayr. Frankfurter Allgemeine Zeitung, Nr. 30, 5. Februar, Seite 37
Sullivan P (2005) Seminal evolutionist Ernst Mayr dies. Washington Post, 4 February
Sulloway FJ (2005) Ernst Mayr, 1904–2005. Remembrances and Tribute. E-Skeptic # 55 for February 11, 2005
Unwin B (2005) A life dominated by birds and Darwinian discovery. The Guardian, February 5
Winsor MP (2005) Ernst Mayr, 1904–2005. Isis 96:415–418
Yoon CK (2005) Ernst Mayr, 100, premier evolutionary biologist. New York Times, February 4
Yoon CK (2005) Ernst Mayr, pioneer in tracing geography's role in the origin of species, dies at 100. New York Times, 5 February

# Index

Printing: Krips bv, Meppel, The Netherlands
Binding: Stürtz, Würzburg, Germany